HERBAL PERSPECTIVES :
PRESENT AND FUTURE

HERBAL PERSPECTIVES : PRESENT AND FUTURE

Editors :
Dr. T. Parimelazhagan
Dr. S. Manian
Dr. M. Pugalenthi

2011

SATISH SERIAL PUBLISHING HOUSE
403, Express Tower, Commercial Complex, Azadpur, Delhi-110033 (India)
Phone : 011-27672852, Fax : 91-11-27672046
E-mail :info@satishserial.com, hkjain1975@yahoo.com
Website : www.satishserial.com

Published by :

SATISH SERIAL PUBLISHING HOUSE
403, Express Tower, Commercial Complex, Azadpur, Delhi-110033 (INDIA)
Phone : 011-27672852 Fax : 91-11-27672046
E-mail : info@satishserial.com, hkjain1975@yahoo.com

ISBN 81-89304-91-7

Typeset at: **Harminder** *for* **Laxmi Art Creations** # 98 11 48 23 28
Printed at: Salasar Imaging Systems, Delhi-110035

PROF M S SWAMINATHAN
Member of Parliament (Rajya Sabha)
Chairman, M S Swaminathan
Research Foundation
Third Cross Street, Taramani
Institutional Area
Chennai - 600 113 (India)

Foreword

Developing countries remain poor inspite of a rich endowment of natural resources including biodiversity. The challenge now is the conversion of bioresources into jobs and income on an environmentally sustainable basis. It is in this context that this book dealing with herbal technology is a very timely one. Dr T. Parimelazhagan has done a superb job in putting together a large number of authentic articles in the form of this publication. I hope the book will stimulate the development of Herbal Biovalleys which can bring together bioresources, biotechnology and business in a mutually reinforcing manner.

I hope the book will be widely read and used for the development of Herbal Biovalleys in our country. If this happens, we can end with the dichotomy of a poor people inhabiting a rich country, a point stressed by Jawaharlal Nehru over fifty years ago. We owe a deep debt of gratitude to Dr Parimelazhagan for this labour of love.

Sd /-

(M.S. Swaminath)

Dr.C.Swaminathan
Vice-Chancellor

Bharathiar University
Coimbatore

Front Line

Since ancient times humanity has depended on the diversity of plant resources for food, clothing, shelter, and traditional medicine to cure myriads of ailments. Early humans recognized their dependence on nature in both health and illness. Medicinal plants have traditionally occupied an important position in the socio-cultural, spiritual and medicinal arena of rural and tribal lives of India. Based on the different publications, it has been estimated that there are about 2000 medicinal plants in India used in different systems of practice such as, Allopathic, Ayurvedic, Unani and Homeopathic and more than 6780 industries are engaged in the production of medicines. It is worthwhile to note that, about 80% of the human populations in India are still dependent on nature for remedies and this can be well understood from the fact that almost all systems of medicine are largely based on drugs of plant origin. Herbal medicine is now globally accepted as a valid alternative system of therapy in the form of pharmaceuticals and functional foods, a trend recognized and advocated by the World Health Organization (WHO).

Renewed interests in alternative medications in general and botanicals in particular have opened new horizons of exploration and debated globally. It is the need of the day to share knowledge and experiences in the field of herbal research. The Department of Botany, Bharathiar University has done commendable research on medicinal plants since its inception. The Department has published good number of scientific research papers and also patented the novel work.

I am happy that an exclusive book, which exemplifies the potential and recent trends on various aspects of herbal science, through manuscripts contributed by eminent researchers has been attempted. This book provides a ground work for natural product chemistry/phytochemistry, quality control and standardization of herbal drugs leading to expansion of knowledge in various selected classes of herbal science. I congratulate the Editors for their tremendous efforts in bringing out this inclusive publication on herbals.

It is earnestly believed that students, learning herbal science will certainly find this text not only useful but also a good companion for further pursuit of higher knowledge. I am sure this book will be welcomed and provide scientific guidelines for the researchers, academicians and students as a reference work for all aspects of herbal research in India.

Sd /-
C.Swaminathan

Preface

India has a rich heritage in the field of herbal medicine covered under the disciplines of Ayurveda, Siddha and Unani. The Indian systems of medicine with the use of plants and other natural sources, are evolved with the practical experiences over hundreds of years. These Indian systems of medicines are not only health care systems but traditional heritage of India. To preserve this heritage, it is necessary to explore them with scientific approaches. Herbal drugs occupy an important place in the movement for achieving the goal of health for all.

Herbal medicine is the use of appropriate plants or plant parts for their therapeutic or medicinal value. It is the oldest form of healthcare known to humanity that has been used by all cultures throughout the history. Many drugs commonly used today in the developing countries are of herbal origin and prescription drugs contain at least one active ingredient derived from plant material either obtained from plant extracts or synthesized to mimic a natural plant compound. Many countries have a national regulation on herbal medicines but the legislative control of medicinal plants has not evolved around a structured model. This is because medicinal products or herbs are defined differently in different countries and diverse approaches have been adopted with regard to licensing, dispensing, manufacturing and trading. The growing market for herbal medicines and other natural health care products also demands an extensive research not only for its great healthcare but also for the commercial benefit.

The principal aim of this book has been to collate the latest information, and to review the viewpoints of eminent investigators, who excel in the areas of herbal research. An attempt has been made to encapsulate scientific information pertinent to ethnomedicine, pharmacology, phytochemistry, post harvest technology and value addition of herbals. In this book, 30 articles by eminent academicians and scientists of the country on different aspect of the subject have been compiled. All recent scientific innovations/advances have been incorporated in a simple and lucid manner. A special emphasis has been laid on screening of herbal drugs for pharmacological activity, antioxidant potential and bioinformatics tools involved in herbal research. A brief discussion on 'Intellectual Property Rights' related to herbal research is also included. This book discusses the WHO protocol, different methods used for standardisation, quality control, standards for herbal extracts and validation of herbal products. The book will be useful to students and professionals in the fields of pharmacy, ayurveda, unani and siddha.

The book **'Herbal Perspectives: Present and Future'** is the outcome of the National symposium on herbal drug research organized by our Department during September, 2008. The chapters have been critically reviewed by the experts to improve the quality. We hope that the present compilation will be useful for the academicians, scientists, researchers and people associated in herbal research.

We appreciate the efforts of all the experts who have contributed quality papers by sharing their current scientific knowledge. The editors would like to convey their sincere thanks to Bharathiar University for the support and also grateful to Dr.C.Swaminathan, Vice-Chancellor, Bharathiar University for his guidance and encouragement. The support rendered by Dr. Ravikumar, Associate Professor, Alagappa University, Thondi campus, Dr.Sunil Jadhav Ekanath, Scientist, Defence Institute of High Altitude Research, DRDO, Ladakh, Dr.Kalidass, Associate Professor, School of Biotechnology, Karunya University, Coimbatore, Dr.R.Ilavarasen, Assistant Director, Captain Srinivasa Murti Drug Research Institute for Ayurveda and Siddha, Govt. of India, Chennai and Dr.V.Vadivel, AvH Post Doctoral Research Fellow, University of Hohenheim, Germany, in critically reviewing the chapters is dully appreciated. Constant support provided by colleagues of our Department of Botany, Bharathiar University, Coimbatore is also gratefully acknowledged.

We would like to express special appreciation for the generous support provided by the publishers and their team and for their sincere efforts in bringing out the book in time. We welcome any suggestion/criticism for further improvement.

Coimbatore **Editors**

List of Contributors

A. Madhusudhana Reddy
Department of Botany
Yogi Vemana University
Kadapa-516 003

A. Rajendran
Department of Botany, School of Life Sciences, Bharathiar University
Coimbatore – 641 046

Ashish R.
Defence Institute of High Altitude Research, Defence Research and Development Organization (DRDO)
C/O 56 APO, Leh, Jammu and Kashmir

Ashutosh A. Murkute
Defence Institute of High Altitude Research, Defence Research and Development Organization (DRDO)
C/O 56 APO, Leh, Jammu and Kashmir

Ayyanar M.
Division of Ethnopharmacology, Entomology Research Institute
Loyola College, Chennai – 600 034
Tamil Nadu

B. Ravi Prasad Rao
Conservation Ecology Division
Department of Botany
Sri Krishnadevaraya University
Anantapur -515 003, A.P.

D. Sathiyan
State Forest College, R.S.Puram, Coimbatore

D. Thiyagaraja Guptha
Patent Office, GST Road, Chennai
Tamil Nadu

Deepu Mathew
Krishi Vigyan Kendra
Kerala Agricultural University
Tavanur – 679 573, Kerala

G. Ramanathan
Department of Microbiology
V.H.S.N. College, Virudhunagar
Tamil Nadu

Hemant Kr. Pandey
Defence Institute of Bio-energy Research
Defence Research and Development Organization, Pithoragarh – 262501
Uttarakhand

Ignacimuthu S.
Division of Ethnopharmacology
Entomology Research Institute
Loyola College, Chennai – 600 034
Tamil Nadu

Indramohan Singh
Centre for Plant Molecular Biology and Biotechnology, Tamil Nadu Agricultural University Coimbatore

J. Dharuman
Department of Pharmaceutical analysis
KMCH, Coimbatore

Janifer Raj X.
Defence Institute of High Altitude Research, Defence Research and Development Organization (DRDO)
C/O 56 APO, Leh, Jammu and Kashmir

Joshni T Chacko
School of Biotechnology
Karunya University, Coimbatore - 641 114

K. Kathiresan
Centre of Advanced Study in Marine Biology, Annamalai University
Parangipettai 608 502, Cuddalore District, Tamil Nadu

M. Geetha Rani
M.S. Swaminathan Research Foundation
Third Cross Street Taramani Institutional Area, Chennai- 600113, Tamil Nadu

M. Geetha
C. L. Baid Metha College of Pharmacy
Chennai-600 097

M. Muthukumar
Central Institute for Subtropical Horticulture, Rahmankhera
Lucknow, U.P.

M.S.M. Rawat
Hemwati Nandan bahu Guna Garhwal University, Srinagar

M. Anbazhagan
Department of Botany
Annamalai University
Annamalai Nagar, Chidambaram

M. Kalpana
Department of Botany, Annamalai University, Annamalai Nagar
Chidambaram

M. Pal Murugan
Defence Institute of High Altitude Research, Defence Research and Development Organization (DRDO)
C/O 56APO

M. Pugalenthi
Department of Botany
Government Arts College
Coimbatore-600 018, Tamil Nadu

Modhumita Ghosh
Division of Plant Biotechnology
Institute of Forest Genetics and Tree Breeding, Forest Campus
R.S. Puram, Coimbatore-641002

N. Kumaravadivel
Centre for Plant Molecular Biology and Biotechnology, Tamil Nadu Agricultural University Coimbatore

N. Anusuya
Department of Botany, School of Life Sciences, Bharathiar University
Coimbatore – 641 046

O.P. Chaurasia
Defence Institute of High Altitude Research, Defence Research and Development Organization (DRDO)
C/O 56 APO, Leh, Jammu and Kashmir

P.N. Sivalingam
Central Institute for Arid Horticulture
Bikaner, Rajasthan

P. Shanmughavel
Department of Bioinformatics
Bharathiar University
Coimbatore- 641 046, Tamil Nadu

Phani Kumar
Defence Institute of High Altitude Research, Defence Research and Development Organization (DRDO)
C/O 56 APO, Leh, Jammu and Kashmir

R. Chandra
Central Institute for Subtropical Horticulture, Rahmankhera, Lucknow, U.P.

R. Ilavarasen
Captain Srinivasa Murti Drug Research Institute for Ayurveda and Siddha (CCRAS, Mino of Health & Family Welfare Govt. of India), Arumbakkam
Chennai -600 106

R. Rajendran
Department of Botany, Annamalai University, Annamalai Nagar
Chidambaram

R.Sathish Kumar
Department of Biotechnology
School of Biotechnology, Bharathiar University, Coimbatore -641 046

Radhika Ramachandran
Department of Bioinformatics
Bharathiar University
Coimbatore- 641 046 Tamil Nadu

S. Anandhan
Defence Institute of Bio-Energy Research
Defence Research & Development Organization, Haldwani -Uttarakhand

S. Jacob Inbaneson
School of Marine Sciences, Centre for Marine Pharmacology, Alagappa University, Thondi Campus - 623 409
Ramanad District, Tamil Nadu

S. Ravikumar
School of Marine Sciences
Centre for Marine Pharmacology
Alagappa University
Thondi Campus-623 409, Tamil Nadu

S. Seal
Central Institute for Subtropical Horticulture, Rahmankhera, Lucknow, U.P.

S. Dhivya
Department of Biotechnology
School of Biotechnology, Bharathiar University, Coimbatore -641 046

S. Girija
Department of Biotechnology
School of Biotechnology, Bharathiar University, Coimbatore -641 046

S. Kalidass
School of Biotechnology, Karunya University, Coimbatore - 641 114

S. Manian
Department of Botany, School of Life Sciences, Bharathiar University
Coimbatore – 641 046

S. Rajeev Kumar
Department of Biotechnology
School of Biotechnology, Bharathiar University, Coimbatore -641 046

S. Sembian
School of Biotechnology, Karunya University, Coimbatore - 641 114

Saji Gomez
College of Agriculture
Keral Agricultural University,
Padannakkad, Kasaragode, Kerala

Seeja Thomachan Panjikkaran
Krishi Vigyan Kendra
Kerala Agricultural University
Tavanur – 679 573, Kerala

Shashi Bala Singh
Defence Institute of High Altitude Research, Defence Research and Development Organization (DRDO)
C/O 56 APO, Leh, Jammu and Kashmir

Sucheta Tripathy
Virginia Bioinformatics Institute
Virginia Polytechnic Institute and State University, Blacksburg, VA-24061, U.S.A.

Sunil E. Jadhav
Defence Institute of High Altitude Research, Defence Research and Development Organization (DRDO)
C/O 56 APO, Leh, Jammu and Kashmir

T. Jeya
School of Biotechnology, Karunya University, Coimbatore - 641 114

T. Parimelazhagan
Department of Botany, School of Life Sciences, Bharathiar University
Coimbatore-641 046, Tamil Nadu

V. Satheesh
National Research Centre on Plant Biotechnology, IARI, New Delhi

V. Vadivel
Department of Biotechnology
Karpagam University
Coimbatore-600 021
Tamil Nadu, India

V.S. Ramachandran
Department of Botany
School of Life Sciences
Bharathiar University
Coimbatore – 641 046

Warghat, G.
Defence Institute of High Altitude Research, Defence Research and Development Organization (DRDO)
C/O 56 APO, Leh, Jammu and Kashmir

Zakwan Ahmed
Defence Institute of Bio-energy Research, Defence Research & Development Organization
Haldwani – Uttarakhand

Contents

Foreword v
Front Line vii
Preface ix
List of Contributors xi

Chapter 1
Role of Traditional Medicinal Plants in Drug Discovery: Ethnobotanical Approaches **1**
M. Ayyanar and S. Ignacimuthu

Chapter 2
***Spilanthes acmella* - A Wonderful Medicinal Herb** **25**
S. Kalidass, S.Sembian, T. Jeya, Joshni T. Chacko

Chapter 3
Dactylorhiza Hatagirea (D. don) A Critically Endangered Medicinal Plant of Trans-Himalaya **37**
Ashutosh A. Murkute, Ashish R. Warghat, G. Phani Kumar G.P. Mishra and Sunil E. Jadhav

Chapter 4
High Altitude Medicinal Plants - A Rich Source of Valuble Phytoconstituents **47**
Janifer Raj, X, O.P. Chaurasia , M. Pal Murugan and Shashi Bala Singh

Chapter 5

Rare Medicinal Plants of Western Ghats in Kerala and their Ayurvedic Uses **63**

Deepu Mathew and Seeja Thomachan Panjikkaran

Chapter 6

Uses of Medicinal Plants to Treat Human Disease Among the Tribals at Pachamalai Hills, Tamil Nadu **81**

M. Geetha Rani

Chapter 7

Some Therapeutic Uses of Biodiversity of the Kolli Hills Eastern Ghats, Tamil Nadu, India **95**

Rajendran, A. & S. Manian

Chapter 8

Ethno-Medico Plants in Courtallum Hills, Southern Western Ghats, India **113**

Sivalingam. R & S.Manian

Chapter 9

Medicinal Plant Wealth of Palakonda Hill Ranges, Kadapa District, Andhra Pradesh, India **125**

A. Madhusudhana Reddy and B. Ravi Prasad Rao

Chapter 10

Pharmaceutical Properties of L-Dopa and its Therapeutic Application **135**

M. Pugalenthi, V. Vadivel and T. Parimelazhagan

Chapter 11

Antioxidant Activity and Hypoglycemic Property of a South Indian Under-Utilized Food Legume *Acacia nilotica* DC. **155**

V. Vadivel, M. Pugalenthi and T. Parimelazhagan

Chapter 12

Post Harvest Handling and Processing of Potential Nutraceutical Crops **177**

Saji Gomez

Chapter 13
Quality Control of Herbal Plant Material 201
M. Geetha

Chapter 14
Toxicity Evaluation of Herbal Drugs in Animal Models .. 211
R. Ilavarasan

Chapter 15
Standardization of Marine Herbal Preparations for the Management of Malaria 223
S. Ravikumar, G. Ramanathan, K. Kathiresan and S. Jacob Inbaneson

Chapter 16
***In vitro* Culture and Studies on Secondary Metabolites in *Nothapodytes nimmoniana* 245**
Singh, Indramohan and N. Kumaravadivel

Chapter 17
Modern Techniques Involving Tissue Culture for Medicinal Plants 259
M. Anbazhagan, R. Rajendran and M. Kalpana

Chapter 18
DNA Barcoding of Indian Medicinal Plants - A Case Study 279
S. Dhivya, S. Rajeev Kumar, V. S. Ramachandran and R. Sathishkumar

Chapter 19
Metabolic Engineering for Therapeutics 291
M. Muthukumar, S. Seal, P.N. Sivalingam, R. Chandra, V. Satheesh and S. Anandhan

Chapter 20
The Role of High Performance Liquid Chromatography in Herbal Research 307
J. Dharuman

Chapter 21
Medicinal and Neutraceutical Properties of Some less Explored *Allium* Species of Uttarakhand Himalayas : A Potential Future Crop **315**
Hemant Kr. Pandey, M.S.M. Rawat and Zakwan Ahmed

Chapter 22
Natural Antioxidant Food Supplements and Human Health **331**
N. Anusuya and S. Manian

Chapter 23
Cancer and Diet **341**
Sucheta Tripathy

Chapter 24
Plant Antifungal Protein as Therapeutic Agents **361**
Modhumita Dasgupta

Chapter 25
Medicinal Plant Databases - A Review **381**
P. Shanmughavel

Chapter 26
***In silico* Structural Analysis of Hypothetical Proteins : A Therapeutic Approach** **391**
R. Radhika and P. Shanmughavel

Chapter 27
Herbs as Green Nanofactories **411**
S.Girija

Chapter 28
Status of Biodiversity and Conservation **423**
D.Sathiyan

Chapter 29

Intellectual Property Rights : Protection of Herbal Products and Plant Varieties **437**

D.Thiyagaraja Guptha

Chapter 30

Role of Plant Secondary Metabolites in Ruminant Nutrition **449**

Jadhav Sunil Ekanath, Guru Charan, Tilak Raj Ashutosh Murkute & Shashi Bala Singh

1

Role of Traditional Medicinal Plants in Drug Discovery: Ethnobotanical Approaches

M. Ayyanar and S. Ignacimuthu

CONTENTS

1.1 Introduction 2

1.2 Traditional Medicine 2

1.3 Medicinal Plants in the Present Scenario 3

1.4 Traditional Medical Systems 4

- 1.4.1 African Traditional Medicine 4
- 1.4.2 American Traditional Medicine (North, Central and South) 5
- 1.4.3 Australian and Southeast Asian Medicine 5
- 1.4.4 Ayurvedic Medicine (Indian Traditional Medicine) 6
- 1.4.5 Chinese Traditional Medicine 6
- 1.4.6 European Medicine 6
- 1.4.7 Classical Arabic and North African Traditional Medicine 7

1.5 Medicinal Plants in Traditional Medicine 7

1.6 Role of Ethnobotanical Research in Drug Discovery 8

1.7 Ethnobotanical Studies on Indian Medicinal Plants 12

1.8 Conclusion 20

References 21

1.1 Introduction

Plants and plant based medications have been man's prime therapeutic tools since the dawn of civilization and are still among the frontline treatments for a variety of ailments. Since time immemorial, man has used various parts of plants in the treatment and prevention of various ailments (Chah *et al.*, 2006). According to the World Health Organization (WHO), about 65-80% of the world's population in developing countries, due to poverty and lack of access to modern medicine, depend essentially on plants for their primary healthcare (Calixto, 2005). Besides serving medical and cultural functions, medicinal plants in developing countries have an important economic role. Of the 252 drugs considered as basic and essential by the World Health Organization, 11% are exclusively of plant origin and a significant number are synthetic drugs obtained from natural precursors (Rates, 2001). About 200 years ago our pharmacopoeia was dominated by herbal medicines (Ernst, 2005). Herbal medicines are comparatively safer than synthetic drugs. Plant-based traditional knowledge has become a recognized tool in search for new sources of drugs and neutraceuticals and the ethnobotanical survey can bring out many different clues for the development of drugs to treat human diseases (Ayyanar and Ignacimuthu, 2005a).

1.2 Traditional Medicine

Traditional Medicine (TM) is a comprehensive term used to refer both to TM systems such as traditional Chinese medicine, Indian Ayurveda and Arabic Unani medicine, and to various forms of indigenous medicine. It is a sum total of the knowledge, skills, and practices based on the theories, beliefs, and experiences indigenous to different cultures, whether explicable or not, used in the maintenance of health as well as in the prevention, diagnosis, improvement or treatment of physical and mental illness. WHO defines traditional medicine as including diverse health practices, approaches, knowledge and beliefs incorporating plant, animal, and/or mineral based medicines, spiritual therapies, manual techniques and exercises applied singularly or in combination to maintain well-being, as well as to treat, diagnose or prevent illness. In countries where the dominant healthcare system is based on allopathic medicine, or where TM has not been incorporated into the national healthcare system, TM is often termed "complementary", "alternative" or "non-conventional" medicine.

The Chinese Materia Medica has been extensively documented over the centuries, with the first record dating from about 1100 BC and the documentation of the Indian Ayurvedic system dates from about 1000 BC; this system formed the basis for the primary text of Tibetan Medicine, Gyu-zhi (Four Tantras), translated from Sanskrit during the eighth century AD. In the ancient Western world, the Greeks contributed substantially to the rational development of the use of herbal drugs (Cragg and Newman, 2001). The practice of traditional medicine is widespread in Asian countries such as China, India, Japan, Pakistan, Sri Lanka and Thailand.

Traditional medical knowledge of medicinal plants and their use by indigenous cultures are not only useful for conservation of cultural traditions and biodiversity but also for community healthcare and drug development in the present and future (Pei, 2001). Several effective medicines used in their natural state such as cinchona, opium, belladonna and aloe were selected as therapeutic agents based on empirical evidence of their clinical application by traditional societies from different parts of the world (Iwu *et al.*, 1999).

In the past decade there has been renewed attention and interest in the use of traditional medicine globally. In China, traditional medicine accounts for about 40% of all healthcares delivered. In Chile 71% of the population, and in Colombia 40% of the population, have used such medicine. In India, 65% of the population in rural areas uses Ayurveda and medicinal plants to help meet their primary healthcare needs. In developed countries, traditional, complementary and alternative medicines are becoming more popular. For example, the percentage of the population that has used such medicines at least once is 48% in Australia, 31% in Belgium, 70% in Canada, 49% in France and 42% in the United States of America (WHO, 2003).

1.3 Medicinal Plants in the Present Scenario

Worldwide, a total of more than 35,000 plant species are used for medicinal purposes. But, there is no reliable figure for the total number of medicinal plants on Earth, and numbers and percentages for countries and regions vary greatly. The value of medicinal plants to human livelihoods is essentially infinite. India and China are the leading countries with more than 40% species of plants market (Table 1) employed in medicinal plant (Hamilton, 2004).

Table 1 : Numbers and percentage of medicinal plant species

Country	Total number of species in flora	Number of medicinal plants	Percentage of flora which is medicinal
China	27100	11146	41
India	17000	7500	44
Mexico	30000	2237	7
North America	20000	2572	13
World	297000	52885	10-18

They obviously make fundamental contributions to human health, and: "Is not health dearer than wealth?". A study of the 25 best-selling pharmaceutical drugs in 1997 found that 11 of them (42%) were either biologicals, natural products or entities derived from natural products, with a total value of US$ 17.5 billion. The total sales value of drugs (such as Taxol) derived from just one plant species (*Taxus baccata*) was US$ 2.3 billion in 2000. The world market for herbal remedies in 1999 was calculated to be worth US$ 19.4 billion, with Europe in the lead (US$ 6.7 billion), followed by Asia (US$ 5.1 billion), North America (US$ 4.0 billion), Japan (US$ 2.2 billion), and then the rest of the world (US$ 1.4 billion). The industrial demand on medicinal plant resources is poised to increase owing to the worldwide buoyancy in the herbal sector engaged in production of herbal health care formulations, herbal based cosmetic products and herbal nutritional supplements (Ved and Goraya, 2007).

The recovery of the knowledge and practices associated with these plant resources are part of an important strategy linked to the conservation of biodiversity, the discovery of new medicines, and the bettering of the quality of life of poor rural communities (Almeida *et al.*, 2006). The importance of medicinal plants is demonstrated by the fact that even in developed countries, about 35% of prescribed drugs are of natural origin and 50% of drugs are of plant origin (Nyiredy, 2004). The primary benefits of using plant derived medicines are that they are relatively safer than synthetic alternatives, offering profound therapeutic benefits and more affordable treatment (Iwu *et al.*, 1999). Safe, effective and inexpensive indigenous remedies are gaining popularity among the people of both urban and rural areas, especially in India and China.

1.4 Traditional Medical Systems

The traditional medicine all over the world is nowadays revalued by an extensive study of results of research on different plant species and their therapeutic principles (Scartezzini and Speroni, 2000). There is an urgent need to inventory and record all ethnobiological information among the diverse ethnic communities before the traditional cultures are completely lost. The various traditional medical systems practiced throughout the world are African, American, Australian, Ayurvedic, Chinese, European, Classical Arabic and North African traditional medicine (Gurib-Fakim, 2006).

1.4.1 African traditional medicine

African traditional medicine is the oldest and perhaps the most diverse of all medicine systems. African traditional medicines in its varied forms, is holistic involving both the body and the mind. The healer typically diagnoses and treats the psychological basis of an illness before prescribing medicines to treat the symptoms. Some important African medicinal plants are: *Acacia senegal* (Gum Arabic), *Agathosma betulina* (Buchu), *Aloe ferox* (Cape Aloes), *Aloe vera* (North African Origin), *Artemisia afra* (African Wormwood), *Aspalanthus linearis* (Rooibos Tea), *Boswellia sacra* (Frankincense), *Catha edulis* (Khat), *Commiphora myrrha* (Myrrh), *Harpagophytum procumbens* (Devil's Claw), *Hibiscus sabdariffa* (Hibiscus, Roselle), *Hypoxis hemerocallidea* (African Potato), *Prunus africana* (African Cherry).

1.4.2 American traditional medicine (North, Central and South)

In the US, just like in any other cultures the indigenous healer or Shaman approaches illnesses by addressing both the physical and spiritual dimension of diseases. These Shamanistic ceremonies involve chanting, dancing and other rituals aimed at expelling evil forces so that the patient or the community as a whole can be healed. Early settlers learnt from native practices and they eventually adopted many of the herbal remedies, which later formed the basis of the Pharmacopeia of the United States. Famous medicinal plants of the United States are Echinacea (*Echinacea purpurea*) and Goldenseal (*Hydrastis canadensis*).

Just like Africa, Central and South American countries have a rich and diverse healing cultures which are poorly known and have not been properly recorded. They will no doubt be a source of new herbal remedies in the years to come. Famous examples of medicinal plants used by them are: *Cinchona pubescens* (Peruvian bark), *Erythroxylum coca* (Coca), *Ilex paraguariensis* (Mate), *Myroxylon balsamum* (Tolu balsam), *Paullinia cupana* (Guarana), *Peumus boldus* (Boldo), *Psidium guajava* (Guava), *Spilanthes acmella* (Brazilian cress), *Tabebuia impetiginosa* (Lapacho) and *Uncarina tomentosa* (Cat's claw).

1.4.3 Australian and Southeast Asian medicine

This region has witnessed a resurgence of interest in traditional medicine and many countries now promote research into medicinal plants as a potential source of new remedies. The Aborigines had a complex healing system but much of the traditional knowledge in Australia was lost before it could be systematically recorded. In contrast, many healing places like Malaysia, Thailand, Vietnam, New Zealand, Borneo, and the Polynesian Islands remain intact and are being recorded and developed. A strong Chinese influence is being observed in most countries. Among the well-known medicinal products originating from this region are *Croton tiglium* (Purging croton), *Duboisia hopwoodii* (Pituri), *Eucalyptus globulus* (Bluegum), *Melaleuca alternifolia* (Tea tree), *Myristica fragrans* (Nutmeg and Mace), *Piper methysticum* (Kava kava), *Strychnos nux-vomica* (Strychnine), *Styrax benzoin* (Benzoin) and *Syzygium aromaticum* (Cloves).

1.4.4 Ayurvedic medicine (Indian Traditional Medicine)

Ayurveda is perhaps, the most ancient of all medicinal traditions and is probably older that the traditional Chinese medicine. It is considered to be the origin of systemized medicine. It is actually a practical and holistic set of guidelines to maintain balance and harmony in the system. Ancient Hindu writings on medicine contain no references to foreign medicines whereas Greek and Middle Eastern texts do refer to ideas and drugs of Indian origin. Ayurveda is derived from the Indian words Ayur (Life) and Veda (Knowledge or Science) and hence means the Science of Life. Ayurveda is similar to Galenical Medicine in that it is based on bodily humours (dosas) and the inner life force (prana) that is believed to maintain digestion and mental activity. The living and the non-living environment, including humans, is composed of the elements earth (prithvi), water (jada), fire (tejac), air (vaju) and space (akasa). Some of the important Ayurvedic medicinal plants are *Azadirachta indica* (Neem), *Centella asiatica* (Gotu Kola), *Cinnamomum camphora* (Camphor), *Elettaria cardamomum* (ela), *Rauvolfia serpentina* (Indian snake root), *Santalum album* (Sandalwood), *Terminalia sp.* (Myrobolan) and *Withania somnifera* (Aswagandha).

1.4.5 Chinese traditional medicine

The civilizations of China and India were flourishing when only modestly sophisticated cultures were developing in Europe. This ancient system of medicine, believed to be more than 5000 years old, is based on two separate theories about the natural laws that govern good health and longevity, namely yin and yang, and the five elements (wu xing). The most complete reference to Chinese herbal prescription is the Modern Day Encyclopedia of Chinese Materia Medica published in 1977. It lists nearly 6000 drugs out of which 4800 are of plant origin. The spread of traditional Chinese medicine to most continents has undoubtedly contributed to the current popularity of herbal medicines throughout the world. Examples of famous Chinese medicinal herbs are *Angelica polymorpha* var. *sinensis* (dang gui), *Artemisia annua* (qing hao), *Ephedra sinica* (ma huang), *Paeonia lactiflora* (bai shao yao), *Panax ginseng* (ren shen) and *Rheum palmatum* (da huang).

1.4.6 European medicine

In the ancient Western world, the Greeks contributed significantly to the rational development of the use of herbal drugs. However, the European healing system is said to have originated with Hippocrates (460–377 BC) and Aristotle (384–322 BC), whose own ideas were rooted in ancient beliefs from India and Egypt. The philosopher and natural scientist, Theophratus (300 BC), in his History of Plants, dealt with the medicinal qualities or herbs, and noted the ability to change their characteristics through cultivation. European tradition has many regional influences on local folk practices and traditions.

A large number of traditional herbal remedies in Europe have become widely known as a result of commercialization and a number of active compounds have been isolated from medicinal plants and are used today as single chemical entities.

1.4.7 Classical Arabic and North African traditional medicine

The Middle East is known as the cradle of civilization and many plants grown nowadays have been domesticated in this region. The Babylonians, Assyrians and Sumerians recorded herbal remedies in cuneiform writing on numerous clay tablets. The Egyptians documented their knowledge in wall paintings of tombs dating from the Old Kingdom and on papyrus, the latter being made from *Cyperus aquaticus*. The Arabs were the first to establish privately owned drug stores in the 8th century, and the Persian pharmacists, physician, philosopher and poet, Avicenna, contributed much to the sciences or pharmacy and medicine throughout the works such as *Canon medicinae*. It includes elements of other healing cultures and forms the basis for a distinct Islamic healing system known today as Unani-Tibb. Some of the important medicinal plants of Middle East and Egypt are: *Allium cepa* (Onion), *Astracantha gummifera* (Tragacanth), *Carthamus tinctorius* (Safflower), *Carum carvi* (Caraway), *Ferula assafoetida* (Asofoetida), *Lawsonia inermis* (Henna), *Papaver somniferum* (Opium poppy), *Peganum harmala* (Syrian rue), *Prunus dulcis* (Almond), *Punica granatum* (Pomegranate), *Rosa damascena* (Damask Rose), *Ricinus communis* (Castor Oil), *Salvadora persica* (Toothbrush tree), *Senna alexandrina* (Senna), *Sesamum indicum* (Sesame), *Trachyspermum ammi* (Ajowan), *Trigonella foenum-graecum* (Fenugreek) and *Vitis vinifera* (Grape).

1.5 Medicinal Plants in Traditional Medicine

Plants have formed the basis of sophisticated traditional medicine systems that have been in existence for thousands of years and continue to provide mankind with new remedies. In recent years, use of ethnobotanical information in medicinal plant research has gained considerable attention in segments of the scientific community (Heinrich, 2000). Medicinal plants are the important part of indigenous pharmaceutical systems. Interest in medicinal plants has been fuelled by the rising costs of prescription drugs in the maintenance of personal health and well-being, and the bioprospecting of new plant-derived drugs (Hoareau and DaSilva, 1999).

Historically, all medicinal preparations were derived from plants, whether in the simple form of plant parts or in the more complex form of crude extracts, mixtures, etc. The vast majority of people on this planet still rely on their traditional Materia Medica for their everyday healthcare needs. Many commercially proven drugs used in modern medicine were initially tried in crude form in traditional or folk healing practices, or for other purposes that suggested potentially useful biological. About 25% of the drugs prescribed worldwide come from plants, 121 such active compounds being in current use.

There are hundreds of significant drugs and biologically active compounds developed from the traditional medicinal plants, a few of which are mentioned here; the antispasmodic agent vasicin isolated from *Adhatoda vasica*, anticancer agents such as vincristine, vinblastine and D-Tubocurarine isolated from *Catharanthus roseus* (Gurib-Fakim, 2006), antibacterial agents isolated from *Diospyros melanoxylon* (Mallavadhani *et al.*, 1998), antimalarial agent isolated from *Sida acuta* (Karou *et al.*, 2006), steroid and lancamarone with cardiotonic properties, lantamine with antipyretic and antispasmodic properties from *Lantana camara* (Ghisalberti, 2000), antimicrobial agents isolated from *Acorus calamus* (Chowdhury *et al.*, 1993), antiviral, antibacterial and anti-inflammatory agents isolated from *Urtica dioica* (Harborne and Buxter, 1993), anticancer agents isolated from *Andrographis paniculata, Phyllanthus amarus, Piper longum, Semecarpus anacardium, Withania somnifera, Moringa oleifera, Aloe vera, Curcuma longa, Allium sativum* and *Tinospora cordifolia* (Balachandran and Govindarajan, 2005), promising and potent antimalarial drug artemisinin isolated from *Artemesia annua* (Dhingra *et al.*, 2000). A large proportion of such drugs have been discovered with the aid of ethno- botanical knowledge of the traditional uses of the plant.

1.6 Role of Ethnobotanical Research in Drug Discovery

Harshberger (1896) defined the purpose of ethnobotany as an aid to elucidate cultural positions of tribes according to the uses of plants, evidence of past distribution and trade routes and useful in suggesting the valuable properties they utilized in their wildlife filling some vacant niche in our own (Wickns, 1990). The ethnobotanical surveys bring out many different clues for the development of drugs to treat human diseases. During the second half of the last century, there has been a tension between more modern approaches to drug discovery including combinational chemistry, rational drug design based upon computer modeling, functional genomics and proteomics, and, traditional approaches based on discovery of novel bioactive molecules from plants (Alan Cox, 2005). Most of the secondary plant compounds employed in modern medicine were first discovered through ethnobotanical investigation (Gurib-Fakim, 2006).

It is estimated that about 75% of the 120 biologically active plant derived compounds (Table 2) presently in use worldwide have been derived through follow-up researches to verify the authenticity of data from folk and ethnomedical uses and there is great scope for new drug discoveries based on traditional plant uses (Pushpangadan and Kumar, 2005). Table 2 provides the important drugs with active compounds and their clinical uses which were developed from the ethnomedicinal source (Fabricant and Farnsworth, 2001).

Table 2 : Drugs derived from plants with their ethnomedical correlations and sources

Plant Source	Action/Clinical Use	Drug/Compound
Adonis vernalis L.	Cardiotonic	Adoniside
Aesculus hippocastanum L.	Anti-inflammatory	Aescin
Agrimonia eupatoria L.	Anthelmintic	Agrimophol
Ammi majus L.	Leukoderma; vitiligo	Xanthotoxin
Ammi visnaga (L.) Lamk.	Bronchodilator	Khellin
Anamirta cocculus (L.) W.&A.	Analeptic	Picrotoxin
Ananas comosus (L.) Merrill	Anti-inflammatory	Bromelain
Andrographis paniculata Nees	Bacillary dysentery	Andrographolide, Neoandrographolide
Anisodus tanguticus Pascher	Anticholinergic	Anisodamine, Anisodine
Ardisia japonica Bl.	Antitussive	Bergenin
Areca catechu L.	Anthelmintic	Arecoline
Artemisia maritima L.	Ascaricide	Santonin
Atropa belladonna L.	Anticholinergic	Atropine
Berberis vulgaris L.	Bacillary dysentery	Berberine
Brassica nigra (L.) Koch	Rubefacient	Allyl isothiocyanate
Camellia sinensis (L.) Kuntze	CNS stimulant Diuretic; bronchodilator	Caffeine Theophylline
Carica papaya L.	Proteolytic; mucolytic	Chymopapain, Papain
Cassia spp.	Laxative	Danthron, Sennosides A & B
Centella asiatica (L.) Urban	Vulnerary	Asiaticoside
Cephaelis ipecacuanha	Amoebicide; emetic	Emetine
Chondodendron tomentosum	Skeletal muscle relaxant	Tubocurarine

Contd...

Table 2 Contd...

Cinchona ledgeriana Trimen	Antimalarial	Quinine
Colchicum autumnale L.	Antitumor agent; antigout	Colchicine
Convallaria majalis L.	Cardiotonic	Convallotoxin
Corydalis ambigua	Analgesic; sedative	Tetrahydropalmatine
Crotolaria sessiliflora L.	Antitumor agent	Monocrotaline
Curcuma longa L.	Choleretic	Curcumin
Cynara scolymus L.	Choleretic	Cynarin
Daphne genkwa Seib. & Zucc.	Abortifacient	Yuanhuacine
Datura metel L.	Sedative	Scopolamine
Digenea simplex Wulf.	Ascaricide	Kainic Acid
Digitalis lanata Ehrh.	Cardiotonic	Acetyldigoxin, Deslanoside, Digoxin, Lanatosides A, B, C
Digitalis purpurea L.	Cardiotonic	Digitalin, Digitoxin, Gitalin
Ephedra sinica Stapf.	Sympathomimetic	Ephedrine
Ephedra sinica Stapf.	Sympathomimetic	Pseudoephedrine
Erythroxylum coca Lamk.	Local anaesthetic	Cocaine
Fraxinus rhynchophylla Hance	Antidysentery	Aesculetin
Glycyrrhiza glabra L.	Sweetener	Glycyrrhizin
Gossypium spp.	Male contraceptive	Gossypol
Helmsleya amabilis Diels	Bacillary dysentery	Hemsleyadin
Hydrangea macrophylla DC	Sweetener	Phyllodulcin
Hydrastis canadensis L.	Hemostatic; astringent	Hydrastine
Hyoscamus niger L.	Anticholinergic	Hyoscamine
Lobelia inflata L.	Smoking deterrent	Lobeline
Lonchocarpus nicou Aubl.	Piscicide	Rotenone
Papaver somniferum L.	Analgesic; antitussive	Codeine, Morphine, Noscapine
Pausinystalia yohimbe	Aphrodisiac	Yohimbine
Physostigma venenosum Balf.	Cholinesterase inhibitor	Physostigmine

Contd...

Pilocarpus jaborandi Holmes	Parasympathomimetic	Pilocarpine
Piper methysicum Forst. f.	Tranquilizer	Kawain
Podophyllum peltatum L.	Antitumour agent	Etoposide, Teniposide
Potentilla fragaroides L.	Haemostatic	(+)-Catechin
Quisqualis indica L.	Anthelmintic	Quisqualic acid
Rauvolfia canescens L.	Antihypertensive	Deserpidine
Rauvolfia serpentina Kurz	Antihypertensive; tranqulizer; Circulatory disorders	Rescinnamine, Reserpine, Ajmalicine
Rhododendron molle G. Don	Antihypertensive	Rhomitoxin
Rorippa indica (L.) Hochr.	Antitussive	Rorifone
Salix alba L.	Analgesic	Salicin
Silybum marianum L.	Antihepatotoxic	Silymarin
Simarouba glauca DC.	Amoebicide	Glaucaroubin
Stephania sinica Diels	Analgesic; sedative	Rotundine
Stevia rebaudiana Bertoni	Sweetener	Stevioside
Strophanthus gratus Baill.	Cardiotonic	Ouabain
Strychnos nux-vomica L.	CNS stimulant	Strychnine
Theobroma cacao L.	Diuretic; bronchodilator	Theobromine
Thymus vulgaris L.	Abortifacient	Trichosanthin
Urginea maritima (L.) Baker	Cardiotonic	Scillarin A
Valeriana officinalis L.	Sedative	Valepotriates
Veratrum album L.	Antihypertensive	Protoveratrines A, B
Vinca minor L.	Cerebral stimulant	Vincamine

Aspirin, atropine, artimesinin, colchicine, digoxin, ephedrine, morphine, physostigmine, pilocarpine, quinine, quinidine, reserpine, taxol, tubocurarine, vincristine, and vinblastine are a few important examples of what medicinal plants have given us in the past. Most of these plant-derived drugs were originally discovered through the study of traditional cures and folk knowledge of indigenous people and some of these could not be substituted despite the enormous advancement in synthetic chemistry (Gilani and Atta-ur-Rahman, 2005).

Drug discovery from medicinal plants on the basis of traditional medicinal practices continues to provide new and important leads against various pharmacological targets including cancer, HIV/AIDS, Alzheimer's, malaria, and pain. Several natural product drugs of plant origin have either recently been introduced to the United States market, including arteether, galantamine, nitisinone, and tiotropium, or are currently involved in late-phase clinical trials (Balunas and Kinghorn, 2005). Plants have always been an exemplary source of drugs and many of the currently available drugs have been derived directly or indirectly from them (Grover *et al.*, 2002). Despite the number of clinical agents developed by the pharmaceutical industry, traditional indigenous phytotherapy is still practiced in many rural areas, using treatments handed down from generation to generation.

1.7 Ethnobotanical Studies on Indian Medicinal Plants

India is one of the twelve mega-biodiversity countries of the World having rich vegetation with a wide variety of plants with medicinal value. The tribal population of the country, as per the 2001 census, is 8.43 crore, constituting 8.2% of the total population (MTA, 2005-06) and has the second largest tribal population in the world after Africa. With enormously diversified living ethnic groups and rich biological resources, India represents one of the great emporia of ethnobotanical wealth (Pal, 2000). Even today, tribals and certain local communities in most of the countries are still collecting and preserving locally available, wild and cultivated plant species and practice herbal medicine to cure a variety of diseases and disorders (Ayyanar and Ignacimuthu, 2005b; Ignacimuthu *et al.*, 2006).

The Indian flora is extensively utilized as source of many drugs mentioned in the traditional systems of medicine. In India, record of the utility of natural and plant products to cure ailments appeared in *Rigveda* dating back to 4500 - 1600 BC., wherein 67 plant species were mentioned which have been botanically equated later on. The *Yajurveda* includes 81 prescriptions for curing several diseases, of which 100 sanskrit names of herbs have been identified botanically. A number of therapeutic recipes, time, method of treatment and trade on herbs have also been given in the *Atharvaveda* and about 290 herbs named in Sanskrit have been botanically equated (Singh *et al.*, 2001).

Medicinal plants have traditionally occupied an important position in the socio-cultural, spiritual and medicinal arena of rural and tribal lives of India. Based on the different publications, it has been estimated that there are about 2000 medicinal plants in India, and more than 6780 industries are engaged in the production of medicines employed in different systems of practice such as, Allopathic, Ayurvedic, Unani and Homeopathic. It is worthwhile to note that, about 80% of the human populations in India are still dependent on nature for remedies and this can be well understood from the fact that almost all systems of medicine are largely based on drugs of plant origin (Singh *et al.*, 2001).

Primarily, Janaki Ammal (1956) conceptualized the importance of ethnobotanical studies in India and initiated such studies. After that from the year 1963, Jain (1964) extended these studies to the forefront through his pioneering research and publications. As a final point, Jain published a book on Dictionary of Indian Folk medicine and Ethnobotany, which was prepared by the compilation of research carried out by different researchers with the different tribal communities in India with documentation of more than 2000 medicinal plants used by tribal communities in India up to 1990 (Jain, 1991). In South India also the ethnobotanical effort was first started by Janaki Ammal with the first publication on ethnomedicinal uses of *Costus speciosus* among the Kani tribals (Janaki Ammal and Nagendra Prasad, 1984). These studies not only formed the foundations of modern ethnobotany but also triggered off a number of ethnobiological studies and inspired a number of research workers from different parts of the country.

In our ethnobotanical survey among the *Kani* tribals and *Paliyar* tribals in Western Ghats of Tamil Nadu, the traditional healers noticed that, most of the medicinal plants found in their environs are used frequently for the treatment of various diseases and most of them have a general knowledge of medicinal plants that are used for first aid remedies, to treat cough, cold, fever, headache, poisonous bites and some other simple ailments (Ayyanar and Ignacimuthu, 2005a, b; Ignacimuthu *et al.*, 2006; 2008).

The plants listed in table 3 are known to contain various active principles of therapeutic value and to possess biological activity against a number of diseases. However, from this small percentage innumerable therapeutically indispensable compounds have been isolated such as alkaloids various glycosides, steroids, vitamins and flavonoids and a large range of antibiotics. This type of research must be promoted as a means for developing countries to understand the potential use of their plant resources, as well as a means to better promote basic healthcare. It is evident that therapeutically interesting and important drugs can be developed from plant sources that are used in traditional systems of medicines. The curative properties of drugs are due to the presence of complex chemical substances of varied composition in one or more parts of these plants. These plant metabolites according to their composition, are grouped as alkaloids, glycosides, corticosteroids, essential oils, etc.

Table 3 : Common medicinal plants used in Indian Traditional Medicine

Botanical Name & Family	Active Compounds*	Reported pharmacological activities**
Abutilon indicum Malvaceae	Leaves and root contain asparagin	Hepatoprotective & hypoglycemic activity
Acalypha indica Euphorbiaceae	Alkaloids acalypus and acalyphine, cyanogenetic glucoside, triacetanamine	Antioxidant, wound healing, anti-fertility and antibacterial activity
Achyranthes aspera Amaranthaceae	Alkaline	Contraceptive spermicidal, cancer chemopreventive, hypoglycaemic, anti-inflammatory and arthritic activity
Adhatoda vasica Acanthaceae	Vasicine, adhatodic acid, 1-pegamine, synthetic 1-pegamine	Hepatoprotective and antituberculosis activity
Allium cepa Apiaceae	Acrid volatile oil, albuminoids, soluble carbohydrates and sugar, catechol and protocatechuic acids	Antidiabetic, antimicrobial and antileishmanial activity
Aloe vera Liliaceae	Aloin, isobarbaloin, emodin, gum, resin, chrysophanic acid, urinic acid, oxidase, catalase and sugars	Anticancer, anti-inflammatory, antidiabetic and wound healing activity
Alternanthera sessilis Amaranthaceae	Protein and iron	Antioxidant and anti-molluscicidal activity
Areca catechu Arecaceae	Choline, isoguvocine, catechu, α-catechin, tannin, gallic acid, alkaloids, arecoline, arecaine, guvacine, arecaidine,guvacoline, guvacine,	Antioxidant, hypoglycemic and active-oxygen scavenging activity
Azadirachta indica Meliaceae	Margosine, margosic acid, margosopicrin, glycerides of fatty acids, butyric acid, valeric acid, neutral resins and acid resins, nimbin, nimbidin, nimbinin, nimbisterin, bakayanin	Anti-plasmodial, anticancer, antioxidant, antisecretory and antiulcer activity

Contd...

Table 3 Contd...

Bambusa arundinacea Poaceae	Silicic acid, peroxides of iron, cholin, betain, nuclease, urease, proteolytic, emulsifying and diastatic enzymes, cyanogenetic glucoside, benzoic acid, reducing sugar, resins and cyanogenetic glucoside	Antifertility, anti-inflammatory and antiulcer activity
Capsicum frutescens Solanaceae	Capsaicin, fatty acid, fixed oil, resin, solanine, volatile alkaloids,volatile oil, oleo-resin capsicin	Antimicrobial & antivenom activity
Cardiospermum halicacabum Sapindaceae	Essential oil and saponin	Antiulcer, anti-inflammatory & antimalarial activity
Carica papaya Caricaceae	Papain or papayotin, chymopapain, carpaine, caricin, carposide, citric, tartaric, malic, palmitic, crystalline, papayic and carica-fat acids, dextrin and resin	Antifertility, antioxidant and contraceptive activity
Carissa carandas Apocynaceae	Alkaloids, salicylic acid	Histamine releasing & cardiotonic activity
Cassia auriculata Caesalpiniaceae	Tannins, di-(2-ethyl) hexyl phthalate	Antioxidant and free-radical-scavenging property, nephroprotective, antibacaterial, antihyperglycemic and antioxidant activity
Centella asiatica Apiaceae	Vallarine, asiaticoside and oxy-asiaticoside, essential oil, fatty oil, sitosterol, tannin, pectic acid,ascorbic acid, alkaline hydrocotyline and resins	Anticancer, antiulcer and antioxidant activity
Cissus quadrangularis Vitaceae	Carotene, calcium oxalate, ascorbic acid	Analgesic, anti-inflammatory and antiulcer activity
Citrus aurantifolia Rutaceae	Citric acid, citrol, limonene, linalool, linalyl acetate, cymene, petitgrain oil, coumatins, iso-pimpinellin, bergapten, citropten, xanthyletin	Antiviral, antibacterial & anticancer activity

Contd...

Table 3 Contd...

Cleome viscosa L. Cleomaceae	glucocapparin and glucocleomin, diterpene lactone, cleomoscosin A, cleomiscosin D, a minor coumarino-lignan and Kaempferide 3-glucuronide	Antimicrobial, antipyretic, analgesic and antidiarrheal activity
Clitoria ternatea Fabaceae	Tannin, resin, fixed oil, tannic acid and starch	Antimicrobial, anti-inflammatory, analgesic and antipyretic activity
Coccinia indica Cucurbitaceae	Glucokenin, alkaloid, fatty acid, amylase, starch, sugar, gum, enzymes and hormones	Antioxidant, antidiabetic and hyperglycemic activity
Eclipta prostrata Asteraceae	Alkaline ecliptine, nicotine	Antihyperlipidemic, antivenom and antimicrobial activity
Elettaria cardamomum Zingiberaceae	Fixed, essential and volatile oil, terpinyl acetate, free terpineol, terpinene, sabinene, limonene, ligneous fibre	Gastro protective, DPPH radical-scavenging and anti-inflammatory activity
Euphorbia hirta Euphorbiaceae	Alkaloid, essential oil, l-inositol and alkaline xanthorhamnin, caoutchouc, resin, tannins, sugar, mucilage, calcium oxalate, carbohydrates, albuminoids, gallic acid, quercitin and a phenol	Antimicrobial and antimalarial activity
Evolvulus alsinoides Convolvulaceae	Neutral fat, alkaloid, organic acid and saline substances	Adaptogenic, anti-amnesic, antioxidant and antharthritic activity
Ficus bengalensis Moracaee	Tannin, wax, caoutchouc, oils, albuminoids, carbohydrates and fibre	Antioxidant and antidiarrhoeal activity
Ficus religiosa Moraceae	Tannin, wax, caoutchouc	Anti-tumor and acetylcholinesterase inhibitory activity
Gymnema sylvestre Asclepiadaceae	Gymnemic acid, resins, tannins, pararapin, glucose, inositol, anthraquinone, carbohydrates, tartaric acid, calcium salts and crystalline concretions	Antimicrobial and antidiabetic activity

Contd...

Table 3 Contd...

Hemidesmus indicus Asclepiadaceae	Coumarin, volatile oil, hemidesmine, smilasperic, essentialoil, saponin, resin, tennins, 2-hydroxy-4-methoxy benzaldehyde, sterols	Antinociceptive, antioxidant, antithrombotic, antiulcer and antidiabetic activity
Hibiscus rosa-sinensis Malvaceae	Aliphatic ethers, fatty acid methyl ethers, Ethyl â-L-arabinopyranoside	Hypoglycemic activity
Lantana camara Verbenaceae	Essential oil containing camarene, isocamarene, micranene, lantanine, lantadene	Antimicrobial and antitumor activity
Lawsonia inermis Lythraceae	Hanno-tannic acid, tannin, olive green resin, fragrant oil and glucoside	Antiparasitic, antibacterial and tuberculostatic activity
Leucas apsera Lamiaceae	Essential oil, glycosides and alkaloid	Antimicrobial
Momordica dioica Cucurbitaceae	Alkaloids	Hypoglycemic, antidiabetic, anticancer and antihyperglycemic activity
Moringa oleifera Moringaceae	Alkaloids moringine and moringinine, resins, organic acids, *Moringa* oil, behenic, myristic, palmatic, oleic, stearic and lignoceric acids	Antioxidant, anticancer, antifungal and hypolipidaemic activity
Murraya koneigii Rutaceae	Koenigin, essential and volatile oils, glucosides	Hypoglycemic, antihyperglycemic, antifungal, antioxidant and radical-scavenging activity
Ocimum canum Lamiaceae	Essential oil contain Camphor, citronellel. l-citronellic acid, eugenol, borneol, citral, methyl-heptenone, linalool, dipentene, terpinelone, crithmene, lomonene, sabinene, camphene, caryophyllene and methylheptenone	Antidiabetic, antibacterial and antifungal activity

Contd...

Table 3 Contd...

Ocimum sanctum Lamiaceae	Basil -camphor, terpenes and mucilage, eugenol, carvacrol, caryophyllene	Antiproliferative, antioxidant and antisecretory activity
Phyllanthus amarus Euphorbiaceae	Phyllanthin, quinine and hypophyllanthin	Antidiabetic, anti-inflammatory, antiallodynic and chemoprotective activity
Phyllanthus emblica Euphorbiaceae	Rich vitamin C, essential oil and fixed oil, phosphatides and tannins	Anticancer activity
Piper betle Piperaceae	Essential volatile oil containing betel-phenol, starch, sugar, tannins, terpene, sesqueterpene, chavicol	Antidiabetic, antioxidant, antibacterial and antifungal activity
Piper nigrum Piperaceae	Piperine, piperidine, piperitine, balsamic, chavicin, essential volatile oil, starch, lignin and insoluble water	Antioxidant, anticancer, antibacterial and antimutagenic activity
Pongamia pinnata Fabaceae	Pongamol (*Pogamia* oil), karanjin, glabrin, alkaloid, resin, mucilage, sugar, acetyl, benzoyl derivatives, myristic, palmitic, stearic, arachidic, lignoceric, dihydroxy stearic, lino lenic, linolic and oleic acids	Anti-inflammatory, antiplasmodial, antihyperglycemic and antilipidperoxidative activity
Punica granatum Punicaceae	Essential oil eugenol, tannins, punico-tannic acid, mannite, sugar, pectin, pelletierine, isopelletierine	Antidiabetic, anti-inflammatory, anticancer and antibacterial activity
Santalum album Santalaceae	A-santanol, B-santanol, isovaleric aldehyde, santanone, santalone, esters and free acids, resins, tannic acid,	Antitumor and antioxidant activity
Solanum nigrum Solanaceae	Solanine, saponine, compound of sugar and solanidine	Antioxidant, antiulcer and antitumor activity
Strychnos nux-vomica Loganiaceae	Strychnine, pseudostrychnine, brucine, vomicine, igasurine, igasuric and strychnic acid, loganine, proteids, starch, sugar and phosphates	Analgesic, anti-inflammatory and antidiarrhoeal activity

Contd...

Table 3 Contd...

Syzygium cumini Myrtaceae	Jamboline, jambosine, resin, gallic acid, resin, tannin, ellagic acid, glucosides, essential oil,	Anti-inflammatory, antidiabetic and antioxidant activity
Tephrosia purpurea Fabaceae	Tephrosin, isotephrosin, rotenone, osyritin, deguelin, quercetin, querritrin, glucoside rutin	Wound healing and Immunomodulatory activity
Terminalia arjuna Combretaceae	Arjunine, arjunetin, lactone, tannin, essential oils, reducing sugars, glucotannic acid, pyrocaechol and phytosterol	Cardio protective, antiatherogenic and antibacterial activity
Terminalia chebula Combretaceae	Chebulinic acid, resin, anthraquinone, tannic acid, gallic acid, mucilage,	Chemomodulatory, antioxidant, antidiabetic and antibacterial activity
Tribulus terrestris Zygophyllaceae	Alkaloid, resin, fat and mineral matter	Antioxidant and aphrodiasic activity
Tridax procumbens Asteraceae	Procumbenetin, dotriacontanol, â-amyrone, 12-dehydrolupen-3-one, â-amyrin, lupeol, fucosterol, 9-oxoheptadecane, 10-oxononadecane and sitosterol	Immunomodulatory and hepatoprotective activity
Vitex negundo Verbenaceae	Essential oil, organic, malic acid, alkaloids and resin, iridoids, lignan, ecdysones, iridoid glucoside, Acetyl oleanolic acid, sitosterol, isomeric flavanones and furanoeremophilane	Antioxidant, anti-inflammatory, analgesic and antibacterial activity
Wedelia calendulacea Asteraceae	Alkaloid ecliptine, kauren doterpenes	Hepatoprotective, antibacterial and antiosteoporotic activity
Zizyphus jujuba Rhamnaceae	Zizyphic acid, saponine, sapogenin, jujubogenin, jujubosides A and B, Acylated flavone-C-glycosides, cyclopeptide alkaloids, frangufoline, mucilage and sugar	Antioxidant and anti-cholinesterase activity

*Nadkarni (1976); Rastogi and Mehrotra (1990 –1994)

**www.sciencedirect.com; Ayyanar and Ignacimuthu (2008a,b)

Research in bioactive products is still in early stages in many countries. There is much information on the chemical constituents of medicinal plants in the literature, where structural determination has been an end in itself and the activity of compounds isolated have not been assessed (Mulholland, 2005). In the last 2000 years of the history of medicine, we can see that for most of this period, mankind had no other source of medicine than plants, either fresh or dried. Several diverse lines of evidence indicate that medicinal plants represent the oldest and most widespread form of medication. Traditional medical knowledge is important not only for its potential contribution to drug development and market values, but also for the people's healthcare (Pei, 2001). Classic examples of phytochemicals in biology and medicine include taxol, vincristine, vinblastine, colchicines as well as the Chinese antimalarial, artemisinin and the Indian ayurvedic drug, forkolin (Hoareau and DaSilva, 1999).

1.8 Conclusion

Research on medicinal plants and the search for plant-derived drugs require a multidisciplinary approach with integrated projects, financial and technical support, and a very carefully planned strategy. Renewed interest in traditional pharmacopoeias has meant that researchers are concerned not only with determining the scientific rationale for the plant's usage, but also with the discovery of novel compounds of pharmaceutical value (Fennell *et al*, 2004). Instead of trying to identify the active components of herbs through massive collection of plants from natural sources, it is better to start investigating the efficacy of the natural product from the traditional use by patients in randomized clinical trials. The curative properties of drugs are due to the presence of complex chemical substances of varied composition (present as secondary plant metabolites) in one or more parts of the traditional medicinal plants used by the traditional healers.

There is still much we can learn from investigating herbals available abundantly in the forests particularly those which are less well known. This type of research must be promoted as a means for developing countries to understand the potential use of their plant resources, as well as a means to better promote basic healthcare. There is an urgent need to inventory and record all ethnobiological information among the diverse ethnic communities before the traditional cultures are completely lost. This type of research needs a multidisciplinary approach and this includes expertise in the fields of ethnobotany, ethnopharmacology and photochemistry.

References

Alan Cox, P. 2005. The seven pillars of ethnomedical wisdom. *Ethnobotany*, 17: 24-34.

Almeida, C.F.C.B.R., de Amorium, E.L.C., de Albuquerque, U.P. and Maia, M.B. 2006. Medicinal plants popularly used in the Xingó region – a semi-arid location in Northeastern Brazil. *Journal of Ethnobiology and Ethnomedicine*, 2: http://www.ethnobiomed.com/content/2/1/15.

Ayyanar, M. and Ignacimuthu, S. 2005a. Traditional Knowledge of Kani tribals in Kouthalai of Tirunelveli hills, Tamil Nadu, India. *Journal of Ethnopharmacology*, 102: 246-255.

Ayyanar, M. and Ignacimuthu, S. 2005b. Ethnomedicinal plants used by the tribals of Tirunelveli hills to treat poisonous bites and skin diseases. *Indian Journal of Traditional Knowledge*, 4: 229-236.

Ayyanar, M. and Ignacimuthu, S. 2008a. Pharmacological actions of *Cassia auriculata* L. & *Cissus quadrangularis* Wall.: a short review. *Journal of Pharmacology and Toxicology*, 3: 213-221.

Ayyanar, M. and Ignacimuthu, S. 2008b. Medicinal uses and Pharmacological actions of five commonly used Indian medicinal plants: a mini-review. *Iranian Journal of Pharmacology and Therapeutics*, 8: (In press).

Balachandran, P. and Govindarajan, R. 2005. Cancer—an ayurvedic perspective. *Pharmacological Research*, 51: 19-30.

Balunas, M.J. and Kinghorn, A.D. 2005. Drug discovery from medicinal plants. *Life Sciences*, 78: 431-41.

Calixto, J.B. 2005. Twenty five years of research on medicinal plants in Latin America: a personal review. *Journal of Ethnopharmacology*, 100: 131-134.

Chah, K.F., Eze, C.A., Emuelosi, C.E. and Esimone, C.O. 2006. Antibacterial and wound healing properties of methanolic extracts of some Nigerian medicinal plants. *Journal of Ethnopharmacology*, 104: 164-167.

Chowdhury, A.K., Ara, T., Muhammed, F.H. and Ahmed, M. 1993. A new phenyl propane derivative from *Acorus calamus*. *Pharmazie*, 48: 786-787.

Cragg, G,M. and Newman, D.J. 2001. Natural product drug discovery in the next millennium. *Pharmaceutical Biology*, 39: 8-17.

Dhingra, V., Vishweshwar Rao, K. and Lakshmi NArasu, M. 2000. Current status of artemisinin and its derivatives as antimalarial drugs. *Life Sciences*, 66: 279-300.

Ernst, E. 2005. The efficacy of herbal medicine - an overview. *Fundamental and Clinical Pharmacology*, 19: 405-409.

Fabricant, D.S. and Farnsworth, N.R. 2001. The value of plants used in traditional medicine for drug discovery. *Environmental Health Perspectives*, 109: 69-75.

Fennell, C.W., Lindsey, K.L., McGaw, L.J., Sparg, S.G., Stafford, G.I., Elgorashi, E.E., Grace, O.M. and Van Staden, J. 2004. Assessing African medicinal plants for efficacy and safety: pharmacological screening and toxicology. *Journal of Ethnopharmacology*, 94: 205-17.

Ghisalberti, E.L. 2000. *Lantana camara* L. (Verbenaceae). *Fitoterapia*, 71: 467-86.

Gilani, A.H. and Atta-ur-Rahman. 2005. Trends in Ethnopharmacology. *Journal of Ethnopharmacology*, 100: 43-49.

Grover, J.K., Yadav, S. and Vats, V. 2002. Medicinal plants of India with anti-diabetic potential. *Journal of Ethnopharmacology*, 81: 81-100.

Gurib-Fakim, A. 2006. Review – Medicinal plants: Traditions of Yesterday and drugs of tomorrow. *Molecular Aspects of Medicine*, 27: 1-93.

Hamilton, A.C. 2004. Medicinal plants, conservation and livelihoods. *Biodiversity and Conservation*, 13: 1477-1517.

Harborne, J.B. and Baxter, H. 1993. *Phytochemical Dictionary. A Handbook of Bioactive Compounds from Plants*. London: Taylor and Francis.

Harshberger. 1896. The purposes of ethnobotany, *Botanical Gazette*, 21: 146-154.

Heinrich, M. 2000. Ethnobotany and its role in drug development. *Phytotherapy Research*, 14: 479-88.

Hoareau, L. and DaSilva, E.A. 1999. Medicinal plants: a re-emerging health aid. *Electronic Journal of Biotechnology*, 2: http://www.ejb.org/content/vol2/issue2/full/2.

Ignacimuthu, S., Ayyanar, M. and Sankarasivaraman, K. 2006. Ethnobotanical investigations among tribes in Madurai District of Tamil Nadu, India. *Journal of Ethnobiology and Ethnomedicine*, http://www. ethno-biomed.com/content/2/1/25.

Ignacimuthu, S., Ayyanar, M. and Sankarasivaraman, K. 2008. Ethno-botanical study of medicinal plants used by Paliyar tribals in Theni district of Tamil Nadu, India. *Fitoterapia*, 79: 562-568.

Iwu, M.W., Duncan, A.R. and Okunji, C.O. 1999. New antimicrobials of plant origin. In: J. Janick (ed.), Perspectives on new crops and new uses. ASHS Press, Alexandria, VA. p. 457–462.

Jain, S.K. 1964. The role of botanist in folklore research. *Folklore*, 5: 145-150.

Jain, S.K. 1991. Dictionary of Indian Folk Medicine and Ethnobotany. Deep publications, Paschim Vihar, New Delhi.

Janaki Ammal, E.K. 1956. Introduction to the subsistence economy of India: William LT Jr (ed.) Man's role in changing face of the earth. University of Chicago Press, Chicago, 324 – 335.

Janaki Ammal, E.K. and Nagendra Prasad, P.N. 1984. Ethnobotanical findings on *Costus speciosus* among the kanikkars of Tamil Nadu. *Journal of Economic and Taxonomic Botany*, 5: 129-133.

Karou, D., Savadogo, A., Canini, A., Yameogo, S., Montesano, C., Simpore, J., Colizzi, V. and Traore, A.S. 2006. Antibacterial activity of alkaloids from *Sida acuta*. *African Journal of Biotechnology*, 5: 195 – 200.

Mallavadhani, U.V., Panda, A.K. and Rao, Y.R. 1998. Pharmacology and chemotaxonomy of *Diospyros*. *Phytochemistry*, 49: 901-951.

Ministry of Tribal Affairs. Annual Report 2005 – 2006.

Mulholland, D.A. 2005. The future of ethnopharmacology: A southern African perspective. *Journal of Ethnopharmacology*, 100: 124-126.

Nadkarni, A.K. 1976. Indian Materia Medica, vol. I. Popular Prakashan, Bombay.

Nyiredy, S. 2004. Review Separation strategies of plant constituents–current status. *Journal of Chromatography B*, 812: 35–51.

Pal, D.C. 2000. Ethnobotany in India. In: Flora of India. Introductory volume - part II, (Eds. Singh, N.P., Singh, D.K., Hajra, P.K. and Sharma, B.D). Botanical Survey of India, Calcutta, India. pp. 303 – 320.

Pei, S.J. 2001. Ethnobotanical approaches of traditional medicine studies: Some experiences from Asia. *Pharmaceutical Biology*, 39: 74-79.

Pushpangadan, P., Kumar, B.M. 2005. Ethnobotany, CBD, WTO and the Biodiversity Act of India. *Ethnobotany*, 17: 2-12.

Rastogi, R.P. and Mehrotra, B.N. 1990. Compendium of Indian Medicinal Plants, Vol. I-V. Central Drug Research Institute, Lucknow and National Institute of Science Communication, New Delhi, India.

Rates, S.M. 2001. Plants as source of drugs. *Toxicon*, 39: 603-613.

Scartezzini, P. and Speroni, E. 2000. Review on some plants of Indian traditional medicine with antioxidant activity. *Journal of Ethnopharmacology*, 71: 23-43.

Singh, N.P., Mudgal, V., Khanna, K.K. and Srivastava, S.C. 2001. Medicinal plants. In: Floristic diversity & Conservation strategies in India. Vol. IV – Angiosperms, Economic & Ethnobotany (Eds. Singh, N.P and Singh, D.K). Botanical Survey of India, Calcautta and Ministry of Environment and Forests, India. pp. 2269 - 2322.

Ved, D.K. and Goraya, G.S. 2007. Demand and Supply of Medicinal Plants in India. NMPB, New Delhi & FRLHT, Bangalore, India.

Wickens, G.E. 1990. What is Economic Botany? *Economic Botany*, 44: 12-28.

World Health Organization, 2003. Traditional medicine. 56th World Health Assembly.

□□□

Spilanthes acmella - A Wonderful Medicinal Herb

S. Kalidass, S.Sembian, T. Jeya, Joshni T. Chacko

CONTENTS

1.1 Introduction 26

1.2 Taxonomic Hierarchy 27

1.3 Geographical Data 28

1.4 Medicinal Active Components of Spilanthes acmella 28

1.5 Medicinal Properties of Spilanthes acmella 29

- 1.5.1 Tingling 29
- 1.5.2 Antidiuretic 31
- 1.5.3 Insecticidal 31
- 1.5.4 Antimicrobial & Immunomodulating 32
- 1.5.5 Anti-inflammatory and Analgesic 32
- 1.5.6 Other Medicinal Properties of the Spilanthes acmella 33

1.6 Conclusion 34

References 34

1.1 Introduction

The uses of medicinal plants for primary health care have steadily increased worldwide in recent years; about 80% of pharmaceuticals are derived from plants. Plants are rich in a wide variety of secondary metabolites that have found anti-microbial properties; scientists are in search of new phytochemicals that could be developed as useful anti-microbials for treatment of infectious diseases. Many of these commercially available compounds are crude preparations administered without performing human clinical trials. In many cases, these substances serve as plant defense mechanisms against predation by microorganisms, insects, and herbivores. Some, such as terpenoids, give plants their odors; others (quinones and tannins) are responsible for plant pigment. Many compounds are responsible for plant flavor (e.g., the terpenoid capsaicin from chili, peppers), and some of the same herbs and spices used by humans to season food yield useful medicinal compounds (Marjorie Murphy cowan, 1999).

Once the local ethnomedical preparations of traditional sources are scientifically evaluated before dispensing they should replace existing drugs commonly used for the therapeutic treatment of infection. This method should be put into practice for future investigations in the field of ethnopharmacology, phytochemistry, ethnobotany and other biological fields for drug discovery. Recent methods are useful to standardize the extraction for scientific investigation of new phytochemicals and anti-microbials of traditionally used plants.

Green plants synthesize and preserve a variety of biochemical products, many of which are extractable and used as chemical feed stocks or as raw material for various scientific investigations. Many secondary metabolites of plants are commercially important and find use in a number of pharmaceutical compounds.

Herbs are staging a comeback and herbal 'renaissance' is happening all over the globe. The herbal products today symbolize safety in contrast to the synthetics that are regarded as unsafe to human and environment. Although herbs had been priced for their medicinal, flavouring and aromatic qualities for centuries, the synthetic products of the modern age surpassed their importance for a while. However, the blind dependence on synthetics is over and people are returning to the naturals with hope of safety and security.

Population rise, inadequate supply of drugs, prohibitive cost of treatments, side effects of several allopathic drugs and development of resistance to currently used drugs for infectious diseases have led to increased emphasis on the use of plant materials as a source of medicines for a wide variety of human ailments. Global estimates indicate that 80% of about 4 billion population cannot afford the products of the Western Pharmaceutical Industry and have to rely upon the use of traditional medicines which are mainly derived from plant material.

India has been known to be rich repository of medicinal plants. The forest in India is the principal repository of large number of medicinal and aromatic plants, which are largely collected as raw materials for manufacture of drugs and perfumery products.

Spilanthes acmella Linn.

The genus *Spilanthes* contains 35 tropical species, of which three of them are reported from India. *Spilanthes acmella* Linn. (Family-*Asteraceae*) is an important medicinal plant commonly known as Akarkara or toothache plant with rich source of therapeutic constituents. It is an annual, spreading plant with bicoloured, red/gold flower buds. In Ayurvedic system of medicine, flower heads and roots are used in treatment of scabies, psoriasis, scurvy, toothache, infections of gums and throat, paralysis of tongue and remedy for stammering in children (Anonymus, 1976). *Spilanthes* is commonly referred to as the "toothache plant, " a name that refers to the analgesic action of its alkylamide constituents (Stacy, 2007). *Spilanthes acmella* has been well documented for its uses as spices, antiseptic, anti-bacterial, anti-fungal, anti-inflammatory and anti-malarial Agents and as remedy for flu, cough, rabies diseases and tuberculosis (Burkill, 1966; Oliver-Bever, 1986; Di Stasi *et al.*, 1994; Akah and Ekekwe, 1995; Ramsewak *et al.*, 1999). Because of its high medicinal value, there is a much demand of this plant in the market.

1.2 Taxonomic Hierarchy

Kingdom	:	Plantae-Planta
Subkingdom	:	Tracheobionta-vascularplants

Division	:	Magnoliophyta- angiosperms
Class	:	Magnoliopsida-dicots
Subclass	:	Asteridae
Order	:	Asterales
Family	:	Asteraceae
Genus	:	*Spilanthes* Jacq. - Spilanthes
Species	:	*Spilanthes acmella* (Linn.) L. - para cress

1.3 Geographical Data

All the species coming under genus *Spilanthes,* are mainly distributed in the tropical and subtropical regions around the world (Jansen, 1985). *Spilanthes acmella* is an annual or short-lived herb, spreads in low open places and requires moist soil (Wilkinson, 1989). It is found in Benguet Subprovince, Cagayan, Nueva Vizcaya, Rizal, and Laguna Provinces in Luzon, and in Mindoro and Balabac, in open waste places, old clearings, etc., at low and medium altitudes. It is pantropic in distribution.

Local names: Biri (Ig.); dila-dilag (If.); gatang-gatang (Sul.); pilet-pilet (Sul.); toothache plant (Engl.)

This is an erect, branched, annual herb which reaches a height of 15 to 60 centimeters. The leaves are opposite, smooth or nearly so, ovate-lanceolate, and 1.5 to 3 centimeters long, with pointed tip and wedge-shaped base, and with toothed or wavy margins. The conical heads occur singly at the ends of long stalks, and are about 1 centimeter in length. The flowers are yellow. The achenes are flattened, oblong, dark brown and enclosed separately in scales. The herb is official in the French Pharmacopoeia.

1.4 Medicinal Active Components of *Spilanthes acmella*

The most abundant active molecules present in *Spilanthes acmella* are the alkamides and especially, (2E, 6Z, 8E) -*N*-isobutyl-2, 6, 8-decatrienamide or spilanthol ($C_{14}H_{25}NO$), (Molecular weight: 221.339 g/mol).

HN

O

(2E, 6Z, 8E)-*N*-isobutyl-2, 6, 8-decatrienamide (spilanthol)

The flowerheads of *Spilanthes acmella* contain drimane sesquiterpenoids (Jansen *et al.,* 2003). Phytochemically, flowers of *Spilanthes acmella* contain amino acids, alkaloids, *N*-isobutylamides (spilanthol, undeca-2*E,* 7*Z,* 9*E*-trienoic acid isobutylamide and undeca-2*E*-en-8, 10-diyonic acid isobutylomide). A number of other *N*-isobutylamides such as 2*E*-*N*-(2-methylbutyl)-2-undecene-8, 10-diynamide;2*E,* 7*Z*-*N*-isobutyl-2, 7-tridecadiene-10, 12-diynamide and 7*Z*-*N*-isobutyl-7-tridecene-10, 12-diynamide from *Spilanthes acmella* have been found. *N*-isobutylamides have also been found from other species such as *Spilanthes alba* and *Spilanthes oleracea*. *N*-2-Phenylethylcinnamamide was isolated from *Spilanthes ocymifolia.*

Myricyl alcohol, α- and -amyrins, -sitosterol, stigmasterol and other compounds have been isolated from the air dried whole plant of *Spilanthes acmella*. Essential oil containing the major constituents as limonene (23.6%), -caryophyllene (20.9%) (Z)-ocimine (14%), germacrene D (10.8%) and myrcene (9.5)% has been obtained from the flower heads of *Spilanthes acmella*. Presence of a mixture of C22 to C35 normal hydrocarbons in the flower heads of *Spilanthes acmella* have been found. Flavonoid glycoside was also isolated from *Spilanthes Calva,* which was characterized as tetrahydroxydihydrochalcone 3'-0-glucoside, the compound possessed hypoglycemic activity in mice where diabetes was induced by streptozotocin. From the aerial parts of *Spilanthes leiocarpa,* a new compound eudesmanolide was found. Secondary volatile metabolites including sesquiterpenes (α-andß-bisabolenes, caryophylleneandc adinenes), nitrogenated (N (isobutyl) 2*E,* 6 *Z,* 8*E* decatrienamide; *N* (2methylbutyl) 2*E,* 6*Z,* 8*E* decatrienamide; dec atrienamide; N (isobutyl) 6Z, 83 decadienamideand *N* (2phenelethyl) 2*E,* 6*Z,* 8*E*decatrienamide) and oxygenated compounds have isolated by simultaneous distillation-solvent extraction (SDE) and supercritical (CO2) extraction (SFE) from flowers, leaves and stems of *Spilanthes Americana* (Mutis) Hieron. SFE extracts from stems were found rich (>40%) in sesquiterpenes, while those from leaves and flowers were abundant in nitrogenated (43 and 27%) and oxygenated (36 and 23%) compounds (Gupta, 2007).

1.5 Medicinal Properties of *Spilanthes acmella*

1.5.1 Tingling

The flowers of this herb when chewed produce a numb sensation in the mouth and thus have been effectively used for the remedy of toothaches, sore mouth, itching, psoriasis, stammering problem and also to regulate the flow of saliva. It is also recommended for paralysis of the tongue and sore mouth in children. *Spilanthes acmella* is used traditionally to treat mouth aliments, stammering, stomatistis and throat complaints. Spilanthol was the most active tingling and mouthwatering compound. (Jakob, 2005). The

Spilanthes acmella is one of the active ingredients in compositions for acute- or long-term treatment of microbial infections, particularly, oral pathogenic microorganisms, dental caries, periodontosis, gum disease, gum bleeding and/or plaque reduction. Caius says that the plant is used as a specific for toothache in Old Calabar. In South Africa the powdered leaf is placed in a carious tooth, and is rubbed on the lips and gums for sore-mouth in children. According to Kirtikar and Basu the flower-heads are by far the most pungent part, and are chewed by the Hindus to relieve toothache, which they do by producing redness of the gums and salvation. A tincture of the flower heads is also recommended for toothache in place of tincture of pyrethrum. It is said to be a specific for inflammation of the periosteum of the jaws.

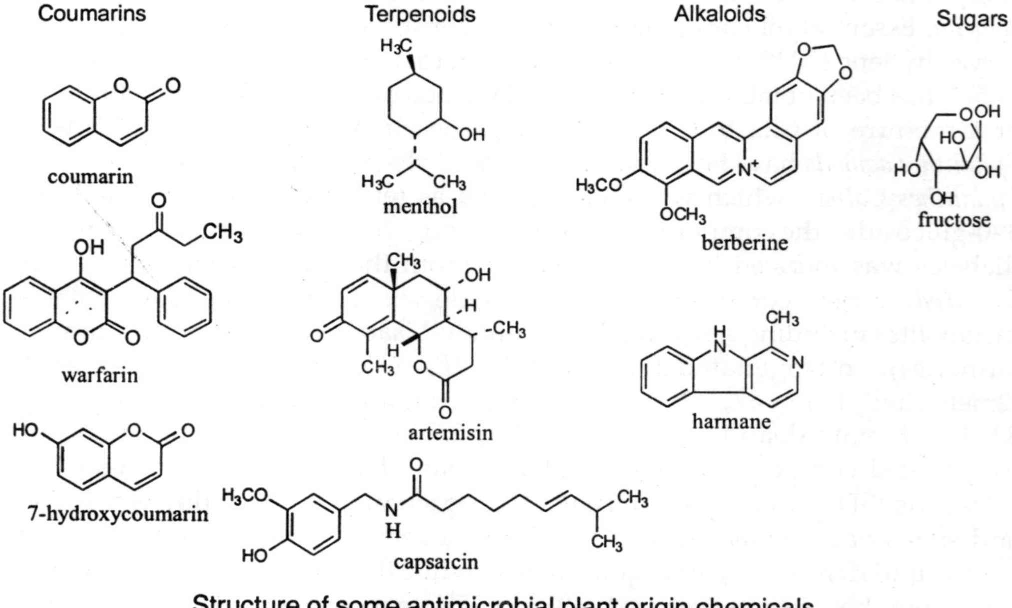

Coumarins: coumarin, warfarin, 7-hydroxycoumarin
Terpenoids: menthol, artemisin
Alkaloids: berberine, harmane
Sugars: fructose
capsaicin

Structure of some antimicrobial plant origin chemicals

The bronze-green leaves and the more potent yellow and red cone shaped flowers have a taste and properties similar to coneflower (Echinacea purpurea). As a sialagogue it stimulates the flow of salvia which cleanses the mouth, tones the gums, and enhances immune function. *Spilanthes acmella* also improves digestion, eases flatulence, improves the appetite, and helps to overcome nausea and vomiting by its stimulating effect on the salivary glands. It is called the toothache plant because when you chew on the leaves or flowers it produces a numbing effect to the tongue and gums. *Spilanthes acmella* can be used in this manner to help ease the pain of a toothache (Orapin, 2008). Moreover, it has been used as anti-toothache formulation for pain relieving within 2-10 min and also useful in treating the swelling and infection of gums.

1.5.2 Antidiuretic

Spilanthes acmella is known for its antidiuretic activities. The flowers have also been shown to have diuretic activity in rats. (Moulick, 2006). Alkylamides from *Spilanthes* have demonstrated strong diuretic (Stacy, 2007). In addition, the crude flower head extracts displayed strong diuretic action which caused marked increase in urine Na^+, K^+ and a reduction in the osmolarity of urine. Its root decoction has been used as a diuretic drug (Perry and Metzger, 1980); (internally it has a diuretic effect and is reputed to be a solvent of vesical calculi). In addition to diuretic activities, they also possess sialagogue (Gupta, 2007).

1.5.3 Insecticidal

Spilanthes acmella is known for its and larvicidal activities. Spilanthol, an Isobutylamide in *Spilanthes* has been isolated and shown to be bioavailable which has similarity in structure with the isobutylamides found in Echinacea. Larvicidal activity has been demonstrated by *Spilanthes acmella* against *Culex quinquefasciatus.*(Moulick, 2006). Spilanthol has been isolated and is regarded as very highly bioactive due to its insecticidal activity. (Moulick, 2006). *Spilanthes* is one of the few herbs, which exhibits antimalarial properties as well. Spilanthol is found effective at extremely low concentrations against blood parasites, and indeed is a poison to most invertebrates while remaining harmless to warm-blooded creatures. The larvicidal potential of *Spilanthes acmella* against *Culex quinquefasciatus* has been shown by Pitasawat *et al.*, 1998. Spilanthol showed interesting bioactivities, e.g. strong local anesthetic, analgesic and insecticidal activities. Crude flower head extracts of the plant exhibited potent ovicidal, marked larvicidal and pupicidal activities. (Orapin; 2008). Alkylamides from *Spilanthes* have demonstrated strong and insecticidal properties. (Stacy, 2007). Spilanthol, a major constituent of ethanolic extract of flower heads of *Spilanthes acmella* Murr. is having potent ovicidal, larvicidal and pupicidal activity. Maximum 7.5 ppm concentration causes 100% motility of eggs, larvae and pupae of Anopheles, Culex and Aedes mosquito. Spilanthol is more effective even at low doses against eggs and pupae. In pupae, it seems to work on nervous system which was evident by abnormal movement like jerks, spinning and uncoordinated muscular activity suggesting thereby that it disturbs nerve conduction (Krishnaswamy *et al.*, 1975; Borges *et al.*, 1984; Jondiko, 1986; Saraf and Dixit, 2002). Spilanthol is effective at extremely low concentrations against blood parasites, and is a poison to most invertebrates while remaining harmless to warm blooded creatures. This is the explanation for its utility against blood parasites, specifically malarial spirochetes, either as a prophylactic or as a treatment for malarial paroxysms.

1.5.4 Antimicrobial & immunomodulating

It is antifungal, antiviral and immunostimulating. The herb also possesses a strong antibacterial activity. Studies have shown strong *in-vitro* activity of *Spilanthes acmella* extracts against common pathogens such as *Escherichia coli, Klebsiella pneumoniae, Proteus vulgaris, Pseudomonas aeruginosa, Salmonella gallinarum* and *Staphylococcus albus. Spilanthes acmella* also inhibits the yeast/ fungal organism such as *Candida albicans,* which is responsible for the nearly epidemic condition known as candidiasis. *S. acmella* also possessed excellent anti-microbial activities against red halophillic cocci from salt cured fish. Researchers have shown preliminary antimicrobial activity in the crude extract from roots and flower heads of *S. mauritiana. Spilanthes acmella* has been used for the treatment of tuberculosis. A study was done on dried *Spilanthes acmella* leaf from Rwanda showing active immunomodulating activity, specifically increased production of mononuclear leukocytes. In case of earache and ear infections, the immune enhancing activity of the Echinacea was enhanced by follow up with anti-bacterial influence of the *Spilanthes acmella* tincture to cure the problem. The plant is used as a snakebite remedy and in the treatment of articular rheumatism. Flowers are used against itching and psoriasis. (Gupta, 2007).

The leaves of *Spilanthes acmella* have been used as a salad ingredient. Additions of a few leaves to salads, sandwiches, steamed vegies, and stirfries were suggested to control the symptoms of various types of infections and stimulation of immunity in general (Gupta, 2007). Alkylamides from other plants (specifically *Echinacea* species) have demonstrated *in vitro* and *in vivo* immunomodulatory activity. Today, alkylamide containing extracts of *Spilanthes acmella* are sold as dietary supplements by a number of herbal extract manufacturers, and the biological activity of these extracts and their constituents is an ongoing subject of research. Validated methods for analysis of alkylamides in complex *Spilanthes acmella* extracts are needed to facilitate this research and to serve as the basis for quality control of commercial *Spilanthes acmella* products. (Stacy, 2007).

1.5.5 Anti-inflammatory and analgesic

On preliminary phytochemical screening, the aqueous extract of *Spilanthes acmella* was found to contain flavonoid compounds. Flavonoids are known to target prostaglandins which are involved in the late phase of acute inflammation and pain perception. The presence of flavonoids may be contributory to the anti-inflammatory and analgesic activities of aqueous *Spilanthes acmella*. Further studies may reveal the exact mechanisms of action responsible for the analgesic and anti-inflammatory activities of *Spilanthes acmella*. (Chakraborty *et al.*, 2003). Several studies have demonstrated biological activity of *Spilanthes acmella* species. Ratnasooriya and Pieris demonstrated that cold water extracts of the flowers of *Spilanthes acmella* possess antinociception activity against persistent pain and anti-hyperalgesic activity. Sukumaran and Kuttan (1995) observed chloroform extracts of *S. calva* inhibited tobacco induced mutagenesis on *Salmonella typhimurium*. *Spilanthe's* biological activity has primarily been attributed to constituents known as alkylamides, which this plant possesses in abundance.

1.5.6 Other Medicinal properties of the *Spilanthes acmella*.

Spilanthes acmella is a popular herbal extract which is used traditionally to treat stammering, stomatistis, and throat complaints. *Spilanthes acmella* Murr. extracts on phenylephrine-induced contraction of rat thoracic aorta as well as their antioxidant activity. Results showed that the extracts exert maximal vasorelaxations in a dose-dependent manner (Orapin, 2008). Constituents found in *Spilanthes acmella* (Murr.) extracts were also shown to exhibit pancreatic lipase inhibition which has potential as candidates for weight loss and obesity control. In addition, its root powder extract is very effective in treating HIV/AIDS infection. Presently, cardiovascular disease is one of the most important public health concerns, accountable for morbidity and mortality. Thus new vasoactive ingredients that influence the endothelial cells, which are key regulating cells in the vessel wall, are the striking areas. So far, the effect of *Spilanthes acmella* (Murr.) on vascular function has not been reported. In this regard, it was of great interest to explore its medicinal value with regards to cardiovascular functions such as regulation of vascular tone (Orapin; 2008). which is a powerful stimulant, sialogogue and local anesthetic. (Prasad and Seenayya, 2000). *Spilanthes acmella* modulates IL-2 production by stimulated human T cells. IL-2 is a critical cytokine required for clonal proliferation of antigen specific T lymphocytes key immune cells involved in adaptive immunity to infectious microorganisms. Since *S. acmella* extract contains high alkylamide contents it suppresses IL-2 production. So, it can be used as a potential therapy for treatment of immune-related medical conditions.The roots, flower heads and whole aerial part yield a compound known as spilanthol, *Spilanthes acmella* Murr., extracts possess vasorelaxant and antioxidant activities (Orapin *et al.*, 2008). Its root decoction has been used as a laxative drug (Perry and Metzger, 1980; Bunyapraphatsara and Chokechareunporn, 1999). In addition, its root powder extract is very effective in treating HIV/AIDS infection (Bajarang, 2007). It is reported that some native herb-doctors use the root as a purgative, giving a decoction of 4 to 8 grams to a cup of water. The infusion is used locally for itches and psoriasis. The leaf juice and the bruised leaves are applied to wound and atonic ulcers.

These leaves, with those of *Blumea balsamifea* and *Tamarndus indica*, are used to prepare aromatic baths for convalescents, rheumatics, and pregnant women. In Cameroons the flowering heads are combined with other plants, it is chewed, and swallowed for snake bite, together with local treatment of the wound. Among the Mundas of Chota Nagpur the crushed plant is used as a fish poison.

1.6 Conclusion

Spilanthes acmella Linn is a valuable medicinal herb containing strong antibacterial and larvicidal activity, toxicity and potential insecticidal activity, anti hepatotoxic activity, analgesic activity, vasorelaxant and antioxidant activities, anti-inflammatory activity, cockroach cercal, anti-malarial, anti-fungal activity, anti-viral, Antiseptic, Immunomodulator, nerve activity, diuretic activity. Besides it is extensively used in treating toothaches, throat infection, gums and urinary tract disease (Vibha *et al.*, 2009). In conclusion, it could be stated that *Spilanthes acmella* wonderful medicinal herb owing to its drug constituents and its herbal significance.

References

Anonymous, 1976. Wealth of India, Raw materials. Publication and Information Directorate, CSIR, New Delhi, India. Vol 2, 11-12.

Akah, P.A., Ekekwe, R.K., 1995. Etnopharmacology of some Asteraceae family used in Nigerian traditional medicine. *Fitoterapia*, 66, 351-355.

Bajarang, B.B., 2007. Invention of herbal drug against HIV infection/AIDS. IN Patent 2005MU01512, *Chem. Abstr.*, 147, 413250.

Borges-Del-Castillo, J., Vazque-Beno, P., Secundino-Lucas, M., Martinez-Maryir, A.I., Joseph-Nathan, P., 1984. The N-2-phenylethylcinnamamide from *Spilanthes ocymifolia*. *Phytochemistry*, 23, 2671-2672.

Bunyapraphatsara, N., Chokechareunporn, O. 1999. Tradition medicinal plants. Prachachon. Bangkok.

Burkill, I.H., 1966. A dictionary of the economic products of the MalayPeninsula. Vol II. Governments of Malaysia and Singapore by the Ministry of Agriculture and Co-operatives, Kuala Lumpur.

Caius, J.F., 1986.The Medicinal and Poisonous Plants of India. Scientific Publishers, Jodhpur, India, 323-325.

Chakraborty, AR., Devi K. B., Rita S., Sharatchandra KH., Singh Th. I., 2003. Preliminary studies on anti-inflammatory and analgesic activities of *Spilanthes acmella* in experimental animal models.

Di Stasi, L.C., Hiruma, C.A., Guimarães, E.M., Santos, C.M., 1994. Medicinal plants popularly used in Brazilian Amazon. *Fitoterapia*, 65, 529-540.

Eellarova, E., Kimakova, K., 1999. Morphoregulatory effects of plant growth regulators on *Hypericum perforatum* L. seedlings. *Acta Biotechnology*, 19, 163-169.

Eisenberg, D.M., Kessler, R.C., Foster, C., Norlock, F.E., Calkins, D.R., Delbanco, T.L., 1993. Unconventional medicine in the United States: prevalence, costs and patterns of use. *N. Engl. J. Med.* 328, 246–252.

Gupta, R.N., 2007. *Pharmbit* Vol.XV, No.1, 2007.

Jakob, P. Ley., 2005. Structure Activity Relationships of Trigeminal Effects for Artificial and Naturally Occuring Alkamides related to Spilanthol.

Jansen, R.K. 1985. The systematics of *acmella* (Asteraceae-Heliantheae). *Syst. Bot. Monogr.* 8, 1-115.

Jansen, B.J.M., de Groot Ae., 2003. Occurrence, biological activity and synthesis of drimane sesquiterpenoids.

Jondiko, I.J.O., 1986. A mosquito larvicide in *Spilanthes mauritiana.Phytochemistry*, 25, 2289-2290.

Joy, P.P., Thomas, J., Samuel Mathew, Baby P. Skaria, 1998. Medicinal plants, Aromatic and Medicinal Plants Research Station, Kerala Agricultural University.

Kiritikar, K.R., Basu, B.P., 1987. *Spilanthes acmella* murr. *Indian medicinal* plants; 2nd edition, Dehradun, International book distributors.

Krishnaswamy, N.R., Prasanna, S., Seshadri, T.R., Vedantham, T.N.C., 1975. a- and b-amyrin esters and sitosterol glucoside from *Spilanthes acmella. Phytochemistry*, 14, 1666-1667.

Marjorie Murphy Cowan, 1999. Plant Products as Antimicrobial Agents. Clinical microbiology reviews, 564-582.

Moerman, D.E., 1996. An analysis of the food plants and drug plants of native North America. J.Ethnopharmacol. 52, 1-22.

Moulick Kamalika, M.S., 2006. Interactions of *Echinacea* and *Spilanthes* with Human Metabolizing Systems.

Oliver-Bever, B., 1986. Medicinal Plants in Tropical West Africa.Cambridge Uni. Press, Cambridge.

Orapin Wongsawatkul, Supaluk Prachayasittikul, Chartchalerm Isarankura-Na-Ayudhya, Jutamaad Satayavivad, Somsak Ruchirawat, Virapong Prachayasittikul, 2008. Vasorelaxant and Antioxidant Activities of *Spilanthes acmella* Murr. *International Journal of Molecular Sciences*, 9, 2724-2744.

Perry, L.M., Metzger, J., 1980. Medicinal plants of East and South-East Asia, MIT Press, Cambridge.

Prasad, M.M., Seenayya, G., 2000. Effects of spices on growth of red halophilic cocci isolated from salt cured fish and solar salt. *Food Research International*, 33, 793-798.

Ramsewak, R.S., Erickson, A.J., Nair, M.G., 1999. Bioactive Nisobutylamides from the flower buds of *Spilanthes acmella*. *Phytochemistry*, 51, 729-732.

Rani, S.A., Murty, S.U., 2006. Antifungal potential of flower head extract of *Spilanthes acmella* Linn. *African Journal of Biomedical Research*, 9, 1, 67-68.

Ratnasooriya, W.D, Pieeris, K.P. Samaratunga, U., Jayakodi, J.K., 2004. Diuretic activity of *Spilanthes acmella* flowers in rats. *Journal of Ethanopharmacology*, 91, 2-3, 317-320.

Sabitha A. Rani, Suryanarayana U. Murty. 2006.African Journal of Biomedical Research, Vol. 9, *67-69*.

Saraf, D.K., Dixit, V.K., 2002. *Spilanthes acmella* Murr. : Study on Its Extract Spilanthol as Larvicidal Compound, *Asian J. Exp. Sci.*, Vol. 16, No. 1&2, 9-19.

Stacy, S. Bae., 2007. A HPLC/ESI-MS Method Developed and Validated to Evaluate the Quantity, Identity, and Stability of the Alkylamides in Ethanolic Extracts of *Spilanthes acmella*, The University of North Carolina, Greensboro.1-2.

Stockwell, C. 1988. *Nature's pharmacy*. Century Hutchinson Ltd., London, United Kingdom.

Storey, C. Salem, J.I., 1997. Lay use of Amazonian plants for the treatment of tuberculosis. *Acta Amazonica*, 27, 175-182.

Sukumaran, K., Kuttan, R. 1995. Inhibition of tobacco-induced mutagenesis by eugenol and plant extracts. *Mutation Research*, 343, 25-30.

Thomson, W.A.R. (ed.). 1978. Medicines from the Earth. McGraw-Hill Book Co., Maidenhead, United Kingdom.

Voravuthikunchai, S., Phongpaichit, S., Subhadhirasakul, S. 2005. Evaluation of Antibacterial Activities of Medicinal Plants Widely Used Among AIDS Patients in Thailand. *Pharmaceutical Biology (Formerly International Journal of Pharmacognosy)*, 43, 8, 701.

Vibha Pandey, Kuldeep Sharma and Veenaagrawal. 2009). Efficient micropropagation protocol of an antimalarial plant - *Spilanthes acmella*. In: *In vitro cellular and developmental Biology* Vol: 45, 491-499.

Wilkinson, J. 1989. Herbs and flowers of the cottage garden, Book Builders: Melbourne.

□□□

Dactylorhiza Hatagirea (D. don) A Critically Endangered Medicinal Plant of Trans-Himalaya

Ashutosh A. Murkute, Ashish R. Warghat,
G. Phani Kumar, G.P. Mishra and Sunil E. Jadhav

CONTENTS

1.1 Introduction 38

1.2 Distribution 39

1.3 Botanical Description and Agro-techniques 40

1.4 Phytochemical Composition 43

1.5 Pharmacology 44

1.6 Future Perspectives 44

References 45

1.1 Introduction

Dactylorhiza hatagirea (D. Don) is a high value medicinal plant species exclusively found in temperate regions of India including certain parts of trans-Himalayan regions of Ladakh. The economic potential of the species could be assessed on the basis of its high market value i.e. Rs 1,500-1,700 kg^{-1} (dry tubers) along with high annual demand (5,000 tons). Its roots were used in the traditionally '*Amchi*' system of medicine as an aphrodisiac, expectorant and nervine tonic.

Dactylorhiza hatagirea syn. *Orchis latifolia* var. *Indica* (Lindley) has been identified as a critically endangered (CAMP status), critically rare (IUCN status) and listed under Appendix II of Convention of International Trade in Endangered Species (CITES) (Kala, 2000; Samant *et al.*, 2001; Uniyal *et al.*, 2002).

It is commonly known as 'Marsh Orchis', it is also referred as Munjataka (Ayurveda), Angmo-lakpa (Ladakh), Salampanja (Kashmir) and Hatajari (Uttaranchal). The name *Dactylorhiza* is derived from Greek words "dactylos" (finger) and "rhiza" (root) referring to the palmately two to five lobed tubers. Locally in Ladakh it is known as Angmo-Lakpa which again means 'Fingers of a Girl'. It belongs to the family Orchidaceae and this is the only plant from this family which grows at high altitudes of Ladakh (Figure 1)

Fig. 1a: Plant of *Dactylorhiza hatagirea*-syn. *Orchis latifolia* var. Indica

Fig. 1b: Inflorescence of *Dactylorhiza hatagirea*-syn. *Orchis latifolia* var. Indica

Table 1: Genetic diversity of *Dactylorhiza* across globe

Species	Country	Species	Country
D. fuchsii	Sweden	*D. sambucina*C	roatia
D. fuchsia	Austria	*D. iberica*	Cyprus
D. foliosa	Madeira	*D. viridis*	UK
D. maculata	Sweden	*D. viridis*	China
D. saccifera	Greece	*D. incarnata*	Germany
D. romana	Italy	*D. incarnata*	Sweden
D .romana	Sicily	*D.euxina*	Turkey
D. sambucina	Sweden	*D. maculata*	Portugal
D. sambucina	Spain	*D. maculata*	UK

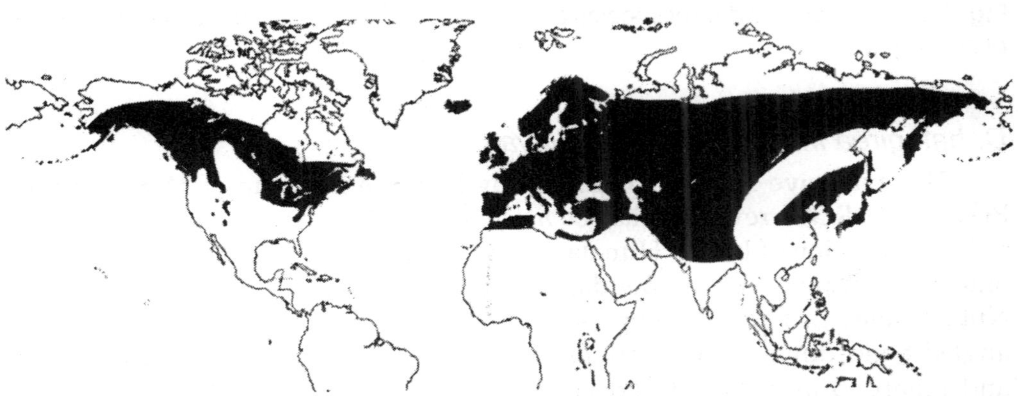

Fig. 2: World distribution of the genus *Dactylorhiza* (shaded areas), modified from Luer (1975) and Pridgeon *et al.*, (2001)

1.2 Distribution

Belongs to family Orchidaceae, *Dactylorhiza* species is distributed in the temperate to alpine regions of Europe, Asia, North Africa, Japan, the Aleutian Islands and northern parts of North America at altitudes of 2800-4000 m (Fig. 2). Marsh Orchis is found in shrubberies, open slopes and marshes in the Himalayas from Pakistan to South - East Tibet and Nepal at an altitude of 2500 - 5000 m (Bhatt *et al.*, 2005). Among different species *D. viridis* is a widely spread species in the new world (Luer, 1975). The world-wide genetic diversity (Table 1) and its distribution (Fig. 3) have been depicted ahead. In India, *D. hatagirea* is an indigenous species and exclusively found in Uttarakhand, Sikkim, Himachal Pradesh and Ladakh region of Jammu and Kashmir (mainly in Nubra valley).

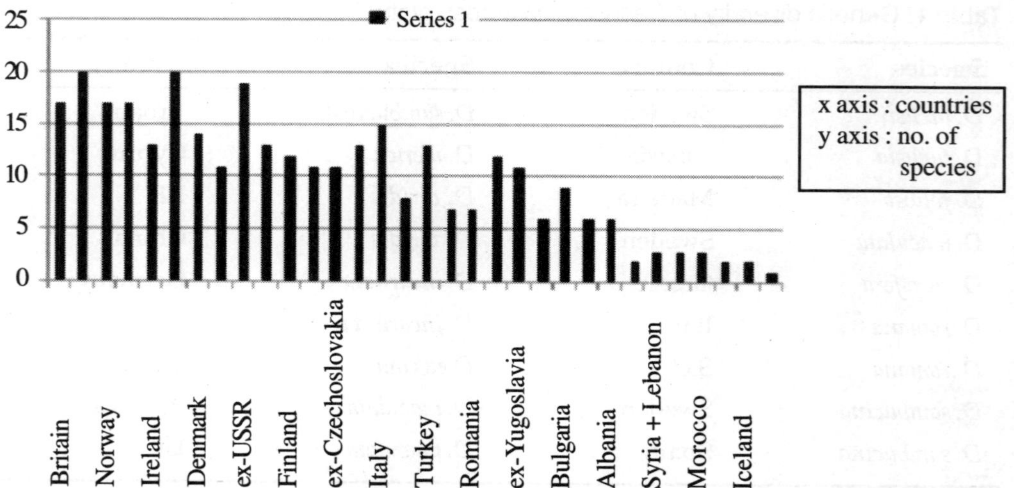

Fig. 3: Distribution of different species among different countries across the globe (Pillon *et al.*, 2006)

D. hatagirea in cold desert himalaya

The extensive survey conducted by Defence Institute of High Altitude Research (DRDO) revealed that *D. hatagirea* is habituated in the Nubra valley of Ladakh region of Indian Himalaya (altitude 2550-5000 m) mainly in Hunder, Skurru, Skampuk, Turtuk, Bogdang, Tirit, Sumur, Tigger etc. villages (Fig. 4). Nubra valley in Ladakh comprises the Valley of Shyok River from its acute-angled bend down to its confluence with Nubra and further towards Indus and regarded as a cold desert climate characterized by high wind velocity continues throughout the year, which causes great variation in temperatures (Temperature varies from -40ºC in peak winters to 35ºC in peak summers)

1.3 Botanical Description and Agro-techniques

Taxonomy of dactylorchids is widely considered to have been complicated by relatively great morphological variability within species and high frequency of hybridization between species (Pillon *et al.*, 2007). However, modern biotechnological tools i.e. allozymes, AFLPs, and plastid DNA data has revealed the patterns of polyploidy evolution (Fig. 5) in *Dactylorhiza* (Hedre *et al.*, 2001 and 2007).

D. hatagirea is a terrestrial ground dwelling perennial herb, up to 60 cm in height. The stem is erect, hollow and obtuse, and bears palmately lobed and lanceolate leaves with sheathing leaf base. The cylindrical and terminal spike bears rosy purple flowers with green bracts. Flowers are 1.7 – 1.9 cm long with curved spur. The inflorescence consists of a compact raceme with

25-50 flowers developed from axillary buds. The dark purple spotted lip of the flower is rounded and lobed (3 lobed). The plants store a large amount of water in their tuberous roots to survive in arid conditions.

Fig. 4: Map of Ladakh (Nubra Valley)

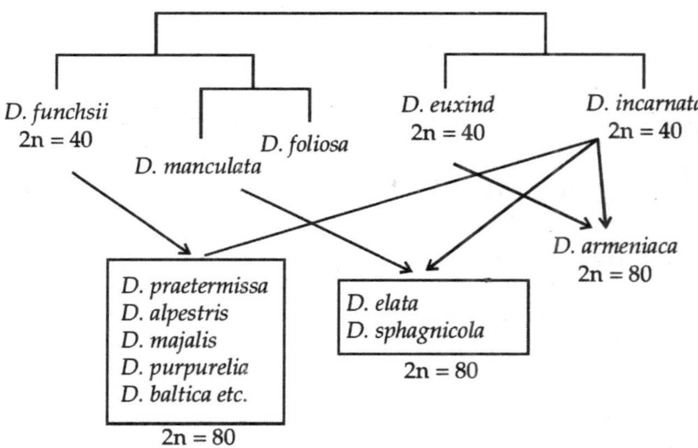

Fig. 5: Hybrid origin of some species and Ploidy level of *Dactylorhiza*

Dactylorhiza favours acidic and sandy loam soil with rich organic manure and sufficient moisture. The healthy plant development and rooting requires 80-90% humidity. The flowering season starts in early June and spreads up to July end. Subsequently the fruiting season starts in August - September (Chaurasia *et al.*, 2007).

Dactylorhiza is being propagated vegetatively using rhizomes, which may be collected after flowering. It may also be propagated by seeds. The mature fruits are collected in September and air dried for 2-3 weeks to extract the seeds. The seeds are stored at low temp until used. Being very minute in size, seeds are mixed with sand before sowing (Lal *et al.*, 2004). Seeds require symbiosis with fungal species for germination and only 20-30% germination is achieved when sown in natural habitat. However, the seed treatment of cytokines has been recommended for higher germination percentage by Tomita (1999).

Tubers may be harvested after seed maturation during mid September to October end. The plants that are developed from the tubers become ready for harvesting in two years. About 3.5 - 4 qt/ ha dry tubers could be harvested from the well maintained field. However, being recognized as a critically endangered plant species it is recommended that only 80% tubers should be harvested as measure of *in situ* conservation (Silori and Badola, 2000).

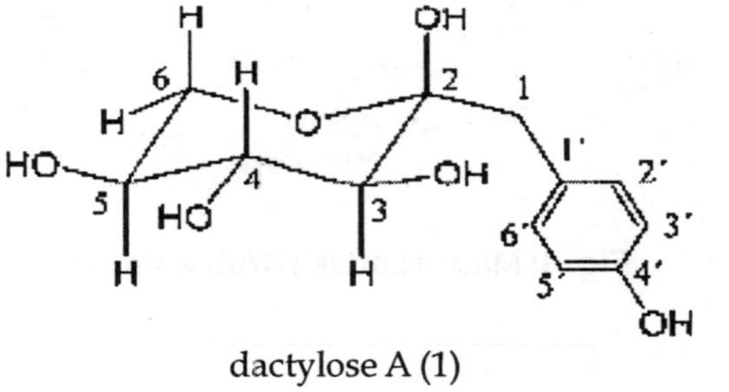

dactylose A (1)

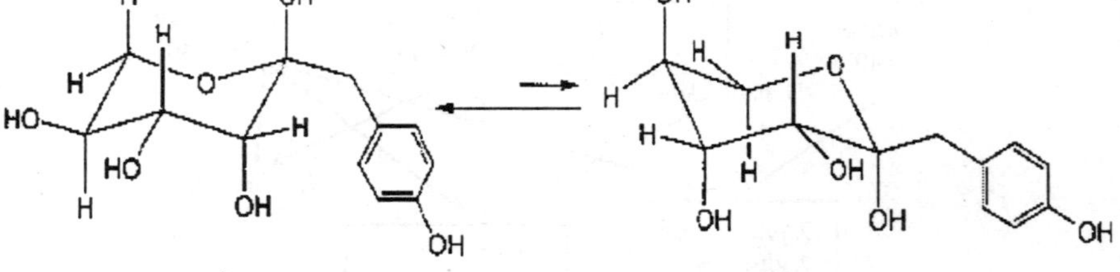

2a

2b

dactylose B (2)
Chart 2

Fig. 6: Structure of dactylose

1.4 Phytochemical Composition

The mature tubers contain mucilage (45%), starch, glucoside, loroglossin, albumen, volatile oil, phosphate (2.7%), chloride and ash contains potassium and lime (Chaurasia *et al.*, 2008). Five new compound known as dactylorhin A-E and two natural compounds known as dactyloses A-B have been reported by Kizu *et al.* (1999). Dactyloses A and B get synthesized from L-ascorbic acid and 4-hydroxybenzyl alcohol via 2-c-(4-hydroxybenzyl)-α-L-xylo-3-ketohexulofuranosono-1, 4-lactone (Fig. 6). However, Dactylorhin A and Dactylorhin E on enzymatic hydrolysis using almond emulsion give Dactylorhin C (Fig. 7). Also, Dactylorhin D and Dactylorhin B on enzymatic hydrolysis using cellulose gives 7a compound (-2-3-dihydroxy-2-2-methylpropyl) butanedioic acid which on hydrolysis gives Loroglossin (Kizu *et al.*, 1999). Leaves contain loroglossin compound which help in peristaltic movement of gastrointestinal tract and tuber contains dactylorhin compound which acts as a neuroprotective agents against dementia, Alzheimer's disease, depression, anxiety, and irritable bowel syndrome and significantly it improves the memory in case of treating with inhibiter such as scopolamine, cycloheximide or alcohol.

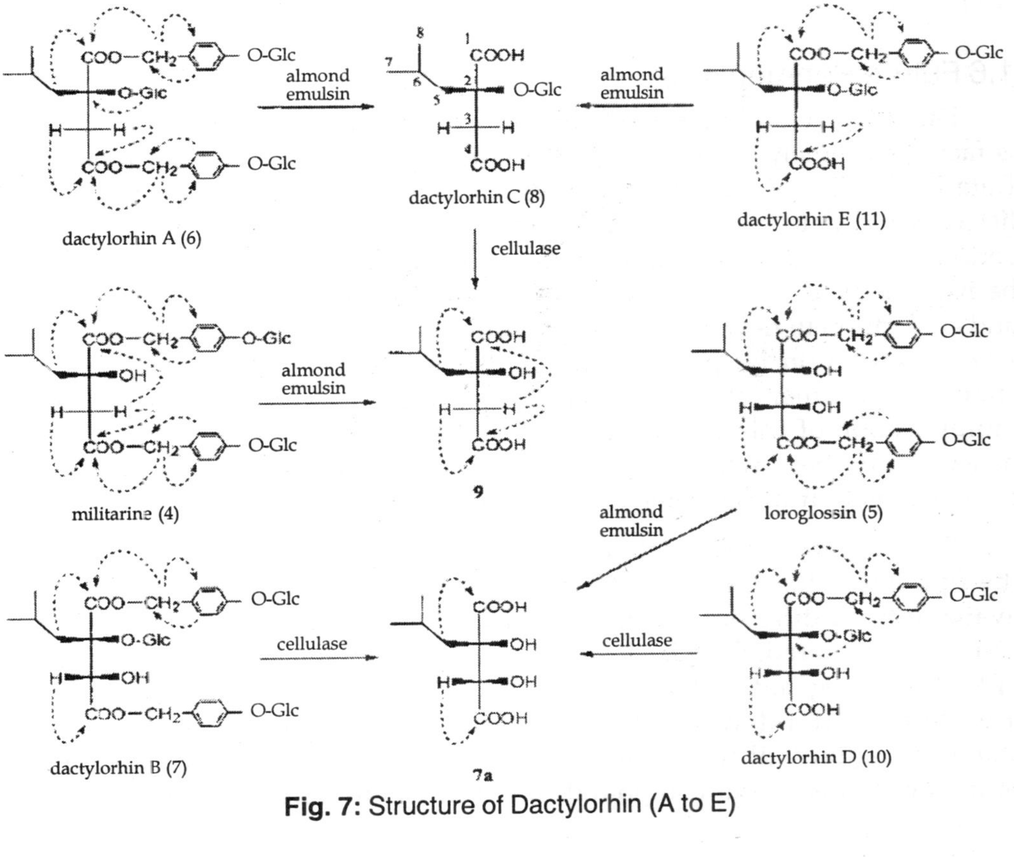

Fig. 7: Structure of Dactylorhin (A to E)

1.5 Pharmacology

Bulbous roots are used traditionally as an aphrodisiac, expectorant and nervine tonic. Studies conducted with lyophilized aqueous extract of roots on the effect of sexual behavior and spermatogenesis in male rats revealed that it increases testosterone production by 2.5 folds and subsequently increases the attraction towards females (Thakur and Dixit, 2007).

The tuber of *Dactylorhiza* yields Salep that is being used as a sizing material in silk industry. It is reputed as expectorant, aphrodisiac and nervine tonic and used in the treatment of dysentery, diarrhea. Salep is also used to increase sperm count and to relieve hoarseness, diabetes, paralysis, convalesance, impotence, malnutrition etc. Salep boiled with milk is being used as a rejuvenating tonic in some parts India. This plant is famous not only in Ayurveda but also in Tibetan Medicine, Unani etc. and about 2.5 gm powder of tubers is considered as a full day's diet in adverse conditions by local people. Decoction of Salep with sugar and flavored with spices has got a tremendous nutraceutical value. Mucilage jelly is also a nutritious byproduct and useful in treating diarrhea, dysentery, chronic fever leucorrhoea etc. (Chaurasia *et al.*, 2008).

1.6 Future Perspectives

Due to over exploitation of rhizomes for medicinal purposes *D.hatagirea* is facing extremely high risk of extinction; hence, listed into the Indian Red Data Book. Near about all the species of *Dactylorhiza* spread world-wide are listed in red data book. Therefore, immediate exclusive steps for its *in situ* conservations including complete ban of its collection from forest areas must be implemented at an early date. The standardization of agro-technologies and cultivation in new areas will give further boost to its conservation. The *ex situ* conservation by plant tissue culture techniques would support the endeavors. Further, of *in vitro* techniques would also prove helpful in mass multiplication of this endangered species. The identification and extraction of secondary metabolites using *in vitro* techniques will definitely strengthen the conservation programme.

The enormous medicinal potential of *Dactylorhiza* has not been tapped to the full extent till today. Using different mass-spectroscopy methods, the availability of different chemical compounds with high medicinal values should be explored. Standardization of agro-technologies and development of high yielding cultivars through conventional breeding methods is also an area left unattended. The efficient use of different DNA marker for characterization will further help the breeding programme for exploitation of important medicinal compounds and other agronomical traits.

Referances

Bhatt, A., Joshi, S. K., Gairola, S. 2005. *Dactylorhiza hatagirea* (D.Don) soo- a west Himalayan Orchid in peril. *Curr. Sci.*, 89: 610-612.

Chaurasia, O. P., Ahmed, Z., Ballabh, B. 2007. Ethnobotany and Plants of Trans-Himalaya. *Satish Serial Publishing House Delhi. pp* 544.

Chaurasia, O. P., Yadav A, Rimando, A.M., Terrill, T.H. 2008. Advances in Agriculture, Environment and Health. *Satish Serial Publishing House, Delhi. pp* 425.

Hedre, M.N., Sofie, N.M., Helena, A., Persson, H., Henrik, A.P. 2007. Patterns of polyploid evolution in greek marsh orchids *Dactylorhiza; orchidaceae* as revealed by allozymes, AFLPS, and plastid DNA data. *American J. Botany*, 94(7): 1205–1218. 2007.

Hedre, N.M., Fay, M.F., Chase, M. W. 2001. Amplified fragment length polymorphisms (AFLP) reveal details of polyploid evolution in Dactylorhiza (Orchidaceae). *American J. Botany*, 88: 1868– 1880.

Kala, C.P. 2000. Status and conservation of rare and endangered medicinal plants in the Indian trans-Himalaya. *Biological Conservation*, 93 (3): 371-379.

Kizu, H., Kaneko, Ei-ichi, Tominori, T., 1999. Studies on Nepalese Crude Drugs. XXVI.[1)] Chemical Constituents of Panch Aunle, the Roots of *Dactylorhiza hatagirea* D. DON. Chem. *Pharma. Bull.*, 47: 1618-1625.

Lal, B., Negi, H.R., Singh, R.D., Ahuja, P.S. 2004. Medicinal uses of *Dactylorhiza hatagirea* among the natives of higher altitudes in Western Himalaya. *J. Orchid Society of India*, 18: 97-100.

Luer, C.A. 1975. The native orchids of the United States and Canada, excluding Florida. *New York Botanical Garden, New york.*

Pillon, Y., Fay M.F., Hedren, M., Bateman, R.M., Devey, D.S., Shipunov A.B., Bank, M.V., Chase, M.W. 2007. Evolutiona dn temoral diversification of western European polylpoid species complexes in *Dactylorhiza* (Orchidaceae). *Taxon*, 56 (4): 1185-1208.

Pillon, Y., Fay M.F., Shipunov A.B., Chase, M.W. 2006. Species diversity versus phylogenetic diversity: A practical study in the taxonomically difficult genus *Dactylorhiza* (orchidaceae). *Bilogical Conservation*, 129: 4-13.

Pridgeon, A.M., Cribb, P.J., Chase, M.W., Rasmussen, F.N. 2001. Genera Orchidacearum. 2. Orchidoideae, 1. *Oxford University Press, Oxford.*

Samant S.S., Dhar, U., Rawal, R.S. 2001. In: Himalayan Medicinal Plants- Potential and Propspects (Eds. Samamt, S.S., Dhar, U. and Palni, L.M.S.) *Gyanodaya Prakashan, Nainital, pp 166-184.*

Silori, C.S., Badola, R. 2000. Medicinal plant cultivation and sustainable development: A case study in the buffer zone of the Nanda Devi Biosphere Reserve, Western Himalaya, India. *Mountain Res. Dev.*, 20: 272-279.

Thakur, M., Dixit, V.K., 2007. Aphrodisiac activity of Dactylorhiza hatagirea (D.Don) soo in male albino rats. *eCAM*, 4: 29-31.

Tomita, M. 1999. The cytokinin preference for immature culture of some terrestrial orchids. *Combined Proc. Int. Plant Propagators' Soc.*, 52: 331-334.

Uniyal, S.K. Awasthi, A., Rawat, G.S. 2002. Current status and distribution of commercially exploited medicinal and aromatic plants in upper Gori Valley, Kumaon Himalaya, Uttaranchal. *Curr. Sci.*, 82: 1246-1252.

□□□

High Altitude Medicinal Plants - A Rich Source of Valuable Phytoconstituents

Janifer Raj, X, O.P. Chaurasia, M.Pal Murugan and Shashi Bala Singh

CONTENTS

1.1	Introduction	48
1.2	Medicinal Plant Resource Base at High Altitude	48
1.3	Amchi System of Medicine	50
1.4	Phytochemical Constituents of Plants	51
1.5	Phytochemical Screening of High Altitude Medicinal Plants	52
1.6	Conclusion	58

1.1 Introduction

Medicinal plants are of great importance to the health of individuals and communities. They have been used for centuries as remedies for human diseases because they contain components of therapeutic values (Nostro *et al.*, 2000). The idea that the natural world, in particular, plants, has benefits for human health is not new. The use of plants as source of remedies for the treatment of many diseases dated back to prehistory and people of all continents have this old tradition. These early attempts used natural substances, usually native plants or their extracts and many of these herbal remedies proved successful (Sofowora, 1982). Seeds from medicinal plants in use today have been found on frozen humans dating back at least 60,000 years. The first known Materia Medica , or book of plant medicines, was written on clay tablets in Sumaria about 2000 BC , and the currently used Chinese Materia Medica was written in approximately the first century AD. About 80 percent of the world population relies on the use of traditional medicine which is predominantly based on plant material (WHO, 1993). Green plants posses the broadest spectrum of synthetic activity and have been the source of many useful compounds (Sofowora, 1986). Many of these indigenous medicinal plants are used as spices and food plants. Coincidentally, the last decade has also witnessed increasing intensive studies on extracts and biologically active compounds isolated from plant species used for natural therapies or herbal medicine (Rýos *et al.*, 2005). . The medicinal value of these plants lies in some chemical substances that produce a definite physiological action on the human body. The most important of these bioactive constituents of plants are alkaloids, tannins, flavonoids, and phenolic compounds (Hill, 1952).

The WHO defines complementary and alternative medicine (CAM) as all forms of health care provision that usually lie outside the official health sector. There are over 100 different therapies available as CAM treatments, but the five discrete clinical disciplines (acupuncture, chiropractic, herbal medicine, homeopathy, and osteopathy) are distinguished by having established foundations of training and professional standards. WHO guidelines define herbal medicines as finished labeled medicinal products containing an active ingredient that is obtained from the aerial or underground parts of botanicals or other plant materials or their combination. Plant materials include juices, gums, fatty oils, essential oils, and any other substances of this nature.

1.2 Medicinal Plant Resource Base at High Altitude

The northern part of India harbours a great diversity of medicinal plants because of the majestic Himalayan range. So far about 8000 species of angiosperms, 44 species of gymnosperms and 600 species of pteridophytes have been reported in the Indian Himalaya (Singh and Hajra,1996), of these 1748 species are known as medicinal plants The maximum medicinal plants (1717 species) have been reported around the 1800 m elevation range. On the regional scale, the maximum species of medicinal plants have been reported from Uttaranchal [3], followed by Sikkim and North Bengal (Samant *et al.*,1998). The trans-Himalaya sustains about 337 species of medicinal plants [8], which is low compared to other areas of the Himalaya due to the distinct geography and ecological marginal conditions (Kala amd Mathur,2002). Several plant species are endemic to the Himalayan region. Out of total known number of higher plants from India, approximately 46% are endemic to the Himalaya. Of the total medicinal plant species, sixty-two species of medicinal plants are endemic to the Himalaya and 208 extend their distribution to the adjacent areas, and are therefore classified as near endemic (Chatterjee,1939). Over 200 species of Himalayan medicinal plants are consumed raw, roasted, boiled, fried, cooked, or they are used in the form of oil, spices, jams or pickles (Dhyani and Dhar,1994). The indigenous communities use some medicinal plant species as a source of food, fodder, timber as well as various other ethnobotanical purposes. For example, apart from the use of Myrica esculenta and Terminalia bellirica as medicines, the fruits of these species are edible, the leaves are used for fodder and the wood is used for fuel. Approximately 81 species of Himalayan medicinal plants are known to be used for the extraction of oil. Of the total 675 species of Himalayan wild edibles, 171 are used for the treatment of diseases (Samant *et al.*, 2001). The crop plants diversity is also a source of traditional medicine (Arora and Nayar,1984).

In India, the rich plant diversity of the Himalaya over 8000 angiosperms, 44 gymnosperms, 600 pteridophytes,1737 bryophytes, 1159 lichens, etc. has been a source of medicine for millions of people in the country and elsewhere in the world (Singh *et al.*,1996). The IHR supports over 1748 (32.2% of India) plant species of known medicinal value[4]. The unique diversity of medicinal plants in the region is manifested by the presence of a number of native (31%), endemic (15.5%) and threatened elements: 14% of total Red Data plant species of the IHR and 3.5% of total medicinal plants in different threat categories of Conservation Assessment and Management Plan (CAMP) (Samant *et al.*, 1998). The economic potential of the MPs in some parts of the IHR5 and their contribution in providing novel biomolecules is recognized Dhawan,1997). The geographical position of this region manifests rich biodiversity of medicinal plants having tremendous potential. Out of 17,500 flowering plants known to occur in India (Sharma *et al.*, 1997), about 3000 are recognized for their medicinal use. They can be placed into the following three categories: (i) plants of codified knowledge (over 1500), used in organized systems of medicine like Ayurveda, Unani and Siddha, (ii)plants of empirical knowledge (3000), used in ethno medicine or folk medicine based on oral (undocumented) information being passed from generation to generation and (iii) plants of scientific knowledge (700), which have been investigated pharmacologically and chemically, and their active principles are used in modern medicine or are providing valuable leads for partial or total synthesis of new drugs (Prakash, 2002).

India is one of the 12 mega biodiversity centres having over 45,000 plant species. Its diversity is unmatched due to the presence of 16 different agroclimatic zones, 10 vegetative zones and 15 biotic provinces. The country has 15,000-18,000 flowering plants, 23,000 fungi, 2500 algae, 1600 lichens, 1800 bryophytes and 30 million micro-organisms. India also has equivalent to 3/4 of its land exclusive economic zone in the ocean harbouring a large variety of flora and fauna, many of them with therapeutic properties. About 1500 plants with medicinal uses are mentioned in ancient texts and around 800 plants have been used in traditional medicine (Kamboj, 2000)

1.3. Amchi System of Medicine

The mountain region in India is spread over 13 states in the Indian Himalayan Region covering a geographical area of 500,000 km (WHO, 1993). Over 51 million people (6% of the total population of India) live in this region (Anon 1992). Majority of them live in villages and belong to diverse cultures and communities. The large human populace with diverse life styles, beliefs, traditions and cultural heritage inhabiting Himalaya has learnt to utilize natural resources and products in various ways.The trans-Himalaya sustains about 337 species of medicinal plants (Kala,2002). Ladakh is a land like no other. It Covers an area of approx. 98,000 sq. km and has sparse population of about 1,35,000. It is bound by two of the world's mightiest mountain ranges, the Great Himalaya and the Karakorum, it lies athwart two others, the Ladakh range and the Zanskar range. In geological terms, this is a young land, formed only a few million years ago by the buckling and folding of the earth's crust as the Indian sub-continent pushed with irresistible force against the immovable mass of Asia. Ladakh lies at altitudes ranging from about 9,000 feet (2,750 m) at Kargil to 25,170 feet (7,672 m) at Saser Kangri in the Karakorum. It is situated between 30° to 36° east latitude and 76° to 79° north longitude Ladakh, the cold desert located in the northern most part of trans- Himalaya in Jammu and Kashmir State is well known for its rich ethno botanical wealth Fig.1. A great deal of traditional knowledge of the use of various plant species is still intact with the indigenous people, and this fact is especially relevant with the mountainous areas such as the Himalaya due to less accessibility of terrain and comparatively slow rate of development (Farooquee, 2004). Health care of the tribal population is mainly dependent on traditional system of medicine which is popularly known as Amchi system of medicine. The system traces its origin to Ayurveda. It was brought to Tibet in the 7th century A.D. Lord Buddha is believed to have founded the original Amchi system of medicine after he meditated for 4 years in the forest, near Budh Gaya. Later in the reign of Tibetan King Streng-Ki-Chan (750 A. D.) one renowned pandit named Barochana was sent to India and he translated the system, into the Tibetan script with the help of Acharya Chandra Dev. The text was then brought to Tibet. Tibetan community in the trans-Himalaya is one of such mountainous communities that had developed their own healthcare system based on the available natural resources. This system of therapy is well known today as the Tibetan medical system. Since amchis are the practitioners of this system of medicine, it is also known as Amchi medical system. There is ample literature available on Tibetan Medicine. (Dash, 1976; Dash,1994; Dommer,1988; Manjrekar, 1998; Namgyal and Phuntsog,1990; Tsarong, 1986).

1.4 Phytochemical Constituents of Plants

"Phyto" is the Greekword for plant. There are many "families" of phytochemicals and they help the human body in a variety of ways. Phytochemicals are non-nutritive plant chemicals that have protective or disease preventive properties. Plant produces these chemicals to protect itself but recent research demonstrates that many phytochemicals can protect humans against diseases. There are many phytochemicals in fruits and herbs and each works differently. A variety of herbs and herbal extracts contain different phytochemicals with biological activity that can be of valuable therapeutic index. The scientific studies available on a good number of medicinal plants indicates that promising phytochemicals can be developed for many human health problems (Gupta, 1994; Dahiru *et al.*, 2005), including diabetes, cancer and infectious diseases.

Different phytochemicals have been found to possess a wide range of activities, which may help in protection against chronic diseases. The phytochemical such as alkaloids, flavonoids, tannins, saponins, phlobatannins, terpenoids etc are known to show medicinal activity as well as exhibit physiological activity (Sofowara, 1993). Phytochemicals such as saponins, terpenoids, flavonoids, tannins, steroids and alkaloids have anti-inflammatory effects (Liu, 2003; Manach *et al.*, 1996; Latha *et al.*, 1998; Akindele and Adeyemi, 2007; Ilkay Orhan *et al.*, 2007). Glycosides, flavonoids, tannins and alkaloids have hypoglycemic activities (Oliver, 1980; Cherian and Augusti, 1995). Tannins are also known antimicrobial agents. Tannins (commonly referred to as tannic acid) are water-soluble polyphenols that are present in many plant food. Tannins have been reported to prevent the development of microorganisms by precipitating microbial protein and making nutritional proteins unavailable for them (Sodipo *et al.*, 1991). The growth of many fungi, yeasts, bacteria and viruses was inhibited by tannins (Chung *et al.*, 1998). Tannins are reported to have various physiological effects like anti-irritant, antisecretolytic, antiphlogistic, antimicrobial and antiparasitic effects. Phytotherapeutically tannin-containing plants are used to treat nonspecific diarrhoea, inflammations of mouth and throat and slightly injured skins (Westendarp, 2006).

Rupasinghe *et al.* (2003) have reported that saponins possess hypocholesterolemic and antidiabetic properties. Saponins are a special class of glycosides which have soapy characteristics (Fluck, 1973). It has also been shown that saponins are active antifungal agents (Sodipo *et al.*, 1991). The terpenoids have also been shown to decrease blood sugar level in animal studies (Luo *et al.*, 1999). Steroids and triterpenoids showed the analgesic properties (Sayyah *et al.*, 2004). The steroids and saponins are responsible for central nervous system activities. Steroids and phlobatannins were found to be of importance and interest in pharmacy due to their relationship with such compounds as sex hormones (Okwu, 2001).

1.5 Phytochemical Screening of High Altitude Medicinal Plants

Considering the importance of the phytochemicals responsible for the medical properties fifteen medicinal plants such as *Achillea millefolium, Bidens pilosa, Carum carvi, Ferula jaeskiana, Gallium pauciflorum, Heracleum pinnatum, Inula racemosa, Nepeta podostachys, Origanum vulgare, Peganum harmala, Rhodiola imbricata, Rosa webbiana, Rubia cordifolia, Tanacetum gracile,* and *Utrica hyperborea* which are extensively used in Amchi system of medicine in the Ladakh region of the Himalaya were investigated for the phytochemicals with biological activity Fig. 3 (a-o). Their various uses in traditional medicine as reported by Chaurasia *et al.* (2004) are reviewed in Table 1. The leaves, stems and roots of the plants were collected and identified by the authors from uncultivated lands and the herbal garden, DIHAR of Ladakh.

The plant samples were air-dried and ground into uniform powder using pestle and mortar. Two types of extracts namely the aqueous and chloroform were prepared. The aqueous extracts were prepared by soaking 5 g of dried powdered samples in 20 ml of distilled water for 12 hours and were filtered using Whatman filter paper No. 42 (125 mm). The chloroform extracts were prepared by soaking 5 g of dried powderd sample. The solution was extracted with dilute H_2SO_4 or dilute HCl and the acid layer was used for tests. Chemical tests were carried out on the aqueous and chloroform extracts and on the powdered specimens using standard procedures to identify the constituents as described by chemical tests were carried out on the aqueous and chloroform extracts, using standard procedures to identify the constituents as described by Sofowara, (1993), Trease and Evansm (1989) and Harborne, (1973). All the biochemical tests to test the qualitative presence of tannins, flavonoids, saponins, steroids, alkaloids and cardiac glycosides were carried out thrice and were found to be replicable.

Table 1 : Review of the various medicinal uses of the high altitude plants studied

Botanical Name	Family	Vernacular Name	Part Used	Traditional medicinal usages
Achillea millefolium	Asteraceae	Yarrow(E) Chuang(L)	Leaves/flowers	Leaves-stomach complaints. Flowers as a tonic
Bidens pilosa	Asteraceae	Local Tea/ Gurgur Chai	Leaves	As a local Tea/Namkin chai
Carum carvi	Apiaceae	Konyot	Seeds	Condiment
Ferula jaeskiana	Apiaceae	Jangli Heeng / Chuklam(L)	Stem/Roots	Roots used in Rheumatism, Stem gum resin used in septic wounds & Toothache
Gallium pauciflorum	Rubiaceae	Phomongo (L)	Stem/Leaves	Used to cure Intestinal parasites & reduce fever
Heracleum pinnatum	Apiaceae	Spru-ma(L)	Roots	Used to treat inflammation & pain caused by Vulnerable fever
Inula racemosa	Asteraceae	Puskarmool(H) Manu (L)	Roots	Asthma
Nepeta podostachys	Lamiaceae	Shangukaram	Aerial parts	Aromatic
Origanum vulgare	Lamiaceae	Wild oregano /Marjoram(E)	Leaves	Flavour agent
Peganum harmala	Zygophyllaceae	Wild Rue(E) / Sepan (L)	Seeds	Diuretic
Rhodiola imbricata	Crassulaceae	Rose root or Stone Crop(E) /Shrolo (L)	Leaves/Roots	Leaves used to making local dish Thukpa. Roots used as anti-stress Health tonic
Rosa webbiana	Rosaceae	Wild Rose(E)/Siah (L)	Flower	As a ornamental
Rubia cordifolia	Rubiaceae	Manjith (H), Indian Maddar (E), Btsod(L)	Stems & roots	Effective against blood diseases & Chest complaints
Tanacetum gracile	Asteraceae	Khamchu(L)	Leaves and flower	Used against Intestinal Worm
Utrica hyperborea.	Utricaceae	Stinging Nettle(E), Dzatsutt or Zozot(L)	Leaves	Used to making local dish Thukpa.

Table 2 : Qualitative analysis of the phyto chemicals of the high altitude medicinal plants

S.No.	Plants	Tannins	Flavanoids	Saponins	Cardiac glycoside	Alkaloids	Steroids
1	*Achillea millefolium*	+	+	+	+	+	+
2	*Bidens pilosa*	+	+	-	+	+	-
3	*Carum carvi*	+	+	+	+	+	-
4	*Ferula jaeskiana*	+	+	+	+	+	+
5	*Gallium pauciflorum*	+	+	+	+	+	+
6	*Heracleum pinnatum*	+	+	+	-	+	+
7	*Inula racemosa*	+	-	+	-	+	-
8	*Nepeta podostachys*	+	+	+	+	+	-
9	*Origanum vulgare*	+	+	+	-	+	+
10	*Peganum harmala*	-	+	+	+	+	-
11	*Rhodiola imbricata*	+	+	+	+	+	-
12	*Rosa webbiana*	+	+	+	+	+	-
13	*Rubia cordifolia*	+	+	+	-	+	-
14	*Tanacetum gracile*	+	+	+	+	+	+
15	*Utrica hyperborea*	+	-	+	+	+	+

+ Presence of constituent - Absence of constituent

(a) *Achillea millefolium*

(b) *Bidens pilosa*

(c) *Carum carvi*

(d) *Ferula jaeskiana*

(e) *Gallium pauciflorum*

(f) *Heracleum pinnatum*

Fig. 3 : (a-f) High Altitude Medicinal Plant of Ladakh

(g) *Inula racemosa*

(h) *Nepeta podostachys*

(i) *Origanum vulgare*

(j) *Poganum pauciflorum*

(k) *Rubia cordifolia*

(l) *Tanacetum gracile*

Fig. 3 : (g-l) High Altitude Medicinal Plant of Ladakh

(m) *Rhodiola imbricata* (n) *Rosa webbiana*

(o) *Utrica hyperborea*

Fig. 3 : (m-o) High Altitude Medicinal Plant of Ladakh

Test for Tannins: 0.5 g of the dried powdered samples was boiled in 20 ml of water in a test tube and then filtered. A few drops of 0.1 percent ferric chloride was added and observed for browrish green or a blue-black colouration.

Test for Flavonoids: Five ml of dilute ammonia solution were added to a portion of the aqueous filtrate of each plant extract followed by addition of concentrated H_2S0_4. A yellow colouration observed in each extract indicated the presence of flavonoids. The yellow colouration disappeared on standing.

Test for Saponins: About 2 g of the powdered sample was boiled in 20 ml of distilled water in a water bath and filtered. 10ml of the filtrate was mixed with 5 ml of distilled water and shaken vigorously for a stable persistent froth. The frothing was mixed with 3 drops of olive oil and shaken vigorously, then observed for the formation of emulsion.

Test for Cardiac glycosides (Keller-Killani test): Five ml of each extracts was treated with 2 ml of glacial acetic acid containing one drop of ferric chloride solution. This was underlayed with 1 ml of concentrated sulphuric acid. A brown ring of the interface indicates a deoxysugar characteristic of cardenolides. A violet ring may appear below the brown ring, while in the acetic acid layer, a greenish ring may form just gradually throughout thin layer.

Test for Alkaloids (Wagner's test): One ml of the chloroform extract was mixed with 1 ml of Wagners reagent and the positive reaction results in a brown coloured precipitate.

Test for Steroids (Salkowski test): Five ml of chloroform extract was mixed concentrated H_2S0_4 (3 ml) and shaken. If the reaction is positive it gives red colouration on standing.

1.6.Conclusion

The preliminary investigation conducted on the high altitude medicinal plants revealed the presence of medicinally active phyto constituents. The phyto constituents of the fifteen medicinal plants investigated are summarized in Table 2. The results indicated that the plants investigated were rich in tannins, flavonoids, saponins, steroids, alkaloids and cardiac glycosides and these phytochemicals were known to exhibit medicinal as well as physiological activity. All the fifteen plants (100%) screened gave a positive reaction for alkaloids. Acetogenins screened included tannins and flavonoids. Fourteen plant materials (93%) gave a positive reaction for tannins which are well documented for physiological astringent properties (Fransworth, 1966) while thirteen samples (87%) tested positive for flavonoids which are known to possess anti-viral and anti-inflammatory properties. Screening for isoprenoids included saponins and steroids which have expectorant, anti-diabetic properties as well as precursors for steroid hormones (Okwu, 2001). Fourteen (93%) and seven (47%) plant samples gave positive results for saponins and steroids respectively. Eleven (73%) plant samples tested positive for cardiac glycosides which are used in treating congestive heart failure and cardiac arrhythmia. Fig 4. The plants studied here can be seen as a potential source of useful drugs. Further studies are going on to isolate, identify, characterize and elucidate the structure of the novel bioactive compounds.

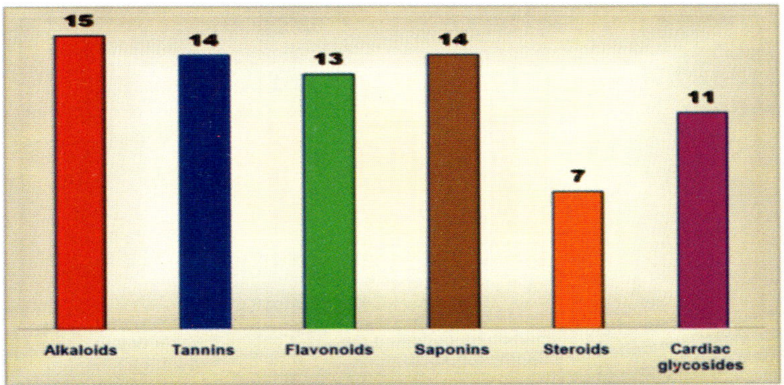

Fig. 4 Distribution of phyto chemicals in the medicinal plants

References

Akindele, A.J. and O.O. Adeyemi, 2007. Antiinflammatory activity of the aqueous leaf extract of *Byrsocarpus coccineus*, *Fitoterapia*, 78: 25-28.

Anon, *Action Plan for Himalaya*, G. B. Pant Institute of Himalayan Environment and Development, Kosi Katarmal, Almora,1992.

Arora RK, Nayar ER: Wild relatives of crop plant in India.*NBPGR Science Monograph* 1984, 7:97.

Chatterjee D: Studies on the endemic flora of India and Burma. *Journal of Royal Asiatic Society Bengal* 1939, 5:19-67.

Chaurasia, O P, Z Ahmed and B Ballabh. 2004. Ethno-botany & Plants of Trans- Himalaya.

Cherian, S. and K.T. Augusti, 1995. Insulin sparing action of leucopelargonidin derivative isolated from *Ficus bengalesis* Linn. *Indian J. Exp. Biol.*, 33: 608- 611.

Chung KT, Wong TY, Wei CI, Huang YW, Lin Y .1998. Tannins and human health: a review. *Crit Rev Food Sci Nutr* 38(6): 421-464.

Dahiru, D., William, E.T. and Nadroi, M.S. 2005. Protective effects of Ziziphus mauritiana leaf extract on carbon tetrachloride-induced liver injury. *African Journal of Biotechnology*, 4 (10):1177-1179.

Dash, B., In *Encyclopaedia of Tibetan Medicine*, Sri Satguru Publications, Delhi, 1994, Vols I–III.

Dash, B., In *Tibetan Medicine*, Library of Tibetan Works and Archives, Dharamsala, 1976.

Dhawan, B. N., *Himalayan Biodiversity – Action Plan* (ed. Dhar,U.), Gyanodaya Prakashan, Nainital, 1997, pp. 111–114.

Dhyani PP, Dhar U: *Myrica esculenta* Box myrtle Kaphal: A promising underexploited tree crop of the Himalaya. Almora: G.B. Pant Institute of Himalayan Environment and Development 1994.

Dommer, T., *Tibetan Medicine and Other Holistic Health CareSystems*, Routledge, Chapman and Hall, New York, 1988.

Farnsworth, N.R. The NAPRALET data base as an information source for application to traditional medicine. In: R.H.Bannerman, J.Burton and C.Wen- Cheih (Eds.), *Traditional medicine and health care coverage*, World Health organization, Geneva,1983, pp.184-193.

Farooquee NA, Majila BS, Kala CP: 2004. Indigenous knowledge systems and sustainable management of natural resources in a high altitude society in Kumaun Himalaya, India. *Journal of Human Ecology.*, 16:33-42.

Fluck H .1973. Medicinal plants and their uses, W. Feulshom and comp. Ltd, New York. pp. 7-15.

Gupta, S.S. 1994. Prospects and perspectives on natural plant products in medicine. *Indian Journal of Pharmacology*, 26:1-2

Harborne JB. *Phytochemical methods*, London. Chapman and Hall, Ltd. 1973, pp. 49-188.

Hill AF.1952. Economic Botany. A textbook of useful plants and plant products. 2nd edn. McGarw-Hill Book Company Inc, New York.

Ilkay Orhan, Esra Kupeli, Bilge Sener and Erdem Yesilada, 2007. Appraisal of anti - inflammatory potential of the clubmoss, *Lycopodium clavatum* L. *J. Ethnopharmacol.*, 109: 146-150.

Kala C.P: 2002 .Medicinal Plants of Indian Trans-Himalaya. Dehradun: Bishen Singh Mahendra Pal Singh.

Kala CP, and Mathur VB: Patterns of plant species distribution in the trans-Himalayan region of Ladakh, India. *Journal of Vegetation Science* 2002, 13:751-754.

Kamboj,V,P. 2000. Herbal Medicine. *Current Science*, 78, (1), 10 35-39.

Latha, R.M., T. Geetha and P. Varalakshmi, 1998. Effect of *Vernonia cinerea* less flower extract in adjuvant induced arthritis. *General Pharmacol.*, 31: 601-606.

Liu, R.H., 2003. Health benefits of fruit and vegetables are from additive and synergistic combinations of phytochemicals. *Am. J. Clin. Nutr.*, 78: 517S-520S.

Luo, J., J. Cheung and E. Yevich, 1999. Novel terpenoidtype quinones isolated from *Pycnanthu angolensis* of potential utility in the treatment of type-2 diabetes. *J. Pharmacol. Exptl. Therapy*, 288: 529-534.

Manach, C., F. Regerat and O. Texier, 1996. Bioavailability, metabolism and physiological impact of 4-oxo-flavonoids. *Nutr. Res.*, 16: 517-544.

Manjrekar, N., *A Study of the Local Health Traditions of Spiti*,Foundation for Revitalization of Local Health Traditions, Bangalore, 1998

Namgyal, G. and Phuntsog, S. T., In *Amchi Pharmaco-therapeutics*, Central Council for Research in Ayurveda and Siddha, Cambridge Printing Works, New Delhi, 1990.

Nostro, A., Germano, M.P., D'Angelo, V., Marino, A. and Cannattelli, M.A. 2000. Extraction Methods and Bioautography for Evaluation of Medicinal plants Antimicrobial activity. *Letters in Applied Microbiology*, 30:379-384.

Okwu, D. E. 2001.Evaluation of the chemical composition of indigenous spices and flavouring Agents. *Global J. Pure Appl. Sci.* 7(3): 455- 459.

Oliver, B., 1980. Oral hypoglycaemic plants in West Africa. *J. Ethnopharmacol.*, 2: 119-127.

Prakash, V., *Himalayan Medicinal Plants Potential and Prospect*(eds Samant, S. S., Dhar, U and Palni, L. M. S.), GyanodayaPrakashan, Nainital, 2002.

Rupasinghe, H.P., C.J. Jackson, V. Poysa, C. Di Berado, J.D. Bewley and J. Jenkinson, 2003. Soyasapogenol A and B distribution in Soybean (Glycine Max L.Merr) in relation to seed physiology, genetic variability and growing location. *J. Agric. Food Chem.*, 51: 5888- 5894.

Rýos J, Recio, M 2005. Medicinal plants and antimicrobial activity. *J. Ethnopharmacol.* 100: 80-84.

Samant SS, Dhar U, Palni LMS: Medicinal Plants of Indian Himalaya: Diversity Distribution Potential Values. Almora: G.B. Pant Institute of Himalayan Environment and Development; 1998.

Samant SS, Dhar U, Rawal RS: Diversity and distribution of wild edible plants of the Indian Himalaya. In *Plant Diversity of the Himalaya* Edited by: Pande PC, Samant SS. Nainital: Gyanodaya Prakashan; 2001:421-482.

Samant, S. S., Dhar, U. and Palni, L. M. S., *Medicinal Plants of Indian Himalaya*, Himavikas Publ., GBPIHED, Kosi Katarmal,Almora, 1998.

Sayyah, M., N. Hadidi and M. Kamalinejad, 2004. Analgesic and anti-inflammatory activity of *Lactuca sativa* seed extract in rats. *J. Ethnopharmacol.*, 92: 325-9.

Sharma, J. R., Mudgal, V. and Hajra, P. K., *Floristic Diversityand Conservation Strategies in India* (eds Mudgal, V. and Hajra,P. K.), Botanical Survey of India, Ministry of Environment andForests, 1997, p. 5.

Singh DK, Hajra PK: Floristic diversity. In *Biodiversity Status in the Himalaya* New Delhi: British Council; 1996:23-38.

Singh, D. K. and Hajara, P. K., *Changing Perspectives of Biodiversity Status in the Himalaya* (eds Gujral, G. S. and Sharma, V.), British Council Division, New Delhi, 1996, pp. 23–38.

Smith, B. W., *Am. J. Bot.*, 1950, 37, 802–815..

Sodipo OA, Akanji MA, Kolawole FB, Odutuga, AA .1991. Saponin is the active antifungal principle in Garcinia kola, heckle seed, *Biosci. Res. Commun.* 3: 171.

Sofowara A. 1993.Medicinal plants and Traditional medicine in Africa. Spectrum Books Ltd, Ibadan, Nigeria, pp. 289.

Sofowora A .1982. Medicinal Plants and Traditional medicine in Africa. Published by John Wiley and Sons Ltd. 1st edition 131: 168 -171.

Sofowora A .1986. The state of medicinal plant research in Nigeria. University of Ife Press Ife, Nigeria.

Trease GE, Evans WC. 1989.Pharmacognsy. 11th edn. Brailliar Tiridel Can. Macmillian publishers.

Tsarong, T. J., In *Handbook of Traditional Tibetan Drugs*, Tibetan Medical Publications, Dharamsala, 1986.

Uniyal, M. R., Bhat, A. V. and Chaturvedi, P. N., Preliminary observations on medicinal plants of Lahaul Spiti forest division in Himachal Pradesh. *Bull. Med. Eth. Bot. Res.*, 1973, 3, 1–26.

Westendarp H .2006. [Effects of tannins in animal nutrition]. *Dtsch Tierarztl Wochenschr* 113(7): 264-268.

WHO, 1993. Regional Office for Western Pacific, Research Guidelines for Evaluating the Safety and Efficacy of Herbal Medicines. Manila. p. 2

WHO/IUCN/WWF, *Guidelines on Conservation of Medicinal Plants,* IUCN Gland, Switzerland, 1993.

□□□

5

Rare Medicinal Plants of Western Ghats in Kerala and their Ayurvedic Uses

Deepu Mathew and Seeja Thomachan Panjikkaran

CONTENTS

1.1	Introduction	64
1.2	Morphological and Therapeutic Uses of Unknown Medicinal Plants	64
1.3	Conclusion	79
	References	80

1.1 Introduction

The Western Ghats region of India and parts of Srilanka is one among the most acclaimed hotspots of biodiversity in the world (Gadgil *et al.*, 1977). The hill chain of the Western Ghats constitutes the Malabar province of the Oriental realm, running parallel to the west coast of India from 8° N to 21° N latitudes, 73° E to 77° E longitudes for around 1600 km.These humid tropics extend at an area of 189,611 square kilometres, harbouring the home for at least 3,049 plant species (Anonymous, 2008). Heavy rainfall especially in the western face of the Ghats, reaching up to 3 metres annually, combined with the stable warm temperatures make this region the most comfortable place for origin and development of several rare species of plants. *Ayurveda* is an ancient Indian system of medicine that literally means the Science of Life. Ayurveda largely makes use of the miraculous curative potentials of these rare plants endemic to the Western Ghats. In this chapter we would like to introduce the comparatively unknown plants and discuss their morphological as well as therapeutic capabilities.

1.2 Morphological and Therapeutic Uses of Unknown Medicinal Plants

1. *Adapathiyan Kizhangu* (Malayalam) Payasvini (Sanskrit)

Holostemma ada-kodien *Schultes.* (Asclepiadaceae)

This plant is a perennial laticiferous woody climber, very common to the evergreen forests of Malappuram, Palakkad, Kozhikode and Kannur districts of Kerala, characterized by cordate simple leaves. Flowers are whitish purple in color, as axillary umbellate cyme. Fruits thick follicles, about 7 cm long and pointed. The tuberous roots of this plant are extensively used in Ayurvedic preparations. As on today this plant has extensive market potential and every kilo of dried roots costs more than 500 Indian Rupees. Plant pacifies vitiated tridoshas, diseases of eye, ear, burning sensation, cough, fever and diabetes. Ayurvedic giants of this state are ready to enter into contract farming with the farmers to assure its availability.

2. Amukkuram (Malayalam) Varahakarni (Sanskrit)

Withania somnifera (Linn.) (Solanaceae)

Amukkuram is a shrub that grows up to 2M with ovate and alternate leaves. Flowers are green arising from the leaf axils and the fruits are globose berries, become reddish orange when mature. Roots and leaves are extensively used in Ayurvedic preparations. Plant pacifies vitiated vata, nerve weakness, impotency, oligospermia and arthritis. The roots are aphrodisiac and stimulant, widely used in rejuvenative ayurvedic formulae. Plant has immuno-stimulant action.

3. Analivegam (Malayalam) Vishagni (Sanskrit)

Alstonia venenata R. Br. (Apocynaceae)

Analivegam is a small tree growing up to 5M height with greyish brown bark, bright yellow hard woody roots and simple leaves. Terminal cyme is white and both roots and fruits are used in medicinal preparations. Plant pacifies vitiated pitha, cobra bite, skin diseases, other venomous bites, epilepsy, fever and otalgia. Fruits are useful in insanity and epilepsy.

4. Arayal (Malayalam) Avatha (Sanskrit)

Ficus religiosa Linn. (Moraceae)

Arayal is a large spreading deciduous tree with cordate, shiny, long petioled and tailed leaves. One constituent of the *Naalpaamaram* in Ayurveda and it pacifies vitiated pitha, kapha, skin diseases, inflammation, poison, neuralgia, constipation and gynecological diseases. Every part of the plant finds use in various preparations.

5. Arogyapacha (Malayalam) Jeevani (San.)

Tricopus zeilanicus Gaertn. (Trichopodaceae)

Arogyapacha was traditionally used by the Kani tribes of Agasthya hills in Southern Kerala for retaining the vitality and to reduce hunger. This plant was introduced to the modern world by the successful interaction of TBGRI (Tropical Botanical Garden and Research Institute) at Thiruvananthapuram. Extensive studies are made on the antioxidant and antiageing properties of this plant and the technology is transferred to a Coimbatore (Tamil Nadu) based company for the development of commercial product named *Jeevani*. This plant is a small perennial herb that grows to 30 cm in height. Leaves are dark green colored, triangular, smooth, arising form underground rhizome. Flowers are purple and very small whereas fruits are dark brown similar to the cardamom fruits and got miraculous properties of suppressing hunger and relieve exhaustion. Plants pacifies vitiated kapha, retains vigor and vitality, relieve anxiety, prevent body changes due to ageing, expels accumulated poisons, immuno-modulatory and aphrodisiac.

6. Adalotakam (Malayalam) Vasa (San.)

Adhatoda beddomei C.B.Clarke (Acanthaceae)

Cheriya Adalotakam is a medium shrub that grows up to 3-4 meters. Leaves simple, opposite, ovate-lanceolate, acute or acuminate and shiny. Flowers arise from axillary condensed spikes, white; fruit capsules. This plant is an essential part of every Kerala households and young leaves are used in home remedy for fever, cough etc. Ayurvedic industry uses the whole plant.

Plant pacifies vitiated pitha, kapha, cough, bronchitis, asthma, inflammation, hemorrhage, hemorrhoids, diseases of eyes, and bleeding diarrhea.

7. Chethi (Malayalam)

Ixora macrothyrsa Linn. (Rubiaceae)

Chethi is a perennial shrub that grows up to 5 meters in height. Leaves simple, opposite, oblong-lanceolate, acuminate, short petioled, and shiny. Flowers arises form terminal umbels, and bright red in color. This plant pacifies vitiated kapha, pitha, burning sensations, eczema, ringworm other skin diseases, menorrhagia, leucorrhea and general weakness. Fruits, seeds, petals and bark are used in Ayurveda. Flowers are used to impart color for herbal preparations.

8. Athi (Malayalam) Falgu (San.)

Ficus carica Linn. (Moraceae)

Athi is a moderate sized tree simple, large and shiny leaves. One of the *Naalpaamaram* in Ayurveda. This plant got the potential to vitiate vata, pitha, burning sensation, burning micturition, sexual weakness, constipation, hemorrhoids, cough, intestinal worms, wounds, ulcers and general debility.

9. Avilpori (Mal.) Sarpari (San.)

Ophiorrhiza mungos Linn. (Rubiaceae)

Avilpori is a small shrub that grows up to 60 cm in height. Leaves are simple, opposite and lanceolate and cymes are white. Roots are used in Ayurvedic preparations. Plant pacifies vitiated kapha, wounds, ulcers, worm infestation, snake poison, skin diseases and cancer.

10. Ayyappana (Mal.) Ajaparna (San.)

Eupatorium ayapana Vent. (Asteraceae)

Ayyappana is an erect annual herb with opposite simple leaves, light blue flowers and reaching 1M height. This plant got the capability to pacify vitiated kapha, pitha, dyspepsia, hemorrhage, hemoptysis, menorrhagia, wounds, edema, ulcers, stomatitis, cardiac debility, skin diseases, poison bites, cough, asthma, bronchitis and general debility.

11. Adakkamanian (Mal.)

Sphaeranthus indicus Linn. (Asteraceae)

Adakkamaniyan is a spreading aromatic herb with hairy stem and branches bearing purple pink globose flowers terminally. The whole plant is used to pacify vitiated vata, pitha epilepsy, migraine, jaundice, fever, cough, hemorrhoids, helminthiasis, skin diseases and in preparing nerve tonics.

12. Chakkarakkolli (Mal.) Meshasringi (San.)

Gymnema sylvestre (Retz.) R.Br. (Asclepiadaceae)

Chakkarakkolli is a large woody climber with simple, glabrous, opposite and elliptic leaves and small yellow flowers in umbellate cymes. Plant pacifies vitiated vata, hepato-splenomegaly, renal and vesical calculi, constipation, asthma and leukoderma.

13. Changalamparanda (Mal.) Vajralatha (San.)

Cissus quadrangualaris Linn. (Vitaceae)

Changalamparanda is a peculiar quadrangular shaped greenish climber common in Kerala and Tamil Nadu. Tendrils are opposite to leaves, long and simple. Leaves are alternate and small. Plant pacifies vitiated kapha, vata, piles, osteoporosis, anorexia and fracture.

14. Chengazhineerkizhangu (Mal.) Hallaka (San.)

Kaempheria rotunda Linn. (Zingiberaceae)

Chengazhineerkizhangu is an aromatic shrub resembling ginger, turmeric and kacholam. Rhizomes are useful in pacifying vitiated vata, kapha, inflammations, wounds, ulcers, blood clots and malignancy.

15. Cherukataladi (Mal.) Markatapippali (San.)

Cyathula prostrata Blume. (Amaranthaceae)

Cherukataladi is an annual herb found in moist localities with the branches ending in inflorescence. Roots and seeds pacify vitiated vata, kapha, diseases of urinary system and uterine disorders and have constipating property.

16. Cherula (Mal.) Gorakshaganja (San.)

Aerva lanata (Linn.) (Amaranthaceae)

This is a semi-erect, many branched shrub that grows up to 60 cm tall. Leaves are simple, alternate, tomentose, and become smaller in the flowering twigs. Flowers are small sessile, greenish or whitish, often found in spikes. Fruits are greenish round compressed utricle, seeds kidney shaped and small. Plant pacifies vitiated pitha, urinary infection, vesical calculi, cough, and boils.

17. Chittamrithu (Mal.) Amrita (San.)

Tinospora cordifolia (Wild.) (Menispermaceae)

This is a spreading perennial deciduous vine with succulent stems on which simple, alternate, cordate leaves are arranged. Flowers yellow in recemes, arising from nodes on the old wood. Stem is used in Ayurveda to pacify vitiated vata, pitha, burning sensation, rheumatoid arthritis, gout, cardiac debility, skin disease, anemia, cough, jaundice, oligospermia and urinary system diseases.

18. Chittaratha (Mal.) Kulanjana (San.)
Alpinia calcarata Rox. (Zingiberaceae)

Chittaratha is a perennial, aromatic, rhizomatous herb reaching up to 2 M height. Flowers are fragrant and greenish white. Rhizomes of this plant is effective to cure vitiated vata, kapha, rheumatoid arthritis, inflammations, cough, asthma, bronchitis, dyspepsia, stomach pain, obesity, diabetes, migraine and intermittent fevers.

19. Chappangam (Mal.) Lohita (San.)
Caesalpinia sapan (Leguminoceae)

A small spiny leguminous tree grows up to 15 meters in height. Flowers yellow in terminal and axillary panicles and fruits small woody pods, with 2-3 seeds. Heartwood of the tree is used for boiling in the drinking water and is proved to have thirst quenching properties. Currently, this plant is having a huge potential for commercial large scale cultivation aimed at its heartwood.

20. Chathuramulla (Mal.) Chaturdharalata (San.)
Myxopyrum serratulum A.W.Hill (Oleaceae)

Chathuramulla is a large woody climbing shrub with quadrangular stems. Leaves are simple, opposite, elliptic and serrate. Flowers yellow in axillary or terminal panicles. Roots and leaves are used to pacify vitiated kapha, vata, cough, asthma, arthritis, headache, low back pain, fever, neuropathy and wounds.

21. Cherupunna (Mal.) Jyotishmati (San.)
Celastrus paniculatus Willd. (Celastraceae)

Cherupunna is a large climbing shrub with woody stem, elongate branches and simple, alternate, ovate leaves. Flowers are greenish white, in terminal panicles. Roots and leaves pacify mental retardation, skin diseases, itching, hemiplegia, arthritis, asthma, insomnia, amnesia, urinary retention and nephritis. Seeds widely used in intelligence promoting medicines.

22. Chethikoduveli (Mal.) Agni (San.)
Plumbago rosea Linn. (Plumbaginaceae)

This plant is under extensive commercial cultivation in Kerala for its tubers. Chethikkoduveli is a perennial shrub with simple alternate, acute leaves.

Flowers are red in long terminal spikes and cylindrical roots are the store house for Plumbagin, the commercially important ingredient. This plant is also very drought hardy. Root tubers when used in Ayurvedic medicines pacify diarrhea, inflammation, fever, nervous palsy, hemorrhoids, skin diseases, irritable bowel disease, amenorrhea and anemia.

23. Amalpori (Mal.) Sarpagandha (San.)
Rauvolfia serpentina (Linn.) (Apocynaceae)

A very familiar plant for the Ayurvedic medicines and this is a small erect shrub grows up to 1 metre in height with white flowers in cymes. Roots are cylindrical, yellowish brown and used to pacify hypertension, insanity, epilepsy, insomnia, wounds, fever, colic and urinary retention, in controlled dose.

24. Ekanayakam (Mal.) Peetika (San.)
Salacia reticulata Wight. (Hippocrateaceae)

This is a woody climber with simple, opposite leaves and yellowish green flowers. Roots pacify urinary infection, diabetes, hemorrhoids, inflammations, leucorrhea, skin diseases, wounds, ulcers, flatulence, colic, and jaundice.

25. Garudapacha (Mal.) Sarpadani (San.)
Selaginella rupestris Spring. (Selaginellaceae)

Garudapacha is a small herb grows up to 20 cm in height with compound very small leaves. Plant pacifies vitiated pitha, skin diseases, snake poison, traumatic edema and general debility.

26. Idampiri valampiri (Mal.) Murva (San.)
Helicterus isora Linn. (Sterculiaceae)

This is a large shrub or small tree grows up to 5 meters in height. Leaves are simple, alternate and obovate and flowers red to grey colored in axillary clusters of 2-6 together. Fruits, roots and bark of this plant are used to pacify pitha, diarrhea, colic, flatulence, wounds, ulcers, hemorrhoids and diabetes.

27. Irattimadhuram (Mal.) Madhuka (San.)
Glycyrrhiza glabra Linn. (Fabaceae)

This is a perennial shrub growing up to 1.5 meters in height. Leaves are compound, pinnate with 5-6 leaflets and flowers are violet in axillary racemes. Underground stems are woody and pale brown with a peculiar sweetish taste. These roots got the capability to pacify inflammation, headache, cough, bronchitis, fever, skin disease, peptic ulcers, graying of hair, ulcer, wounds and oligospermia.

28. Kacholam (Mal.) Gandhamoolaka (San.)

Kaempferia galanga Linn. (Zingiberaceae)

Kacholam is very familiar plant for the farmers of Kerala. In growth and morphology it resembles ginger, turmeric and chengazhineer kizhangu. Rhizomes of this perennial plant are very fragrant and compressed round leaves arise from the rhizome. Flowers are white with purplish spots in axillary fascicles. Rhizomes pacify vitiated vata, kapha, skin diseases, wounds, fever and nasal obstruction.

29. Kadukka (Mal.) Rohinee (San.)

Terminalia chebula Retz. (Combretaceae)

This is a medium to large deciduous tree with spreading branches and yellowish white flowers in the terminal simple spikes. This is widely used in Ayurveda as a constituent of *triphala* recommended for children. Fruits of this plant when administered pacifies tridoshas, wound, ulcer, inflammation, anorexia, flatulence, haemorrhoids, jaundice, hiccough, cough, vesical and renal calculi, epilepsy, skin diseases, leprosy, intermittent fever, cardiac disorders, stomatitis, neuropathy and general debility.

30. Kalloorvanchi (Mal.) Asmabheda (San.)

Rotula aquatica Lour. (Boraginaceae)

This is a much sought after but difficult to get magical shrub that grows up to 120 cm in height and found usually on riverbanks. Few years back, this shrub was very common in Achankovilar and few other rivers of Pathanamthitta district of Kerala and uncontrolled exploitation for its capability to cure the vesical calculi has resulted in almost extinction of this species. Leaves are simple, almost sessile, hairy and crowded on the branches. Roots of this plant pacify piles, diabetes, renal and venereal diseases and vesical calculi.

31. Kanjiram (Mal.) Vishatindukam (San.)

Strychnos nux-vomica Linn. (Loganiaceae)

This plant is known as a synonym for bitterness. Skin of the tree is highly poisonous. This is a tall growing deciduous tree reaching up to 15-20 meters in height. Leaves are opposite, oblong and dark green with smooth cuticle surfaces. Fruits become orange red on ripening and contain many seeds. Seed, bark, root, leaves of the plant are used for various demands in ayurveda. Plant got the capability to pacify cardiomyopathy, hypertension, arthritis, and dementia.

32. Karunochi (Mal.) Neelamanjari (San.)

Vitex negundo Linn. (Verbenaceae)

This is an aromatic large shrub reaching up to 5 meters in height with quadrangular branches. Plant is used in the treatment of arthritis, inflammations, lumbago, dyspepsia, colic, flatulence, dysentery, bronchitis, cough, malarial fever, hemorrhoids, dysmenorrhoea and leprosy.

33. Kasturivenda (Mal.) Latakasturi (San.)

Abelmoschus moschatus Medikus (Malvaceae)

This is an erect growing hairy annual herb with palmate hairy leaves. Flowers are large and yellow with purple center and fruits hairy capsules with many grayish brown and kidney shaped and scented seeds. Seeds are mainly used in Ayurvedic preparations and plant pacifies vitiated kapha, pitha, asthma, bronchitis, burning sensation, nausea, dyspepsia, flatulence, colic, diarrhea, calculi, vomiting, and nervous system disorders.

34. Kayyonni (Mal.) Bringaraja (San.)

Eclipta prostrata Linn. (Asteraceae)

Kayyonni is a common weed in the wetlands and garden lands of Kerala. It is an erect prostrate annual herb grows up to 70 cm height. Branches are few, stems are soft and fleshy, leaves are simple and opposite and white flowers are in head inflorescence. Plant pacifies vitiated kapha, vata, reduces pain, promote hair growth, stimulate the functions of liver, heal and clean ulcers and increase eye sight.

35. Keezharnelli (Mal.) Bhoomyamalaki (San.)

Phyllanthus amarus Schum. & Thonn (Euphorbiaceae)

This is a much known and profusely available medicine to cure the abnormalities of liver. Keezharnelli is an annual herb with slender leaf-bearing branches. Leaves are numerous, with small, globose, 3 seeded fruits underneath the branches. Plant pacifies jaundice, intermittent fevers, urinary retention, constipation, ulcers and wounds.

36. Kiriyath (Mal.) Kiratatikta (San.)

Andrographis paniculata (Burm.f.) Wall. ex Nees (Acanthaseae)

Kiriyath is a wasteland plant commonly seen in the northern districts of Kerala. This annual herb grows up to 50 cm in height and leaves are opposite and lanceolate. This plant has active anti-feedant property against insects and this is commercially exploited in organic farming. Plant pacifies tridoshas, skin diseases, burning sensation, cough and bronchitis. Plant stimulates liver and purifies vitiated blood.

37. Koovalam (Mal.) Vilva (San.)
Aegle marmelos Linn. Corr. (Rutaceae)

This tree is believed to be a holy tree and its leaves are used for garland preparations in Hindu temples of Kerala. As a vitamin C rich fruit tree, it is known as bael and is commercially grown in North India. Tree reaches up to 10 metres in height with trifoliate and aromatic leaves. Globose fruits resemble oranges. Leaves, roots and fruits of the plant are used in various medications and it cures body pain, poison, diarrhea, dysentery, vomiting, and intermittent fever. Pulp of unripe fruit is constipating whereas that of ripened fruit is laxative. Leaves cure diabetes, cough, inflammation and asthma.

38. Kallurukki (Mal.) Asmaghni (San.)
Scoparia dulcis Linn. (Scrophulariaceae)

Kallurukki is an erect annual herb grows up to 30 cm in height. Leaves simple, opposite and serrate. Small, dull white flowers are seen hanging in leaf axils and fruits are small round. Plant pacifies renal and vesical calculus, inflammation, fever, wounds, ulcer, skin diseases, diarrhea, dysmenorrhea and leucorrhea.

39. Karalayam (Mal.) Garudi (San.)
Aristolochia indica Linn. (Aristolochiaceae)

This is a perennial woody climbing shrub with simple, alternate leaves and round 6 chambered fruits containing numerous winged compressed seeds. Plant pacifies vata, poison, skin diseases, intestinal worms, colic, arthritis and ulcers.

40. Kaippanpadavalom (Mal.) Patola (San.)
Trichosanthes lobata Roxb. (Cucurbitaceae)

This is the most demanded medicinal plant of commerce as on today in Kerala. As a close relative of snakegourd, it shares all the morphological features. Fruits are green stripped with white line when immature, and scarlet red when ripe. Many Ayurvedic industries including *Kottakkal Arya Vaidyasala* are ready to take up commercial contract with farmers for its cultivation. Plant pacifies vitiated tridosha, skin diseases, ulcers, constipation, stomach pain, anorexia, flatulence, hyperacidity, hemorrhoids, cough, fever and general debility.

41. Kodapunna (Mal.) Brihatpunnaga (San.)
Dillenia pentagyna Roxb. (Dilleniaceae)

This is a large deciduous tree grows up to 40 meters in height. Leaves are large, alternate and ovate-rhomboid. Flowers are yellowish, fragrant and arise from the nodes of the fallen leaves. Leaves, bark and fruits of this plant are used for various medicinal preparations and it pacifies anal fistula, wounds, diabetes, diabetic carbuncle, neuritis, pleurisy, pneumonia, and burning sensation.

42. Kurumthotti (Mal.) Vadyala (San.)
Sida retusa (Linn.) Borssum (Malvaceae)

This is again a much sought after medicinal plant which is exclusively used for the treatment of arthritis. Kurumthotti is a semi-woody, erect under-shrub growing up to 50 cm in height. Leaves short petioled, obovate and serrate and flowers are yellow and solitary borne in the leaf axils. Roots are the economic part. Once this plant was profuse in the wastelands, garden lands and roadsides of Kerala. But indiscriminate harvest for its highly valuable roots has left the species extinct.

43. Moovila (Mal.), Chitraparni (San.)
Pseudarthria viscida W&A (Fabaceae)

This is again a well known medicinal plant in Ayurveda. Moovila is a semi erect, pubescent, perennial shrub growing up to 1 meter in height. Branches are covered with whitish hairs and leaves are hairy, compound and trifoliate. Flowers are purple, small and numerous, found on terminal and axillary spikes. Root is the commercial part and the plant pacifies tridoshas, cough, asthma, fever, dysentery, cardiac ailments, rheumatoid arthritis and aid in fast healing of fractured bone.

44. Mukkutti (Mal.), Peethapushpa (San.)
Biophytum sensitivum Linn. (Oxalidaceae)

Once a common plant to all the garden lands of Kerala, Mukkutti is now getting extinct, due to the overexploitation. Morphologically it resembles a coconut bonsai. It is a slender erect annual with abruptly pinnate leaves and opposite 6-12 pairs of leaflets. Flowers are yellow. Plant pacifies vitiated kapha, pitha, urinary calculi, wounds, abscesses, asthma and stomachalgia.

45. Kudangal (Mal.), Mandookaparni (San.)
Centella asiatica Linn. (Apiaceae)

A perennial runner common to the wetlands of the entire Western Ghats. Leaves are simple with long petioles. This plant is known in Chinese medicine for its blood purifying and brain nourishing properties and in North India, this is called *Brahmi*. Plant pacifies vitiated pitha, insomnia, epilepsy, asthma, bronchitis, abdominal disorders, mental retardation, insanity and fever.

46. Maramanjal (Mal.), Darvi (San.)
Coscinium fenestratum (Gaertn.) Colebr. (Menispermaceae)

Maramanjal is a large woody climber grows with cylindrical and yellowish stem. Leaves are simple, oblong, alternate and minutely tomentose beneath and smooth above. Flowers yellow, unisexual in supra-axillary inflorescence. Plant pacifies skin diseases, diseases of the eye, inflammations, wounds, jaundice, diabetes, fever and general debility.

47. Mayilellu (Mal.), Tilakam (San.)
Vitex alata Heyne, (Verbenaceae)

This is a large tree grows up to 40 meters in height. Leaves are compound, palmate, 3-5 foliate with winged rachis. Leaflets are acute and glabrous. Flowers are whitish, arising from the terminal panicles. Leaves and roots pacify vitiated kapha, vata, inflammation, wounds, ulcers, allergy, eczema, pruritus, worm infestations, urinary system diseases, stomatitis, emaciation, and ailments after parturition.

48. Nannari (Mal.), Anantamoola (San.)
Hemidesmus indicus (Linn.) (Asclepiadaceae)

This is a perennial, slender, laticiferous twinning shrub with woody stock and numerous slender stems. Leaves are simple and opposite and flowers are greenish purple. The tuberous root is the economic part and is dark brown, with transversely cracked and longitudinally fissured bark. It has a strong central vasculature and pleasant smell and taste. Plant pacifies skin diseases, burning sensation, leukoderma, pruritus, bronchitis, asthma, hyperdipsia, diarrhea, dysentery, hemorrhoids, leucorrhoea, arthritis, fever and general debility.

49. Neermaruthu (Mal.), Arjuna (San.)
Terminalia arjuna (Roxb) W&A (Combreteaceae)

Neermaruthus is a moderate sized deciduous tree growing up to 15 meters in height. Skin shedding is a characteristic feature of the plant. Bark is the economic part and plant pacifies vitiated skin diseases, poison, fever, wound and promote the re-union of fractured bone.

50. Neermathalam (Mal.), Varuna (San.)
Crataeva magna (Lour.) DC. (Capparaceae)

This is a deciduous tree reaching up to 7-9 m. Leaves are trifoliate and flowers are large, greenish white. Fruits are fleshy, ovoid berries. Leaves and bark are used in many Ayurvedic preparations and it got the capability to pacify vitiated vata, kapha, renal and vesical caliculi, asthma and bronchitis.

51. Neelamari (Mal.), Neelini (San.)

Indigofera tinctoria Linn. (Fabaceae)

This is again a much used plant in Ayurvedic preparations. *Neelibhringadi* is a medicinal oil prepared from this plant and is highly used to promote the hair growth. This is a highly branched shrub growing up to 3 meters. The bluish tinted leaves are compound with 6-14 leaflets. Flowers are many and small. Plant pacifies vitiated pitha, hair fall, splenomegaly, epilepsy, nervine disorders, headache, cardiac ailments and in stray dog intoxication.

52. Naikkurunam (Mal.), Markati (San.)

Mucuna pruriens (Linn.) DC. (Fabaceae)

This plant is known as herbal Viagra. This is a perennial climber with hairy branches. Leaves are trifoliate with leaflets ovate. Leaflets are pubescent above and flowers are purple in axillary drooping pendulous racemes. Fruits are curved pods, densely clothed with pale brown irritant hairs and seeds are black, ovoid, 4-6 per pod. Seeds are commercially and extensively used in treatment of impotency, Parkinsonism, kidney diseases, neuropathy, worm infestations, fever and general debility.

53. Neervalam (Mal.), Dandibeeja (San.)

Croton tiglium Linn. (Euphorbiaceae)

This is a small evergreen tree growing up to 10 meters height. Leaves are simple, alternate, ovate, acute and membranous. Flowers are unisexual, found in slender terminal panicles. Seed is the economic part and is used for the extraction of the oil also. Plant pacifies constipation, worm infestation, inflammations, insanity, fever, ascites, and generalized edema. In larger dose or without purification, the plant is toxic.

54. Njerinjil (Mal.), Gokshura (San.)

Tribulus terrestris Linn. (Zygophyllaceae)

This is a creeping perennial herb growing parallel to the ground and is very common to the rice fallows in Kerala and Tamil Nadu. Leaves are simple, pinnate, opposite and leaflets are rounded, sessile. Fruits are spiny, 5 angled, woody schizocarp containing one or more seeds in each unit. Plant pacifies urinary retention, vesical and renal calculi, urinary tract infections, sexual weakness, impotency, cough, asthma, anemia, general weakness and constipation.

55. Orila (Mal.), Sthira (San.)

Desmodium gangeticum DC. (Fabaceae)

A very common plant for all the Ayurvedic preparations and is common to the wastelands of Kerala. This is an erect, branched shrub grows up to 1 meter in height. Stem is woody with numerous prostrate branches and covered with soft hairs. Leaves are single, ovate-lanceolate, and covered with numerous gray hairs. Flowers are white, purple or lilac found in elongated terminal or axillary recemes. Fruits are moniliform 6-8 joined pods. Joints separate when ripe into one seeded segments. Root is the economic parts. Plant pacifies vitiated tridoshas, edema and cardiomyopathy. Plant has detoxifying property.

56. Poovaam kurunnila (Mal.), Sahadevi (San.)

Vernonia cinerea (Linn.) Less. (Asteraceae)

This is a very familiar plant in all the garden lands of Kerala. Poovaam kurunnila is an erect annual herb growing up to 75 cm in height. Stems are cylindrical, striate, and somewhat pubescent. Leaves are simple and alternate. Flowers are many, pinkish violet in small head inflorescences. Plant pacifies tonsillitis, stomach pain, diarrhea, intermittent fever, eczema, herpes, ringworm, and elephantiasis. Leaves are useful in conjunctivitis, and in lacrimation. Seeds are useful in worm infestation, cough, psoriasis, leukoderma and in other skin diseases and plant possess anti cancerous property.

57. Paalmuthukkin kizhangu (Mal.), Ksheeravidari (San.)

Ipomoea mauritiana Jacq., (Convolvulaceae)

This is a perennial twiner with broad palmately compound 5-7 lobed leaves, used mainly in Ayurveda in the preparations for increasing the breast milk production. Tuberous root is the economic part and roots are large, fleshy and tuberous. Plant pacifies emaciation, skin diseases, anorexia, colic, fever, burning sensation, vomiting, inflammation, menorrhagia and general debility.

58. Paanal (Mal.), Asvasakhota (San.)

Glycosmis pentaphylla (Rutaceae)

This plant is well known for its anti-viral properties. This is a moderate sized shrub growing up to 1 meter. Leaves are imparipinnately compound, alternate, gland doted and entire. As a preventive measure against the chikunguinea viral fever, daily intake of emerging leaves pulped in milk is recommended. Also the chicken pox patients are directed to lie on the leaves. Plant pacifies inflammations, traumatic inflammations, fever, cough, arthritis, skin diseases, and jaundice.

59. Punna (Mal.)
Calophyllum inophyllum Linn. (Clusiaceae)

This is a moderate evergreen tree growing up to 25 meters in height. Leaves are simple opposite and elliptic and flowers white. Bark, leaves and seeds are used in various preparations. Plant pacifies arthritis, ulcers, skin diseases, pruritus, migraine and urinary retention.

60. Putharichunda (Mal.), Simhi (San.)
Solanum anguivi Lam. (Solanaceae)

This is a much demanded medicinal plant for the Ayurvedic industry in Kerala, third only to *Adapathiyan kizhangu* and *kaippan padavalam*. Putharichunda is a prickly shrub growing up to 2 meters. A close relative of brinjal and hence having all the morphological identities. Fruits are greenish yellow globular berries, contain large number of small smooth compressed seeds. Roots and fruits are commercially important and it pacifies skin diseases, urinary retention, fever, cough, asthma, dysmenorrhea, colic, flatulence and constipation.

61. Raktachandanam (Mal.), Raktasara (San.)
Pterocarpus santalinus Linn. f. (Fabaceae)

A highly valued medicinal plant used for the preparation of facial products. This is a very tall growing tree reaching almost 15 meters in height. Leaves are compound with alternate leaflets. Heartwood is the commercial part and apart from the extensive use in cosmetics, it is also used to cure burning sensation, skin discoloration, fever, headache, diarrhea, dysentery, skin diseases, hemorrhage, inflammation, psychic disorders and general debility.

62. Ramacham (Mal.), Useera (San.)
Vetiveria zizanioides (Linn.) Nash (Poaceae)

This is an extensively grown perennial aromatic grass that may reach up to 2 meter in height. This plant is mainly grown in degraded and highly eroded soils to prevent the soil erosion due to its capability to bind large volume of soil. Root is the economic part that pacifies hyperdipsia, ulcer, skin diseases, vomiting nausea, flatulence, dyspepsia, colic, cough, fever, low back pain, headache and general debility.

63. Thazhuthama (Mal.), Punarnava (San.)
Boerhaavia diffusa Linn (Nyctaginaceae)

This is a succulent trailing perennial familiar for all the wetlands and rice fallows of Kerala and Tamil Nadu. Leaves are thick and in opposite unequal pairs. Leaves are ovate-oblong, acute or obtuse, rounded at base, glabrous above and whitish beneath. Plant pacifies fever, constipation, leucorrhoea, lumbar pain, myalgia, skin diseases, cardiac disorders, urinary infection, vesical stone, anemia, dyspepsia, constipation, and general debility.

64. Thiruthali (Mal.), Klitani (San.)

Ipomoea sepiaria Roxb. (Convolvulaceae)

This is a slender twinning perennial with tuberous roots. Plant pacifies vitiated pitha, sterility in women, urinary retention, constipation, and gynecological disorders.

65. Thakara (Mal.), Prapunnada (San.)

Cassia tora Linn. (Caesalpiniaceae)

This is an erect annual herb, familiar to the roadsides and wastelands and grows up to 90 cm in height. This plant is well known for its allelopathic effects against Parthenium, the noxious weed. Leaves and seeds are commercially used in the preparations to treat skin diseases, dandruff, constipation, cough, hepatitis, fever, and hemorrhoids.

66. Thanni (Mal.), Vibheetaka (San.)

Terminalia bellirica (Gaertn.) Roxb. (Combretaceae)

One constituent of Triphala, Thannikka is extensively used in Ayurveda. This is a large tree growing up to 15 meters in height. Leaves are simple, alternate, long petioled and crowded at the tip of branches. Flowers are small, pale green, found in axillary spikes. Fruits are ovoid drupes containing single stony hard seed. Bark and fruits are commercially used in the preparations for treating cough, bronchitis, allergic cough, insomnia, edema, dyspepsia, flatulence, hemorrhage, skin diseases, diarrhea, dysentery, leukoderma, and graying of hair.

67. Uzhinja (Mal.), Indravalli (San.)

Cardiospermum helicacabum Linn. (Sapindaceae)

This is an annual spreading herb with tendril hooks. Leaves are bicompound and leaflets are acuminete at apex. Plant pacifies vitiated pitha, constipation, fever, arthritis, amenorrhea, low back pain, neuropathy, chronic rheumatism and mania.

68. Vankadaladi (Mal.), Apamargah (San.)

Achyranthes aspera Linn. (Amaranthaceae)

Vankadaladi is shrub that grows up to one meter height. Leaves are simple, opposite and obovate. Plant pacifies asthma, bronchitis, flatulence, painful inflammations, vomiting, skin diseases, renal and vesical calculi.

69. Vallippala (Mal.), Lataksheeri (San.)

Tylophora indica (Burm.f.) Merrill. (Asclepiadaceae)

This is a week climber with round, dark green opposite leaves. Flowers are greenish white and borne in umbel. Roots and leaves are commercially used to pacify cough, asthma, bronchitis, dysentery, diarrhea, wounds, ulcer, hemorrhoids, malignant tumor, and leukemia.

70. Vaathamkolli (Mal.), Vataghni (San.)

Justicia gendarussa Burm.f. (Acanthaceae)

This is a very common shrub in all the garden lands of Kerala, locally known as Odichukuthi. This small erect shrub grows up to 1 meter in height. Leaves are simple, opposite, lanceolate, glabrous, short petioled and dark violet green above. Plant pacifies inflammation, arthritis, headache, hemiplegia, facial paralysis, migraine, cough, bronchitis, dysmenorrhea, amenorrhea, fever, ascites, and general debility.

71. Venga (Mal.), Beejaka (San.)

Pterocarpus marsupium Roxb. (Fabaceae)

This is a large sized softwood tree growing up to 20 meters in height. Heart wood is generally the economic part of this plant. It pacifies diabetes, skin disease, leucoderma, hemorrhoids, diarrhea, arthritis cough and graying of hair. The gum Kino extracted from the bark is useful in many diseases including diabetes. Gum, leaves and bark are also used in few preparations.

72. Vishapacha (Mal.), Sophaghni (San.)

Clinacanthus siamensis (Acanthaceae)

This is a medium sized perennial shrub growing up to 2 meters in height. Leaves are simple, opposite, acute and slightly serrated. Leaves, root and stem are used in various Ayurvedic preparations and plant pacifies vitiated kapha, pitha, poison bites, inflammation, traumatic edema and swelling due to poison stings.

1.3 Conclusion

The Western Ghats is richly credited with varied kinds of vegetation and unimaginable topographical features. Our discussion throws a definite idea on the miraculous curative properties of the plants that are endemic to the Western Ghats. More importantly, the enormous responsibility of the present generation to protect these rare medicinal plants should be conceived. India's Biological Diversity Act 2002 aims to promote conservation, sustainable use and equitable sharing of benefits of India's biodiversity resources. Further, huge amount of knowledge on the traditional therapeutic uses of these plants is being transmitted through generations and poor documentation leads to its gradual depletion. Though the emergence and acceptance of Ayurveda as a frontier medicinal field is an answer to this problem to some extent, serious efforts should be diverted for the identification of rare plants and clinical examination on their properties, conservation of important plants and documentation of the traditional wisdom.

References

Anonymous, 2008, http://www.biodiversityhotspots.org/xp/hotspots/ghats/Pages/default.aspx retrieved on 24 November 2008.

Gadgil, M., Prasad, S.N., Nair, P.V.K. and Sharatchandra, H. 1977 A system of biosphere reserves for Western Ghats of Karnataka CTS technical report 13.

□□□

6

Uses of Medicinal Plants to Treat Human Disease Among the Tribals at Pachamalai Hills, Tamil Nadu

M. Geetha Rani

CONTENTS

1.1	Introduction	82
1.2	Study Area	84
1.3	Methodology	85
1.4	Observations	85
1.5	Conclusions	92
	References	92

1.1 Introduction

The concept of indigenous knowledge (IK) or traditional knowledge (TK) refers to the complete bodies of knowledge, know-how, practices and representations that are maintained and developed by people with long histories of close interaction with nature. These sets of understandings, interpretations and meanings are part of a cultural complex that encompasses language, naming and classification systems, ways of using resources, rituals, spirituality and a world view. Such knowledge is embedded in the local situation that makes it so valuable and provides the basis for local decision making about many fundamental aspects of day-to-day life in these societies. This body of knowledge, which is transferred orally from generation to generation and is seldom, if ever, documented, is classified as non-formal knowledge in contrast to the documented formal knowledge.

A major component of this IK in many developing countries continues to be the traditional methodologies and practices of using biological and genetic resources for different day-to-day applications in life, particularly for human health care. It also involves a range of sectors, such as agriculture (intercropping techniques, animal husbandry, pest control, crop diversity, animal health care, seed varieties), biology (botany, fish-breeding techniques), the use and management of natural resources (soil conservation, irrigation and other forms of water management), education (oral traditions, local languages) and poverty alleviation in general. This knowledge is conceived and evolved by trial and error over several years and conveyed across generations. Hence, most of the IK is collectively evolved and conserved by communities, although a few of them are also innovations by individuals in the recent past and are relatively new.

During the process of intergenerational transfer and repeated applications, this knowledge undergoes modifications or improvements in efficacy, specialized applications or process of formulation. As most of the indigenous knowledge is closely linked with various biological and genetic resources, conservation of this knowledge goes hand in hand with conservation of these biodiversities. Innovation of IK might have origin in individuals or groups within indigenous communities. Some of this knowledge, largely individual owned and a few owned by the community, may be kept confidential to the innovators and their descendants and access to such knowledge is not only difficult but comes with restrictions too. However many of the IK are disseminated locally, placed in public domain of the community concerned and hence are easily accessible to those within the community and outside. Nevertheless, in many poor countries and to poor communities, this knowledge and biological resources constitute an important desideratum for their well-being and livelihood.

Different communities may have varying capability to produce, reproduce improve and conserve the IK for individual or common good. There is a view that wider the sharing of the IK, greater the opportunity for its improvement. When such improvement by individuals leads to substantial modification of the IK in a single step, it may either end in continuing in public domain or getting shifted to private individual domain. The communities, nonetheless, benefit from the specialized knowledge of individuals with expertise in standardizing the extraction, combination of herbal drugs and calibrating the dose according to the status of the health of the subject. This expertise is recognized in practitioners of traditional medicine or traditional healers and therefore the community revere these IK experts. It is also important to appreciate that when a given knowledge is shared within or across the communities, such transfer may often be critically lacking in these individual expertise or the know-how mastered by traditional healers.

Two major increasing dangers to the practices and conservation of IK and the bio-resources are their erosion due to the changing lifestyles and cultural value systems and their misappropriation under the global intellectual property rights regime. The erosion is enhanced by weakening intergeneration linkages, higher rates of migration of the youth from rural areas, increasing urbanization and new value systems pushed by market economy, technological advances and its faster diffusion as in the case of modern crop varieties, and monoculture of species in forests and terrestrial or aquatic farming. The indifference of public policy makers in national domains in establishing incentives or institutions to promote the conservation of TK and bio-resources also hastens their erosion.

Although IK has proven its value in many cases it cannot be promoted without first being critically assessed. Not all IK offers sustainable solutions to today's pressing problems. Furthermore, most local solutions are very context specific. It is in danger of disappearing not only under influence of global processes of rapid change, but also because the capacity and facilities needed to document, evaluate, validate, protect and disseminate such knowledge are lacking in the south. More research needs to be done on IK systems, and methods need to be developed for dealing with it. Only then can claims of solutions to specific problems by certain IK systems be backed up, and attempts made to improve or adapt those systems. This research must of course be done together with the people who possess the IK and with the local communities involved.

The present study is an effort to document community knowledge available in the public domain of Malayali tribal community in Pachamalai hills in Tamil Nadu. Such documentation of the IK in public domain is an important step to mitigate its increasing loss and concurrently create awareness on conserving the bio-resources associated with these IK. During this study, it was considered very important to seek the consent of concerned individuals or communities before documenting the IK and also to ascertain their concurrence on the knowledge documented. In this context it is relevant to mention the recent concern expressed by the Intergovernmental Committee on "Matters Concerning Intellectual Property and Genetic Resources, Traditional Knowledge and Folklore" constituted by the World Intellectual Property Organization (WIPO 2000) on the purpose and the implications of documentation of TK and the inclusion of TK into databases. Given the wide range of IK (or TK) documentation projects currently under way in India and elsewhere with the aim of preservation as well as positive and defensive protection of intellectual property of the TK, the Committee endorsed the development of a documentation system which strengthens the ownership right of the traditional communities on the documented TK (WIPO 2000).

The IK documented by this study have been in existence over generations among the Malayali tribe and they are still practicing with occasional innovative improvements. Although it is recognized that compared with the modern health care systems the IK-based health remedies take a long time to achieve favorable results, the community values the latter for its ready accessibility with least or no economic cost and no side effect. It is found that for practicing the IK-based health care, it is not only the knowledge that is important but also the ready access to all required plant species resources. The ethno-botanical knowledge of the indigenous community on the medicinal value of different plants, the importance of different morphological parts of these plants for treating different health problems of cattle, human and crop plants along with taxonomy and distribution pattern of species is discussed in detail in this chapter. This study highlights and also touches upon the significance of this study in relation to the sustainable management of bio-resources of the Eastern Ghats.

1.2 Study Area

The Pachamalai Hills (Fig. 1) spread over two districts namely, Salem and Tiruchirapalli in Tamil Nadu. It is about 75 km away from Tiruchirapalli city. This hill is situated in the southern parts of the Eastern Ghats. Its elevation ranges from 400 metres to 1200 metres above Mean Sea Level. The forest comprises of tropical thorny, dry deciduous and moist deciduous types of vegetation. Total geographical area of hill Pachamalai is 14,277 ha. This survey covers Kombai, Thenparanadu, Vannadu (Tiruchirapalli Dt) and Pachamalai (Salem Dt). Out of 81 villages in this region, only 30 villages selected for this study.

Fig. 1 : Pachamalai Hills

1.3 Methodology

The methods adopted for collections of data were interviews with the tribals, observations, interaction with local Panchayats, Forest Officials and NGOs working in and around Pachamalai Hills. Information on the use of medicinal plants to cure human ailments was documented as it was disclosed by the individuals along with the methods of preparation, prescribed dose as well. The plant voucher specimens were prepared and deposited with the Community Herbarium located at MSSRF, Chennai.

1.4 Observations

Under this study, the use of medicinal plants to treat 40 different ailments came to light which covers physical, physiological as well as chemical aspects of human disease management. The survey showed the use of 65 medicinal plant species to cure various human diseases.

The following are the various ailments: Abscess; Anaemia; Anaemia & constipation; Anaemia & hair fall; Anaemia, jaundice, liver disease & loss of appetite; Anaemia & swelling (Oedema); Arthritis; Bite - Dog bite, Insect bite,Poison bite, Pooran bite (millipede / centipede), Scorpion bite; Burning feet & fungal infections of the feet; Burning urination, scanty urination & white discharge; Cold, cough, asthma, wheezing, throat pain, gastritis, ulcer, fever, etc; Cold; Cold & asthma; Cold & cough; Cold, cough & fever; Cold, cough, asthma, wheezing & throat pain; Cold, cough, difficulty in breathing & throat pain; Cough, cold, gastritis & ulcer; Fever; Constipation; Constipation & gastritis; Constipation, excess bleeding, piles & iron tonic; Constipation & uterine problems; Control bleeding at the time of menstrual period; Diabetes; Diarrhoea; Dysentery; Fits; Headache – one-sided; Enhancing health; Increasing milk at the time of breast feeding; Indigestion; Indigestion, bleeding and piles; Indigestion & loss of appetite; Mouth ulcer; Nervous problems & weakness; Pain & sprain- Body, Join, Eye; Sprain; Piles; Reduce swelling & abscess; Respiratory disorders & problems; Skin disease, painful menstruation & Intestinal worms; Stomach ulcer; Tooth; Vomiting, flatulence & diarrhoea; Weakness ; White discharge; White discharge & loose motion; Wounds. The distribution of use of medicinal plants in treatment is depicted in the Figure 2.

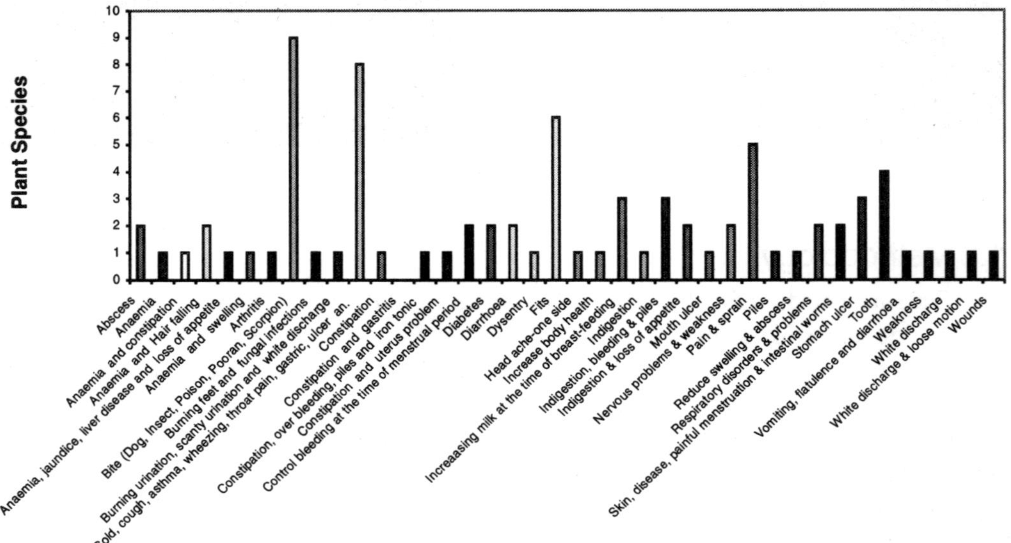

Fig. 2 Distribution of use of medicinal plants in curing human ailments

It was noted that 65 species put in use to various human diseases were found to belong to 36 families and 58 genera. Mimosaceae, Solanaceae, Amaranthaceae, Asclepiadaceae, Caesalpiniaceae, Fabaceae, Liliaceae, Rutaceae and Verbenaceae were found to be the top 9 families. Among the species, the per centage of tree was 29.2 %, shrub 27.7%, herb 18.5 % and climber 18.5 % (Table 1).

Table 1 : Growth habit of plant species used in management of human diseases

S.No.	Vernacular name	Plant species	Plant habit	Family name
1	Athi	*Ficus racemosa*	Tree	Moraceae
2	Ayila	*Holoptelea integrifolia*	Tree	Ulmaceae
3	Etti	*Strychnos nux-vomica*	Tree	Loganiaceae
4	Kaddukai	*Terminalia chebula*	Tree	Combretaceae

Contd...

Table 1 Contd...

5	Konna	*Cassia fistula*	Tree	Caesalpiniaceae
6	Malaivembu	*Melia azaderach*	Tree	Meliaceae
7	Murungai	*Moringa oleifera*	Tree	Moringaceae
8	Nuna	*Morinda pubescence*	Tree	Rubiaceae
9	Pala	*Artocarpus heterophyllus*	Tree	Moraceae
10	Puli	*Tamarindus indica*	Tree	Caesalpiniaceae
11	Seetha	*Annona squamosa*	Tree	Annonaceae
12	Usila	*Albizia amara*	Tree	Mimosaceae
13	Vagai	*Albizia lebbeck*	Tree	Mimosaceae
14	Velam	*Acacia Arabica*	Tree	Mimosaceae
15	Vembu	*Azadirachta indica*	Tree	Meliaceae
16	Viluthi	*Cadaba fruticosa*	Tree	Capparaceae
17	Vilvam	*Aegle marmelos*	Tree	Rutaceae
18	Kumala	*Gmelina arborea*	Tree	Verbenaceae
19	Naval	*Syzygium cumini*	Tree	Myrtaceae
20	Karuveppilai	*Murraya koenigii*	Shrub/Tree	Rutaceae
21	Pappali	*Carica papaya*	Shrub	Caricaceae
22	Aavaaram	*Cassia auriculata*	Shrub	Caesalpiniaceae
23	Adathoda	*Justicia adhatoda*	Shrub	Acanthaceae
24	Adutheendapalai	*Aristolochia bracteolata*	Shrub	Aristolochiaceae
25	Amanakku	*Ricinus communis*	Shrub	Euphorbiaceae
26	Anathalai	*Clausena dentate*	Shrub	Rutaceae
27	Erukku	*Calotropis gigantea*	Shrub	Asclepiadaceae
28	Kalakka	*Carissa carandas*	Shrub	Apocynaceae
29	Kandankathiri	*Solanum surattense*	Shrub	Solanaceae
30	Kodikali	*Sarcostemma acidum*	Shrub	Asclepiadaceae
31	Maruthani	*Lawsonia inermis*	Shrub	Lythraceae
32	Nochi	*Vitex negundo*	Shrub	Verbenaceae
33	Peenarisangu	*Clerodendron inerme*	Shrub	Verbenaceae
34	Semparuthi	*Hibiscus rosa-sinensis*	Shrub	Malvaceae
35	Sundai	*Solanum torvum*	Shrub	Solanaceae
36	Thuthi	*Abutilon indicum*	Shrub	Malvaceae
37	Virali	*Dodonaea angustifolia*	Shrub	Sapindaceae
38	Pasalai	*Basella alba*	Herbacious climber	Basellaceae

Contd...

Table 1 Contd...

39	Pirandai	*Cissus quadrangularis*	Herbacious climber	Vitaceae
40	Thaneervittan kizhangu	*Asparagus racemosus*	Herbacious climber	Liliaceae
41	Sirukurinjan	*Gymnema sylvestre*	Herbacious climber	Asclepiadaceae
42	Thalaisurili	*Aristolochia indica*	Herbacious climber	Aristolochiaceae
43	Thuthuvalai	*Solanum trilobatum*	Herbacious climber	Solanaceae
44	Poonaikali	*Mucuna pruriens*	Herbacious climber	Fabaceae
45	Kundumani	*Abrus precatorius*	Herbacious climber	Fabaceae
46	Kovakka	*Coccinia grandis*	Herbacious climber	Cucurbitaceae
47	Kumati	*Citrullus colocynthis*	Herbacious climber	Cucurbitaceae
48	Mudakkattan	*Cardiospermum halicacabum*	Herbacious climber	Sapindaceae
49	Molagaranai	*Stephania japonica*	Herbacious climber	Menispermaceae
50	Thottalsinungi	*Mimosa pudica*	Herb	Mimosaceae
51	Karisalangani	*Eclipta alba*	Herb	Asteraceae
52	Manathakkali	*Solanum americanum*	Herb	Solanaceae
53	Manjal	*Curcuma longa*	Herb	Zingiberaceae
54	Nithyakalyani	*Catharanthus roseus*	Herb	Apocynaceae
55	Sirukeerai	*Amaranthus tricolor*	Herb	Amaranthaceae
56	Sothukatrazhai	*Aloe vera*	Herb	Liliaceae
57	Avuri	*Tephrosia purpurea*	Herb	Fabaceae
58	Mullukeerai	*Amaranthus spinosus*	Herb	Amaranthaceae
59	Nayuruvi	*Achyranthes aspera*	Herb	Amaranthaceae
60	Puthina	*Mentha arvensis*	Herb	Lamiaceae
61	Seruppadai	*Coldenia procumbens*	Herb	Boraginaceae
62	Siriyanangai	*Andrographis paniculata*	Herb	Acanthaceae
63	Thumbai	*Leucas aspera*	Herb	Lamiaceae
64	Vasambu	*Acorus calamus*	Herb	Araceae
65	Vengkayam	*Allium cepa*	Herb	Liliaceae

Total of 68 formulations under 40 subjects matter used leaves in 3 instances, roots in 11 instances, bark and fruit in 8 and 9 instances and bulb in 3 instances. The whole plant is used only in two formulations and they possessed shrub and climber habits. Only leaf is used in 14 formulations, while fruit alone is used in 3 formulations, seed alone in 2 formulations, flower alone in 2 formulations and tuber, rhizome, and bark alone in 1 formulation each (Table 2).

Table 2 : Parts of different species used for human disease management

S. no.	Management	Plant species used	Plant parts used
1	Abscess	Clausena dentata	Leaf
		Ricinus communis	Seed
2	Anaemia	Moringa oleifera	Leaf
3	Anaemia & constipation	Carica papaya	Fruit
4	Anaemia & hair fall	Murraya koenigii	Leaf
		Tamarindus indica	Fruit
5	Anaemia, jaundice, liver disease & loss of appetite	Eclipta alba	Leaf
6	Anaemia & swelling	Aegle marmelos	Leaf
7	Arthritis	Strychnos nux-vomica	Seed
8	Bite		
	By Dog	Achyranthes aspera	Leaf
		Allium cepa	Bulb
	By Insect	Leucas aspera	Leaf
	Poisonous	Azadirachta indica	Leaf
		Andrographis	Leaf
		paniculata	Root
		Aristolochia indica	Bulb
		Allium cepa	
	By Pooran	Leucas aspera	Leaf
	By Scorpion	Morinda pubescence	Leaf
9	Burning feet & fungal infections	Lawsonia inermis	Leaf
10	Burning urination, scanty urination & white discharge	Basella alba	Leaf
11	Cold, cough, asthma, wheezing, throat pain, gastritis, ulcer & fever	Solanum trilobatum	Leaf
		Justicia adhatoda	Leaf
		Solanum surattense	Whole plant
		Solanum americanum	Leaf
		Amaranthus spinosus	Leaf + Root
		Amaranthus tricolor	Leaf
		Ricinus communis	Bark
		Acacia Arabica	Root

Contd...

Table 2 Contd...

12	Constipation	Terminalia chebula	Fruit
13	Constipation & gastritis	-	-
14	Constipation, over bleeding, piles & iron tonic	Ficus racemosa	Fruit + Bark
15	Constipation & uterine problem	Citrullus colocynthis	Fruit
16	Control bleeding at the time of menstrual period	Syzygium cumini	Bark
		Holoptelea integrifolia	Bark
17	Diabetes	Gymnema sylvestre	Leaf
		Cassia auriculata	Flower
18	Diarrhoea	Stephania japonica	Root
		Syzygium cumini	Bark
19	Dysentery	Mimosa pudica	Leaf
20	Fits	Tephrosia purpurea	Root
		Albizia amara	Root
		Sarcostemma acidum	Root
		Calotropis gigantea	Root
		Acacia arabica	Root
		Leucas aspera	Leaf
21	Head-ache one-sided	Leucas aspera	Leaf
22	Increase body health	Catharanthus roseus	Flower
23	Increasing milk at the time of breast feeding	Carissa carandas	Flower
		Coldenia procumbens	Leaf
		Artocarpus heterophyllus	Seed
24	Indigestion	Moringa oleifera	Leaf
25	Indigestion, bleeding & piles	Cissus quadrangularis	Stem
		Allium cepa	Bulb
		Tamarindus indica	Fruit
26	Indigestion & loss of apetite	Mentha arvensis	Leaf
		Tamarindus indica	Fruit
27	Mouth ulcer	Abrus precatorius	Leaf
28	Nervous problems & weakness	Mucuna pruriens	Seed
		Solanum torvum	Fruit
29	Pain & sprain	Cardiospermum	Whole plant
		halicacabum	Leaf
		Cadaba fruticosa	Leaf
		Coccinia grandis	Root
		Stephania japonica	Leaf
		Tamarindus indica	
30	Piles	Abutilon indicum	Leaf
31	Reduce swelling & abscess	Dodonaea angustifolia	Leaf

Contd...

Table 2 Contd...

32	Respiratory disorders & problems	Calotropis gigantean	Flower
33	Skin, disease, painful menstruation & intestinal worms	Melia azaderach	Bark
		Aristolochia bracteolate	Leaf
34	Stomach ulcer	Solanum americanum	Leaf
		Tamarindus indica	Fruit
		Cassia fistula	Bark
35	Tooth	Achyranthes aspera	Root
		Gmelina arborea	Leaf
		Vitex negundo	Leaf
		Solanum surattense	Seed
36	Vomiting, flatulence & diarrhoea	Acorus calamus	Rhizome
37	Weakness	Asparagus racemosus	Tuber
38	White discharge	Hibiscus rosa-sinensis	Flower
		Clerodendron inerme	Leaf
		Aloe vera	Leaf
39	White discharge & loose motion	Albizia lebbeck	Bark
40	Wounds	Annona squamosa	Seed

Predominant use of leaves reflects possibly a conscious practice of non-destructive harvesting from plants. King (2005) and Kumari (1998) also made similar observations in the Kolli hills and the Jawadi hills, respectively where the Malayali tribe uses leaf as their major plant part for traditional medicinal preparations. Kala (2005) reveled that, in the case of Apatani tribals, use different underground plant parts such as root, tuber, rhizome, bulb and pseudo stem are also used for traditional medicines. Hence, constructive use of medicinal plants and use of parts, which do not demand destructive harvesting, are important for use and concurrent conservation of species (Etkin 2002). In this context, the conservation sense of Malayali tribe is commendable and their practices do not threaten the wild and naturally growing species.

Literature survey shows that some of the knowledge on use of medicinal plants species identified with the Malayali tribe in the current study is also prevalent among the other tribes living in the Eastern Ghats. It was earlier reported that the *Kandhas* of Kandamal (Behera *et al.* 2006), the local people living at the Narayanapatna hills, Koraput (Dash & Mishra 2003), the tribals at Jajpur in Orissa (Satapathy & Brahman 2003), the *Konda reddis* (Prasad *et al.* 2002), the *Chenchus*, the *Sugalis* and the *Yerkuals* of Kurnool (Goud *et al.* 2003) and in Mahabunagar (Kumar & Pullaiah 2003), in Anantpur, Andhrapradesh (Reddy *et al.* 2003), the *Paliyans* and the *Pulayan* (Ganesan *et al.* 2004), the *Badagas*, the *Chetties*, the *Kotas*, and the *Kattunayaka* tribals at

Nilgiri, Tamil Nadu (Baburaj *et al.* 2003) used some of the plant species which has been documented at Pachamalai hills for various treatment.

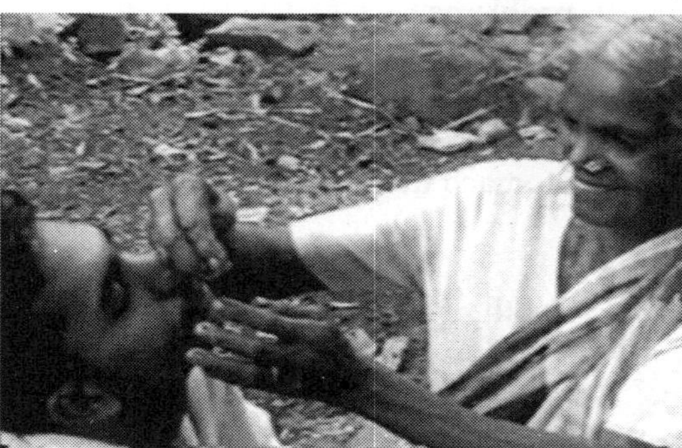

Fig. 3 : Resource persons treating ailments

Diagnosis of diseases and administration of each of the formulation documented under various disease management category was based on the expertise and the knowledge of the individuals who disclosed it (Figure 3). Diseases are classified under the subject matter and the plant species are described with their vernacular name, common name, scientific name, habit as well as the particular plant parts used.

1.5 Conclusion

In India, about 5000 species of plants are being used as sources of drugs since ancient times in indigenous systems of medicine as well as folklore. However less than 100 species of these are accepted so far in modern system of medicine which established its lead over the indigenous systems in the present century (Ved 1997). It is well recognized that the knowledge on the plant resources are often constrained by location specificity, with great bearing on the local flora and fauna. There is reciprocity in the conservation of the knowledge and the flora by the community. Das *et al.* (2004) after conducting validation studies on several indigenous practices related to the use of medicinal plants documented from all over India, reported that the treatment are very effective in many cases. In the present study, validation of the indigenous knowledge was done through triangulation. It is there for evaluate them scientifically for the development of new modern drugs and to validate the efficacy of traditional usages.

References

Baburaj, Suresh D., S. John Britto, G. K. Mathew and S. Rajan. 2003. Cultivated Medicinal Plants Useful in Homoeopathy Found in Nilgiri District, Tamil Nadu. Ethnobotany and Medicinal Plants of Indian Subcontinent. (ed.) J. K. Maheswari. Scientific Publishers, Jodhpur. 381-389.

Behera Soumit K., Anima Panda, Susanta K. Behera and Malaya K. Misra. 2006. Medicinal Plants Used by the Kandhas of Khandhamal District of Orissa. *Indian Journal of Traditional Knowledge*. 5(4): 519-528.

Das, P., S.K. Das, A. Mishra, R.P.Singh Ratan, G.Subba Reddy, H.P.S. Arya, M.Geetha Rani, L.R.Verma and D.P.Ray. (comp). 2004. Cross - Sectoral Validation of Indigenous Knowledge in Agriculture. Document 4. Indian Council of Agricultural Research, New Delhi. 230

Dash, S. S. and M. K. Misra. 2003. Taxonomic Survey and Systematic Census of Economic Plants of Narayanapatna Hills of Koraput District, Orissa. Ethnobotany and Medicinal Plants of Indian Subcontinent. (ed.) J. K. Maheswari. Scientific Publishers, Jodhpur. 473-498.

Etkin, N. L. 2002. Local Knowledge of Biodiversity and its Conservation in Rural Hausaland, Northern Nigeria. *Economic Botany*. 56: 73-88.

Ganesan, S., N. Suresh and L. Kesavan. 2004. Ethnomedicinal Survey of Lower Palni Hills of Tamil Nadu. Indian J. of Traditional Knowledge. 3(3): 305-313.

Geetha Rani, M. 2008. Documentation of Medicinal Plant Genetic Resources of Pachamalai Hills, Tamil Nadu, India. In: Proceedings of National Seminar on Forest Medicinal Plants & People. Organized by Dept. of Natural Resource Management College of Forestry, Sirsi & UAS, Dharwad. Sponsored by Medicinal Plants Board, New Delhi. Sirsi 21-22 Feb.

Goud, P. S., T. Pullaiah and K. Sri Rama Murthy. 2003. Native Phytotherapy for Fever and Malaria from Kurnool District, Andhra Pradesh. 2003. Ethnobotany and Medicinal Plants of Indian Subcontinent. (ed.) J. K. Maheswari. Scientific Publishers, Jodhpur. 337-340.

Kala, C. P. 2005. Ethnomedicinal Botany of the Apatani in the Eastern Himalayan Region of India. J. of Ethnobiology and Ethonomedicine. 1(1): 11. www. ethnobiomed. com/ content/1/1/11.

King, E. D. Israel Oliver. 2005. Sacred Forests of Kolli Hills, Tamil Nadu, India: A Study on Botany, Ecology and Community Interventions. Ph.D. Dissertation, MSSRF.Chennai

Kumari, R.M. 1998. Ethnobotany of Malayalis of Nellivasal Nadu and its Neighbourhood, Jawadi Hills, Vellore District, Tamil Nadu. M.Phil. Dissertation. Madras Christian College, Chennai

Kumar, Dharma Chandra T. and T. Pullaiah. 2003. Ethno-Medicinal Uses of Some Plants of Mahabubnagar District, Andhra Pradesh, India. Ethnobotany and Medicinal Plants of Indian Subcontinent. (ed.) J. K. Maheswari. Scientific Publishers, Jodhpur. 341-345.

Prasad, Krishna, V., T. Rajagopal and K. V. S. Badarinath. 2002. Ethnobotany of Kondareddis of Rampa Agency, Eastern Ghats, Andhra Pradesh. Ethnobotany An Over View. (ed.) P. C. Trivedi . Aavishkar Publishers, Jaipur. 205-220.

Reddy, K. N. and R. R. Venkata Raju. 2003. Plants in Ethnoveterinary Practices in Anantapur District, Andhra Pradesh. Ethnobotany and Medicinal Palnts of Indian Subcontinent. (ed.) J. K. Maheswari. Scientific Publishers, Jodhpur. 347-357.

Satapathy, K. B. and M. Brahmam. 2003. Some Interesting Phytotherapeutic Claims of Tribals of Jajpur District (Orissa), India. Ethnobotany and Medicinal Palnts of Indian Subcontinent. (ed.) J. K. Maheswari. Scientific Publishers, Jodhpur. 467-4372.

Ved, D.K. 1997. Trade in Medicinal Plants: The State of Our Ignorance. *Amruth.* 1 (10): 2-8.

□□□

7

Some Therapeutic Uses of Biodiversity of the Kolli Hills Eastern Ghats, Tamil Nadu, India

Rajendran, A. & S. Manian

CONTENTS

1.1 Introduction 96

1.2 Biogeography of the Kolli Hills 96

1.3 Methodology 97

1.4 Therapeutic Uses of Potential Medicinal Plants 98

1.5 Conclusion 109

References 110

1.1 Introduction

India is bestowed with a treasure of medicinal plants. It is also rich in its ethnic diversity of which many indigenous culture have retained traditional knowledge concerning the medicinal utility of the native plants. Tribal communities living in biodiversity rich area possess a wealth of knowledge on the local utilization and conservation of food and medicinal plants. (Taothathiri *et al.*, 1989; Jain 1991). Traditional knowledge systems, including various medicinal plant utilities appear to vary according to local population domain (Tumbrook, 2004).

India has over 537 different aboriginal and other ethnic groups constituting approximately eight percent of the country's population (Jain 1991; Nazer, 2000). The traditional knowledge which developed over years of observation, trial and error, inference and inheritance has largely remained with the aboriginal people (Ragupathy and Mahadevan, 1991; 1996). In every culture, there were men and women especially knowledgeable in the properties of edible and medicinal plants. Southeastern Indians have been known to put a great emphasise on traditional systems and practice which is supported by their vast intra-ethnic diversity (Ganesan *et al.*, 2004).

Medicinal folklore over the years has proved to be an invaluable guide in the present day to the screening for important modern drugs (Anyinam, 1995). In view of this, there is a need to document the traditional knowledge of these communities, mainly because the majority of such communities are rapidly losing their socio-economic and cultural characteristics. Studies on the indigenous healthcare practices among the wide Malayali tribals of the Eastern Ghats of Tamil Nadu have been carried out as scattered or isolated efforts.

The Kolli Hills is noted for its rich biodiversity and tribal populations. Although a good number of ethnobotonical investigations has been carried out in Kolli Hills (Lakshmanan and Kalyani, 1980; Rajendra, 1981; Ansari, 1997; Ansari *et al.*, 1993; Alagesaboopathi, 1992; Dwarakan and Ansari, 1996; Geetha, 1996; Geetha *et al.*, 1996; Ansari, 1997; Dwaraka, 1999; Dwarahan and Alagesaboopathi, 1999; Dwarakan *et al.*, 1999; Rajendran, 2005) not much information has been collected from the tribe Malayalis.

1.2 Biogeography of The Kolli Hills

The Kolli Hills forms a major range in the Eastern Ghats that is rich in biodiversity and indigenous population. It is located (11°10' - 11° 30' N Latitude; 75°15' - 75° 30' Longitude) in the Namakkal district of the southeastern Tamil Nadu. It extends to an area of about 418.5 km^2. Elevation of the hills ranges between 1000 and 1500 m above MSL. The average annual rainfall is about 1200 mm and the temperature varies between 10° and 30°C

(Lakshminarayanan, 1990). The vegetation is predominantly dry deciduous forest to tropical riparian forest occuring along the rivers and streams. The ecosystems and aboriginal community of the Kolli Hills already has been described by Lakshmi (1995) and Geetha (1996). The 'Kolli Malai' means Killer Mountain. It is also called Chathuragiri Malai, because of its square shaped topography. In Ramayana it is quoted differently as Rishymukha, Parvatham, Sugriva Parvatham and Kolachalam. The Arappaleswar Temple, Masiperianan temple, Kongu Ayi Temple and the cave of Korakkar are popular pilgrimages within the Kolli Hills. It is believed to be the sacred abode of the great saint Kalalginathar and the heaven of medicinal plant resources.

The Malayalis (etymology in Tamil - Malai = hill areas; Yalis = people who live in and depend on the hills) are an aboriginal community who reside in the forest of the Kolli Hills. They appear to have migrated from Kanchipuram between seventh and eleventh centuries. They are supposed to be descendents of 'Kanchipuram Vellalor'. They are also believed to be descendents of proto-Australoid group who lived here before Harappan civilization. The seclusion provided by the hills and unapproachable forests retained some forms of old dialects of the Tamil language. They have accumulated comprehensive knowledge of plant and animal utility, which may be accredited to their long association with rich flora of the hills. They are cultivating food crops, fruits and vegetables. More recently many young Malaiyalis work as casual laborers in coffee estates.

1.3 Methodology

Ethnobotanical surveys were conducted repeatedly in different seasons and areas for 3 years. Good rapport was inveterated with one or two persons preferably the Chief, guidance sought and contact was then established with other tribals of the locality before actually commencing into the field work. The local informants were the medicine men, men and women working in the fields, village headman and other community leaders. Mostly persons in the age group of 50 – 60 years and above had accurate information regarding their old traditions. Field sites were visited with the local medicine men. The purpose of the study was explained and interviews of both individuals and groups were taken. In group interviews, more than two individuals were approached. To determine the authenticity of information collected during field work, repeated cross verification of data from different informants and in different areas at different times was done. Thus, only the specific and reliable information, cross-checked with many informants, has been incorporated in the study. The collected medicinal plants were identified upto species level with help of floras (Gamble, 1915-1936; Matthew, 1970). All the collected specimens were deposited in the Herbarium of Department of Environmental and Herbal Science of the Tamil University, Thanjavur.

1.4 Therapeutic Uses of Potential Medicinal Plants

The plant species of therapeutic interest are arranged alphabetically. The data on the medicinal properties of plants is given with a mention of their local names, part/(s) used, mode of their administration and uses.

Ananas comosus (L.) *Merr. (Bromeliaceae)*

Annasipazham. The plant is a short-stemed terrestrial. Plantlings distributed free of cost to the tribal farmers of the Kolli Hills from the government research station. Widely cultivated. Immature fruit juice taken as an abortifacient. Powdered leaves mixed with water and taken early in the morning to control blood sugar in antidiabetic.

Anisomeles malabarica (L.) *R.Br. ex Sims (Lamiaceae)*

Paeimerrati. Densely white-woolly, gregarious, aromatic, subshrub, clumps in wastelands, stream banks and along fences. Leaf tied in front of the house is believed to keep off the evil spirit and also used in black magic.

Anoectochilus elatus *Lindl. (Orchidaceae)*

The plant is a terrestrial, baby herbs, on moist floor of shoals in humus and in deep shade. Keeping the dried plant is believed to fulfil one's desire.

Argemone mexicana (L.) *(Papaveraceae)*

Naai Kadugu. A short herb. Gregarious alien weed, locally abundant in wastelands, near human settlements and along river banks. Yellow sap of the plant is mixed with cow's milk and taken against ulcer complaint. Root bark decoration used for spermatorrhoea.

Acalypha indica (L.) *(Euphorbiaceae)*

Kuppaimaeni. A subshrubby on wastelands, in moist and shady places and in river banks. Fresh leaf juice is used as regular mouthwash for 3-5 days for tooth ache.

Azadirachta indica *A. Juss. (Meliaceae)*

Semievergreen tree. Cultivated and naturatised. Leaf juice is used to control stomach troubles. Leaf juice mixed with salt and turmeric and applied gently for skin disease.

Amorphophallus sylvaticus *(Roxb.) Kunth. (Araceae)*

Kattukkilangu. Cormous herb. Foothills to 1200 m in the shade. The plant rhizome along with garlic (*Allium sativum*), pepper (*Piper nigrum*) and pork's ghee consumed for piles complaints.

Abrus precatorius (L.) *(Fabaceae)*

Kattukundumani. Straggler in serals, jungles and thickets. Seed paste is mixed with cow's milk and taken against snake bite.

Aloe barbadensis *Mill. (Liliaceae)*

Sottrukartazhai. Erect herbs with radical leaves in rosettes succulent, spiny. Plains and foothills, scrubs jungles and wastelands. Inner fleshy portion of the leaves applied for rheumatic pain and inflammation.

Asparagus racemosus *Willd. (Liliaceae)*

Armed vine. Plains in scrubs jungles, slopes and forest border. Root bark decoction prepared with turmeric (*Curcuma longa*) and taken as general tonic.

Alternanthera sessilis (L.) *R.Br. ex. DC (Amaranthaceae)*

Ponnankanni. A prostrate branched herb. Leaf decoction prepared with garlic (*Allium sativum*) and taken twice a day for pneumonia fever. Maccerated leaves tied on snake bitten spot as an antidote.

Arisaema tortuosum *(Wall.) Schott. (Araceae)*

Karunai kizhangu. Cormous herb. Shady places in the foot hills. The tubers are crushed and applied on snake biten spot.

Alstonia scholaris (L.) *R. Br. (Apocynaceae)*

Palaimarum. Tree. Occurs in foot hills. Bark extract is used to control blood sugar level in diabetics.

Argyria cuneate *(Wild.) Ker. -Gawl. (Convolvulaceae).*

Vettaicheddi. Bushy erect shrubs. Locally abundant on the degraded slopes. Leaf juice and paste is used to cure burns.

Aglaia elaegnoidea *(Juss) Benth. (Meliaceae)*

Erramulli. An evergreen shola tree by forest border. Leaf juice of this plant is believed to cure cancer.

Asystasia gangetica (L.) *T. Anders. (Acanthaceae)*

Gungulacheddi. A straggling perennial herb. Hilly areas. Root is taken internally for easy child birth.

Aegle marmelos (L.) *Corr. Serr. (Rutuceae)*

Vilvamaram. A deciduous armed tree, apparently truly wild. Often planted in plains. Leaf juice prepared with *Acorus calamus* together with cow's milk and taken for burns.

Andrographis paniculata *Wall. Ex Nees. (Acanthaceae)*

Serriyanangai. A subshrub, in dry forest under growth. Often planted by households as a putative snake repellent. Leaf powdered, mixed with cow's milk and given internally for snake bite. Leaf paste applied on snake bitten spot.

Aristolochia bracteolata *Lam. (Aristolochiaceae)*

Perumarundhu kodi. A perennial herb in dry areas. Leaf juice and root paste alongwith black pepper is taken orally and also applied externally for snake bite.

Aristolochia tagala *Cham. (Aristolochiaceae)*

Kozhikkudu cheddi. An extensive climbing shrub. Chewing the whole plant or taking its juice internally acts as an antidote for snake poison.

Achyranthes aspera (L.) *(Amaranthaceae)*

Naayuruvi. A subshrub. Weed of wasteland. Leaf juice mixed with honey in equal ratio and taken against asthma complaints. Juice of the whole plant is given orally as antidote for snake bite.

Aerva lanata (L.) *Juss. (Amaranthaceae)*

Poonaippo. Erect to decumbent herb. Wayside weed. Chewing of leaf raw is thought to have effective antidote action in cases of snake bite.

Butea monosperma *(Lamk.) Thunb. (Fabaceae)*

Porasu. A deciduous tree, very conspicuous when flowering. Occasional. Root decoction mixed with cow's milk and taken as an aphrodisiac. Gum together with rice washed water with salt taken against diarrhoae and dysentery.

Begonia malabarica *Lam. (Begoniaceae)*

Chenthandu. Subshrubby. Occur in deep shade of ravines and rocks. Locally abundant. Leaf paste greatly applied externally over body to reduce fever.

Begonia integrifolia *Dalz. (Begoniaceae)*

Chenthandu. A succulent herb with short stem. Occassional clumps in moist forest. Stem paste together with cow's milk taken as a blood purifier.

Bambusa arundinacea *(Retz.) Wild. (Poaceae)*

Mungil. A long thorny bamboo. Leaf paste is applied externally on body to relieve body pain.

Basella alba (L.) *(Basellaceae)*

Pasalai keerai. Rhizomatous succulent herbaceous twiner. Boiled young leaves and stems consumed against constipation.

Bauhinia racemosa *Lam. (Fabaceae)*

Aathi. A deciduous tree. In most of the hilly areas. Bark decoction together with cow's milk is used against epilepsy. Bark gently heated and applied externally over the stomach for stomach disorders.

Boerhavia diffusa (L.) *(Nyctaginaceae)*

Saaranai, Mukurrattai. A diffuse herb with red flowers, in wastelands and along roadsides. Root juice together with roots of *Pavetta indica* taken internally for epilepsy. Root juice mixed with honey for jaundice.

Bacopa monnieri (L.) *Penn. (Scrophulariaceae)*

Brahmi. A profusely branched succulent herb. Very abundant in low-lying marshy places, along water courses. Whole plant made into paste, mixed with honey and used against nervous disability. Leaf paste mixed with honey take for spermatozzhoea.

Balanophora fungosa *Forst. & Forst. (Balanophoraceae)*

Nilagunguligam. A thick deciduous herb, spadicles club-shaped in moist and shady places. Fried stem together with onion made into fine paste and applied for rheumaticism.

Bidens pilosa (L.) *(Asteraceae)*

Naethirum Poondu. A stout herb. Gregarious weed of fallow fields. Whole plant made into paste together with cow's milk and taken orally to cure ulcer complaints.

Boswella serrata *Colebr. (Burseraceae)*

Kungli. A deciduous tree with papery bark, mostly in dry forest. The plant is used to keep off evil spirit and also for black magic.

Centella asiatica (L.) *Urban. (Apiaceae)*

Vallarrai. Prostrate herb, in moist and sheltered places. Powdered leaves together with cow's milk taken internally for ulcer complaints.

Calotropis gigantea (L.) *R.Br. (Asclepiadaceae)*

Errukku. A shrubby in wastelands, ready colonizer by new roads. Latex applied for tooth ache and rabies. Root paste prepared with *Acorus calamus* and applied over chest to control pneumonia fever.

Cipadessa baccifera *(Roth.) Miq. (Meliaceae)*

Peenathazhai. A bushy shrub or small tree in the hilly area. Stem paste is taken orally for stomach disorders.

Cajanus cajan (L.) *Mellsp. (Fabaceae)*

Thuvarai. An erect shrub, cultivated for its seeds. Leaf paste applied over breast of feeding mother to increase breast milk.

Cissus quadrangularis (L.) *(Vitaceae)*

Pirandai. Rambling shrub in scrub jungles and wastelands on thickets. Chatiney prepared from the fried stem in gingley oil (*Sesamum indicum*) and taken internally to reduce fever.

Cardiospermum helicacabum (L.) *(Sapindaceae)*

Mudakattan. A twining herb, abundant in wasteland and abundance on disturbed ground. Decoction prepared from the leaves and stem and taken internally to treat fever as well as paralysis.

Cynodon dactylon (L.) *Pers.- Poaceae*

Arugampullu. A perennial rhizomatous herb. Leaf juice prepared with *Mentha piperata* and taken as general tonic.

Chroloxylon swietenia *DC. (Rutaceae)*

Porasu. A deciduous unarmed tree. Very common on the lower slopes. Stem juice is used for eye diseases.

Coccinia grandis (L.) *Voight. (Cucurbitaceae)*

Kovai. A deciduous vine. Very common on thickets and wastelands. Leaves and mucilage from immature fruits given twice or thrice a day till cure for diabetes. Leaf juice is used to treat eye complaints.

Curcuma neilgherrensis *Wight. (Zingiberaceae)*

Kattu manjal. A tuberous herb. Common but often not recognized as it is ephemeral. Flowers appear when the plant is leafless. Tuber juice is taken in diarrhoea and dysentery problems.

Commelina erecta (L.) (Commenlinaceae)

Mayura singu. A spreading herb in hills. Leaf juice is administered orally for children with constipation.

Cochlospermum religiosum (L.) Als. (Cochlospermaceae)

Tanaku, Kongillium. A deciduous tree in dry forest on stony hills. Gum of this plant tied on fractured bone for bone setting.

Canthium dicoccum (Gaertn.) Teijin & Binnend. (Rubiaceae)

Allumba maram. A small tree in hills. Leaf juice mixed with palm jaggery consumed for easy delivery.

Dodonaea angustifolia (L.) (Sapindaceae)

Viraali. A shrub with extensive distribution on the hills even in rocky, gravelly or limestone ground. Leaf juice is used for epilepsy. Leaf paste applied on fractured bone for bone setting.

Drosera peltata Sm. (Droseraceae)

An insectivorous herb with a reddish tuber, in moist ground, especially on the thin layer of soil on exposed rocks. It is believed that the whole plant is used to cure cancer. Plants in demand by pharmacists.

Derris scandens (Roxb.) Benth. (Fabaceae)

Theykil. A strangler in shrub jungles, along river banks. Leaves and flowers juice taken once in a day for diabetes.

Dalbergia lanceolaria (L.f.) (Fabaceae)

Velangiri. A deciduous tree, densely foliaceaous with a crown of equal spread. Locally abundant on the deciduous slopes. Stem bark paste prepared with roots of *Ficus benghalensis,* mixed with water and internally taken for jaundice.

Datura metel (L.) (Solanaceae)

Oomathai. Subshrub. Leaf paste mixed with cow's milk taken against rabies.

Drynaria quercifolia (L.) Sm. (Drynariaceae)

Aattukkal kilangu. Common on branches of tress and also in crevices of rocks. Rhizomes made into paste applied for rheumatism.

Eclipta prostrata (L.) (Asteraceae)

Karisalankanni. A perennial herb with ascending branches, very common in clayey soil by arable lands. Leaf juice prepared together with onion (*Allium alpha*) in 2:1 ratio, taken twice a day for spermatorrhoea.

Evolvulus alsinoides (L.) (Convolvulaceae)

Vishnu kiranthi. A prostrate to ascending herb, the floor or shrub jungles even in poor soil, more abundant on hills on the bare, exposed slopes. Whole plant made into paste, mixed with cow's milk, taken against nervous disability.

Entada pursaetha DC. (Mimosaceae)

Papparan kottai. Unarmed liana. Occassional in sholas. Paste prepared from the cotyledons applied for inflammation.

Enicostemma axillare (Lam.) Raynal. (Gentianaceae)

Vellarugu. A perennial erect herb, in gravelly wastelands, fallow fields, river banks. Leaf juice prepared with pepper (*Piper nigram*), orally given to reduce fever.

Elaeocarpus tuberculatus Roxb. (Elacocarpaceae)

Ruddratsum. A tree in sholas in evergreen forests. Bark decoction used as abortifacient.

Ficus racemosa (L.) (Moraceae)

Athimaram. A tree, leaf galls copious and characteristic. Bark decoction taken orally after menstrual period to get fertility.

Ficus religiosa (L.) (Moraceae)

Arasamaram. A tree. Often planted near temples. The leaf juice is used as general tonic.

Ficus hispida (L.f.) (Moraceae)

Paeipperukku. A tree, common in wastelands. Fruit paste applied externally over throat against tonsil.

Filicium decipiens Thw. (Sapindaceae)

Paeimeraatti. A evergreen shola tree, in moist ravines, shady floor of sholas, copious regeneration. Leaves used to keep off evil spirits and also used for black magic.

Gloriosa superba (L.) (Liliaceae)

Kalappai kizhangu. A climber on thickets. Getting scare owing to collection of rhizome for medicine. Leaf juice is used to cure eye disease.

Gmelina arborea (L.) (Verbenaceae)

Kumilamaram. An unarmed tree. Common in deciduous belt. Leaf juice orally taken to control diabetes.

Hemionitis arifolia (Burn.f.) Moore. (Hemionitidaceae)

Kambithamarai. A small herb, in rock shady places. Whole plant juice is used to control irregular menstruation.

Hydrocotyle javanica Thunb. (Apiaceae)

Kamalakkodi. A prostrate shrub in shady and moist places like shola floor, stream banks and rotting wood. The whole plant paste prepared with turmeric (*Curcuma longa*) together with milk, taken orally for jaundice.

Ipomoea obscura (L.) Ker. -Gawl. (Concolvulaceae)

Serruthali. A slender twiner on fences, in arable lands, less on the hills. Leaf paste prepared with onion (*Allium ceba*) applied on blisters.

Justicia adhatoda (L.) (Acanthaceae)

Aadathodai. A shrub, often planted along fences. Widely used medicinally. Leaf decoction prepared with garlic (*Allium sativum*) and pepper (*Piper nigrum*), taken internally to cure asthma and bronchitis.

Leucas aspera (L.) (Lamiaceae)

Thumbaichedi. A scabrid erect herb. Common wayside weed. Locally abundant in fallow fields, wastelands, smaller size in the hills. Leaves made into paste together with cow's milk, taken internally to induce lactation in nursing mothers.

Moringa olifera (L.) (Moringaceae)

Murungaimaram. A tree cultivated. Gum mixed with cow's milk taken internally to control spermatorrhoea.

Mucuna pruriens (L.) DC. (Fabaceae)

Poonaikkali. A vine, plains and lower hills along river banks. Occasional. Powdered seeds mixed with cow's milk, internally taken for leucorrhoea.

Myristica dactyloides Gaert. (Myristicaceae)

Kattujathikkaei. Densly foliaceous shola tree. Immature fruit paste applied on forehead to control headache.

Mangifera indica (L.) (Anacardiaceae)

Mangaai. A tree, often cultivated. Decoction prepared with immature fruits, given internally to control fever.

Manilkara zapota (L.) P. Royon. (Sapotaceae)

Sapotamaram. A tree, in plains and lower slopes of hills, often planted. Fruits taken to control diarrhea and dysentery.

Madhuca longifolia (Koen.) Macbr. (Sapotaceae)

Kattueluppai. Tree with crown of equal spread. Stem bark decoction regularly used to maintain normal blood sugar level.

Nilgirianthus foliosus (Wight) Bremek. (Acanthaceae)

A large gregarious shrub, in sholas. Root paste mixed with honey and cow's milk taken for aphrodisiac.

Oxalis corniculata (L.) (Oxalidaceae)

Serrupulichan. A diffuse herb. Abundant in fallow fields, optimum size in shade at the higher altitude. Whole plant paste externally over body to reduce body pain.

Orthosiphon thymiflorus (Ruth.) Sleensen. (Lamiaceae)

Kutchi Tulasi. Slightly aromatic subshrub on the slopes, in crevices of rocks, more numerous by arable lands or as a colonizer along new roads. Leaf juice prepared with paddy crop (*Oryza sativa*), given orally after menstrual period for fertility.

Passiflora foetida (L.) (Passifloraceae)

Ponnaikuttipazham. A vine, foetid when bruised. Mature fruits used to reduce stomach pain.

Pergularia daemia (Forssk.) Chiov. (Asclepiadaceae)

Veliparuthi. A straggler in wastelands, wayside thickets, less on hills. Leaf juice is used for rheumatic pain.

Psidium guajava (L.) (Myrtaceae)

Guaiyamaram. A small tree cultivated for its fruits. Mature fruits taken to reduce stomach pain.

Pyrrosia mollis (Kuntz.) Ching – Polypodiaceae

Paraipatchai. A small herb, on rock surface and rock crevices in moist localities, and shola florest. Whole plant mixed with onion (*Allium cepha*), made into paste, applied for rheumatic pain.

Peperomia tetraphylla (Forst. F) Hook. & Arn. (Piperaceae)

Alingucheddi. An epiphytic herb often on rotting wood, in shady places. Leaf juice is used for mouthsore.

Pedalium murex (L.) (Pedaliaceae)

Yannai nerunji. A herb. Leaf paste applied over forehead to reduce headache. Leaves taken with sugar orally for leucorrhoea.

Pterocarpus marsupium (L.) (Fabaceae)

Vengaimaram. Semievergreen tree on the slopes. Even in dry and fully exposed habitats. Very showy and recognizable from great distance. Bark decoction given internally to control dysentery and diarrhoea.

Rhaphidophora laciniata (Burm.) Merr. (Araceae)

Pannipirandai. Evergreen stout root climber, in shade on shola trees, on vertical rocks and in ravines. Leaf paste applied on fractured bone for bone setting. Leaf smoke inhaled to cure asthma.

Solanum anguivi Lamk. (Solanaceae)

Sundai. A shrub on the slopes. Fruits regularly used as vermicide.

Solanum nigrum (L.) (Solanaceae)

Kunni. An unarmed shrub, especially by the river bank, river bed. Boiled leaves taken to control fever and mouthsore.

Schefflera wallichiana Harms. (Araliaceae)

Paeimerratti. A deciduious tree, branches radiating, crown hemispherical, among crevices of rocks by ravines in shade and in full sun, often by forest border. Leaves are used for black magic and keep off evil spirit.

Scoparia dulcis (L.) (Scophulariaceae)

Nilavembu. A robust erect herb in waste lands – clearing and fallow fields. Leaf juice used to cure eye diseases.

Spermacoce hispida (L.) (Rubiaceae)

Nathaisoori. A spreading hispid herb, an aggressive colonizer, abundant in fallow fields. Leaf juice mixed with cow's milk given orally as aphrodisiac.

Sanseviera roxburghiana Schult. & Schult. (Agavaceae)

Nagathali. A stout herb, firm rooted, on the dry slopes, crevices of rocks. Stem juice is used for ear complaints.

Sida cordata (Burm. f) Bress. (Malvaceae)

Vellurum. A subshrub. Common by the wayside wastelands, locally abundant in forest clearings. Bioled leaf juice is given internally to control diarrhoea and dysentery.

Syzygium cumini (L.) Skeels. (Myrtaceae)

Naaval. Evergreen tree in a variety of habitats, sholas, river banks, scrub jungles or planted as avenue tress. Decoction of stem bark given orally as an anti-diabetic. Immature fruits eaten raw.

Sonerila clarkei Cogn. (Melastomatoceae)

A woody herb in the hills. Leaf juice mixed with cow's milk taken internally as an aphrodisiac.

Tephrosia purpurea (L.) Pers. (Fabaceae)

Kolungi. A subshrub. Gregarious in wastelands and fallow fields. Powdered roots is used as an abortifacient.

Trichosanthes curcumerina (L.) (Cucurbitaceae)

Therupudal. A vine, in the lower hill slopes. The whole plant paste is used as an abortifacient and also for asthma complaints.

Tinospora cordifolia (Wild.) Merr. (Menispermaceae)

Seenthil kodi. A deciduous woody climber, common in scrub jungles, on thickets, fences and tress. Stem decoction taken internally daily in the morning for diabetic.

Terminalia chebula Retz. (Combretaceae)

Kadukkaie. A deciduous tree with dark brown bark. In deciduous forests. Decoction prepared from the mature fruits is used to control diarrhoea and dysentery.

Terminalia bellerica (Gaert.) Roxb. (Combretaceae)

Thani. A large deciduous tree with bluish grey, tissued bark. Common in hilly areas. Fruit decoction is used as general tonic.

1.4 Conclusion

Malaiyalis preferred to use a variety of plant species with medicinal utility. A total of 100 species distributed in 94 genera belonging to 54 families were identified for medicinal and general healthcare purpose during the study. Indigenous medicinal utility in the present study supports much of their traditional knowledge as reported in Geetha's (1996) study of illness, remedies and mode of action. However, there are few disagreement among remedies between the present study and that of Geetha's study (l.c.)

Malayalis prefer to utilize species from primary and secondary semi-evergreen rainforests of Kolli Hills rather than the weedy species from disturbed areas. The most common families in the study were Fabaceae (9 species), Acanthaceae (4 species), Amaranthaceae (4 species), Asclepiadaceae (3 species), Convolvulaceae (3 species), Meliaceae (3 species), Lamiaceae (3 species), Moraceae (3 species), Sapindaceae (3 species) and Solanaceae (3 species). Herbs (38 species), were the most common used group of plants followed by tress (31 species), herbs (20 species) and climbers (13 species).

An assortment of plant parts were utilized as medicine by the Malaiyalis of which the leaves were used most frequently followed by roots, bark, seeds, whole plants, flowers, fruits and latex/sap. The number of internal application of phytomedicines is visibly higher than external applications. There are indications that the plants, much employed for curing purpose in other parts of India are currently used as herbal medicines in the Kolli Hills as well.

The simultaneous usages of a particular plant for multiferous treatment of a number of diseases are observed in the present study for the species like *Annona squamosa, Argemone mexicana, Azadirachta indica, Bauhinia racemosa, Boerhavia diffusa, Bacopa monnieri, Calotropis gigantea, Coccinia grandis, Dodonea angustifolia, Jatropha curcas, Pedalium muxex, Rhaphidophora pertusa* and *Trichosanthes cucumerina.* This may also be considered encouraging with regard to their positive biological activity.

The numbers of antidiabetic and antidysentric plant species are fairly high indicating that diabetic and dysentery are the commonly encountered ailments among the Malayalis tribe.

The plant species such as *Aglaia elaegniodea, Acorus calamus, Artocarpus hirsutus, Bacopa monnieri, Boerhavia diffusa, Drosera peltata, Dalbergia lanceolaria, Gloriosa superba, Gymnema sylvestris, Justicia adhatoda, Pterocarpus marsupium, Syzygium cumini* and *Terminalia chebula* are highly valuable medicinal plants found in Kolli Hills area.

The present study identified plant species such as *Anectochilus elatus, Begonia integrifolia, Begonia malabarica, Drosera peltata, Soneritia clarkei* and *Curcuma neilgherrense* which are rare and endemic. Malayalis tribes collect and preserve these locally available wild species, unaware of the fact that they are endemic to India.

References

Alagesaboopathi, C., Dwarakan, P. and Balu, S. 1999. Plants used as medicine by tribals of Shervarayan Hills, Tamil Nadu. *J. Econ. Tax. Bet.* 49:391-93.

Alagesaboopathi,C. 1994. Medico- Botanical survey of plants in Kangmalai Hills of Salem. *Anc. Sci. Life* 149:112-116.

Ansari, A.A. 1997. Less known wild edible plants of Shervarayan and Kolli Hills of South India. In: S.K.Jain *(Ed). Contribution of India Ehtinobotany. Scientific Publishers, Jodhpur, India.* Pp. 243-247.

Ansari, A.A., Diwahar, P.G. and Dwarakan, P. 1993. Less known edible plants of Shevarayan and Kolli Hills. *J. Econ. Tax. Bot.* 17: 245-247.

Anyinam, C. 1995. Ecology and Ethnomedicine: Exploring links between current environmental crisis and indigenous medicinal practices. Sec. *Sci.Med.*

Dwarakan, P. 1999. Less known plants of Kolli Hills in Namakkal District, Tamil Nadu in National Seminar on Siddha system of medicine, "Focus on current status", organized by Coimbatore District Siddha Assistant Medical Officer's Association, Coimbatore.

Dwarakan, P., and Ansari, A.A. 1982. Ethnobotanical notes of Valikadupatti and surroundings to Killimalais of Salem Dsitrict, Tamil Nadu. *J. Econ. Tax. Bot* (Addl. Ser) 10: 495-499.

Dwarakan, R., Rajasekharan, K. and Selvam, A.B. 1999. Hitherto unreported medicinal uses of plants from Kolli Hills. *Anc. Sci. Life* 12: 259-260.

Dwarakan. P. and Alagesaboopathi, C. 1999. Traditional crude drug resources used for human and live stock diseases in Salem District, Tamil Nadu. *J. Econ. Tax. Bot.* 47: 421-424.

Dwaraken, P. and Ansari, A.A. 1996. Less known uses of plants of Kollimalai (Salem District) Tamil Nadu in South India. *J. Econ. Tex. Bot.* (Addl. Ser.) 12: 284-286.

Gamble, J.S. 1915-1936. Flora of the Presidency of Madras, London.

Ganesan, S., Suresh, N. and Kasavan, L. 2004. Ethnobotanical survey of lower Palani Hills of Tamil Nadu. *Ind. J. Trad. Knowled* 3: 299-304.

Geeka, S., Lakshmi, G. and Ranjitakani, P. 1996. Ethnobotanical review. Wild Fibre yielding Plants of Kolli Hills, Tamil Nadu. *J. Econ. Tax. Bot.* (Add. Ser) 12: 250-252.

Geetha, S. 1996. Ethnobotanical studies of Kolli Hills, Salem District, Ph.D, Thesis, Bharathiar University, Coimbatore.

Jain, S.K. 1991. Dictionary of India Folk Medicine and Ethnobotony. Deep Publications, New Delhi.

Lakshmanan, K.K. and Kalyani, K. 1980. Kollimalaiyum athen Mooligai selvangalum. In: Tamil Nadu *(Ed.) Salem Movattam Sila Aaivakal Kaavya Publications*, Bangalore.

Lakshmi, G. 1995. Ecological studies on the vegetation of Kolli hills, Salem District, Tamil Nadu, Ph.D, Thesis, Bharathiar University, Coimbatore.

Lakshminarayanan, 1990. Management Plan for Salem Forest Division (From 1.4.1990 to 31.3.2000) Part – I.

Matthew, K.M. 1983. The Flora of the Tamil Nadu Carnatic Vol 3 (1-2): Rapinet Herbarium, Tiruchirapalli.

Nazer, M. 2000. Tribal people in India: *Report. A Study of Land Alienation and Indebetedness among Tribals; Tamil Nadu, Kerala and Karnataka States,* submitted to Planning Commission of India, New Delhi.

Ragupathy, S. and Mahadevan, A. 1996. Traditional medicine and ethno-botany among Dravidian tribal communities. In: Menon, T.M. *(Ed.) Encyclopedia of Dravidian Tribes; Thiruvananthapuram, India.* International School of Dravidian Linguistic 1-5.

Ragupathy, S. and Mahadevan, A. 1991. Ethnobotony of Kodaikkarai Reserve Forest, Tamil Nadu, South India. *Ethnobotany* 3: 79-82.

Rajendran, A. 1980. Kalanginathar Coorum Kollimalai ragasiam. In: Tamilnadu *(Ed.) Salem Mavattam sila Aaivukal. Kaavya Publications*, Bangalore.

Rajendran, A. 2005. Wild Plants utilization in Kolli Hills, Tamil Nadu. Tamil University Publication, Thanjavur.

Thothathri, K., Sen, R. and Pal, D.C. 1989. Status of the medicinal plants in Indian Flora. *J. Econ. Tax. Bot.* 13(3): 605-614.

Timbrook, J. 2004. *Ethnobiota. J. Ethnobiol* 24(2) 5-6.

□□□

Ethno-Medico Plants in Courtallum Hills, Southern Western Ghats, India

Sivalingam, R. & S. Manian

CONTENTS

1.1 Introduction 114

1.2 Study Area 115

1.3 Methodology 115

1.4 Observation 116

1.5 Conclusion 122

References 123

1.1 Introduction

Plants have been used by human race from time immemorial to treat several diseases. Phyto-medicines appear to be safer and now there is a growing trend of acceptance of this medicine worldwide. According to World Health Organization (WHO), 80% of the world's population rely on plants for their primary health care (Lindsay, 1993), because it is easily available, less costly and safer. This means that 3300 million people use medicinal plants on a regular basis (Fransworth, 1994). Caniago and Siebert (1998) stated that many rural people throughout the world rely on medicinal plants because of their effectiveness, lack of modern healthcare alternatives and cultural preferences. Of late more and more scientific and commercial activities are being centered on medicinal plants.

India being a tropical country is rich in species diversity. India ranks tenth among the plant rich countries of the world and sixth among the centers of diversity (Gadgil, 1996). It is estimated that India accounts for 8% of the global biodiversity existing in only 2.4% of land area of the world. It harbours about 33,000 medicinal taxa. Among these, 8,000 species are the most common medicinal plants (Swaminathan, 2000).

Traditional ethnomedicinal studies have in recent years received much attention due to their local acceptability and clues for new or lesser-known medicinal plants (Tripathi, 2000). Over 16,000 species of higher plants are reported to occur in India, of which about 9,000 are economically useful, where as, 7,500 of these species are reported to be used for health care by various ethnic communities in India, 3,900 are reported to be edible, 700 culturally important, 525 used for fiber, 400 for fodder, 300 for pesticides and insecticides, 300 for gums, resins and dyes and 100 provide incense and perfume (Anon., 1994; Arora, 1997).

Only about 2,500 plant species of medicinal value have been well documented in Indin systems of medicines and under ethnobotany (Jain, 1991). Ethnobotanical investigations were carried out in Tamil Nadu only in late seventies. Several researchers (Nair & Henry, 1983; Henry *et al.*, 1987; Binu *et al.*, 1992; Geetha *et al.*, 1996) reported the plants and plant parts used by different tribes of Tamil Nadu for ethno-medicine purposes.

Due to over exploitation, habitat destruction and various developmental activities, the potential medicinal plants may be lost for ever. It is necessary to undertake the ethnobotanical exploration to identify and record the information possessed by rural people, particularly in the Western Ghats of India. Hence, the present study is an attempt to ascertain preferences of rural communities on traditional phytomedicines and to enlist the medicinal plants of the Courtallam hills.

1.2 Study Area

The Courtallum hills (8° 50' - 9° 0' N and 77° 10' - 77° 22' E) are one of the important forest reserves of Tirunelveli forest division, an offshoot of Agastyamalai Biosphere Reserve of southern Western Ghats. This biosphere reserve is one of the 18 hot spots in the world, (Nayar, 1986) and has a high floristic diversity attracting many botanists (Nair & Daniel, 1986). The hills of Courtallum, one of the major tourist and pilgrimage centers, are 5 km, east of Tenkasi, which is known as South Kasi. The forest covers an area of approximately 70 sq. km and is extending to the Kulathupuzha forest reserve of Kerala state in the West, Papanasam and Kadayam ranges in the south, and Tenkasi and Shencottah taluks of Tirunelveli district, in the east and north respectively.

The physiography of this range differs much from the adjacent portions of southern Western Ghats of Tamil Nadu state due to the varying altitude, valleys, ridges and its height and breadth. The altitude of the hilly rugged terrain ranges from 150 to 1600 m with the highest peak in Valrattimottai (1600 m). This area is traversed by three rivers namely, Chittar, Aindalaiar and Palayacourtallar which form a few and cheerful sanitoriums, namely Mainfalls, Honeyfalls, and Five falls. These attract several thousands of tourists during the months of June- September which is known as charal season.

Geologically, the soil is red-black sandy loam intermingled with occasional granitoid genessic rock. The ghats of Courtallum experience both south-west (July-September) and north-east (October-December) monsoon and occasional showers, which promote the vegetations in the hills and its adjoining plains. The annual rainfall varies from 80 to 150 cm. The temperature is maximum (38°C) during the months of April-June and minimum (17°C) in the rainy season (November-January).

The various topographic, edaphic and climatic factors display heterogenous type of vegetation viz., hydrophytic, mesophytic and rheophytic flora in which, the mesophytes manifest five southern tropical forests such as 1. Thorny forests (150-300m) 2. Dry-deciduous forests (300-500m) 3. Moist-deciduous forests (500-700m) 4. Wet evergreen forests (700-1600m) and 5. Grasslands of both low altitude (150-500m) and high altitude (900-1600m).

1.3 Methodology

More than 10 systematic field trips were conducted throughout the Courtallum hills in different seasons. The information about medicinal plants with reference to local name, medicinal uses, parts of the plant used, method of plant crude drug preparation and their composition, mode of application, dosage and food restriction if any during the period were recorded with the help of local inhabitants who have knowledge about plants and prescribe medicine to cure various ailments. Repeated queries were raised and cross-checked to get correct information. The method of gathering information was the same as suggested by Schultes (1960;1962) and Jain (1963; 1965; 1967). Bilateral discussion and open ended interviews with traditional healers, herbalists and local people have been held in various villages to collect the maximum traditional phytotherapeutic information from them within the limited period. The plants were identified using the Flora of British India (Hooker, 1887), Flora of the Presidency of Madras (Gamble, 1918) and Flora of Tamil Nadu (Nair *et al.,* 1989).

1.4 Observation

The plants used as traditional medicine in Courtallum hills, southern western Ghats, India are presented in Table 1. Plants are arranged in alphabetical order of their botanical names followed by vernacular name and family with the data regarding medicinal uses, ingredients used, mode of application, dosage etc.

Table. 1: List of plants used by rural people of Courtallum Hills, Western Ghats.

Botanical name	Vernacular name	Family	Part(s) used	Uses
Abrus precatorius L.	Kundumani	Fabaceae	Leaf	The leaves are made into paste and applied for skin diseases like scabies and itch.
Acalypha fruticosa Forssk.	Kaatukuppaimeni	Euphorbiaceae	Leaf	Leaf paste applied on skin for scabies, itch and insect bite.
Alpinia galanga Sw.	Chitharathai	Alpiniaceae	Rhizome	The rhizome is boiled in a water to make a decoction. The decoction is given orally during labour pain for easy delivery.
Andrographis paniculata (Burm.f.) Wall.ex Nees	Nilavembu	Acanthaceae	Whole plant	100 ml of whole plant decoction given 3 times a day for cold and fever.

Contd...

Table 1 contd...

Anisomeles malabarica (L.) R.Br. ex Sims.	Peymirati	Lamiaceae	Leaf	Fresh leaf paste is applied on the leg bone joints to cure rheumatism.
Arisaema tortuosum (Wall.) Schott.	Kaatu seanai	Araceae	Rhizome (Corms)	Corms boiled in water are taken for 10 days to cure piles
Asparagus racemosus Willd.	Thannervitan	Liliaceae	Tuber	The tubers are mixed with honey and taken orally twice in a day to cure male sterility
Azima tetracantha Lam.	Mulsangu	Salvadoraceae	Leaf	25 g of leaves with 25 g of *Ocimum tenuiflorum* leaves are boiled well in 250 ml of water for 10 minutes and the orally for dry cough
Bambusa arundinacea (Retz.) Roxb.	Moongil	Poaceae	Leaf	50 ml decoction from fresh leaves is taken orally for fever with chillness.
Canthium parviflorum Lam.	Kaarai	Rubiaceae	Leaf	The leaves are crushed and applied over the cuts and wounds to heal.
Capparis divaricata Lam.	Athandai chedi	Capparaceae	Fruit	The fruit juice is applied over the place of scabies.
Cardiospermum halicacabum L.	Mudakathan	Sapindaceae	Leaf	50 ml of decoction of fresh leaves is taken orally for gastritis.
Cayratia pedata (Lam.) Juss. ex Gagnep.	Kalalai kodi	Vitaceae	Leaf	The leaf juice is administered twice a day to ladies until cured the body heat with white discharge.
Citrus limon (L.) Burm.f.	Elimitchai	Rutaceae	Fruit	Fruit juice is applied on the head for removing body heat.

Contd...

Table 1 contd...

Commelina benghalensis L.	Theankaraipattu	Commelinaceae	Leaf	25 g of leaf decoction given orally for profuse bleeding during menstruation.
Crateva magna (Lour.) DC.	Maavilangam	Capparaceae	Leaf	The fresh leaves are crushed with gingely oil and the paste is externally applied for piles.
Curculigo orchioides Gaertn.	Nilappanai	Hypoxidaceae	Tuber	The tubers are made into powder, then mixed with honey and is given orally twice in a day for body muscle improvement.
Didymocarpus ovalifolia Wight.	Punnakkukulai	Gesneriaceae	Leaf	Leaf paste mixed with neem oil and applied externally for rheumatism.
Dioscorea esculenta Burk.	Boomi sarkarai kizhangu	Dioscoreaceae	Tuber	Tubers made into a paste and mixed with boiled cow milk and given orally for chest pain.
Dodonaea viscosa (L.) Jacq.	Viraali	Sapindaceae	Leaf	The leaf paste is used using externally at the affected site to treat bone fractures.
Drymaria cordata (L.) Roem. ex Schult.	Aatukal kizhangu	Caryophylla-ceae	Bulb	The bulb paste is applied to the naval region of pregnant woman to quicken labour.
Emilia scabra DC.	Kaanakkaduku	Asteraceae	Leaf	Leaf paste is prepared with castor oil and applied on the paining bone joints of legs.

Contd...

Table 1 contd...

Evolulus alsinoides (L.) L.	Vishnukaranthi	Convolula-ceae	Whole plant	The whole plant is used to prepare decoction and 20 ml of decoction given orally for heavy fever.
Gymnema sylvestre (Retz.) . R.Br	Sakaraikolli	Asclepiadaceae	Leaf	The dried leaf powder (10 g) is taken daily early in the morning in empty stomach to cure diabetes.
Habenaria decipiens Wight.	Vellaipattu	Orchidaceae	Whole plant	Decoction of whole plant is used to control pain during menstruation.
Hemidesmus indicus (L.) R.Br.	Nannari	Periplocaceae	Root	The dry root powder (5 g) is added with honey (one tea spoon) and all are mixed with cold water, then given orally to promote free urination.
Hemionitis arifolia (Burm.) Moore.	Kalpaasi	Hemionitida-ceae	Frond	10 ml of frond extract mixed with boiled cow milk and given orally for pitha problems.
Hydrocotyle javanica Thunb.	Malai valarai	Apiaceae	Leaf	10 g of leaf powder mixed with honey is taken orally for curing madness.
Impatiens leschenaultii (DC.) Wall. ex Wight & Arn.	Kannadichedi	Balsaminaceae	Whole plant	The whole plant juice is applied to body to get rid of burning sensations.
Lantana indica Roxb.	Unni	Verbenaceae	Leaf	Leaf paste is applied on the place of bone fracture.

Contd...

Table 1 contd...

Melia azadirach L.	Malai vembu	Meliaceae	Leaf	The leaves with flowers are made into paste and applied on skin diseases.
Mimosa pudica L.	Thotarsuringi	Mimosaceae	Leaf	The leaves are made into powder, mixed with cow's milk and given orally twice a day to control diabetes.
Mucuna pruriens (L.) DC.	Poonikaali	Fabaceae	Leaf	Leaf paste is applied on cuts and wounds twice a day until cure.
Orthosiphon comosus Wight ex Benth.	Pakkurinji	Lamiaceae	Leaf	Leaf made into paste, and then applied externally to cure muscle contraction.
Oxalis corniculata L.	Puliyarai	Oxalidaceae	Leaf	The cooked leaves are orally given for stomach problems and blood piles.
Pandanus fascicularis Lam.	Thazhai	Pandanaceae	Pollen grain	The pollengrains mixed with coconut oil is applied on the penis for strength.
Phyllanthus emblica L.	Perunelli	Euphorbiaceae	Fruit	Fruits are soaked in the honey and administered orally twice in a day to cure anemia.
Plumbago indica L.	Chitramoolam	Plumbagina-ceae	Root	10 of root powder with boiled cow milk taken orally after men struation for birth control.
Pterocarpus marsupium Roxb.	Veangai	Fabaceae	Bark	The bark is made into paste and applied on burns.

Contd...

Table 1 contd...

Rubus elipticus L.	Kaaturosa	Rosaceae	Fruit	Mature fruits are eaten to cure stomachulcer.
Santalum album L.	Santhanam	Santalaceae	Wood	The hard wood paste is applied externally to the body as refrigerant.
Securingia virosa (Roxb. ex Willd.) Baill.	Karumpoolathi	Euphorbiaceae	Leaf	A handful of leaves are ground the juice obtained is mixed with 100 ml of buttermilk and administered orally twice a day to control diarrhoea.
Strychnos potatorum L.f.	Thetran	Loganiaceae	Seed	Seed paste is applied on the place of millipede bites to cure.
Syzygium cumini (L.) Skeels.	Naaval	Myrtaceae	Seed	10 g of seed powder taken orally on every morning with empty stomach to cure diabetes.
Vallaris solanaceae (Roth.) O. Ktze.	Manipushpam	Apocyanaceae	Leaf	The leaves are crushed and made into a paste. The paste is applied on the bone fracture and tied with a cloth for 5 days.
Wrightia tinctoria (Roxb.) R.Br.	Veppalai	Apocynaceae	Leaf	Small amount of leaves are chewed in the morning and evening for tooth ache.
Zanthoxylum limonella (Dennst.) Alston	Kaatu elumitchai	Rutaceae	Leaf	Leaf paste is applied externally for tooth-ache.
Ziziphus mauritiana Lam.	Elanthai	Rhamnaceae	Fruit	The fruit juice is taken orally for curing whooping cough.

1.5 Conclusion

A total of 47 Angiosperms (39 dicotyledonous and 8 monocotyledonous) and 1 Pteridophyte affiliated to 39 families distributed over 48 genera have been documented in this study (Fig.1). The present investigation has brought to light some popular and frequently used prescriptions available for minor ailments, such as treatment for fever, cold and cough, wounds and cuts, piles, leg pain, diarrhoea, diabetes, skin diseases, mensuration problems, rheumatism, insect bite, tooth ache, gastritis, bone fracture, and ulcer. In a majority of cases, fresh plants are used for alleviating complaints.

The Euphorbiaceae and Fabaceae each with 3 species are copiously represented in the overall collections followed by Rutaceae, Apocynaceae and Sapindaceae. These plants show wide diversity in their habit such as trees, shrubs, under-shrubs, sub-shrubs, herbs and climbers. Out of the 48 plants, the herbaceous medicinal plants recorded 19 individuals, trees 13 individuals, climbers 11 individuals and shrubs 5 individuals (Fig. 2).

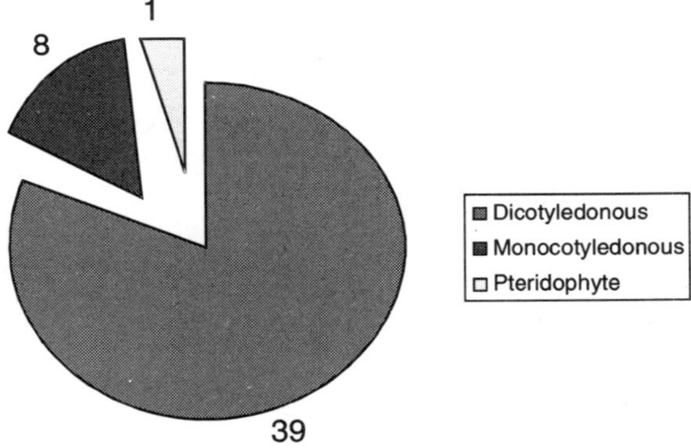

Fig.1 : Class Diversity

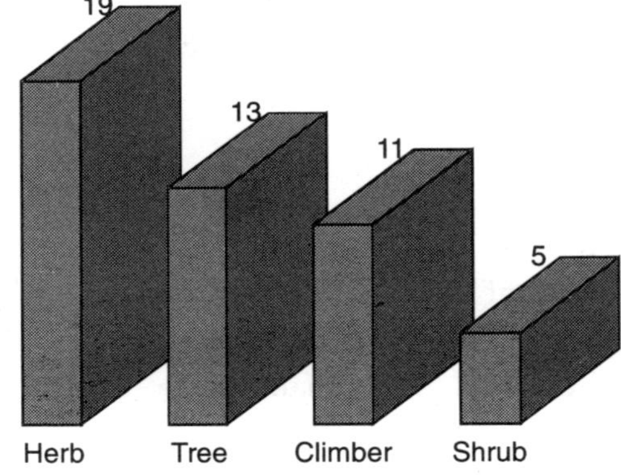

Fig. 2 : Habit Diversity

Leaves, flowers, fruits, roots, seeds, bark, bulb, rhizome, tuber, frond, wood, pollengrains and whole plant are used as medicine. Among the various morphological parts leaves are the most useful part in medicine.

The method of using these plants varied according to the nature of ailments. In the majority of cases, a decotion / juices of leaf, stem, root, frond, fruit, seed, wood, pollengrain, bulb, tuber and rhizome is drunk or rubbed on the affected area to cure ailments. Mostly decoction is extracted by just grounding the plant parts or boiled with water and the liquid decanted. In some ailments (eg., skin diseases), paste of plants is applied to the affected areas.

The commonly used medicinal plants are: *Alpinia galanga, Andrographis paniculata, Cardiospermum halicacabum* and *Gymnema sylvestre*. Some of the plants may have greater medicinal value as *Acalypha fruticosa,* hence they are used to treat different kinds of diseases like scabies, itch and insect bite; *Canthium parviflorum* is used to treat cuts and wounds. The number plant species curing, skin diseases, cough and fever are fairly high in the data indicating that skin diseases, cough and fever are the commonly encountered ailments among the rural people of this area.

The present study observed that overeating of the fruits of *Rubus elipticus* and prolonged contact with the roots of *Plumbago indica* can be dangerous to human beings. This study will promote further research on the potentially useful medicinal plants and their development for new therapeutic agents. It may also shed light on the relationship between plant species and cultural and other existing practices in rural population in the Courtallum Hills of the southern Western Ghats of Tamil Nadu.

References

Anonymous 1994. *Ethnobotany in India:* A status Report. All India Coordinated Research Project in Ethnobotany, Ministry of Environment & Forest. Govt of India. New Delhi.

Arora, R.K. 1997. Ethnobotany and its role in the conservation and use of plant genetic resources in India. *Ethnobotany.* **9:** 6-15.

Binu, S., Nayar, T.S. and Pushpangadan, P. 1992. An out line of ethnobotanical research in India. *J. Econ.Tax.Bot (Addl Ser.)* **10:** 405-428.

Caniago, I. and Siebert, S. 1998. Medicinal plants ecology, knowledge and conservation in Kalimantan, Indonesia. *Econ. Bot.* **52:** 229-250.

Fransworth, N.R. 1994. Ethnopharmacology and drug development. *In: G.T. Prance (Ed.) Ethnobotany and the search for new drugs.* Wiley, Chichester Ciba Foundation symposium. **185:** 42-59.

Gadgil, M. (1996). Documenting diversity: An experiment. *Crop Science*. **70:** 36-44.

Gamble, J.S. and Fischer, C.E.C. 1915-1936. Flora of Presidency of Madras Vol. 1-3.

Geetha, S., Lakshmi, G. and Ranjithakani, P. 1996. Ethnobotanical review: Wild fibres yielding plants, Kolli Hills, Tamil Nadu. *J.Econ.Tax.Bot. (Addl. Ser.)* **12:** 250-252.

Henry, A.N., Chitra, V. and Balakrishnan N.P. 1989. Flora of Tamil Nadu, India. Series I: Analysis. Vol. III. Botanical Survey of India, Coimbatore.

Henry, A.N., Kumari, G.R. and Chitra, V. 1987. Flora of Tamil Nadu, India. Series I: Analysis. Vol. II. Botanical Survey of India, Coimbatore.

Hooker, J.D. *(edi.)* 1872-1897. The Flora of British India (7 vols.) London.

Jain, S.K. (Ed.). 1991. Dictionary of Indian Folk-medicine and Ethnobotany. Deep Publications, Paschim Vihar, New Delhi.

Jain, S.K. 1963. Studies in Indian Ethnobotany-Plants used in medicine by the tribals of Madhya Pradesh. *Bull. RRL.* **1:** 126-129.

Jain, S.K. 1965. Medicinal plant-lore of the tribals of the Bastar. *Econ. Bot.* **19:** 236-250.

Jain, S.K. 1967. Plant of Indian Medicine and folk-lore associated with healing of bones. *Ind. J. Orthopaed.* **1:** 95-104.

Lindsay, B. (1993). Guidelines on the conservation of medicinal plants. International Union of Conservation on Nature and Natural Resources (IUCN) Publication. *Glan Switzerland.* 1-11.

Nair, D.C. and Daniel, D. 1986. The floristic diversity of the Western Ghats and its conservation: a review. *Proc. Indian. Acad. Sci: (Animal Sci./Plant Sci.)* Supl. 127-167.

Nair, N.C. and Henry, A.N. 1983. Flora of Tamil Nadu, India. Series I: Analysis Vol I. Botanical Survey of India, Coimbatore.

Nayar, M.P. 1986. "Hot spots" of Endemic plants of India, Nepal and Bhutan. Tropical Botanical Garden and Research Institute, Trivandrum, India.

Schultes, R.E. 1960. Tapping our heritage of ethnobotanical lore. *Econ. Bot.* **14:** 257-262.

Schultes, R.E. 1962. The role of the ethnobotanist in the search for new medicinal plants. *Lloydia.* **25:** 257-266.

Swaminathan, K. (2000). Herbal wealth of India. *In : Kisan world. Ed.* Mahalingam. N. **27** (9) 52.

Tripathi, Y.C. 2000. Ethnomedicinal Treasure of Tribal Rajasthan. *J. Non. Timber For. Products.* **7 (1/2):** 77-84.

□□□

Medicinal Plant Wealth of Palakonda Hill Ranges, Kadapa District, Andhra Pradesh, India

A. Madhusudhana Reddy and B. Ravi Prasad Rao

CONTENTS

1.1 Introduction 126

1.2 Study Area 126

1.3 Methodology 126

1.4 Observations 127

1.5 Conclusion 134

References 134

1.1 Introduction

Medicinal plants, the world's oldest known health care products, play a key role in traditional medicine. They are the major source of drugs for the treatment of various health disorders especially in rural areas. The use of plant based medicines dates back to 4000-5000 BC. Plants contain thousands of chemicals which act against diseases and infections of humans and animals when properly used. In the past few decades, there is a shift of using allopathic medicine rather than herbal medicine derived from plants, although majority of the allopathic medicines contain plant based ingredients. However, there are communities still using the medicinal plants collected from wilderness for curing various diseases. In the context of erosion of Traditional Botanical Knowledge (TBK) over the decades, the documentation of the existing knowledge regimes with the local communities is of prime concern.

1.2 Study Area

Palakonda ranges are situated in kadapa district of Andhra Pradesh. They are naturally formed hills bordering kadapa town. The Palakonda hill ranges lies between 13° 50′- 14° 20′ N latitude and 77° 51′- 78° 50 E longitudes and comprises an area of 916 sq.km. They represent Seshachalam hills in Kadapa district. In this range lies, the beautiful gorge called the Vempalle Gandi of Ramayana fame where the river Papaghni cuts through the range. The vegetation of the hills is varied depending upon the climate, altitude and edaphic factors. According to Champion and Seth (1968) the forests are of South Indian dry mixed deciduous types, Red Sanders forests type, Hardwickia forest type and scrub.

1.3 Methodology

Intensive explorations were held in Palakonda hill ranges during 2007-2009 to document medicinal plant resources. The community representatives of Sugali, Yanadi and Yerrukala tribes living in and around the hills were consulted and data were collected through using semi-structured questionaires on types of ailments cured by the traditional use of medicinal plants and plant parts used in curing different ailments. Crosschecking of data was made with the help of group discussions among different age classes of local peoples. Local herbal healers and knowledgeable elders are the target groups in the interviews. Voucher specimens representing medicinal plant species were collected for taxonomic identification with the help of standard floras and confirmed with the authenticated specimens in the Herbarium of Sri Krishnadevaraya University (SKU). All the representative specimens were documented at Yogi Vemana University Herbarium.

1.4 Observations

The present study documented a total of 56 medicinal plant species belonging to 34 families. All the species are alphabetically arranged with correct names, local names, family to which it belongs and uses with mode of application.

1. *Acalypha indica* Klein ex Willd. (Euphorbiaceae)

Local Name: Muripinda

Uses: Leaf juice used as a lotion for blood cancer and skin eruptions

2. *Achyranthes aspera* Linn.

Family: Amaranthaceae. Local Name: Uthareni

Uses: Leaf paste with honey used as an external application for deep cuts of iron tools. Dry root powder with common salt used as a tooth powder for toothache.

3. *Alangium salvifolium* (Linn.f.) ang.

Family: Alangiaceae. Local Name: Udaga

Uses: Root bark (4-5 inches) soaked in overnight in a glass of water and same is taken orally for next day morning for stomach problems.

4. *Aloe vera* (Linn.) Burm.f.

Family: Liliaceae. Local Name. Manchi Kalabanda

Uses: Tender leaf pulp used as a lotion for foot cracks and skin ulcers.

5. *Alternanthera sessilis* (Linn.) R.Br.ex DC.

Family: Amaranthaceae. Local Name: Ponnagantiaku

Uses: Tender twigs used for leafy vegetable.

6. *Anacardium occidentale* Linn.

Family: Anacardiaceae. Local Name: Jeedimamidi

Uses: Seed coat paste used as a lotion for foot cracks.

7. *Anisomeles malabarica* (Linn.) Kuntze

Family: Lamiaceae. Local Name: Magabeera

Uses: Dried leaves are burnt and fumes spreads out in home for evil spirits and cold.

8. *Annona squamosa* Linn.

Family: Annonaceae. Local Name: Sita palamu

Uses: Leaf paste applied over the head before taking bath twice in a week for fortnight free from lice.

9. *Asparagus racemosus* Willd.

Family: Liliaceae. Local Name: Pillitigalu.

Uses: Fresh peeled tubers (1-2) consumed daily once in a one month, for female potency. The plant is used for treating nervous disorders, tumors, scalding of urine, throat infections, tuberculosis and bronchitis.

10. *Borassus flabellifer* Linn.

Family: Palmaceae. Local Name: Taati chettu

Uses: Male cones powder (one teaspoonful) administered daily once for a week free from cold & cough.

11. *Borreria hispida* (Linn.) K. Schum.

Family: Rubiaceae. Local Name: Madanakattu aku

Uses: Whole plant powder (one teaspoonful) administered daily twice for a week for minimizing excess of heat.

12. *Calophyllum inophyllum* Linn.

Family: Clusiaceae. Local Name: Ponnaga Chettu

Uses: Seed paste used as an external application for body swellings.

13. *Cassia alata* Linn.

Family: Caesalpiniaceae. Local Name: Taamara

Uses: Leaf and flower paste used as an external application for skin eruptions.

14. *Cassia tora* Linn.

Family: Caesalpiniaceae. Local Name: Tanthemu

Uses: Leaf juice (3-4 teaspoonfuls) mixed with 8-10drops of lemon juice taken orally free from stomachache.

15. *Cleome gynandra* Linn.

Family: Cleomaceae. Local Name: Kukka Vaminta

Uses: Leaf juice used as a lotion for cuts and wounds.

16. *Cleome viscosa* Linn.

Family: Cleomaceae. Local Name: vaminta

Uses: Leaf aroma inhale daily 10-12 times for three days as free from fever.

17. *Cocculus hirsutus* (Linn.) Diels

Family: Menispermaceae. Local Name: Dussaritiga

Uses: Leaf paste contains cloth kept over the eyes for free from reddening of eyes.

18. *Corollocarpus epigaus* Hook..F.

Family: cucurbitaceae. Local Name: Nagadonda

Uses: Tuber paste (half teaspoon) administered with the same quantity of *Tinospora cordifolia* (Tippatiga) for poisonous bites.

19. *Dalbergia sissoo* Roxb.

Family: Papilionaceae. Local Name: Sarkari jitregi

Uses: Leaf juice is used as eye drops to reduce the reddening of eyes.

20. *Emilia sonchifolia* (Linn.) DC.

Family: Asteraceae. Local Name: Kundeti cheviaku

Uses: Tender tips decoction (two teaspoonfuls) administered daily once for fortnight for eye diseases.

21. *Ficus benghalensis* Linn.

Family: Moraceae. Local Name: Marri chettu

Uses: Latex used as a lotion for foot cracks and skin eruptions.

22. *Ficus racemosa* Linn

Family: Moraceae. Local Name: Medi chettu

Uses: Few fresh fruits consumed daily to free from gynecological problems.

23. *Ficus religiosa* Linn.

Family: Moraceae. Local Name: Ravi chettu

Uses: Fresh tender leaves with 4-5 drops of honey consumed daily for easy fertilization of women.

24. *Gmelina asiatica* Linn.

Family: Verbenaceae. Local Name: Chundrukaya

Uses: Fresh fruit paste applied over the scalp, free from falling of hair.

25. *Helicteres isora* Linn.

Family: Sterculiaceae. Local Name: Gubatada

Uses: Fruit powder (half teaspoon) administered daily once a week for stomach problems.

26. *Heliotropium indicum* Linn.

Family: Boraginaceae. Local Name: Telukondichettu

Uses: Leaf paste smeared over the sting area for scorpion.

27. *Hibiscus rosa-sinensis* Linn

Family: Malvaceae. Local Name: Mandara

Uses: Flower decoction (two teaspoonfuls) administered orally for cold and cough.

28. *Hybanthus enneaspermus.* (Linn.) F.V. Muell.

Family: Violaceae. Local Name: Ratnapurusha.

Uses: Whole plant Paste (half teaspoon) administered daily once in a month for male potency.

29. *Hyptis suaveolens* (Linn.) Poit.

Family: Lamiaceae. Local Name: Seema tulasi

Uses: Aerial parts are burnt and fumes are inhaled for cold and cough.

30. *Ipomoea repatans* Poir.

Family: Convolvulaceae. Local Name: Tootikaada

Uses: Slightly warmed leaf paste used as an external application for body pains.

31. *Ixora arborea* Linn.

Family: Rubiaceae. Local Name: Korivi

Uses: Leaf decoction (two teaspoonfuls) administered daily twice for one week to stomach problems.

32. *Leucas aspera* (Willd.) Link

Family: Lamiaceae. Local Name: Tummi

Uses: Leaf aroma inhaled for headache and cold.

33. *Manihot esculenta* Muell.Arg.

Family: Euphorbiaceae. Local Name: Karrapendalam.

Uses: Burnt tubers are consumed for general debility.

34. *Morinda pubescens* J.E.Smith

Family: Rubiaceae. Local Name: Togaru

Uses: Leaf decoction (2-3 Full teaspoon) given orally for loose motions.

35. *Ocimum canum* Linn.

Family: Lamiaceae. Local Name: Kukka tulasi

Uses: Leaf juice (half teaspoon) administered with 3-4 drops of honey for dry cough.

36. *Ocimum tenuiflorum* Linn.

Family: Lamiaceae. Local Name: Tulasi

Uses: Fresh leaf juice (half teaspoon) administered with 3-4 drops of honey for winter allergy.

37. *Mimosa Pudica* Linn.

Family: Mimosaceae. Local Name: Attipatti.

Uses: Whole plant burnt and fumes spread over the home which free from cold.

38. *Opuntia stricta* (Haw.) Syn.

Family: Cactaceae. Local Name: Nagajamudu

Uses: Flower paste with pinch of turmeric and salt, mixture used as a lotion for ulcers.

39. *Passiflora foetida* Linn.

Family: Passifloraceae. Local Name: Jumiki

Uses: Leaf paste used as an external application for skin eruptions.

40. *Pergularia daemia* (Forsk.) Chiov.

Family: Asclepiadaceae. Local Name: Jutapaku

Uses: Slightly warmed leaf paste applied over the swellings for minimizing pains.

41. *Phyla nodiflora* (Linn.) Greene

Family: Verbenaceae. Local Name: Bokkinaaku.

Uses: Tender parts juice (1-2 teaspoonfuls) administered daily once for three days for stomach problems.

42. *Physalis angulata* Linn.

Family: Solanaceae. Local Name: Buda busara

Uses: Few fruits are consumed for general debility and also dissolving stones in kidneys.

43. *Plumeria alba* Linn.

Family: Apocynaceae. Local Name: Devaganneru.

Uses: Latex used as a lotion for skin eruptions.

44. *Pongamia pinnata* (Linn.) Pierre

Family: Papilionaceae. Local Name: Kanuga

Uses: Root bark juice (two teaspoonfuls) administered orally for insect bite.

45. *Portulaca oleracea* Linn.

Family: Portulacaceae. Local Name: Paeaelaaku

Uses: Leaves used as a curry for general debility.

46. *Pterocarpus marsupium* Roxb.

Family: Papilionaceae. Local Name: Yegisa

Uses: Wood extract (3 teaspoonfuls) administered daily twice for minimizoy diabetes.

47. *Sida acuta* Burm.f.

Family: Malvaceae. Local Name: Parasikam

Uses: Whole plant paste is applied externally for healing of cuts and wounds.

48. *Solanum xanthocarpum* Schrad. & Wendl.

Family: Solanaceae. Local Name: Nelavakudu

Uses: Seeds burnt and fumes are pulling for free from rotting of teeth.

49. *Strychnos nux–vomica* Linn.

Family: Loganiaceae. Local Name: Musti

Uses: Stem bark paste (2-3g) administered daily once in fortnight for body swellings.

50. *Strychnos potatorum* L.f.

Family: Strychnaceae. Local Name: Chilla ginja

Uses: Leaf paste applied over the swellings free from pains.

51. *Syzygium cumini* (Wt.) Walp.

Family: Myrtaceae. Local Name: Neredu

Uses: Ripen fruits are consumed.Leaf decoction (2-3 teaspoonfuls) administered thrice a day for loose motions.

52. *Thespesia populnea* (Linn.) Sol. ex Corr.

Family: Malvaceae. Local Name: Ganga ravi

Uses: Flower paste used as an eyetex for free from reddening of eyes.

53. *Tridax procumbens* Linn.

Family: Asteraceae. Local Name: Gayapaku

Uses: Leaf paste (half teaspoon) with a pinch apepper powder, administered daily once a week for stomach disorders.

54. *Tragia involucrata* Linn.

Family: Euphorbiaceae. Local Name: Duradagondi

Uses: Leaf paste mixed with a pinch of turmeric powder, the mixture used as an external application for skin eruptions.

55. *Triumfetta rhomboidea* Jacq.

Family: Tiliaceae. Local Name: Kushtumokka

Uses: Leaf paste are used as an external application for skin eruptions.

56. *Xanthium strumarium* L.

Family: Asteraceae. Local Name: Marulamatangi

Fruit paste are used as an external application for skin eruptions.

1.5 Conclusion

The present study documented a total of 56 medicinal plant species belonging to 34 families used by the tribals against 26 diseases. Plant parts used ranged from leaves, root, stem bark, tubers, fruits, rhizome, root bark and in some cases the whole plant. Medicines were prepared in the form of paste, decoction, infusion, juice, powder, pills and water extract. Both external application and internal consumption are involved in the treatment of various ailments. In our study, we have compared our data with others published works on Indian Medicinal plants. The study recommends further investigations on chemical and pharmacological actions to validate the claims. Generally, the people of the study area still have a strong belief in the efficacy and success of herbal medicine. The results of the present study reveal that medicinal plants continue to play an important role in the primary healthcare system of this tribal community.

References

Champion, H.G. and Seth, S.K. 1968. A revised survey of forest types of India. Delhi: Manager of Publications.

Handa, S.S. and Kaul, M.K. 2001. *Supplement to Cultivation and Utilization of Medicinal Plants.* Regional Research Laboratory, Jammu - Tawi, India.

Parrotta, J.A. 2001. *Healing Plants of Peninsular India.* CABI Publishers, New York, USA.

Pullaiah, T. 1998. Medicinal Plants of Andhra Pradesh.

Rama Rao, N. and Henry, A.N. 1996. The Ethnobotany of Eastern Ghats in Andhra Pradesh, India. Botanical Survey of India, Calcutta.

Shankar Gopal Joshi, 2000. *Medicinal Plants.* Oxford and IBH Publishing Co.Pvt.Ltd., New Delhi, India.

Trivedi, P.C. 2007. *Medicinal plants Traditional Knowledge.* I.K. International Publishing House Pvt. Ltd.

□□□

10

Pharmaceutical Properties of L-Dopa and its Therapeutic Application

M. Pugalenthi, V. Vadivel and T. Parimelazhagan

CONTENTS

1.1	Introduction	136
1.2	L-Dopa: The Non-protein Phenolic Amino Acid	138
1.3	Pharmaceutical Properties of L-Dopa	141
1.4	Conclusion	149
	References	149

1.1 Introduction

One of the serious problems facing most of the developing countries is the scarcity of good quality protein for the teeming human population and dwindling livestock industry. Although the common legume seeds have been playing a key role as an economic protein source throughout the world, their production is not sufficient to meet the increasing protein requirements (Janardhanan *et al.*, 2003a). Further, stiff competition exists between the increasing human population and expanding livestock industries for the consumption of legume seeds as a protein source, leading to an inadequate supply of protein, which ultimately results in Protein-Energy-Malnutrition (PEM) in many developing countries, including India (Pugalenthi *et al.*, 2005).

Hence, at present there is an increasing interest among the researchers to explore other alternative, additional or novel protein foods for the world of tomorrow. In this context, recent research trends have been directed to identify and evaluate the under-utilized legume seeds, whose potential remains untapped to the extent that they deserve, in view of the pressure on conventional legumes which arises due to increasing multiple uses (Vadivel and Janardhanan, 2005).

Among the various under-utilized legumes, velvet bean [*Mucuna pruriens* var. *utilis* (Wall ex Wight) Baker ex Burck], a tropical wild legume having good nutritional properties, merits a wider use as a food/fodder legume in South Asian countries (Pugalenthi *et al.*, 2005). Velvet bean belongs to the family Fabaceae and is widely distributed in Southern and Southeastern Asian regions and in some parts it is cultivated as a green manure and cover crop (Buckles, 1995). This little-known legume has a wider adaptation and exhibits more tolerance to many adverse environmental conditions such as drought, low soil fertility, high soil acidity and wide range of climatic conditions (Pugalenthi and Vadivel, 2007a).

The velvet bean seeds have been reported to contain higher level of protein (26-30%), starch (34-40%), desirable amino acid, fatty acid and mineral composition with good nutritional properties (Vijayakumari *et al.*, 2002; Vadivel and Janardhanan, 2005; Siddhuraju and Becker, 2005; Pugalenthi *et al.*, 2005; Vadivel and Pugalenthi, 2008a; 2008b; 2008c) and their nutritional quality is comparable with that of soybean and other common legumes in relative nutrient profiles (Pugalenthi *et al.*, 2005). Velvet bean seeds have been traditionally used as food by peoples living in rural areas of India, Sri Lanka, Philippines, Nigeria, Ghana, Brazil, Malawi and other countries. In India, certain ethnic groups, particularly, the Northeastern tribes and Kanikkar tribes in Kerala State and Dravidian tribals in Tamil Nadu State have been consuming this legume grains, after repeated boiling, either as such or mixed with other cereals (Janardhanan *et al.*, 2003b; Pugalenthi *et al.*, 2005).

The seeds of velvet bean have been used in indigenous ayurvedic medicine as a uterine stimulant (Jayaweera, 1981; Lorenzetti *et al.*, 1998), powerful aphrodisiac, emetic (Amin *et al.*, 1996; Tripathi and Upadhyay, 2001) and used to treat nervous disorders and arthritis and also have been used against snakebite in traditional medicine in India and West Africa (Houghton and Skari, 1994; Aguiyi *et al.*, 1999; Guerranti *et al.*, 2004). Gastroprotective actions of L-Dopa was studied by Mercer *et al.* (1998). The effect of L-Dopa on indomethacin-induced intestinal ulceration in rat was studied by Kiro *et al.* (2002) and demonstrated that L-Dopa has a protective effect on indomethacin-induced small bowel injury in rat. The hypoglycemic activity of velvet bean with relation to anti-diabetic potential was demonstrated by Grover *et al.* (2002). A mild antimicrobial activity of L-Dopa was reported by Zhao *et al.* (2007). L-Dopa is used to provide symptomatic relief in Parkinson's disease (Siddhuraju *et al.*, 2000; Pugalenthi *et al.*, 2005).

The available information on L-Dopa often shows confusing and conflicting data concerning its medicinal properties and side effects. There are different opinions among the researchers with respect to the toxicity of this amino acid, because, although the effectiveness of L-Dopa in the treatment of Parkinson's disease- a progressing disabling disorder associated with a deficiency of dopamine in the brain- over the chemically synthesized L-Dopa was proved (Hussain and Manyam, 1997), the treated patients often reported to be severely affected by some serious side effects such as hallucinations (Infante *et al.*, 1990) in addition to gastrointestinal disturbances such as nausea, vomiting and anorexia (Reynolds, 1989) and it has also been proved to induce favism in individuals with glucose-6-phosphate dehydrogenase deficiency in their erythrocytes (Nachama and Edward, 1967).

Thus, the lack of consensus in the scientific literature on L-Dopa is scattered and conflicting information are available on the pharmacological properties of L-Dopa. Additionally, in the available literature there is a apparent disagreement with regarding to L-Dopa content that varies depending on geographic origin (Laurent *et al.*, 2002; Janardhanan *et al.*, 2003b), the particular species and part of the plant examined (Prakash and Tewari, 1999; Szabo and Tebbett, 2002), the time of harvest (Prakash and Tewari, 1999) and climatic considerations (Lorenzetti *et al.*, 1998), all of which added further confusion. Moreover, to our knowledge, there is only meager information is available on the antioxidant and hypoglycemic properties of L-Dopa. Even though many researchers working with L-Dopa throughout the world in both a nutritional and medicinal context, there is no scientific summary of this information till date. Hence, the present review emphasizes and evaluates the information available concerning the pharmacology of L-Dopa.

1.2 L-Dopa: The Non-Protein Phenolic Amino Acid

The non-protein amino acids are not the normal components of proteins but occur in a free state in many plants, particularly in the Leguminosae. This kinds of amino acids are usually concentrated in seeds but can also be found in the whole plant. Some examples of such non-protein amino acids are L-canavanine, indospicine, homoarginine, mimosine, β-cyanoalanine, β-N-oxalyl-α-β-diaminopropionic acid, β,γ-diaminobutyric acid, djenkolic acid, among others (Duffus and Duffus, 1991).

L-Dopa was one such non-protein amino acid, first isolated over 90 years ago from the fruits of *Vicia faba* (Brain, 1976). Since then its occurrence has been noted in a number of other plants, including cereals like wheat and oats, although it is confined to leguminous species (Barbeau and McDowell, 1970; Duffus and Duffus, 1991). Its presence in plants remained of only academic significance until the discovery that the oral administration of L-Dopa relieves Parkinson's disease. The widespread application of this therapy created a demand for large quantities of L-Dopa at an economic price level and this has led to the reinvestigation of various plant resources.

Among the various plants, the seeds of the genus *Mucuna* have been found to contain the highest levels of L-Dopa, but even the levels found in this species necessitate the processing of large quantities of material, a difficulty, which is accentuated by the high dosage rate of the drug. The relatively large and consistent L-Dopa concentration in *Mucuna* beans has been well verified. Daxenbichler *et al.* (1971, 1972) surveyed a total of 1062 plant species in 160 families and determined that the L-Dopa content was sufficient only in the velvet bean seeds, merits for extraction of this compound in commercial scale for medicinal purposes. In this survey, they found that the L-Dopa concentration of velvet bean seeds varied between 3.1 and 6.7%. Similarly, Bell and Janzen (1971) measured L-Dopa content in six *Mucuna* accessions collected from different locations and found that the L-Dopa concentration ranged between 5.9 and 9.0%.

Table 1 : L- Dopa content in velvet beans and other leguminous seeds.

S. No.	Name of the legume seed	L- Dopa content (g/100g seed flour)	Reference
1	*Mucuna pruriens* var. *utilis*	3.62- 6.98	Vadivel and Janardhanan (2005) Siddhuraju and Becker (2001a; 2001b; 2001c; 2003; 2005) Siddhuraju *et al.* (2000) Vadivel and Janardhanan (2000) Diallo *et al.* (2002) Szabo and Tebbett (2002) Myhrman (2002) Vadivel and Pugalenthi (2006; 2008a; 2008b)
2	*Mucuna pruriens* var. *pruriens*	4.08- 9.1	Vadivel and Janardhanan (2005) Adebowali *et al.* (2005) Mary Josephine and Janardhanan (1992) Del Carmen *et al.* (1999) Carew *et al.* (2003)
3	*Mucuna monosperma*	4.52- 4.56	Arulmozhi and Janardhanan (1992)
4	*Mucuna atropurpurea*	2.6	Mohan and Janardhanan (1991)
5	*Mucuna gigantean*	1.5	Rajaram and Janardhanan (1991)
6	*Mucuna cochinchinensis*	6.1- 6.2	Diallo *et al.* (2000) Adebowali *et al.* (2005) Myhrman (2002)
7	*Mucuna rajada*	5.35	Adebowali *et al.* (2005)
8	*Mucuna veracruz*	6.35- 7.12	Adebowali *et al.* (2005)
9	*Mucuna deeringeana*	3.87	Adebowali *et al.* (2005)
10	*Canavalia ensiformis*	2.64	Vadivel and Janardhanan (2005)
11	*Canavalia gladiata*	2.83	Vadivel and Janardhanan (2005)
12	*Cassia floribunda*	1.57	Vadivel and Janardhanan (2005)
13	*Cassia obtusifolia*	1.34	Vadivel and Janardhanan (2005)

Several studies have focused on L-Dopa quantification in plant parts of various *Mucuna* species. In India, the L-Dopa content of *Mucuna* species have reported as much as 9% in some *Mucuna* seeds (Janardhanan and Lakshmanan, 1985; Mary Josephine and Janardhanan, 1992; Mohan and Janardhanan, 1995), and as little as 1.5% in the seed of *M. gigantea,* which grows wildly in Southern India (Rajaram and Janardhanan, 1991) (Table- 1). The L-Dopa content of velvet bean seeds collected from five different agro-ecological regions of South India was investigated to evaluate the effect of agroclimatic conditions of different growing locations on the L-Dopa concentration in the velvet bean seeds by Janardhanan *et al.* (2003b) and their findings revealed that the existence of diversity in the L-Dopa content (5.60 to 6.56%) among the different accessions.

Apart from the seed materials, various plant parts of *Mucuna* have also investigated for L-Dopa content, which revealed that the L-Dopa was largely accumulated in the seeds. Prakash and Tewari (1999) measured L-Dopa concentrations in velvet bean seeds and various parts of the plant collected from different agroclimatic zones of India, which ranged between 0.17-0.35% in leaves, 0.19-0.31% in stems, and 0.12-0.16% in roots. In another study, L-Dopa was present most notably only in the seeds (4.93-5.39%) with the stems and roots containing the next highest levels (0.49 and 0.27%, respectively). Lower than roots, no difference in concentration was found among pods and leaves (0.15%) of velvet beans growing in Benin and Gainesville samples (<0.3% in stems, <0.2% in leaves) (Szabo and Tebbett, 2002).

Evidences suggested that the existence of high variability in L-Dopa concentration among various accessions of velvet beans collected from different places. The L-Dopa content in velvet bean seeds ranged from 2.2 to 7.2% in a survey of 36 accessions collected from India, Benin, Honduras, Mexico and Georgia (Lorenzetti *et al.*, 1998). On the basis of their results, they reported that the *Mucuna* plants growing within 6° 40' of the equator contained significantly higher levels of L-Dopa than the plants growing far away from the equator regions (Table 2).

Laurent *et al.* (2002) also examined the L-Dopa concentration in the seed materials of 38 accessions of *M. pruriens* and two accessions of *M. brachycarpa* using High Performance Liquid Chromatography (HPLC). The stability of velvet beans across all the environments indicated that all the accessions may be widely adapted for L-Dopa production. Since many species of *Mucuna* have been shown to produce L-Dopa, its presence and quantification could eventually be used as an interesting systematic biochemical marker.

On the other hand, many studies reported the levels of apparent genetic variability in L-Dopa content explains the genotypic diversity leads to variations in L-Dopa production in velvet beans (Pugalenthi and Vadivel, 2007b). It must be noted that although the studies appear to be heterogeneous in terms of methodologies used, description of materials tested, plant parts examined, age/stage of the tissues, and the geographic location from which the seeds were obtained, the final point is very important, because while some studies have clearly showed an interaction of L-Dopa level in the velvet bean seed with the environmental conditions, under which *Mucuna* was grown.

Table 2 : Variability in the L- Dopa content of velvet beans growing at different places.

S. No.	Name of the location	Latitude	L- Dopa content (g/100 g seed flour)
1	Hawaii	20°03′	2.26
2	Honduras	15°07′	4.34
3	Benin	6°40′	7.64
4	Florida	26°39′	3.39
5	Israel	32°08′	5.65
6	Colombia	3°40′	4.52
7	Georgia	33°25′	4.71

Source: Laurent *et al.* (2002)

1.3 Pharmaceutical Properties of L-Dopa

Despite certain negative attributes of the non-protein amino acids, there is increasing evidence that their toxic properties may be harnessed for therapeutic and other economic purposes. Thus, mimosine is endowed with anti-tumour activity and the use of BAPN (β-aminopropionitrile) in reversing the abnormalities of collagen metabolism, which occur in diabatic subjects. Among the various non-protein amino acids, the mimosine and L-Dopa are aromatic amino acids and are considered as an analogue of tyrosine and its neurotransmitter derivatives, dopamine and norepinephrine found in the brain. The L-Dopa can also act as a precursor of neurotransmitter dopamine and is used in the treatment of Parkinson's disease (Barbeau and McDowell, 1970). Different preparations of velvet bean seeds containing L-Dopa are used for the management of several free radical mediated diseases such as rheumatoid arthritis, diabetes, atherosclerosis, male infertility and nervous disorders in the ayurvedic system of medicine (Siddhuraju and Becker, 2003).

In a screening survey of more than 1000 species in 135 plant families, *Mucuna* was the only species with sufficient L-Dopa to suggest a possible use for commercial production (Daxenbichler *et al.*, 1971) and used to alleviate symptoms of Parkinson's disease. However, their commercial exploitation is hampered by the herbaceous twining habit of these plants, which creates some difficulties in large-scale cultivation. Hence, efforts were taken regarding the production of L-Dopa from plant sources, which are undesirable in view of the amounts of drug required today. Moreover, Brain (1976) observed that due to the requirement of large quantities of raw materials for drug production, a problem exacerbated by the high dosage required in the treatment and by the difficulty of cultivating the twining plants, research trend had emerged to investigate synthetic and microbial transformation methods of production (Pugalenthi and Vadivel, 2007b).

However, all these synthetic procedures generally produce a racemic mixture, which must be resolved, whilst the microbial methods are generally restricted to conversion of close precursor of L-Dopa. Hence, the production of L-Dopa through tissue culture of *Mucuna pruriens* was investigated by Brain (1976). The yield of L-Dopa was surprisingly higher and it was interesting to note that the maximum growth and L-Dopa accumulation occurred at 25 mg/L 2,4-Dichlorophenoxy acetic acid, a level toxic to many other cell lines. Also of particular interest was the observation that in old cultures the majority of the L-Dopa was found in the medium. In terms of practical exploitation on a large-scale, extraction of L-Dopa from the medium could be carried out relatively simply by an ion-exchange procedure. The presence of L-Dopa at high levels in the medium is presumably due to active excretion from the cells, which maintain a low internal level so that feed back control of the synthesizing enzymes does not occur. Alternatively the high level could be due to the release of material that cause cell lysis in old cultures (Brain, 1976).

The large L-Dopa content in *Mucuna* has given rise to a number of medicinal uses throughout the history. Chief among them has been Parkinson's disease that apparently was already been treated with *Mucuna* in the ancient Indian medical system *Ayurveda* (Manyam, 1995); in fact, out of the 35 formulations for Parkinson's disease reported in *Ayurveda*, at least 18 contained *Mucuna* seeds as a major ingredient (Manyam and Sánchez-Ramos, 1999). Later, this ancient use was adopted at global level as a consequence of three findings: (1) Isolation of L-Dopa from *Vicia faba* over 90 years ago; (2) Isolation of L-Dopa from *Mucuna*, and (3) Invention of L-Dopa provides symptomatic relief for Parkinson's disease by decreasing the dopamine deficiency of certain areas of the brain, and the normalization of such treatment during the 1960's (Yahr, 1973).

Though Manyam (1995) stated that L-Dopa was first isolated from *Mucuna* by Indian scientists, in fact, Alabama scientist, Miller (1920) has also reported the same amino acid, dihydroxyphenylalanine, which had been previously identified in *Vicia faba*. Due to the resulting demand for large quantities of L-Dopa at an economic price, reinvestigation of plant sources for L-Dopa was initiated (Brain, 1976), including extensive plant surveys and the development of extraction methods (Daxenbichler *et al.*, 1972). Even, certain viable methods for the extraction of L-Dopa from *Mucuna* were patented (Wysong, 1966).

Although L-Dopa medicines are synthetically manufactured today, *Mucuna* as a source of L-Dopa continues to receive attention from two sources: from those interested in developing low-cost medical treatments for Parkinson's disease and from those interested in natural and alternative medicines (Hussain and Manyam, 1997; Manyam, 1995). Clinical trials and other research works have been conducted on the effect of L-Dopa extracted from *Mucuna* on Parkinson's disease suggested a higher effectiveness of *Mucuna* powder over synthetic L-Dopa (Hussain and Manyam, 1997). Now-a-days, *Mucuna*-derived L-Dopa products, such as *Kapi Kacchu*, are available through the Internet for various medicinal purposes.

In addition to relieving symptoms of Parkinson's disease, L-Dopa has been used for various other medicinal purposes. It has been used for treating snakebites in India (Siddhuraju *et al.*, 1996) and used against snakebites in West Africa (Houghton and Skari, 1994). It has also been used for increasing sexual potency (Siddhuraju *et al.*, 1996), it is used for deworming purposes (Hanum and Maesen, 1996) and as an expectorant to treat cough, asthma and tongue infection (Prakash and Misra, 1987). Even criminal purposes have been reported in the use of L-Dopa along with the pod hairs: they have been secretly fed to persons whose death has been desired. *Mucuna* containing L-Dopa has also been used as a uterine stimulant and aphrodisiac (Lorenzetti *et al.*, 1998). Interestingly, Ajiwe *et al.* (1997) reported that in Nigeria *M. sloanei* has been cooked for pregnant women to avoid miscarriage. Amin *et al.* (1996) also proved that the *Mucuna* seeds increase the sexual potential in rats. However, the role of L-Dopa in velvet beans in increasing the sexual potency is not yet revealed and remains to be investigated.

The mechanism of pharmacological treatment of Parkinson's disease during L-Dopa administration was described by Szabo and Tebbett (2002). Approximately 33% of an orally administered dose of L-Dopa will be absorbed from the gastrointestinal tract, primarily from the jejunum in the small intestine. Peak plasma concentrations will occur within 1-3 h with levels that may vary as much as ten-fold among the individuals. A majority of the successfully absorbed L-Dopa will be converted into dopamine in the periphery (the body excluding the brain) via action of the enzyme L-aromatic amino acid decarboxylase (LAAD).

Less than 1% of the administered dose will actually enter into the brain, where it will be converted to dopamine in the basal ganglia. In addition to dopamine, peripheral L-Dopa will also be metabolized to melanin, norepinephrine, 3-methoxytyramine, methyldopa, 3,4-dihydroxyphenylacetic acid, and 3-methoxy-4-hydroxyphenylacetic acid (homovanillic acid). These metabolites are rapidly excreted in the urine with about 80% of an administered dose being excreted within 24 h and less than 1% of which will be unchanged (Moffat, 1986; Dollery, 1999).

Currently in the treatment of Parkinson's disease, L-Dopa is the almost always prescribed drug regimen with a peripheral decarboxylase inhibitor (usually carbidopa or benserazide) to prevent the loss of L-Dopa by the LAAD metabolism that begins in the intestinal mucosa. The presence of an inhibitor increases the amount of L-Dopa available to the brain by 75-80% (Jankovic and Caine, 1987), thereby decreasing the dosage necessary to achieve a therapeutic effect. This, in turn, resulted in a corresponding decrease in dose-dependent peripheral side effects.

Treatment levels for Parkinsonism typically ranged from starting doses of 0.25 g to maintenance doses of 1-3 g per day when an inhibitor is used upto 8 g of L-Dopa per day in the absence of an inhibitor (Moffat, 1986; Spratto and Woods, 1996; Dollery, 1999), a situation occasionally caused by the lack of availability or by intolerance for the inhibitor on the part of the patient. When treatment is first initiated, administration typically follows a gradual increase over a period of weeks to decrease the likelihood of side effects while the body adjusts to higher levels of dopamine in the periphery. Thus, a starting dose of 0.25 g per day would be increased by 0.5 g every 7-10 days until the desired level was reached.

Since L-Dopa transport across the intestinal membrane is facilitated, the bioavailability of L-Dopa is highly dependent on competitive factors such as the presence of other amino acids (especially phenylalanine and leucine) that utilized the same carrier and the gastric emptying rate (longer stay, lower availability due to metabolism in the mucosal lining), which is affected by pH and the presence of protein (Standaert and Young, 1996; Dollery, 1999). Thus, nutritional status and metabolic characteristics of an individual were influenced the absorption and conversion rates of L-Dopa, which assists in explaining the wide disparities observed in peak plasma levels (up to ten-fold).

It is also interesting to note that the treatment regimen in cases of overdose on L-Dopa consists of the administration of vitamin B_6 (pyridoxine) as a means of increasing peripheral decarboxylation to quickly lower the drug level and resolve the overdose (Dollery, 1999). This approach is not effective when vitamin B_6 is given in the presence of an LAAD inhibitor (Klawans *et al.*, 1971); in this case, general supportive measures are taken to stabilize the patient during the course of the overdose with arrhythmia, especially, being guarded against Parkinsonism.

Patients taking L-Dopa for Parkinson's are counseled against eating large quantities of protein and vitamins or consuming foods prepared with vitamin

B_6, as the protein will limit absorption of L-Dopa and a single vitamin B_6 dose of 10-25 mg can effectively cancel the therapeutic effect of L-Dopa (Spratto and Woods, 1996). Studies with the safe dosage level of L-Dopa in human beings during treatment for Parkinson's disease indicated that prior to the use of LAAD inhibitors the maximum human exposure during supervised treatment period was 8 g of L-Dopa/day for an average body weight of 70 kg (or 0.1 g L-Dopa/kg/day). This level was known to have fairly severe side effects; current maximum dosage levels with the inhibitors have decreased to 2-3 g/day (0.03-0.04 g L-Dopa/kg/day).

L-Dopa itself has very little pharmacological effect, since it is rapidly converted to dopamine; thus, ingestion of L-Dopa results in significantly elevated levels of systemic dopamine. Most side effects arise directly from dopamine's activity as a neurotransmitter involved in the regulation of the heart, vascular system, digestive tract and excretory system, rather than from its well-known effect on receptors in the brain (Standaert and Young, 1996; Dollery, 1999; Metman and Mouradian, 1999).

Some consequences of increased peripheral dopamine including orthostatic hypotension resulting in dizziness and in some cases staggering and increased heart rate. Although tolerance have developed in individuals ingesting L-Dopa over prolonged periods of a few months, the heart-related effects could pose a serious dilemma to those have a predisposing cardiac condition. Nausea, vomiting and anorexia are also common side effects of L-Dopa and are associated with the effects of peripherally produced dopamine on the dopamine receptors in the area of postrene of the brain stem. Nausea and vomiting may occur on consumption of as little as 250 mg of L-Dopa, if the patient is unaccustomed to such an exposure.

Numerous psychiatric disturbances have also been reported in a significant number of patients receiving high doses of L-Dopa, in a dose-dependent manner. These are typically mild in nature and include nervousness, anxiety, agitation, insomnia and vivid dreams. About 15% of Parkinson's patients have experienced serious psychiatric manifestations such as confusion, delirium, depression as well as psychotic reactions with hallucinations have been reported in extreme cases as a result of L-Dopa therapy. L-Dopa should be avoided by patients with a history of glaucoma, asthma, psychosis, peptic ulcers, epilepsy, diabetes and endocrine, renal, hepatic, cardiovascular or pulmonary disease (Dollery, 1999; Spratto and Woods, 1996).

Since it is a precursor of melanin, it may promote the formation of skin cancers and should therefore be avoided by patients with pigmented lesions or pre-existing melanomas (Spratto and Woods, 1996; Pfutzner and Przybilla, 1997; Dollery, 1999; Letellier *et al.*, 1999). L-Dopa should be avoided during pregnancy and by nursing mothers, since the drug can appear in the milk. It is worth noting here that the anti-Parkinsonian effects of L-Dopa on movement and muscle tone have not appeared in healthy individuals.

The main drug interactions of this compound were associated with cardiovascular and antipsychiatric drugs (i.e., clonidine, digoxin and tricyclic antidepressants). L-Dopa ingestion should be avoided by patients receiving these medications (Standaert and Young, 1996). With these limitations in mind, the inadvertent ingestion of L-Dopa as a result of consuming *Mucuna* that has not been adequately prepared could have serious consequences for individuals with any of the pre-existing conditions mentioned above. One incidence of *Mucuna*-related poisoning has been reported by Infante *et al.* (1990), in which over two hundred cases of acute psychosis were diagnosed in Mozambique. Due to drought conditions, villagers who were prepared *Mucuna* seeds (detoxified by repeated boiling combined with frequent water exchanges) drank rather than discarded the water, in which the seed had been boiled. The most common symptoms observed in them includes severe head ache, palpitations, confusion, agitation, hallucinations and paranoid delusions, all of which could be attributed to L-Dopa exposure.

Even though, L-Dopa, a compound chiefly used in the treatment of Parkinson's disease, has been reported to have the serious side effects of causing hallucination (Infante *et al.*, 1990) in addition to causing gastrointestinal disturbances such as nausea, vomiting and anorexia in treated patients (Pugalenthi and Vadivel, 2007b). It has also been shown to be toxic in individuals with glucose-6-phosphate dehydrogenase deficiency in their erythrocytes to induce favism (Nechama and Edward, 1967). It also affects the neurological system including aggression, paranoid delusions, delirium, severe depression, and unmasking of dementia (Reynolds, 1989).

In human beings, it has been found that the chronic L-Dopa administration could reduce pyridoxine deficiency, particularly, in persons who have a higher requirement for this vitamin (Friedman, 1970). Studies conducted by Ordonez and Wurtman (1974) in rat model demonstrated that folate deficient animals show a significant decrease in brain S-adenosylmethionine when treated with L-Dopa. Some authors also suggested that during chronic L-Dopa treatment of patients, excessive utilization of the *de novo* pathway for methyl group synthesis could increase nutritional requirements for folic acid and vitamin B_{12}. The inhibitory effect of higher protein intake on the therapeutic efficacy of L-Dopa has been attributed to competitive absorption of dietary amino acids and L-Dopa in the gastrointestinal tract (Cotzias, 1968). On the other hand, L-Dopa induces the reduction of NADPH and Glutathione (GSH) where the glucose-6-phosphate dehydrogenase enzyme is deficient in blood cells and ultimately it causes favism (haemolytic anemia) (Liener, 1994).

Lorenzetti *et al.* (1998), based on European clinical recommendations regarding the use of L-Dopa in the treatment of Parkinson's disease (OVP, 1995) as well as on observations following the ingestion of *Mucuna* preparations over a study period of 12 weeks (Manyam, 1995) estimated that 1500 mg of L-Dopa was the maximum daily dose that could be tolerate by an individual without experiencing any side effects. Typically, this level would be exceeded with less than 100 g of unprocessed velvet bean (2-6% L-Dopa). This study used to set the L-Dopa guideline lasted only 12 weeks and severely limits extrapolation of the conclusions with regarding to long-term health effects for a healthy adult consuming a diet includes 1500 mg of L-Dopa per day. The considerations of children, pregnant women and of those with various common ailments also were not addressed.

While it is doubtful that *Mucuna* will find wide acceptance as a source of L-Dopa, since the drug is readily produced by synthetic means, it may prove to be an affordable alternative for Parkinson's patients living in developing countries (Kempster and Wahlqvist, 1994; Manyam and Sánchez-Ramos, 1999). Manyam (1995) observed only mild effects in sixty Parkinson's patients treated for twelve weeks with a mean daily dose of *Mucuna* containing 1500 mg of L-Dopa. However, there has heretofore been no reason for anyone to study the effects of regular, long-term consumption of L-Dopa by persons not afflicted by Parkinson's disease. It is therefore not surprising that extensive literature searches have uncovered the definitive reports with respect to the health implications of chronic consumption of L-Dopa by persons not so afflicted. In addition, people with certain medical conditions can respond more adversely than the general population for the consumption of small quantities of L-Dopa (Szabo and Tebbet, 2002).

Currently, products containing L-Dopa are being sold through the Internet for a number of purposes including increased muscle mass and strength, male vitality, enhancement of mental alertness and coordination, as well as for an alleged anti-aging effect. But no reliable evidences are available to confirm these benefits of L-Dopa. However, on the other hand, the antioxidant property of L-Dopa was proved by Siddhuraju and Becker (2003).

The polyphenolics constituents of various legume seeds have been reported to have potential medicinal properties, including antioxidant activity. Effective hydroxyl scavenging ability of a non-protein amino acid β-N-Oxalyl-L-α,β-diaminopropionic acid (ODPA) in *Lathyrus sativus* was reported by Gongke *et al.* (2001). When considering the radical scavenging effect on DPPH (α,α-diphenylamine-β-picrylhydrazyl) radicals, L-Dopa was found to be more potent than all other compounds of *M. pruriens* tested including the reference standards. Siddhuraju and Becker (2003) also studied the non-enzymatic generation of superoxide anion radical scavenging activity of L-Dopa by

Nitrobluetetrazolium (NBT) reduction assay. Superoxide is a biologically important substance, which can be decomposed to form stronger oxidative species such as singlet oxygen and hydroxyl radicals (Korycka-Dahl and Richardson, 1978). The highly reactive hydroxyl radicals can cause oxidative damage to DNA, lipids and proteins (Spencer *et al.*, 1994). The results of Siddhuraju and Becker (2003) revealed that the L-Dopa inhibit superoxide radicals in a dose-dependant manner.

Spencer *et al.* (1996) reported the pro-oxidant and antioxidant actions of L-Dopa and its metabolites such as dopamine, melanine, norepinephrine, 3-methoxytyramine, 3-O-methyldopa, 3,4-dihydroxyphenyl acetic acid and 3-methoxy-4-hydroxy phenyl acetic acid were promotes oxidative DNA damage and could also be harmful to tissues damaged by neurodegenerative disease, namely Parkinsonism. Moreover, a study through *in vitro* models revealed that L-Dopa significantly increases the levels of oxidized glutathione in rat brain striatal synaptosomes (Spina and Cohen, 1988). Such depleted glutathione could be due to the generation of reactive semiquinones from L-Dopa. The reducing power of L-Dopa might be due to their di- and mono-hydryl substitutions in the aromatic ring, which possess potential hydrogen-donating abilities. Thus, this non-protein amino acid might contain higher amounts of reducing reductones and could react with free radicals to stabilize and terminate radical chain reactions.

Table 3 : Antioxidant activity of L-Dopa compound extracted from the seed materials of velvet bean

Antioxidant properties	L-Dopa in velvet bean	
	White seeds	Black seeds
L-Dopa content (g/100 g seed flour)	$5330^{a} \pm 0.07$	$4020^{b} \pm 0.42$
Free radical inhibition activity (%)	$93.56^{b} \pm 0.13$	$86.32^{b} \pm 0.15$
Antioxidant activity (%)	$48.16^{a} \pm 0.14$	$48.75^{a} \pm 0.15$
Oxidation Rate Ratio	$0.92^{a} \pm 0.14$	$0.87^{a} \pm 0.16$
Antioxidant Activity Coefficient	$269.85^{a} \pm 0.21$	$242.07^{a} \pm 0.14$
α-amylase inhibition activity (%)	$28.48^{a} \pm 0.08$	$22.82^{a} \pm 0.08$
α-glucosidase inhibition activity (%)	$65.76^{a} \pm 0.21$	$58.66^{a} \pm 0.12$

[1]Values are mean and ± standard deviation.
[2]Values in the same row with different roman superscript are significantly different ($p<0.05$).

In our present investigation also we have found excellent antioxidant activity and free radical scavenging property of L-Dopa compound in addition to potential hypoglycemic effect (Table 3). Spencer *et al.* (1994) reported that L-Dopa, which is rich in the nervous system, and its metabolites dopamine and 3-O-methyl-dopa was found to cause extensive oxidative DNA damage in the presence of H_2O_2 and traces of copper ions via promotion of lipid peroxidation and highly reactive hydroxyl radicals. Nonetheless, chronic treatment with L-Dopa was reported to decrease lipid peroxidation in the cerebral cortex of healthy mice, but to worsen peroxidation in mice with neuronal injury and thus L-Dopa appears to have a mixture of pro- and antioxidant effects *in vivo,* which means that it could be harmful to tissues damaged by neurodegenerative diseases.

1.4 Conclusion

The available reports showed that the L-Dopa is present at high concentration in the seeds of *Mucuna pruriens* var. *utilis* (velvet bean) and many factors were found to influence their levels. The L-Dopa compound was found to possess many medicinal properties including antioxidant potential. When considering the pharmaceutical properties of this compound, although it plays an important role in treating Parkinson's disease, often some treated patients were reported to receive many side effects. Hence, extensive long-term clinical trials are required to determine the safe dosage concentration of L-Dopa used for the Parkinson's therapy. Due to heavy demand for this compound for various medicinal purposes, certain novel biotechnological approaches are required to improve the L-Dopa content in the velvet bean seeds and new strategies should also be developed for the production of this pharmaceutically active factor through viable *in vitro* techniques.

References

Aguiyi, J.C., Igweh, A.C., Egesie, U.G. and Leoncini, R. 1999. Studies on possible protection against snake venom using *Mucuna pruriens* protein immunization. *Fitoterapia,* 70: 21-24.

Ajiwe, V.I.E., Okeke, C.A., Nnabuike, B., Ounleye, G.A. and Emeka, E. 1997. Applications of oils extracted from African start apple (*Chrysophyllum africanum*), horse eye bean (*Mucuna solanei*) and African pear (*Dacryodes edulis*) seeds. *Bioresource Technology,* 59: 259-261

Amin, K.M.Y., Khan, M.N., Zillur-Rehman, S. and Khan, N.A. 1996. Sexual function improving effect of *Mucuna pruriens* in sexually normal male rats. *Fitoterapia,* LXVII: 53-56.

Barbeau, A. and McDowell, F.H. 1970. L-Dopa and Parkinsonism. F.A. Davis Company, Philadelphia, PA, USA. p 403

Bell, E.A. and Janzen, D.H. 1971. Medical and ecological considerations of L- Dopa and 5-HTP in seeds. *Nature,* 299: 136-137.

Brain, K.R. 1976. Accumulation of L- Dopa in cultures from *Mucuna pruriens. Plant Science Letters,* 7: 157-161.

Buckles, D. 1995. Velvet bean: A "new" plant with a history. *Economic Botany*, 49: 13-25.

Cotzias, G.C. 1968. Parkinsonism and Dopa. *Trans Association of American Physiologists*, 81: 171-183

Daxenbichler, M.E., VanEtten, C.H., Earle, F.R. and Tallent, W.H. 1972. L-Dopa Recovery from *Mucuna* seed. *Journal of Agriculture and Food Chemistry*, 20: 1046-1048.

Daxenbichler, M.E., VanEtten, C.H., Hallinan, E.A., Earle, F.R. and Barclay, A.S. 1971. Seeds as sources of L-Dopa. *Journal of Medicinal Chemistry*, 14: 463-465.

Dollery, C. 1999. Therapeutic drugs, 2 Ed. Churchill Livingstone: New York, pp. 39-43.

Duffus, C.M. and Duffus, J.S. 1991. Non-protein toxic amino acids. In: Toxic substances in crop plants, D'Mello, F.J., Duffus, C.M. and Duffus, J.H. (Eds.), Thomas Graham House, Science Park, Cambridge: The Royal Society of Chemistry. pp. 1-21.

Friedman, S.A. 1970. Levodopa and pyridoxine deficient states. *Journal of American Medical Association*, 214: 1563-1570.

Gongke, Z., Yingzhen, K., Kairong, C., Zhixiao, L. and Yafu, W. 2001. Hydroxyl radical scavenging activity of β-N-oxalyl-L-α,β-diaminopropionic acid. *Phytochemistry*, 58: 759-762.

Grover, J.K., Yadav, S. and Vats, V. 2002. Medicinal plants of India with anti-diabetic potential. *Journal of Ethnopharmacology*, 81: 81-100.

Guerranti, R., Aguiyi, J.C., Ogueli, I.G., Onorati, G., Neri, S., Rosati, F., Buono, F.D., Lampariello, R., Pagani, R. and Marinello, E. 2004. Protection of *Mucuna* pruriens seeds against *Echis carinatus* venom is exerted through a multiform glycoprotein whose oligosaccharide chains are functional in this role. *Biochemistry and Biophysics Research Communications*, 323: 484-490.

Hanum, F.I. and Maesen, L.J.G. 1996. In: Prosea handbooks No. 11, Auxilliary plants. Prosea/Bogor, Wagening, Netherlands.

Houghton, P.J. and Skari, K.P. 1994. The effect on blood clotting of some West African plants used against snakebite. *Journal of Ethnopharmacology*, 44: 99-108.

Hussain, G. and Manyam, B.V. 1997. *Mucuna pruriens* proves more effective than L-Dopa in Parkinson's Disease animal model. *Phytotherapia Research*, 11: 419-423.

Infante, M.E., Perez, A.M., Simao, M.R., Manda, F., Baquete, E.F. and Fernandez, A.M. 1990. Outbreak of acute psychosis attributed to *Mucuna pruriens*. *Lancet*, 336: 1129-1131.

Janardhanan, K. and Lakshmanan, K.K. 1985. Studies on the pulse, *Mucuna utilis*: Chemical composition and antinutritional factors. *Journal of Food Science and Technology*, 22: 369-371.

Janardhanan, K., Gurumoorthi, P. and Pugalenthi, M. 2003b. Nutritional potential of five accessions of a South Indian tribal pulse, *Mucuna pruriens* var. *utilis* Part. I The effect of processing methods on the contents of L- Dopa, phytic acid, and oligosacaharides. In: *Proceedings of an International workshop "Increasing Mucuna's potential as a food and feed crop"*. Eilitta, M., Mureithi, J., Muinga, R., Sandoval, C. and Szabo, N. (Eds.), Mombosa, Kenya, 23-26 September, 2002. *Trop. Subtrop. Agroecosyst.* 1: 141-152. Available online: http://www.veterinaria.uady.mx/publicaciones / journal/ 2002-2-3/indice-1(2-3).html

Janardhanan, K., Vadivel, V. and Pugalenthi, M. 2003a. Biodiversity in Indian under-exploited/tribal pulses. In: Improvement strategies for Leguminosae Biotechnology, Jaiwal, P.K. and Singh, R.P. (Eds), Kluwer Academic publishers, Britain. pp. 353-405.

Jankovic, J. and Caine, D.B. 1987. Parkinson's disease: etiology and treatment. *Curr. Neurol.* 7: 193-234.

Jayaweera, D.M.A. 1981. In: *Medicinal plants used in Ceylon.* National Science Council of Sri Lanka, Colombo, Sri Lanka.

Kempster, P.A. and Wahlqvist, M.L. 1994. Dietary factors in the management of Parkinson's disease. *Nutrition Reviews,* 52: 51-58.

Kiro, A., Zahavi, I., Mercus, H. and Dinari, G. 2002. L-Dopa is protective against indomethcin-induced small intestinal ulceration in the rat: Possible role of a á-2-adreagic mechanism. *Life Science,* 51: 1151-1156.

Klawans, H.L., Ringel, S.P. and Shenker, D.M. 1971. Failure of vitamin B6 to reverse the L-Dopa effects in patients on a dopa decarboxylase inhibitor. *Journal of Neurology and Neurosurgery Psychology,* 34: 682-686.

Korycka-Dahl, M. and Richardson, M. 1978. Photogeneration of superoxide anion in serum of bovine milk and in model systems containing riboflavin and amino acids. *Journal of Dairy Science,* 61: 400-407.

Latellier, S., Garnier, J.P., Spy, J., Stoitchkov, K., Le Bricon, T., Baccard, M., Revol, M., Kerneis, Y. and Bousquet, B. 1999. Development of metastases in malignant melanoma is associated with an increase in the plasma L-Dopa/L-tyrosine ratio. *Melanoma Research,* 9: 389-394.

Laurent, L., Livesey, J., Arnason, J.T. and Bruneau, A. 2002. Variation in L-Dopa Concentration in Accessions of *Mucuna pruriens* (L.) DC. and in *Mucuna brachycarpa* Rech. In: *Proceedings of an International Workshop "Food and Feed from Mucuna: Current Uses and the Way Forward"*. Flores, M., Eilitta, M., Myhrman, R., Carew, L.B. and Carsky, R.J. (Eds.), Tegucigalpa, Honduras, 26-29 April, 2000, pp. 352- 375. Available online: www.cidicco.hn.

Liener, I.E. 1994. Implications of antinutritional components in soybean foods. *CRC Critical Reviews in Food Science and Nutrition,* 34: 31-67.

Lorenzetti, F., MacIsaac, S., Arnason, J.T., Awang, D.V.C. and Buckles, D. 1998. The phytochemistry, toxicology and processing potential of the cover crop velvet bean (Cow(h)age, Cowitch) (*Mucuna* Adans. spp., Fabaceae). In: *Cover Crops in West Africa Contributing to Sustainable agriculture.* Buckles, D., Etèka, A., Osiname, O., Galiba, M. and Galiano, N. (Eds.), IDRC, Ottawa, Canada pp. 67-84.

Manyam, B.V. 1995. An alternative medicine treatment for Parkinson's disease: results of a multicenter clinical trial- HP- 200 in Parkinson's disease study group. *Journal of Alternative and Complementary Medicine,* 1: 249-255.

Manyam, B.V. and Sánchez-Ramos, J.R. 1999. Traditional and complementary therapies in Parkinson's disease. In: Parkinson's disease- Advances in Neurology, Williams S.L. and Wilkins, G.M. (Eds.), Philadelphia. pp. 565-574.

Mary Josephine, R. and Janardhanan, K. 1992. Studies on chemical composition and antinutritional factors in three germplasm seed materials of the tribal pulse, *Mucuna pruriens* (L.) DC. *Food Chemistry,* 43: 13-18.

Mercer, D.M., Cross, J.M., Castaneda, A.A. and Guinter, J.A. 1998. Gastroprotective actions of bombsin, L-Dopa and mild irritants- Role of prostaglandins and sensory neurons. *Surgery,* 124: 864-870.

Metman, L.V. and Mouradian, M.M. 1999. Levodopa therapy of Parkinson's disease and associated long-term motor reverse complications. In: Parkinson's disease: The Treatment Options. LeWitt, P.A. and Oertel, W.H. (Eds.), Martin Dunitz, London. pp.117-130.

Miller, E.R. 1920. Dihydroxyphenylalanine, a constituent of the velvet bean. *Journal of Biological Chemistry,* 44: 481-486.

Moffat, A.C. 1986. Clarke's Isolation and Identification of Drugs, 2 ed. Pharmaceutical Press, London, pp. 702-714.

Mohan, V.R. and Janardhanan, K. 1995. Chemical determination of nutritional and antinutritional properties in the tribal pulses. *Journal of Food Science and Technology,* 32: 465-469.

Nachama, S.I. and Edward, R.K. 1967. Does 3,4- dihydroxyphenylalanine play a part in favism? *Nature,* 215: 285-286.

Ordonez, L.A. and Wurtman, R.J. 1974. Folic acid deficiency and methyl group metabolism in rat brain: effects of L- Dopa. *Archieves in Biochemistry and Biophysics,* 160: 372-376.

OVP, 1995. Office de Vulgarisation Pharmaceutique. Dictionnaire Vidal (71st ed.). OVP Paris, France, Editions du Vidal.

Pfutzner, W. and Przybilla, B. 1997. Malignant melanoma and levodopa: Is there a relationship? Two new cases and a review of the literature. *Journal of American Acadamy of Dermatology,* 37: 332-336.

Prakash, D. and Misra, P.S. 1987. Protein and amino acid composition of some wild leguminous seeds. *Plant Foods for Human Nutrition,* 37: 29-32.

Prakash, D. and Tewari, S.K. 1999. Variation on L-Dopa contents in *Mucuna* species. *Journal of Medicinal and Aromatic Plant Science,* 21: 343-346.

Pugalenthi, M., Vadivel, V. and Siddhuraju, P. 2005. Alternative food/feed perspectives of an under-utilized legume *Mucuna pruriens* var. *utilis*-a review. *Plant Foods for Human Nutrition,* 60: 201-218.

Pugalenthi, M. and Vadivel, V. 2007a. Agrobiodiversity of eleven accession of *Mucuna pruriens* (L.) DC. var. *utilis* (Wall. ex Wight) Baker ex Burck (velvet bean) collected from four districts of South India. *Genetic Resources and Crop Evol*usion, 54: 1117-1124.

Pugalenthi, M. and Vadivel, V. 2007b. L-Dopa (L-3,4-Dihydroxyphenylalanine): A non-protein toxic amino acid in *Mucuna pruriens* seeds. *Food,* 1: 322-343.

Rajaram, N. and Janardhanan, K. 1991. The biochemical composition and nutritional potential of the tribal pulse, *Mucuna gigantea* (Willd) DC. *Plant Foods Human Nutrition,* 41: 45-51.

Reynolds, J.E.F. 1989. The extra pharmacopoeia; The pharmaceutical press: London, UK, pp. 1015-1020.

Siddhuraju, P. and Becker, K. 2003. Studies on antioxidant activities of *Mucuna* seed (*Mucuna pruriens* var. *utilis*) extract and various non-protein amino/imino acids through *in vitro* models. *Journal of the Science of Food and Agriculture,* 83: 1517-1524.

Siddhuraju, P. and Becker, K. 2005. Nutritional and antinutritional composition, *in vitro* amino acid availability, starch digestibility and predicted glycemic index of differentially processed mucuna beans (*Mucuna pruriens* var. *utilis*): an under-utilized legume. *Food Chemistry,* 91: 275-286.

Siddhuraju, P., Becker, K. and Makker, H.P.S. 2000. Studies on the nutritional composition and antinutritional factors of three different germplasm seed materials of an under-utilized tropical legume, *Mucuna pruriens* var. *utilis*. *Journal of Agricultural and Food Chemistry,* 48: 6048- 6060.

Siddhuraju, P., Vijayakumari, K. and Janardhanan, K. 1996. Chemical composition and protein quality of the little-known legume, velvet bean (*Mucuna pruriens* (L.) DC.).*Journal of Agricultural and Food Chemistry,* 44: 2636-2641.

Spencer, J.P.E., Jenner, A., Aruoma, O.I., Evans, P.J., Kaur, H., Dexter, D.T., Jenner, P., Lees, A.J., Marsden, D.C. and Halliwell, B. 1994. Intense oxidative DNA damage promoted by L- Dopa and its metabolites, implications for neurodegerative disease. *FEBS Letters,*. 353: 246-250.

Spencer, J.P.E., Jenner, A., Butler, J., Aruoma, O.I., Dexter, D.T., Jenner, P. and Halliwell, B. 1996. Evaluation of the pro-oxidant and antioxidant actions of L- Dopa and dopamine *in vitro*: implications for Parkinson's disease. *Free Radical Research,* 24: 95-105.

Spina, M.B. and Cohen, G. 1988. Exposure of school synaptosomes to L-Dopa increases levels of oxidized glutathione. *Journal of Pharmacology and Experimental Therapy,* 247: 502-507.

Spratto, G.R. and Woods, A.L. 1996. Delmar's NDR-96: Nurse's Drug Reference. Delmar Publishers, Boston, pp. 779-782.

Standaert, D.G. and Young, A.B. 1996. Treatment of central nervous system degenerative disorders. In: Goodman & Gilman's The Pharmacological Basis of Therapeutics.

Hardman, J.G., Limbird, L.E., Molinoff, P.B., Ruddon, R.W. and Gilman, A.G. (Eds.), 9 Ed.. McGraw Hill, New York. pp. 503-519.

Szabo, N.J. and Tebbett, I.R. 2002. The chemistry and toxicity of *Mucuna* species. In: *Proceedings of an International Workshop "Food and Feed from Mucuna: Current Uses and the Way Forward"*. Flores, M., Eilitta, M., Myhrman, R., Carew, L.B. and Carsky, R.J. (Eds.), Tegucigalpa, Honduras, 26-29 April, 2000, pp. 120-141. Available online: www.cidicco.hn.

Tripathi, Y.B. and Upadhyay, A.K. 2001. Antioxidant property of *Mucuna pruriens* Linn. *Current Science*, 80: 1377-1378.

Vadivel, V. and Janardhanan, K. 2005. Nutritional and antinutritional characteristics of seven South Indian wild legumes. *Plant Foods for Human Nutrition*, 60: 69-75.

Vadivel, V. and Pugalenthi, M. 2008a. Removal of antinutritional/toxic substances and improvement in the protein digestibility of velvet bean seeds during various processing methods. *Journal of Food Science and Technology*, 45: 242-246.

Vadivel, V. and Pugalenthi, M. 2008b. Effect of various processing methods on the antinutritional constituents and protein digestibility of velvet bean seeds. *Journal of Food Biochemistry*, 32: 795-812.

Vadivel, V. and Pugalenthi, M. 2008c. Effect of soaking in sodium-bi-carbonate solution followed by autoclaving on the levels of antinutritional compounds and *in vitro* protein digestibility of velvet bean seeds. *Journal of Food Processing and Preservation*, 33: 60-73.

Vijayakumari, K., Smitha, K.B. and Janardhanan, K. 2002. Biochemical characterization of the tribal pulse, *Mucuna utilis* Wall ex. Wight. Seeds. *Journal of Food Science and Technology*, 39: 650-653.

Wysong, V. 1966. U.S. Pattern, 3, 253, 23.

Yahr, M.D. 1973. Advances in neurology. In: Treatment of Parkinsonism, : The role of dopa decarboxylase inhibitors. Vol. 2. Yahr, M.D. (Ed.), Raven Press Publishers, New York.

Zhoa, P., Li, J., Wang, Y. and Jiang, H. 2007. Broad-spectrum antimicrobial activity of the reactive compounds generated *in vitro* by *Manduca Sexta* phenoloxidase. *Insect Biochemistry and Molecular Biology*, 37: 952-959.

□□□

11

Antioxidant Activity and Hypoglycemic Property of a South Indian Under-Utilized Food Legume *Acacia nilotica* DC.

V. Vadivel, M. Pugalenthi and T. Parimelazhagan

CONTENTS

1.1	Introduction	156
1.2	Acacia nilotica	158
1.3	Bioactive Compounds	158
1.4	Reducing Power	161
1.5	Free Radical Inhibition Activity	163
1.6	Antioxidant Activity	165
1.7	α-Amylase Inhibition Activity	168
1.8	α-Glucosidase Inhibition Activity	169
1.9	Conclusion	171
	References	172

1.1 Introduction

Poverty and wealth have a profound effect on the global dietary intake. While chronic malnutrition resulted from poverty, economic growth gives rise to the nutrition transition (Drewnowski and Popkin, 1997). As the income rises and population becomes more urban, consumption of diets rich in complex carbohydrates gave a way to more energy-dense diets (Drewnowski, 1998). These global dietary trends were accompanied by changes in disease patterns, which move away from infectious and nutrient-deficiency diseases towards higher rates of childhood obesity, chronic heart diseases, cancer and diabetes (Ranilla *et al.*, 2008).

Diabetes mellitus is one of the most common chronic diseases, which affected approximately 150 millions of people worldwide and its incidence rate is expected to be doubled in the next 20 years (Cohen and Goedert, 2004). Epidemiological studies and clinical trials strongly supported the notion that hyperglycemia is the main cause of complications related with coronary artery disease, cerebro-vascular disease, renal failure, blindness, limb amputation, neurological complications and pre-mature death (Lopez-Candales, 2001). Recent studies suggested that postprandial hyperglycemia could induce the non-enzymatic glycosylation of various proteins and resulted in the development of various chronic complications.

In diabetes mellitus, the predominant form is Type 2 (non-insulin dependent), which accounted for more than 90% of the total diabetes occurrence (Nathan *et al.*, 1997). Although, the cause for Type 2 diabetes depends on several factors (genetic and environment), the scientific literature emphasized the effect of specific dietary macronutrients on the risk of Type 2 diabetes. The recent research reports indicated that the consumption of higher level of refined carbohydrates with the concomitant decrease of fiber intake in the diets could be related with the prevalence of Type 2 diabetes and may also play an important role in the epidemic of obesity (Apovian, 2004; Gross *et al.*, 2004).

The postprandial phase in diabetes is characterized by a rapid increase in the blood glucose level, which is reported to contribute the onset of oxidation-linked cardiovascular complications through the over production of free radicals at the mitochondrial level (Ceriello, 2005). Due to high oxidation stress, the Type 2 diabetes has been clearly associated with the prevalence of hypertension and the accelerated development of atherosclerosis (Sowers *et al.*, 1993).

The synthetic therapeutics available for diabetes was reported to control the Type 2 diabetes by retarding the breakdown of starch and absorption of glucose through the inhibition of α-amylase in the pancreas and α-glucosidase enzyme in the intestinal tract. Acarbose and miglitol were certain examples of commercially available inhibitors for clinical use in the treatment of Type 2 diabetes (Bailey, 2003). However, these drugs were reported cause various side effects such as abdominal distention, flatulence and possibly diarrhea due to the excessive inhibition of pancreatic α-amylase, which resulted in the abdominal bacterial fermentation of undigested carbohydrates in the colon (Ranilla *et al.*, 2008).

Hence, at present there is an increasing interest among the Food Scientists to find out an alternative natural source α-amylase and α-glucosidase inhibitors with potential antioxidant activity without any side effects for the dietary management of Type 2 diabetic patients. The natural α-amylase and α-glucosidase inhibitors from certain fruits, vegetables and legume grains were reported to offer a good strategy to control postprandial hyperglycemia and provide benefit without any side effects (Matsui *et al.*, 2001; McDougall *et al.*, 2005; Kwon *et al.*, 2006). In this context, the seed materials of *Acacia nilotica* DC., a promising tropical under-utilized legume, having good nutritional properties received more attention.

This wild legume is distributed throughout India, especially, in the forests of Deccan plateau, Western Himalayas and Khasi hills (Janardhanan *et al.*, 2003; Vijayakumari *et al.*, 2007). Cooked young pods and seeds of *A. nilotica* have been consumed by certain ethnic groups, particularly the Garo, Naga, Khatkharis and Gonds living in Northeastern and central India. It exhibited many favourable agro-botanical traits such as fast vegetative growth, early flowering, high fertility index with high pod weight (8.3 g) and number of seeds per pod (8-9) (Vijayakumari *et al.*, 2007). The plant parts have been used in the indigenous medicine for curing body pain, fever, cancerous growth in stomach and indigestion in human beings (Janardhanan *et al.*, 2003).

The seeds of this non-conventional legume constituted a rich source of crude protein (25.6-27.2%), crude lipid (12.3-14.3%), fibers (4.6-5.8%), carbohydrates (51-57%), minerals and essential amino acids (Vijayakumari *et al.*, 2007). It was also found to possess 66% of unsaturated fatty acids, which is more desirable feature from the nutritional point of view for human food. The raw seeds exhibited a low biological value, however, heat treatments significantly improved the true protein digestibility and net protein utilization of this wild legume. From our earlier study, we have reported the presence of certain bioactive compounds like total free phenolics, tannins, phytic acid and L-Dopa (L-3,4-Dihydroxyphenylalanine, a non-protein phenolic amino acid) in the seed samples of *A. nilotica* apart from its nutrient profiles (Vijayakumari *et al.*, 2007).

Although a few reports were available on the nutritional value and certain bioactive compounds of seed materials of *A. nilotica*, to our knowledge, there is no information available regarding the antioxidant activity and hypoglycemic potential of this wild legume grain. Hence, the present study was carried out to analyze the antioxidant activity and hypoglycemic property of certain bioactive compounds extracted from raw and cooked seed samples of *A. nilotica* with a view to recommend it as a natural source of á-amylase and α-glucosidase inhibitors with potential antioxidant activity for the dietary management of Type 2 diabetes.

1.2 *Acacia nilotica*

The seed materials of *Acacia nilotica* L. were collected from Karpagam Botanical Garden, Karpagam University, Coimbatore, Tamil Nadu, India. After removing the immature and damaged seeds, the mature seeds were dried in the open sun light for 2 days and stored in plastic containers in the refrigerator (5°C) until further use.

The whole seeds of *A. nilotica* were randomly divided into two batches (25 g each) and the first batch of seeds was taken in a metal container and cooked at 90-95°C for 30 min in the bean to water ratio of 1:10 (w/v) (Vadivel and Pugalenthi, 2008). After processing, the seed samples were rinsed with distilled water and dried at 55°C for 6 h in a hot air oven. The second batch of seeds was stored as such without any treatment and regarded as raw seeds. The cooked as well as raw seeds of *A. nilotica* were powdered in a Willey Mill to 60-mesh size and used for further analysis.

1.3 Bioactive Compounds

The phenolic compounds in raw and cooked *A. nilotica* seed flour (200 mg) were extracted with 10 ml of 80% aqueous ethanol in the dark at room temperature on a magnetic stirrer for 2 h. Then the contents were centrifuged at 13,000 x g for 10 min and the supernatant was recovered and used for the analysis of phenolic compounds according to the method of Mazza *et al.* (1999). Briefly, the method consists of mixing 10 μl of extract with 240 μl of a solution of 2% HCl in 75% ethanol in a 96-well flat-bottom microtitration plate. The absorbance of the solution was monitored at 280, 500, 320, 360 and 520 nm with a spectrophotometer (Make: Systronics, Model: 119) using (+)-catechin (0-200 μg/ml), tannic acid (0-50 μg/ml), caffeic acid (0-30 μg/ml), quercetin (0-80 μg/ml) and cyanidin-3-glucoside (0-50 μg/ml) as standards for total phenolics, tannins, tartaric esters, flavonols and anthocyanins, respectively. Standards were prepared in 80% (v/v) aqueous ethanol except for quercetin, which was prepared in 95% (v/v) aqueous ethanol. From the absorbance values, the levels of various phenolic compounds were calculated and the results of three separate determinations were expressed in mg/100 g seed flour on dry weight basis.

The flavonoid content of raw and processed *A. nilotica* seed samples was extracted and estimated according to the method of Hairi *et al.* (1991). 50 μl of the ethanolic extract was mixed with 180 ìl of distilled water and 20 μl of a solution of 2-aminoethyldiphenylborate (10 g/l) in a 96-well flat-bottom microtitration plate. The absorbance of the solution was monitored at 404 nm in a spectrophotometer and compared with that of a rutin standard at concentrations ranging from 0 to 50 μg/ml and the flavonoid content was estimated and expressed as mg rutin equivalent/100 g sample.

The L-Dopa content of raw and cooked *A. nilotica* seeds was quantified according to the method of Brain (1976). One g of air-dried seed flour was extracted with 5 ml of 0.1 N HCl over a boiling water bath for 5 min. After cooling, an equal volume of ethanol was added and the mixture was shaken mechanically for 10 min. The contents were centrifuged at 5000 x g for 10 min and the supernatant was collected and made upto a known volume with ethanol. L-Dopa content was quantified by measuring the ultra-violet light absorption at 282 nm in a UV-Visible Spectrophotometer (Make: Systronics, Model- 119) using L-Dopa (Sigma Chemicals) as a standard. The content of L-Dopa was calculated and expressed as mg/100 g seed flour on dry weight basis.

Table 1 : Levels of various bioactive compounds in the raw and processed seed materials of *Acacia nilotica* [1, 2].

S. No.	Bioactive compounds (mg/100 g seed flour)	Raw Seeds	Cooked Seeds
1	Total phenolics	$3900^b \pm 0.21$	$4260^a \pm 0.06$
2	Tannins	$66^b \pm 0.41$	$98^a \pm 0.04$
3	Flavonols	$386^a \pm 0.32$	$316^b \pm 0.11$
4	Anthocyanins	$65^b \pm 0.23$	$78^a \pm 0.14$
5	Tartaric esters	$81^b \pm 0.15$	$106^a \pm 0.23$
6	Flavonoids	$172^a \pm 0.24$	$144^b \pm 0.16$
7	L-Dopa	$1330^a \pm 0.07$	$920^b \pm 0.42$
8	Phytic acid	$934^a \pm 0.22$	$837^b \pm 0.31$

[1]Values are mean and ± standard deviation of three separate determinations.
[2]Values in the same row with different roman superscript are significantly different (p<0.05).

The phytic acid content of *A. nilotica* seed samples was determined by following the method of Wheeler and Ferrel (1971). 200 mg of seed flour was homogenized with 5 ml of 3% TCA and centrifuged at 5000 x g for 10 min. The collected supernatant was mixed with 4 ml of 1% $FeCl_3$ solution and heated on a boiling water bath for 45 min and centrifuged at 10,000 x g for 10 min. To the residue, 2 ml of distilled water and 3 ml of 1.5 N NaOH were added and diluted to 30 ml with distilled water and heated on a boiling water bath for 10 min. Then the solution was filtered through a hot Whatman No. 1 filter paper. The residue was washed with hot water and 3.2 N HNO_3 and the washings were collected and made upto 100 ml with distilled water. From this extract, 2 ml of aliquot was taken with 4 ml of 1.5 N KSCN and diluted to 20 ml with distilled water. After 10 min of incubation at room temperature, the absorbance was measured at 480 nm. The phytic acid content was calculated and expressed in mg/100 g seed flour on dry weight basis.

The raw seed materials of *A. nilotica* of the present study were found to contain appreciable levels of various bioactive compounds such as total phenolics (3900 mg/100 g seed flour), tannins (66 mg/100 g seed flour), flavonols (386 mg/100 g seed flour), anthocyanins (65 mg/100 g seed flour), tartaric esters (81 mg/100 g seed flour), flavonoids (172 mg/100 g seed flour), L-Dopa (1330 mg/100 g seed flour) and phytic acid (934 mg/100 g seed flour) (Table 1). Among the various bioactive compounds, the total phenolics registered the highest level in both raw and processed *A. nilotica* seeds.

The levels of various bioactive compounds of the present analysis were comparable with that of an earlier report on certain common legume grains such as *Phaseolus vulgaris,* soybean, pea, mung bean, horse gram, cowpea and moth bean (Yen and Duh, 1993; Tsuda *et al.*, 1994; Duh *et al.*, 1997; Lee *et al.*, 2000; Siddhuraju, 2006; Siddhuraju and Becker, 2007; Siddhuraju and Manian, 2007). The levels of total phenolics, flavonols and anthocyanins in the raw seeds of *A. nilotica* were found to be higher and the content of tartaric esters and flavonoids were lower when compared to an earlier report on *Lupinus angustifolius* (Oomah *et al.*, 2006).

During cooking treatment, a significant level ($p < 0.05$) of increase on the levels of phenolics, tannins, anthocyanins and tartaric esters was noticed, which might be due to their release from the complex bounded form with other nutrients like proteins (Pugalenthi and Vadivel, 2007). This is in agreement with that of an earlier study on certain antioxidant compounds of *Vigna radiata* var. emerald, *Glycine max* var. jutro and *G. max* var. merit (Fernandez-Orozco *et al.*, 2008).

On the other hand, the cooking treatment caused a significant level ($p < 0.05$) of reduction of certain bioactive compounds such as flavonols (18%), flavonoids (16%), L-Dopa (30%) and phytic acid (10%) (Table 1). Such drastic level of reduction of L-Dopa during cooking treatment of the present study was in consonance with that of an earlier report on *Abrus precatorius* (Pugalenthi *et al.*, 2007); *Mucuna pruriens* var. *utilis* (Vadivel and Pugalenthi, 2008) and *Entada scandens* (Vadivel *et al.*, 2008). Such substantial level of removal of bioactive compounds might be due to leaching out into the cooking medium and subsequent degradation by heat.

1.4 Reducing Power

The reducing power of different bioactive compounds extracted from raw and cooked seed materials of *A. nilotica* was determined according to the method of Oyaizu (1986). All the extracts were lyophilized and dissolved in respective solvents. Various concentrations (20-100 µg/ml) of each extract were taken along with 2.5 ml of 0.2 M phosphate buffer (pH 6.6) and 2.5 ml of 1% potassium ferricyanide. The control contain all the reagents except the sample extract. The reaction mixture was incubated at 50°C for 20 min and then 2.5 ml of 10% TCA was added and centrifuged at 650 x g for 10 min at 23°C. The upper layer of the supernatant (2.5 ml) was collected and mixed with 2.5 ml of distilled water and 0.5 ml of 0.1% ferric chloride and stand for 5 min at room temperature and the absorbance was measured at 700 nm and the results were recorded.

Figures 1 & 2 showed the absorbance values as a measure of reducing power of bioactive compounds extracted from raw and processed *A. nilotica* seeds. All the bioactive compounds present in the raw seed materials of *A. nilotica* exhibited excellent reducing power with the highest values recorded for L-Dopa in a dose-dependent manner (Fig. 1). The reducing power results revealed that all the bioactive compounds of the present study have a good ability to donate electrons, which involved in the antioxidant activity.

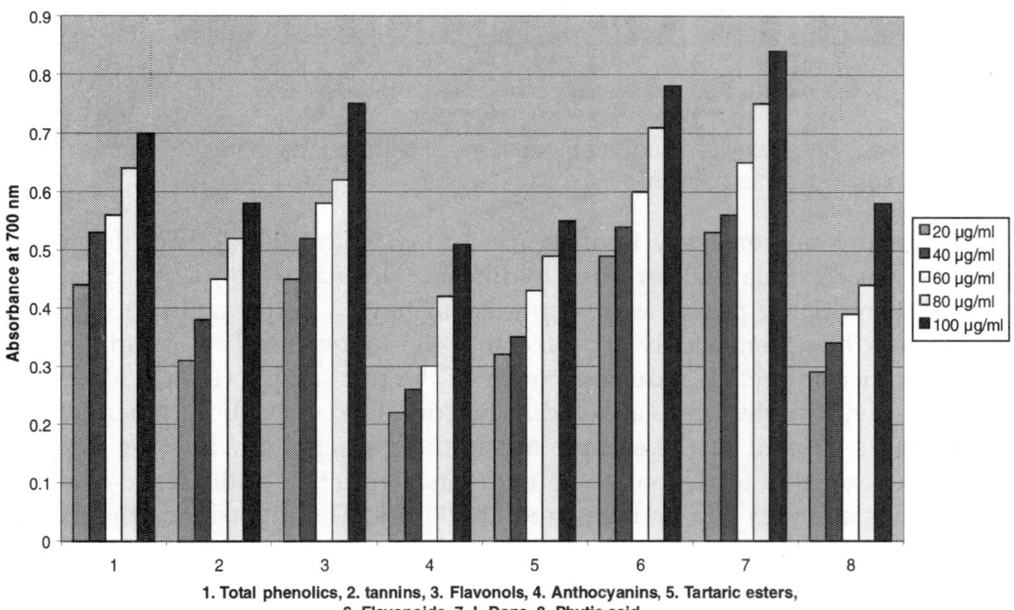

Fig. 1 : Reducing power of various bioactive compounds extracted from raw seed samples of Acacia nilotica

The cooking processing method was found to improve the reducing power of all the bioactive compounds of the present analysis, except phytic acid (Fig. 2). This result was in agreement with an earlier report of Kim *et al.* (2006) that the heat treatment enhances the release of antioxidant compounds. The antioxidant potential of bioactive compounds was reported to associate with their reducing power, which is related with the presence of reductones (Jayaprakasha *et al.*, 2000).

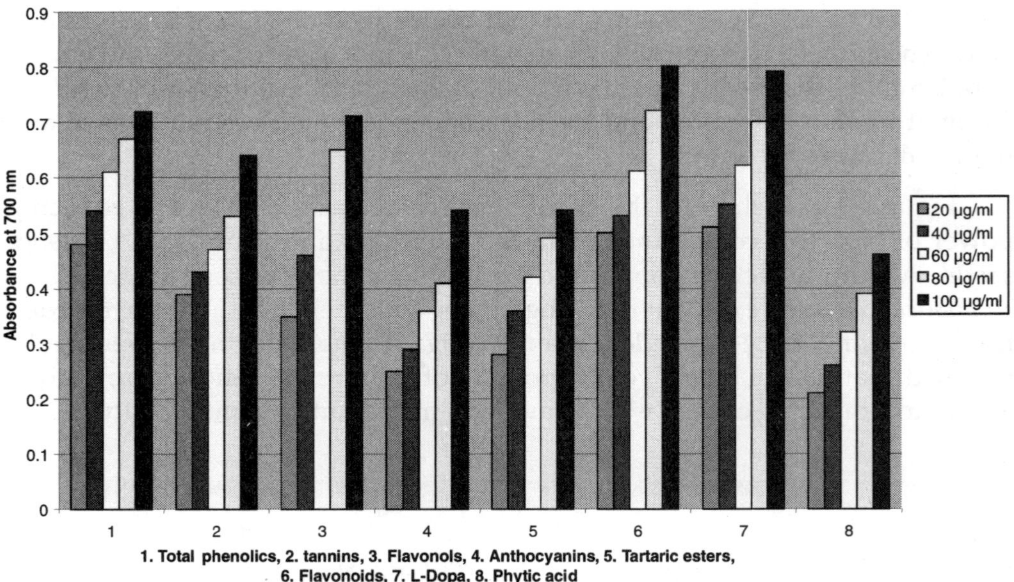

Fig. 2 : Reducing power of various phytochemical compounds extracted from cooked seed samples of Acacia nilotica

Reducing power assay is often used to evaluate the ability of natural antioxidants to donate electron or hydrogen (Dorman *et al.*, 2003). Samples with high reducing power were reported to have a better ability to donate electrons. It has been widely accepted that the higher level of absorbance at 700 nm indicates greater reducing power of the test sample (Duh *et al.*, 1997). In this assay, samples were assayed in a redox-linked reaction, whereby the antioxidants present in the sample act as the reductants, while the reagent containing excess of ferric ions act as oxidants. Reduction of the ferric-chloride into ferrous complex can be measured at 700 nm. The intensity of the colour is related to the amount of antioxidant reductants present in the sample. Many research efforts have been revealed that there was a direct correlation between antioxidant activity and reducing power (Duh *et al.*, 1997; Siddhuraju, 2006). Free radicals form stable substances by accepting the donated electrons and the free radical chain reactions were thus interrupted.

The reducing ability of the samples assessed in this study suggested that the extracts were able to donate electrons to free radicals in actual biological or food systems and convert the free radicals into stable and unreactive species. Among the various bioactive compounds, L-Dopa recorded a higher level of reducing power, which is followed by total phenolics in both raw and cooked seed samples of *A. nilotica*. This is in agreement with that of an earlier report on the reducing power of L-Dopa compound extracted from *Mucuna pruriens* var. *utilis* (Siddhuraju and Becker, 2003).

The reducing power of L-Dopa might be due to its di- and mono-hydryl substitutions in the aromatic ring, which possess potential hydrogen-donating abilities. Thus, this non-protein amino acid might contain higher levels of reducing reductones and could react with free radicals and stabilize the free radicals and also terminate the free radical chain reactions (Pugalenthi and Vadivel, 2007). Recently, apart from phenolic compounds, the protein hydrolysates from Alfalfa leaves (Xie *et al.*, 2008) and proline content of *Doryştoechas hasta* (Karagözler *et al.*, 2008) were also reported to exhibit remarkable reducing power.

1.5 Free Radical Inhibition Activity

The free radical inhibition activity of various bioactive compounds extracted from raw and processed *A. nilotica* seeds was measured by using DPPH assay (Kwon *et al.*, 2006). 1 ml of aliquot of each extract (100 µg/ml) was mixed with 5.0 ml of 1 mM DPPH in methanol. The control contains all the reagents except the sample extract or positive control antioxidant (BHT). The mixture was vortexed for 1 min and allowed to stand for 30 min in the dark at 25°C. Then the absorbance was measured at 517 nm using a Spectrophotometer. The percentage of free radical inhibition activity was calculated by using the following equation,

$$\text{Free radical inhibition activity (\%)} = \frac{A^{control} - A^{extract}}{A^{control}} \times 100$$

All the presently studied bioactive compounds extracted from both raw and processed seeds of *A. nilotica* were found to exhibit more effective free radical inhibition activity against DPPH free radicals (Table 2). The α,α-diphenyl-β-picrylhydrazyl (DPPH), a stable nitrogen centered free radical, has been used to evaluate the antioxidant property of natural compounds by measuring their radical quenching capacity in a relatively short period of time. The absorbance decreases as a result of colour change of DPPH solution from purple to yellow as the radical is scavenged by antioxidant compounds through the donation of protons to form the stable DPPH-H molecule (Gadow *et al.*, 1997).

The free radical scavenging activity recorded by all the bioactive compounds in raw *A. nilotica* seeds (36.25-86.32%) was in consonance with that of a previous report on bioactive compounds in soybean (67.74-92.78%) (Astadi *et al.*, 2009); *Phyllanthus neruri* (85%) (Sabir and Rocha, 2008); *Anacardium occidentale* (74%) (Razali *et al.*, 2008), *Mimordica charantis* (36-79.39%) (Kubola and Siriamornpun, 2008) and *Phaseolus vulgaris* (Ranilla *et al.*, 2007), but higher than that of *Phaseolus vulgaris* (16.6-44.1%) (Cardador-martiänez *et al.*, 2002). The observed *in vitro* free radical inhibition activity suggested that all the investigated bioactive compounds of the present analysis will exert protective effects under *in vivo* system also against oxidative and free radical injuries appeared during different pathological conditions.

Table 2 : Free radical inhibition activity (%) of various bioactive compounds extracted from raw and processed *Acacia nilotica* seeds[1,2].

S. No.	Bioactive compounds	Free radical inhibition activity (%)	
		Raw Seeds	Cooked Seeds
1	Total phenolics	$58.17^{b} \pm 0.24$	$72.12^{a} \pm 0.12$
2	Tannins	$54.46^{b} \pm 0.18$	$59.32^{a} \pm 0.15$
3	Flavonols	$71.36^{b} \pm 0.16$	$80.19^{a} \pm 0.21$
4	Anthocyanins	$67.47^{a} \pm 0.14$	$69.35^{a} \pm 0.06$
5	Tartaric esters	$36.25^{b} \pm 0.23$	$41.66^{a} \pm 0.22$
6	Flavonoids	$75.14^{a} \pm 0.17$	$69.38^{b} \pm 0.08$
7	L-Dopa	$86.32^{b} \pm 0.13$	$93.56^{a} \pm 0.15$
8	Phytic acid	$58.42^{b} \pm 0.11$	$65.40^{a} \pm 0.20$
9	Standard (BHT)	98.4 ± 0.12	98.4 ± 0.12

[1]Values are mean and ± standard deviation of three separate determinations.
[2]Values in the same row with different roman superscript are significantly different ($p<0.05$).

The cooking treatment was found to significantly ($p < 0.05$) improve the free radical scavenging ability of all the bioactive compounds, except anthocyanins, in which only a very little level of increase was noticed (Table 2). But, the free radical inhibition activity was decreased in the case of flavonoids after subjected to cooking treatment. Among the various bioactive compounds, L-Dopa was found to register higher level of free radical scavenging activity in the processed seed samples (93.56%), which is found to comparable with that of standard BHT (98.4%). Similarly, effective hydroxyl scavenging ability of an another non-protein amino acid, β-N-Oxalyl-L-α,β-diaminopropionic acid (β-ODPA) in *Lathyrus sativus* was also previously reported (Pugalenthi and Vadivel, 2007).

Yamaguchi *et al.* (2001) reported the increase of free radical scavenging activity in several vegetables after hydrothermal treatments, because of the suppression of oxidation by the antioxidant compounds, which is mainly attributed to thermal inactivation of oxidative enzymes. In addition, the destruction of cell walls and sub-cellular compartments of the samples during boiling caused the release of potent radical scavenging antioxidants. Similarly, Gahler *et al.* (2003) reported a significant level of improvement on the antioxidant activity of tomatoes after heat treatment due to the increased release of phytochemicals, such as lycopene from the matrix. Boiling, microwave cooking and steaming treatments were reported to increase the total antioxidant activity in pepper, green beans, broccoli and spinach at significant level (Turkmen *et al.*, 2005).

Conversely, the levels of antioxidant compounds were decreased after hydrothermal treatment in broccoli (Zhang and Hamauzu, 2004) and certain cruciferous vegetables (Sikora *et al.*, 2008). Puupponen-Pimia *et al.* (2003) reported that the DPPH free radical scavenging activity of cauliflower decreased by 23% but increased by 9% in the case of cabbage during blanching in water.

1.6 Antioxidant Activity

The antioxidant activity of various bioactive compounds of the present study was evaluated by using β-carotene/linoleic acid model according to Velioglu *et al.* (1997) method. 20 µl of aliquot of each extract (100 µg/ml) or BHT (1000 µM) and 200 µl of β-carotene solution were taken in a 96-well flat-bottom microtitration plate. The sample mixture was diluted by transferring 30 ìl to another plate containing air-sparged distilled water (210 µl). Dilutions were made in triplicate, since the β-carotene bleaching reaction was subject to noticeable variations. 20 µl of ADIBA (0.3 M) was added to each well to initiate the reaction. Absorbance was monitored at 450 nm with a spectrophotometer at regular time intervals of 10-90 min and plotted against time. Plates were kept in the dark at ambient temperature between the readings. Antioxidant activity was calculated relative to the control using the following equation,

$$\text{Antioxidant activity (\%)} = \frac{R^{control} - R^{sample}}{R^{control}} \times 100$$

Where, $R_{control}$ and R_{sample} are the degradation rates of â-carotene in the reaction mixer with BHT (control) and sample extract, respectively. The absorbance values for different times were averaged to give the antioxidant activity value for the sample.

In the present study, the antioxidant activity of various bioactive compounds was determined by using β-carotene bleaching method. Inhibition of breakdown of the lipid hydroperoxides into unwanted volatile products allowed us to determine the antioxidant activity of the natural compounds in related mechanisms. In the absence of antioxidants, oxidation products of linoleic acid (lipid hydroperoxides, conjugated dienes and volatile byproducts) were simultaneously attack the β-carotene and resulted in bleaching of its characteristic yellow colour in ethanolic solution. But, in the presence of antioxidant compounds, oxidation products were scavenged and bleaching was prevented.

Table 3 : Antioxidant activity of various bioactive compounds extracted from raw and processed *Acacia nilotica* seeds by β-carotene method/linoleic acid model[1,2].

S. No.	Bioactive compounds	Antioxidant activity (%) Raw Seeds	Cooked Seeds
1	Total phenolics	$61.13^{b} \pm 0.12$	$66.22^{a} \pm 0.15$
2	Tannins	$48.25^{a} \pm 0.13$	$50.65^{a} \pm 0.04$
3	Flavonols	$53.68^{a} \pm 0.20$	$55.82^{a} \pm 0.10$
4	Anthocyanins	$15.06^{b} \pm 0.15$	$17.40^{a} \pm 0.06$
5	Tartaric esters	$11.38^{b} \pm 0.17$	$14.53^{a} \pm 0.05$
6	Flavonoids	$56.83^{a} \pm 0.22$	$53.98^{b} \pm 0.12$
7	L-Dopa	$80.16^{b} \pm 0.14$	$88.75^{a} \pm 0.15$
8	Phytic acid	$45.75^{b} \pm 0.02$	$51.57^{a} \pm 0.14$
9	Standard (BHT)	92.8 ± 0.12	92.8 ± 0.12

[1]Values are mean and ± standard deviation of three separate determinations.
[2]Values in the same row with different roman superscript are significantly different (p<0.05).

The antioxidant activity analyzed through β-carotene/linoleic acid system showed that the L-Dopa (80.16%) and total phenolics (61.13%) extracted from the raw seed samples of *A. nilotica* exhibited higher level of antioxidant activity next to that of BHT standard (92.8%) (Table 3). The results of the present analysis was in agreement with that of antioxidant activity reported for certain bioactive compounds in velvet bean (Siddhuraju and Becker, 2003) and methanolic extract of pecan (79%) and walnut (70%) (Miraliakbari and Shahidi, 2008). Apart from L-Dopa and total phenolics, all other compounds of the present study exhibited moderate antioxidant activity of 11.38-56.83%, which is comparable with that of an earlier report on *Phaseolus vulgaris* (16.6-44.1%) (Cardador-martiänez *et al.*, 2002). The antioxidant capacity of all the presently investigated bioactive compounds was significantly improved by the cooking processing method, except in flavonoids.

The data presented in this study have demonstrated that almost all the bioactive compounds possess remarkable antioxidant and free radical inhibition activity. Indeed, the bioactive compounds showed a considerable level of variation in the inhibition of linoleic acid oxidation and DPPH radical scavenging activity under *in vitro* assays. Such different behaviour probably due to the different mechanisms involved in the steps of oxidation process. The antioxidant activity measured by an individual assay reflects only the chemical reactivity under specific conditions applied in that assay. Hence it is inappropriate and misleading to use a single method to evaluate the antioxidant capacity of a sample (Huang *et al.*, 2005).

Several studies have been reported the relationship between the antioxidant activity of plant materials and their phenolic contents. Some authors found a correlation between the phenolic content and antioxidant activity, whilst others found no relationship. Velioglu *et al.* (1997) found a strong relationship between total phenolic content and antioxidant activity in selected fruits, vegetables and grain products. On the other hand, no correlation was observed between the antioxidant activity and phenolic content in some plant extracts by Kähkönen *et al.* (1999).

The findings of the present study, have not showed any relationship between the antioxidant activity and total phenolic content. Even though, the total phenolics were present at higher level, the maximum level of antioxidant activity was noticed in L-Dopa in both raw and processed *A. nilotica* seeds. The relatively higher levels of antioxidant and free radical inhibition activities of L-Dopa extract suggested that the type of phenolics has a direct impact on the antioxidant activity rather than their amounts.

This is in agreement with those of Kähkönen *et al.* (1999) and Shahidi and Marian (2003), who reported that the differences in antioxidant activities of plant extracts could be due to different qualitative and quantitative composition of their phenolic constituents, from phenolic acids to flavonoids and their derivatives. For instance, the antioxidant activity of phenolic acids and their derivatives, such as esters depends on the number of hydroxy groups present in the molecules (Soobrattee *et al.*, 2005). Hence, the higher level of antioxidant activity exhibited by the L-Dopa may be due to its aromatic ring structural characteristics with two hydroxy groups (Pugalenthi and Vadivel, 2007).

The intake of L-Dopa was reported to inhibit both the endogenous and exogenous plasma lipid peroxidation in patient's body. By using single cell gel electrophoresis ("Comet") assay it was demonstrated that the oxidation products of L-Dopa prevent the H_2O_2-induced oxidative DNA damage in cultured tissue cells. The L-Dopa was also reported to stimulate the expression of antioxidant enzymes such as NADPH-Quinone-Oxidoreductase (Pugalenthi and Vadivel, 2007).

1.7 α-Amylase Inhibition Activity

The α-amylase inhibition activity of various bioactive compounds extracted from raw and processed *A. nilotica* seeds was determined according to the assay described by Worthington (1993a). 500 μl of each extract (100 μg/ml) and 500 μl of 0.02 M sodium phosphate buffer (pH 6.9) containing α-amylase solution (1 unit liberates 1.0 mg of maltose from starch in 3 min at pH 6.9 and temperature 25°C) were taken and incubated at 25°C for 10 min. After the pre-incubation, 500 μl of 1% starch solution in 0.02 M sodium phosphate buffer (pH 6.9) was added to each tube at regular time intervals. After the incubation at 25°C for 30 min, the reaction was stopped with 1.0 ml of dinitrosalicylic acid reagent. The test tubes were then incubated in a boiling water bath for 5 min and cooled to room temperature. The reaction mixture was then diluted with the addition of 15 ml of distilled water and the absorbance was measured at 540 nm. The readings were compared with the control, which contain buffer instead of sample extract. The results were calculated by using the following equation,

$$\alpha\text{-Amylase inhibition activity } (\%) = \frac{A^{control} - A^{extract}}{A^{control}} \times 100$$

The α-amylase enzyme inhibition activity of various bioactive compounds such as total phenolics (17.45%), tannins (12.73%), flavonols (9.21%), anthocyanins (10.28%), tartaric esters (8.14%), flavonoids (19.16%) and L-Dopa (28.48%) extracted from the raw seeds of *A. nilotica* was given in the Table 4. These values were found to be higher than that of an earlier report on *Mucuna pruriens* var. *utilis* (5.1%) (Vadivel and Pugalenthi, 2008) and *Canavalia virosa* (3.77%) (Siddhuraju and Becker, 2001) and comparable to that an earlier report on *Abrus precatorius* (26.01%) (Pugalenthi *et al.*, 2007); *Entada scandens* (16.23%) (Vadivel *et al.*, 2008); different varieties of chickpea, green gram and pea (14-30%) (Mulimani *et al.*, 1994; Salunke *et al.*, 2006) and strawberry fruits (8-26%) (Pinto *et al.*, 2008).

α-Amylase inhibitor is a proteinaceous substance, which binds with α-amylase enzyme and forms the α-amylase enzyme-inhibitor complex and thereby inhibit the α-amylase activity, which resulted in slow digestibility of starch. Among the various bioactive compounds, the L-Dopa was found to exhibit maximum level of α-amylase inhibition activity (28.48%), whereas, the α-amylase inhibition activity was not detected in phytic acid. A significant level ($p < 0.05$) of improvement was noticed on the α-amylase inhibition activity of all the presently analyzed bioactive compounds, except phytic acid, when subjected to cooking treatment.

Table 4 : α-Amylase inhibition activity of various bioactive compounds extracted from raw and processed *Acacia nilotica* seeds[1,2].

S. No.	Bioactive compounds	α-Amylase inhibition activity (%)	
		Raw Seeds	Cooked Seeds
1	Total phenolics	17.45^b ± 0.14	20.51^a ± 0.15
2	Tannins	12.73^b ± 0.16	15.24^a ± 0.06
3	Flavonols	09.21^b ± 0.14	10.14^a ± 0.15
4	Anthocyanins	10.28^b ± 0.05	11.46^a ± 0.24
5	Tartaric esters	08.14^b ± 0.24	09.15^a ± 0.15
6	Flavonoids	19.16^b ± 0.15	21.33^a ± 0.21
7	L-Dopa	28.48^a ± 0.08	32.82^a ± 0.08
8	Phytic acid	ND	ND

[1]Values are mean and ± standard deviation of three separate determinations.
[2]Values in the same row with different roman superscript are significantly different (p<0.05).
ND: Not Detected.

α-amylase and α-glucosidase are the well-known enzymes playing a key role in the management of hyperglycemia-linked Type 2 diabetes (Bailey, 2003). Cheplick *et al.* (2007) reported that the α-amylase inhibition activity of certain fruit samples such as raspberry might be due to some specific phenolics. Many fruits including strawberry, raspberry and grape were known to contain higher levels of soluble tannins with considerable α-amylase inhibition property (McDougall *et al.*, 2005).

1.8 α-Glucosidase Inhibition Activity

The α-glucosidase inhibition activity of various bioactive compounds of the present investigation was analyzed according to the method of Worthington (1993b). 50 μl of each extract (100 μg/ml) and 100 ìl of 0.1 M phosphate buffer (pH 6.9) containing α-glucosidase solution (1 unit/ml) were taken in a 96-well flat-bottom microtitration plate and incubated at 25°C for 10 min. After the pre-incubation, 50 μl of 5 mM *p*-nitrophenyl-α-D-glucopyranoside solution in 0.1 M phosphate buffer (pH 6.9) was added to each well at regular time intervals and the reaction mixture was incubated at 25°C for 5 min. After the incubation period, the absorbance readings were recorded at 405 nm and compared to a control that had 50 ìl of buffer solution in place of the extract. The results were calculated according to the following equation and expressed as percentage of α-glucosidase inhibition.

$$\alpha\text{-Glucosidase inhibition activity } (\%) = \frac{A^{control} - A^{extract}}{A^{control}} \times 100$$

The analysis of α-glucosidase enzyme inhibition activity revealed that except phytic acid, all other bioactive compounds such as total phenolics (52.32%), tannins (46.41%), flavonols (45.32%), anthocyanins (48.33%), tartaric esters (41.24%), flavonoids (55.19%) and L-Dopa (65.76%) extracted from the raw seed samples of *A. nilotica* possess higher level of enzyme inhibition capacity (Table 5). This is in agreement with that of an earlier report on α-glucosidase inhibition activity of certain bioactive compounds in strawberry fruits (8-26%) (Pinto *et al.*, 2008).

Table 5 : α-Glucosidase inhibition activity of various bioactive compounds extracted from raw and processed *Acacia nilotica* seeds[1,2]

S. No.	Bioactive compounds	α-Glucosidase inhibition activity (%) Raw Seeds	Cooked Seeds
1	Total phenolics	$52.32^{ab} \pm 0.16$	$55.12^{a} \pm 0.18$
2	Tannins	$46.41^{a} \pm 0.25$	$48.33^{a} \pm 0.14$
3	Flavonols	$45.32^{b} \pm 0.06$	$48.56^{a} \pm 0.11$
4	Anthocyanins	$48.33^{b} \pm 0.17$	$52.62^{a} \pm 0.12$
5	Tartaric esters	$41.24^{a} \pm 0.20$	$38.49^{b} \pm 0.15$
6	Flavonoids	$55.19^{a} \pm 0.15$	$53.38^{a} \pm 0.18$
7	L-Dopa	$65.76^{a} \pm 0.21$	$68.66^{a} \pm 0.12$
8	Phytic acid	ND	ND

[1]Values are mean and ± standard deviation of three separate determinations.
[2]Values in the same row with different roman superscript are significantly different (p<0.05).
ND: Not Detected.

McDougall *et al.* (2005) studied certain high-anthocyanin-containing fruit extracts from blueberry, currant, raspberry and strawberry and observed a better level of α-glucosidase inhibition activity. Similarly, Cheplick *et al.* (2007) reported high α-glucosidase inhibition activity for a yellow raspberry cultivar among red, black and yellow coloured raspberries, which suggested that the á-glucosidase inhibition may be influenced by specific anthocyanins rather than the actual amount of the overall total plant phenolics.

The cooking processing method was found to cause a significant level ($p < 0.05$) of improvement on the α-glucosidase inhibition activity of total phenolics, flavonols and anthocyanins, while the tannins, flavonoids and L-Dopa showed only a slight increase on their α-glucosidase inhibition activity.

But, the α-glucosidase inhibition activity of tartaric esters was decreased after cooking treatment. Among the various bioactive compounds, the L-Dopa registered higher level of inhibition of α-glucosidase activity in both raw (65.76%) and processed (68.66%) seed samples. This is in agreement with that of an earlier report on α-glucosidase inhibition activity of methanolic extract of *Tournefortia hartwegiana* (64%) (Ortiz-Andrade *et al.*, 2007).

The recent research efforts are being focused to identify a natural source of α-glucosidase inhibitors. Inhibitors of this enzyme have been recently developed from various natural sources. In Morelos, Mexico, 51 plant species have been registered from traditional medicine were found to possess α-glucosidase inhibition capacity and were employed as anti-diabetic agents (Monroy-Ort´ýz and Castillo, 2000).

Comparison of α-amylase and α-glucosidase inhibition activities in the same sample indicated that the *A. nilotica* seeds have a lower level of α-amylase inhibition activity when compared to α-glucosidase inhibition activity. It is a well-known fact that the dietary management of hyperglycemia-linked Type 2 diabetes can be targeted through whole foods that have a high α-glucosidase and moderate level of α-amylase inhibition (Matsui *et al.*, 2001; Kwon *et al.*, 2006). Because, excessive α-amylase inhibition leads to accumulation of undigested starch in the intestine and consequent stomach distension and discomfort (Bailey, 2003). Hence, the cooked *A. nilotica* seeds with high α-glucosidase and moderate á-amylase inhibition activities could be considered as a potential candidate for further *in vivo* studies as a part of more comprehensive dietary designs to manage early stages of hyperglycemia-linked Type 2 diabetes.

1.9 Conclusion

Results of the present study indicated that the seed materials of *A. nilotica* constitute a rich source of various bioactive compounds. By virtue of the hydrogen donating ability, all the presently investigated compounds exhibited potential antioxidant and free radical scavenging activities with the highest values recorded for L-Dopa. Except phytic acid, all other compounds showed a higher level of α-glucosidase inhibition activity and moderate level of α-amylase inhibition activity, which is more desirable for the dietary management of hyperglycemia-linked Type 2 diabetes. The cooking processing method was appeared to improve both the antioxidant as well as hypoglycemic properties of various bioactive compounds in *A. nilotica* seeds. Hence, such processed seed materials of *A. nilotica* could be recommended as a potential natural source of dietary ingredient with desirable antioxidant and hypoglycemic activities, particularly for the management of Type 2 diabetic patients, after conducting an extensive trial with human subject. *In vivo* studies on evaluation of antioxidant and hypoglycemic properties of various bioactive compounds in processed seed materials of *A. nilotica* and their bioavailability are in progress.

References

Apovian, C.M. 2004. Sugar-sweetened soft drinks, obesity, and type 2 diabetes. *The Journal of American Medical Association,* 292: 978-979.

Astadi, I.R., Astuti, M., Santoso, U. and Nugraheni, P.S. 2009. *In vitro* antioxidant activity of anthocyanins of black soybean seed coat in human low density lipoprotein (LDL). *Food Chemistry,* 112: 659-663.

Bailey, C.J. 2003. New approaches to the pharmacotherapy of diabetes. In J. C. Pickup & G. William. Textbook of Diabetes, 3rd ed., Vol. 2, (pp 73.1-73.21). Blackwell Science, Oxford, U.K.

Brain, K.R. 1976. Accumulation of L-Dopa in cultures from *Mucuna pruriens. Plant Science Letters,* 7: 157-161.

Cardador-Martiänez, A., Loarca-Pinta, G. and Oomah, B.D. 2002. Antioxidant Activity in Common Beans (*Phaseolus vulgaris* L.). *Journal of Agricultural and Food Chemistry,* 50: 6975-?6980.

Ceriello, A. 2005. Perspective in diabetes: postprandial hyperglycemia and diabetes complications: is it time to treat? *Diabetes,* 54: 1-7.

Cheplick, S., Kwon, Y., Bhowmik, P. and Shetty, K. 2007. Clonal variation in raspberry fruit phenolics and relevance for diabetes and hypertension management. *Journal of Food Biochemistry,* 31: 656-679.

Cohen, P., and Goedert, M. 2004. GSK3 inhibitors: development and therapeutic potential. *Nature Reviews,* 3: 479-487.

Dorman, H.J.D., Peltoketo, A., Hiltunen, L. and Tikkanen, M.J. 2003. Characterization of the antioxidant properties of de-odourised aqueous extracts from selected Lamiaceae herbs. *Food Chemistry,* 83: 255-262.

Drewnowski, A. 1998. Energy density, palatability, and satiety: implications for weight control. *Nutrition Reviews,* 56: 347-353.

Drewnowski, A. and Popkin, B.M. 1997. The nutrition transition: new trends in the global diet. *Nutrition Reviews,* 55: 31-43.

Duh, P.D., Yen, W.J., Du, P.C. and Yen, G.C. 1997. Antioxidant activity of mung bean hulls. *Journal of American Oil Chemists Society,* 74: 1059-1063.

Fernandez-Orozco, R., Frias, J., Zielinski, H., Piskula, M.K., Kozlowska, H. and Vidal-Valverde, C. 2008. Kinetic study of the antioxidant compounds and antioxidant capacity during germination of *Vigna radiata* cv. emerald, *Glycine max* cv. jutro and *Glycine max* cv. merit. *Food Chemistry,* 111: 622-630.

Gadow, A., Joubert, E. and Hansmann, C.F. 1997. Comparison of the antioxidant activity of aspalathin with that of other plant phenols of rooibos tea (*Aspalathus linearis*), á-tocopherol, BHT and BHA. *Journal of Agricultural and Food Chemistry,* 45: 632-638.

Gahler, S., Otto, K. and Bohm, V. 2003. Alterations of vitamin C, total phenolics, and antioxidant capacity as affected by processing tomatoes to different products. *Journal of Agriculture and Food Chemistry,* 51: 7962-7968.

Gross, L.S., Li, L., Ford, E.S. and Liu, S. 2004. Increased consumption of refined carbohydrates and the epidemic of type 2 diabetes in the United States: an ecologic assessment. *American Journal of Clinical Nutrition,* 79: 774-779.

Hairi, B., Sall´e, G. and Andary, C. 1991. Involvement of flavonoids in the resistance of two poplar cultivars to mistletoe (*Viscum album* L.). *Protoplasma,* 162: 20-26.

Huang, D., Ou, B. and Prior, L.R. 2005. The chemistry behind antioxidant capacity assays. *Journal of Agricultural and Food Chemistry,* 53: 1841-1856.

Janardhanan, K., Vadivel, V. and Pugalenthi, M. 2003. Biodiversity in Indian under-exploited/tribal pulses. In P. K. Jaiwal & R. P. Singh. Improvement strategies for Leguminosae Biotechnology. (pp. 353-405). Kluwer Academic Publishers, The Netherlands

Jayaprakasha, G.K., Negi, P.S., Sikder, S., Rao, L.J. and Sakariah, K.K. 2000. Antibacterial activity of *Citrus reticulate* peel extracts. *Zeitschrift fur Naturforschung C-A Journal of Bioscience,* 55: 1030-1034.

Kähkönen, M.P., Hopia, A.I., Vuorela, H.J., Rauha, J.P., Pihlaja, K. and Kujala, T.S. 1999. Antioxidant activity of plant extracts containing phenolic compounds. *Journal of Agricultural and Food Chemistry,* 47: 3954-3962.

Karagözler, A.A., Erdag, B., Emek, Y.C. and Uygun, D.A. 2008. Antioxidant activity and proline content of leaf extracts from *Dorystoechas hastate. Food Chemistry,* 111: 400-407.

Kim, S.Y., Jeang, S.M., Park, W.P., Nam, K.C., Ahn, D.U. and Lee, S.C. 2006. Effect of heating conditions of grape seeds on the antioxidant activity of grape seed extracts. *Food Chemistry,* 97: 472-479.

Kubola, J. and Siriamornpun, S. 2008. Phenolic contents and antioxidant activities of bitter gourd (*Momordica charantia* L.) leaf, stem and fruit fraction extracts *in vitro. Food Chemistry,* 110: 881-890.

Kwon, Y.I., Vattem, D.V. and Shetty, K. 2006. Evaluation of clonal herbs of Lamiaceae species for management of diabetes and hypertension. *Asia Pacific Journal of Clinical Nutrition,* 15: 107-118.

Lee, K.G., Mitchell, A.E. and Shibamoto, T. 2000. Determination of antioxidant properties of aroma extracts from various beans. *Journal of Agricultural and Food Chemistry,* 48: 4817-4820.

Lopez-Candales, A. 2001. Metabolic syndrome X: a comprehensive review of the pathophisiology and recommended therapy. *Journal do Medico*, 32: 283-300.

Matsui, T., Ueda, T., Oki, T., Sugita, K., Terahara, N. and Matsumoto, K. 2001. á-Glucosidase inhibitory action of natural acylated anthocyanins. I. Survey of natural pigment with potent inhibitory activity. *Journal of Agricultural and Food Chemistry*, 49: 1948-1951.

Mazza, G., Fukumoto, L., Delaquis, P., Girard, B. and Ewert, B. 1999. Anthocyanins, phenolics, and color of Cabernet Franc, Merlot, and Pinot Noir wines from British Columbia. *Journal of Agricultural and Food Chemistry*, 47: 4009-4017.

McDougall, G.J., Shpiro, F., Doboson, P., Smith, P., Black, A. and Stewart, D. 2005. Different polyphenolic compounds of soft fruits inhibit á-amylase and á-glucosidase. *Journal of Agricultural and Food Chemistry*, 53: 2760-2766.

Miraliakbari, H. and Shahidi, F. 2008. Antioxidant activity of minor components of tree nut oils. *Food Chemistry*, 111: 421-427.

Monroy-Ort´ýz, C. and Castillo, P. 2000. Plantas medicinales utilizadas en el Estado de Morelos. Centro de Investigaciones Biol´ogicas, Universidad Aut´onoma del Estado de Morelos M´exico, Cuernavaca, M´exico, p. 260.

Mulimani, V.H., Rudrappa, G. and Supriya, D. 1994. Alpha amylase inhibitors in chickpea (*Cicer arietinum* L.). *Journal of the Science of Food and Agriculture*, 64: 413-415.

Nathan, D.M., Meigs, J.B. and Singer, D.E. 1997. The epidemiology of cardiovascular disease in type 2 diabetes mellitus: how sweet it is . . . or is it?. *Lancet*, 350: I4-I9.

Oomah, B.D., Tiger, N., Olson, M. and Balasubramanian, P. 2006. Phenolics and antioxidative activities in narrow-leafed Lupins (*Lupinus angustifolius* L.). *Plant Foods for Human Nutrition*, 61: 91-97.

Ortiz-Andrade, R.R., Garc´ýa-Jim´enez, S., Castillo-Espa˜na, P., Ram´ýrez-´Avila, G., Villalobos-Molina, R. and Estrada-Soto, S. 2007. á-Glucosidase inhibitory activity of the methanolic extract from *Tournefortia hartwegiana*: An anti-hyperglycemic agent. *Journal of Ethnopharmacology*, 109: 48-53.

Oyaizu, M. 1986. Antioxidant activity of browning products of glucosamine fractionated by organic solvent and thin layer chromatography. *Nippon Shokuhin Kogyo Gakkaishi*, 35: 771-775.

Pinto, M.S., Kwon, Y., Apostolidis, E., Lajolo, F.M., Genovese, M. and Shetty, K. 2008. Functionality of bioactive compounds in Brazilian strawberry (*Fragaria × ananassa* Duch.) cultivars: Evaluation of hyperglycemia and hypertension potential using *in vitro* models. *Journal of Agricultural and Food Chemistry*, 56: 4386-4392.

Pugalenthi, M. and Vadivel, V. 2007. L-Dopa (L-3,4-Dihydroxyphenylalanine): A non-protein toxic amino acid in *Mucuna pruriens* seeds. *Food*, 1: 322-343.

Pugalenthi, M., Vadivel, V. and Janaki, P. 2007. Comparative evaluation of protein quality of raw and differentially processed seeds of an under-utilized food legume, *Abrus precatorius* L. *Livestock Research for Rural Development*, 19: Article No. 168. http://www.cipav.org.co/lrrd/lrrd19/12/ puga19179. htm

Puupponen-Pimia, R., Hakkinen, S.T., Aarni, M., Suortti, T., Lampi, A.M. and Eurola, M. 2003. Blanching and long-term freezing affect various bioactive compounds of vegetables in different ways. *Journal of the Science of Food and Agriculture,* 83: 1389-1402.

Ranilla, L.G., Genovese, M. and Lajolo, F.M. 2007. Polyphenols and antioxidant capacity of seed coat and cotyledon from Brazilian and Peruvian bean cultivars (*Phaseolus vulgaris* L.). *Journal of Agricultural and Food Chemistry,* 55: 90?98-9104.

Ranilla, L.G., Kwon, Y.I., Genovese, M.I., Lajolo, F.M. and Shetty, K. 2008. Antidiabetes and antihypertension potential of commonly consumed carbohydrate sweeteners using *in vitro* models. *Journal of Medicinal Foods,* 11: 337-348.

Razali, N., Razab, R., Junit, S.M. and Aziz, A.A. 2008. Radical scavenging and reducing properties of extracts of cashew shoots (*Anacardium occidentale*). *Food Chemistry,* 111: 38-44.

Sabir, S.M. and Rocha, J.B.T. 2008. Water-extractable phytochemicals from *Phyllanthus niruri* exhibit distinct *in vitro* antioxidant and *in vivo* hepatoprotective activity against paracetamol-induced liver damage in mice. *Food Chemistry,* 111: 845-851.

Salunke, B.K., Patil, K.P., Wani, M.R. and Maheshwari, V.L. 2006. Antinutritional constituents of different grain legumes grown in North Maharastra. *Journal of Food Science and Technology,* 43: 519-521.

Shahidi, F. and Marian, N. 2003. Phenolics in food and nutraceuticals, Vol. 1 (pp. 144-150). Boca Raton, FL, CRS Press LLC.

Siddhuraju, P. 2006. The antioxidant activity and free radical-scavenging capacity of phenolics of raw and dry heated moth bean (*Vigna aconitifolia* Jacq. Marechal) seed extracts. *Food Chemistry,* 99: 149-157.

Siddhuraju, P. and Becker, K. 2001. Species/variety differences in biochemical composition and nutritional value of Indian tribal legumes of the genus *Canavalia. Nahrung/Food,* 45: 224–233

Siddhuraju, P. and Becker, K. 2003. Studies on antioxidant activities of *Mucuna* seed (*Mucuna pruriens* var. *utilis*) extract and various non-protein amino/imino acids through *in vitro* models. *Journal of Science of Food and Agriculture,* 83: 1517-1524.

Siddhuraju, P. and Becker, K. 2007. The antioxidant and free radical scavenging activities of processed cowpea (*Vigna unguiculata* (L.) Walp.) seed extracts. *Food Chemistry,* 101: 10-19.

Siddhuraju, P. and Manian, S. 2007. The antioxidant activity and free radical-scavenging capacity of dietary phenolic extracts from horse gram (*Macrotyloma uniflorum* (Lam.) Verdc.) seeds. *Food Chemistry,* 105: 950-958.

Sikora, E., Cieslik, E., Leszczynska, T., Filipiak-Florkiewicz, A. and Pisulewski, P. 2008. The antioxidant activity of selected cruciferous vegetables subjected to aquathermal processing. *Food Chemistry,* 107: 55-59.

Soobrattee, M.A., Neergheen, V.S., Luximon-Ramma, A., Aruoma, O.I. and Bahorun, T. 2005. Phenolics as potential antioxidant therapeutic agents: Mechanism and actions. *Mutation Research,* 579: 200-213.

Sowers, J.R., Standley, P.R., Ram, J.L., Jacober, S., Simpson, L. and Rose, K. 1993. Hyperinsulinemia, insulin resistance, and hyperglycemia: contributing factors in the pathogenesis of hypertension and atherosclerosis. *American Journal of Hypertension*, 6: 260-270.

Tsuda, T., Ohshima, K., Kawakishi, S. and Osawa, T. 1994. Antioxidant pigments isolated from the seeds of *Phaseolus vulgaris* L. *Journal of Agricultural and Food Chemistry*, 42: 248-251.

Turkmen, N., Sari, F. and Velioglu, Y.S. 2005. The effect of cooking methods on total phenolics and antioxidant activity of selected green vegetables. *Food Chemistry*, 93: 713-718.

Vadivel, V. and Pugalenthi, M. 2008. Removal of antinutritional/toxic substances and improvement in the protein digestibility of velvet bean seeds during various processing methods. *Journal of Food Science and Technology*, 45: 242-246.

Vadivel, V., Pugalenthi, M. and Megha, M. 2008. Biological evaluation of protein quality of raw and processed seeds of gila bean (*Entada scandens* Benth.). *Tropical and Subtropical Agroecosystems*, 8: 125-133.

Velioglu, Y.S., Mazza, G., Gao, L. and Oomah, B.D. 1997. Antioxidant activity and total phenolics in selected fruits, vegetables and grain products. *Journal of Agricultural and Food Chemistry*, 45: 304-309.

Vijayakumari, K., Pugalenthi, M. and Vadivel, V. 2007. Effect of soaking and hydrothermal processing methods on the levels of antinutrients and *in vitro* protein digestibility of *Bauhinia purpurea* L. seeds. *Food Chemistry*, 103: 968-975.

Wheeler, E.L. and Ferrel, R.E. 1971. A method for phytic acid determination in wheat and wheat fractions. *Cereal Chemistry*, 48: 312-320.

Worthington, V. 1993a. Alpha amylase. In *Worthington Enzyme Manual*; Worthington Biochemical Corp. (pp 36- 41). Freehold, NJ.

Worthington, V. 1993b. Maltase-R-glucosidase. In *Worthington Enzyme Manual*; Worthington Biochemical Corp. (p. 261). Freehold, NJ.

Xie, Z., Huang, J., Xu, X. and Jin, Z. 2008. Antioxidant activity of peptides isolated from alfalfa leaf protein hydrolysate. *Food Chemistry*, 111: 370-376.

Yamaguchi, T., Mizobuchi, T., Kajikawa, R., Kawashima, H., Miyabe, F. and Terao, J. 2001. Radical-scavenging activity of vegetables and the effect of cooking on their activity. *Food Science and Technology Research*, 7: 250-257.

Yen, G.C. and Duh, P.D. 1993. Antioxidant properties of methanolic extracts from peanut hulls. *Journal of American Oil Chemists Society*, 70: 383-386.

Zhang, D. and Hamauzu, Y. 2004. Phenolics, ascorbic acid, carotenoids and antioxidant activity of brocolli and their changes during conventional and microwave cooking. *Food Chemistry*, 88: 503-509.

□□□

12

Post Harvest Handling and Processing of Potential Nutraceutical Crops

Saji Gomez

CONTENTS

1.1 Introduction 178

1.2 Ginger 180
  1.2.1 Post harvest technology of ginger 180

1.3 Turmeric 182
  1.3.1 Post harvest technology of turmeric 182

1.4 Garlic 183
  1.4.1 Post harvest technology of garlic 184

1.5 Aonla 185
  1.5.1 Post harvest technology of aonla 185

1.6 Kokum 187
  1.6.1 Post harvest technology of kokum 187

1.7 Black Pepper 188
  1.7.1 Post harvest technology of pepper 189

1.8 Pomegranate 190
  1.8.1 Post harvest technology of pomegranate 190

1.9 Bael 192
  1.9.1 Post harvest technology bael 192

1.10 Passion Fruit 193
1.10.1 Post harvest technology of passion fruit 194
1.11 Saffron 195
1.11.1 Post harvest technology of saffron 195
1.12 Long Pepper 196
1.13 Conclusion 196
References 197

1.1 Introduction

Post harvest technology has immense potential to meet the food and nutritional requirements of an ever burgeoning population. Rapid urbanization and enhanced standard of living will have an impact on reduction of post harvest losses and improve net availability of quality produce to consumers. In spite of the rapid strides made in the expansion of area and production of several crops, efficient post harvest management of agricultural commodities is still a neglected area. Though many of our neighbouring Asian nations process 70-80% of their produce, India is still lagging behind in this front. It is estimated that 30-40% of the horticultural production goes waste which amounts to Rs 23,000 crore every year. This huge loss can be minimized by adopting appropriate post harvest measures namely, development of infrastructure for handling, grading, packing, storage, transportation, marketing and processing.

About ¼ of India's population is malnourished especially, people living in rural areas. The present per capita availability of nutrients is far below the ICMR recommendation. High incomes and improved life styles have made people becoming more health conscious and nutritionally aware. There is always demand for healthful and nutritious food. Quality enhancement is possible through value addition of agricultural commodities. Horticultural crops, especially fruits, vegetables and spices have significant role in human nutrition. They are considered as "protective food" as they are rich sources of vitamins and minerals. They are also important sources of dietary fibre and many phytochemicals. The antioxidants present in these food materials play an important role in scavenging the free radicals, which are responsible for various degenerative diseases like cancer, arthritis, premature ageing, cardiac problems etc.

The tendency is to avoid chemicals and synthetic foods and choose therapy and nutrition through natural resources. There are several fruits which are nutritive, therapeutic and of immense medicinal values. Many fruits are highly perishable and storing them for long periods, without being spoiled is a difficult task. Besides, some of them are not easy to eat out of hand. The bael fruit has a hard shell, mucilaginous pulp with numerous seeds and therefore, is not much popular as dessert fruit. Similarly the kokum and aonla fruits are not relished in their fresh form because they are highly acidic and astringent. The passion fruit is another example where the rind, the mucilaginous and acidic pulp pose problems in its fresh consumption. However, these fruits have immense potential in the processed form and consumers can enjoy these delicately flavoured and nutritious products during off season as well.

Post harvest technology is emerging as an important area linking agriculture with industry, there by offering scope for setting up cottage and small scale enterprises. Several agricultural commodities remain underutilized and by adopting simple and hygienic techniques, various nutritive, value added products can be prepared from these commodities. Lele (2001) has given a list of plants clinically proven as immunomodulatory drugs. He also exclaims the wisdom of ancient Indians to choose the most appropriate anti-oxidants for routine use to maintain positive health through drugs like turmeric, garlic, ginger, gooseberry etc. which are a part of the normal Indian vegetarian diet. Although many normal diets are culturally motivated, others are chosen with the intention of benefiting health. The ingestion of high amounts of "junk foods" in the population as a whole, but particularly among children seriously compromises the intake of nutritious alternatives (Subar *et al*, 1998).

Now-a-days, interest in the beneficial effects of antioxidants in vitamins and in foods from plant sources rich in flavanoids has increased. In a normal diet, most people consume sufficient amounts of antioxidants to satisfy nutritional requirements. In intervention studies, the use of antioxidants has been shown to reduce the risk of coronary heart disease (Mojzisova and Kuchta, 2001). Based on modest and statistically insignificant results, other studies actually promote the ingestion of certain foods to promote health (e.g., apples and onions to prevent heart attacks or lung cancer). Others generally suggest that the consumption of fruits, vegetables and red wine have value (Coursin, 1970).

Although several crops have been known to possess beneficial effects by virtue of their nutraceuticals, this chapter deals with some prominent crops which are reckoned as potential nutraceutical crops and their post harvest handling and processing.

1.2 Ginger

(*Zingiber officinale* Rosc., F. Zingiberaceae)

Ginger is a herbaceous perennial grown as an annual crop. Ginger of commerce constitutes the rhizome of *Zingiber officinale*. India is the largest producer of ginger. Kerala and Meghalaya are the major ginger producing states. Rhizomes contain carbohydrates 40-70%, protein 6-20%, fibre 3-8%, ash 8%, moisture 9-12%, essential oil 1.5-3%, pungent principles, minerals etc. Rhizomes contain numerous biologically active compounds. The major compound responsible for pungency is gingerol. Other compounds include zingiberol, zingiberrenol, zingiberine etc.

Ginger is a natural preservative, and hence widely used as a bactericide. It is found effective against the typhoid causing bacteria, *Salmonella typhi*. Ginger extracts are also used in anticarcinogenic preparations. It may protect the blood vessels from blockage such as atherosclerosis, which can lead to a heart attack or stroke (Furhmann *et al.*, 2000). Ginger juice is very effective against constipation, stomach upset and abdominal cramps. Ginger is used in traditional medicines to increase memory power. Ginger aids in blood circulation and is used against cold. Ginger oil is widely used in perfumery too.

1.2.1 Post harvest technology of ginger

For green ginger purpose, the crop is harvested from six month onwards. The rhizomes are repeatedly washed in clean water until the adhering dirt and dust are removed. The rhizomes are then dried in sun for a day.

Storage of rhizomes

Storage of seed rhizomes faces many problems due to rotting, sprouting, rooting, shriveling etc.Trials conducted in Kerala showed that storing ginger in 200 gauge thick polythene covers of size 35x25 cm with 125 punch holes (each hole with 4 mm diameter) was an effective method (Jayachandran *et al.*, 1992). There are three traditional methods of seed rhizome storage, namely, storage in soil pits, storage in a dry shady place, and storage in field involving delayed harvesting. The first method is the best for small scale growers, but expensive and laborious for large scale growers. Storage in a dry, shady place is economic for large growers but there is a problem of rhizome drying. Storage in field by delayed harvesting is not to be encouraged as it harbours rhizome rot causing organisms and pests (Rai and Hossain, 1998).

Dry ginger

For preparing dry ginger, the crop is harvested between 240-260 days after planting. The maturity is indicated by yellowing and subsequent drying up of leaves. The clumps are lifted carefully using a spade. The rhizomes are washed thoroughly 2-3 times in clean water to remove the adhering soil and dirt. These rhizomes are then soaked overnight in clean water. After soaking, the rhizomes are removed from water and their outer skin is peeled off using bamboo splinters having pointed ends. Peeled rhizomes are washed and dried in sun for a week. Dried rhizomes are then rubbed together to remove traces of adhering skin pieces. These rhizomes are known as 'unbleached ginger'.

Bleached ginger

Bleached ginger is obtained by soaking the peeled ginger in 2% lime solution for 6 hours and then drying.

Ginger powder

Ginger powder is made by pulverizing dry ginger to a mesh size of 50 to 60.

Ginger oil

Dried, cracked and comminuted ginger when subjected to steam distillation yields 1.0 to 3.0% oil. The oil prepared this way possesses the aromatic odour but lacks the pungent flavour of the spice. Ginger oil can also be distilled from fresh ginger and from ginger scrapings (Nybe *et al.*, 2007).

Ginger oleoresin

Ginger oleoresin is obtained by solvent extraction of powdered dried ginger. Suitable organic solvents are alcohol or acetone. It is known as 'gingerine' in trade. Oleoresin obtained from fresh ginger is known as green ginger oleoresin.

Grading

Ginger is sorted and graded based on size of rhizome, colour, shape, extraneous matter, presence of light pieces. Some well known grades are Garbled Non bleached Calicut (NGK), Ungarbled Non bleached Calicut (NUGK), Garbled Non bleached Cochin (NGC), Garbled Bleached Cochin etc.

Packaging

Packaging of ginger in proper containers can reduce post harvest losses considerably. Dry ginger is normally packed in gunny bags. Fresh ginger is packed in brace boxes or in conical bamboo baskets.

Other products

Ginger is an essential ingredient in many food products like curry powder, gingerbread, ginger ale, sauces etc. It is also used in soft drinks like cordial, ginger cocktail, carbonated drinks etc. It finds application in the preparation of ginger oil, oleoresin, tinctures etc. Ginger candy, a universally known product is prepared from green and fresh ginger. Spent ginger finds application in the preparation of several soft drinks. Alcoholic beverages from ginger include ginger brandy, ginger wine, ginger beer etc.

1.3 Turmeric

(*Curcuma longa* L.F. Zingiberaceae)

The rhizome of *Curcuma longa* L., belonging to the family Zingiberaceae, is used as spice. India is the largest producer and exporter of turmeric. The yellow pigment which gives turmeric rhizomes its natural colour is due to 'curcumin'. It is gaining importance as a natural food colourant as people are becoming increasingly health conscious and the use of synthetic colourants in food products is often looked down with apprehension. Curcumin is an antioxidant and is being increasingly used in several antiseptic, anticancerous and anti-inflammatory preparations. The neuroprotective effect of curcumin was investigated by Vajragupta *et al.*, (2003) and reported its efficacy against vascular dementia. They also evaluated the free radical scavenging ability of curcumin and its derivatives. Role of turmeric in Indian dishes for imparting flavour is unsubstitutable. It is mainly used as a food adjunct. It is also used widely in cosmetic preparations.

1.3.1 Post harvest technology of turmeric

The crop will be ready for harvest 7-9 months after planting. The rhizomes are collected by hand after ploughing the field. The rhizomes are cleaned to remove the adhering mud and dirt. The first step in processing is curing. For this, the fingers are separated from mother rhizomes. In the traditional method of curing, fresh rhizomes are boiled with water just enough to immerse them in copper or earthern vessels. Boiling is continued till froth appears and white fumes emanate with a typical odour. At this point (about 45-60 minutes) rhizomes become soft. If the rhizomes are over cooked it will spoil the colour of final product where as under cooking makes the product too brittle. In the improved method of curing, cleaned fingers (about 50kg) are taken in perforated troughs of 0.9m x 0.5m x 0.4m size made of GI or MS sheet with extended parallel handle. These perforated troughs containing fingers are then immersed in a pan, 100 litters of water is poured into the through to immerse fingers. Cooked fingers are taken out of the pan by lifting the trough so as to drain water into the pan. Water used for boiling can be used for curing fresh samples (Sreenivasan and Rema, 2002) Rhizomes should be processed either two or three days after harvest. If there is delay in processing, rhizomes should be stored under shade or covered with saw dust or coir dust.

Cooked fingers are dried in sun by spreading them in 5-7 cm thick layers on bamboo mats or on drying floor. During night, rhizomes should be heaped or kept covered with any material that provides aeration. The normal drying period is 10-15 days. Drying can also be carried out in dehydrators at a temperature of 60 ^{0}C. Yield of dried product varies from 10-30%.

The dried rhizomes are then subjected to polishing to improve their appearance by removing the scales and root bits on their outer surfaces. This is achieved by manual or mechanical rubbing. Polishing can also be done on power operated dreams. Colour of final product influences its price. In order to make the product attractive and bright, turmeric powder may be sprinkled during last phase of polishing. Half polished rhizomes can be given a much better colour by shaking with turmeric powder @ 20g per kg of turmeric and again drying under sun (Regeena, 2006). Agmark grades are prescribed for whole turmeric and turmeric powder separately for trading abroad and internally. Alleppey Finger Turmeric, Rajapore Finger Turmeric etc. are some well known grades.

Other processed products

Turmeric powder, volatile oil and turmeric oleoresin are other processed products of turmeric. India exports all these products .The curcumin content of turmeric oleoresin is 30-35%. Other value added products of turmeric include spice emulsions or aqua resins and encapsulated water- dispersible powders.

1.4 Garlic

(*Allium sativum* L., F. Amaryllidaceae)

Garlic is an important spicy bulb valued as food and medicine. The compound bulb consists of numerous smaller bulblets called cloves, covered by thin white or pinkish papery membrane. Garlic is a native of Western Asia and Mediterranean area. China, Korea, India, USA, Spain, Argentina and Egypt are the major garlic growing countries, China being the largest producer (Pandey, 2004). Garlic has a strong characteristic odour and an acrid taste. Active principle in garlic is an antibiotic 'allicin', which is produced from its precursor 'allin', a natural component of garlic. Pungency of garlic is due to diallyl disulphide.

Garlic is mainly used as a condiment for flavouring dishes. It acts as an antiseptic and also improves digestion. Recent research findings have proved the efficiency of garlic in reducing blood cholesterol, blood sugar and hyper tension. Allicin, the natural antibiotic in garlic has been found very effective against tumors. Further, it also increases the bioavailability of thiamine in humans.

1.4.1 Post harvest technology of garlic

Garlic is harvested when the top turns yellow and starts drying up, 5-6 months after planting. The bulbs are lifted carefully, adhering mud is removed and leaves are tied at top. Common practice is to bunch 15 to 25 bulbs and hang in well ventilated room. High humidity during storage increases microbial spoilage. The practice of smoking bulbs by hanging them from kitchen roof with tops intact is followed in certain hilly tracts. In cold storage, bulbs without tops are kept in crates or in clean gunny bags. The ideal storage condition is 0-1.6^{0}C temperature and 70% relative humidity. Pre harvest spray of maleic hydrazide 2500- 3000 ppm, 2-3 weeks before harvest inhibits sprouting of bulbs during storage.

Dehydrated garlic, garlic powder, garlic paste, garlic flake, garlic oil and oleoresin are some of the important processed products of garlic, which are described below.

Dehydrated garlic slices or rings

This product includes garlic pieces larger than 4mm slices by cutting garlic cloves and removing broken pieces smaller than 4 mm by sieving.

Garlic flake or garlic in pieces

Dehydrated garlic passing through a sieve of 4 mm aperture size but retained on a sieve of 1.25mm aperture size, which yields product without definite shape.

Garlic grits

Dehydrated garlic passing through a sieve of 1.25mm aperture size but retained on a sieve of 2.50 mm aperture size.

Garlic powder

It consists of dehydrated garlic passing through a sieve of 0.25mm aperture size. About 95% of the product passes through the sieve.

Garlic salt

This product comprises of 20% garlic powder, 78% table salt and 2% desiccant.

Oil of garlic

It is obtained by steam distillation of crushed fresh garlic cloves. Its yield is very low (around 0.1%) but is an excellent substitute for fresh garlic. It is used in pharmaceutical preparations and also for flavouring various dishes.

Garlic oleoresin

It is manufactured by extraction of garlic with water by vacuum concentration, the expressed juice and the aqueous extract of the press- cake, which yields a dark brown extract. It has been reported that one part of garlic oleoresin is equivalent to 50 parts of fresh garlic. Methyl alcohol was found to be an effective solvent in the preparation of garlic oleoresin.

1.5 Aonla

(*Emblica officinalis.*, G. Euphorbiaceae)

Aonla or Indian gooseberry is exclusively cultivated in the drier tracts of India. It is an important ingredient in the Unani and Ayurvedic systems of medicines. Aonla is well known for its nutritional qualities and pharmacological properties, as it is very rich in ascorbic acid and polyphenols. It is valued as an antiscorbutic, diuretic, alterative (Nadkarni, 1927). The fruits are used either in fresh or powdered form in various preparations such as *Chyavanaprash, Triphala* and *Arishta.* Besides ascorbic acid and tannins, it also contains gallic acid, tannic acid, catechol etc.The chemical constituents of the fruit include diterpenes, triterpenes, polyphenols, phyllanthine, alkaloids and low molecular weight hydrolysable tannins, emblicanin A and emblicanin-B are vitamin C like molecules. The antioxidant property of amla resides in the tannoid principles-emblicanin A and emblicanin B of the fruits (Benny *et al.,* 2006).

1.5.1 Post harvest technology of aonla

Storage, Grading and Packaging

The main season of harvest is during November -December. The mature fruits show change of colour from light green to dull, greenish yellow or yellowish with red blush. Bajpai (1969) observed that the best time of harvesting aonla fruits is February, when the fruits have maximum vitamin C content. Proper grading is not standardized in aonla; even then the harvested fruits are graded into three groups i.e. (i) large-sized, (ii) small-sized and (iii) blemished fruits. Large sized fruits are used for candy and preserve; small sized fruits are used for *chyavanprash* and *triphala* and the blemished fruits are used for making powder. For packaging aonla, wooden crates of 20-25 kg capacity with polythene liners have been found better. However, traditionally aonla is packed and sent to markets in gunny bags.

Ventilated cardboard cartons can also be used. Pigeon pea basket of 40-50 kg capacity with newspaper liner has also been found better. Aonla fruits cannot be stored beyond a week at ambient temperature. However, steeping in salt solution is a traditional practice followed to extend the storage period.

Due to its limited shelf life, aonla suffers huge post harvest losses during transportation and marketing. Pre- treatment of aonla fruits of cv. Chakaiya with chlorination (100ppm) and virosil agro (2.5%) in combination with low temperature storage in cool chamber and cool store prolonged the shelf life up to 16 days (Gomez and Khurdiya,2007). Pre-treatment with suitable chemicals and subsequent storage at optimum temperature can reduce the rate of respiration and thereby maintain the quality of aonla fruits. Kumar and Nath (1993) reported that physiological weight loss, decay and decrease in vitamin C content were reduced in *E.officinalis* cv. Chakaiya fruits stored in zero-energy cool chamber compared with fruits stored in ventilated polythene bags at room temperature. In the zero-energy cool chamber, fruits could be stored for up to 12 days with acceptable minimum decay and quality loss, compared with 4 days at room temperature. Aonla fruits can be stored at 0 to 1.7^0 C and 85-90% Relative Humidity for 8 weeks (Pantastico, 1975). The physiological loss in weight was minimum in aonla fruits of cv. Chakaiya when stored at zero energy cool chamber in combination with modified atmosphere storage (Singh and Kumar, 1997).

Processing

The acidity and astringency are two inherent characteristics of aonla that prevent consumers from relishing the fruits in the fresh form. Value added processed products of aonla which are nutritious and delicately flavoured can meet the demands of a health conscious population. Diversification of processed aonla products with altered palatability and enhanced nutrition has been attempted at the IARI, New Delhi. Heat processed and pasteurized, pure aonla juice can be stored safely up to 4 months at ambient temperature and up to 6 months low temperature (cool chamber and cool store) without much deterioration in quality. Similarly, pressure cooked and hermetically sealed aonla pulp yields shelf stable, intermediate product which can take care of post harvest wastage of aonla during gluts. Addition of 2000 ppm potassium metabisulphite can prevent spoilage of pulp. Dehydration of aonla shreds, after addition of 4% salt and subsequent packing in laminated aluminum foil-polythene pouches yielded a crispy and palatable product which can be used for masticatory purposes. Fruit bars prepared from pure aonla and aonla pulp enriched with apple, banana, carrot, papaya and mango pulps resulted in nutritious products with optimum blends of colour, flavour and texture. All the fruit bars, irrespective of their proportion of blending gave an organoleptic score above 6.0, which reflected on their acceptability. The processing techniques and storage conditions (cool chamber and cool store) employed helped in retention of quality in terms of ascorbic acid, tannins, sugars etc. and retarded microbial proliferation (Gomez and Khurdiya, 2007)

Fruits can be made into *murabba*, candy, pickles, jellies etc. Fruits are also used to prepare aonla powder which is superior to synthetic vitamin C in treating deficiencies. An easy way to prepare aonla *murabba* with high vitamin C content has been reported by Gupta and Bopaiah (1986). A comparative evaluation of different products revealed that dried aonla (flakes) had maximum nutritive value followed by *chyavanprash*, aonla preserve (*murabba*), pickle and brined aonla fruits (Naik and Chundawat, 1993).

1.6 Kokum

(*Garcina indica* Choisy. F. Clausiaceae)

Kokum or kokam an evergreen tree, is mainly confined to Southern India especially, along the Western Ghats. The kokum of commerce is the sun dried rind of this fruit after repeatedly soaking in its juice. It is used to impart flavour to curries. The rind of ripe fruits and juice are used in culinary for preparing 'Solkadi', a traditional preparation in Konkan region. The syrup prepared from ripe, fresh kokum fruits is known as 'Amrit kokam', which has attained commercial status in Konkan region and down south up to Kasaragod district of Kerala. The fat obtained from the seed kernels of this fruit is edible and is used in the preparation of kokum butter. It is used in several pharmaceutical preparations and cosmetic products. Fruits contain 15% hydroxyl citric acid, which has anti obesity properties. Ripe rind of kokum fruits is one of the richest sources of anthocyanin pigment. Studies have shown that anthocyanin has potential antioxidant properties. Garcinol, a fat soluble yellow pigment has also been isolated from the fruit rind (Sampathu & Krishnamurthy, 1982).

1.6.1 Post harvest technology of kokum.

Harvesting of kokum fruits starts from March and continues up to June. On ripening, the fruit changes its colour from green to red. Several processed products are prepared from kokum fruits as described below.

Kokum syrup

It is popularly known as 'Amritkokum'. The fresh, ripe kokum rind is made into pulp. Sugar syrup is added to this pulp i.e. for every 1 kg of kokum pulp, 2kg of sugar dissolved in 1 L of water is required to make the syrup. This mixture can be strained to improve the clarity of syrup and subsequently stored in glass bottles. For long term storage, 600 ppm of sodium benzoate can be added. Syrup is diluted with water before consumption as per one's taste. A pinch of common salt and cumin powder are sometimes added for improving the flavour (Joshi *et al.*, 1999).

Dried kokum rind

Rind of kokum fruits is dried under sun or in a dehydrator at 60⁰ C. It is treated with brined kokum juice and again dried. This process can be repeated for about five times. This dried product is used as an acidulant in curries.

The rind of green kokum fruits is also dried and used. The fruit is cut longitudinally into four pieces and dried under sun.

Kokum agal (Brined kokum juice)

It is prepared from either rind of ripe fruits and pulp or only from pulp after treating with 15-16% common salt. This mixture is kept for about 8 days and occasional string is done. After 8 days, the brined kokum juice is extracted and stored in glass bottles. This brined juice is used for making "Solkadi", a refreshing drink consumed after lunch.

Other products

Kokum rind contains about 2-3% anthocyanin pigment, a potential antioxidant. It is used as a natural food colourant. Kokum butter is another commercial product obtained from kernels. It is extracted mostly on a cottage industry basis by crushing the kernels, boiling the pulp in water and skimming off the fat from the top; or by churning the crushed pulp with water (Pruthi, 2001).

1.7 Black Pepper

(*Piper nigrum* L. F. Piperaceae)

It is also known as "King of spices" or "Black gold". It occupies a prime place in international trade and is widely used in cuisines all over the world. The black pepper of commerce is the dried mature fruit of the tropical perennial climbing plant *Piper nigrum*. It possesses antimicrobial and anti-oxidant properties. Pepper originated in the tropical evergreen forests of the Western Ghats. In India, pepper is grown mainly in Kerala, Karnataka, Tamil Nadu and to a small extent in Goa, Orissa, Assam and Andaman group of islands. Kerala accounts for 95% of area and production of black pepper in India. Quality of pepper is determined by two components viz.piperine, the alkaloid that contributes to pungency and volatile oil that imparts aroma and flavour.The essential oil of pepper contains a large number of volatile compounds. The main groups are monoterpene hydrocarbons (69.4-84%), sesquiterpene hydrocarbons (15-27%) and the rest were oxygenated constituents (Lewis, 1973). The major monoterpene hydrocarbons were a-pinene (5.9-12.8%), b-pinene (10.6%) and limonine (22-31.1%). The major sesquiterpene hydrocarbon was b-caryophylline (10.3-22.4%).

1.7.1 Post harvest technology of pepper

Harvesting of pepper is usually done in December to January. Stage of harvest is determined when one or two berries in a spike turn orange or red. Harvested spikes are spread on a clean floor and threshed manually by trampling with legs or by using mechanical thresher. The various processed products of pepper are described below.

Black pepper

It is produced from ripe, fully developed berries. After threshing, they are sun dried on drying floors. Occasional raking of berries is necessary to ensure uniform drying and to prevent mould growth. The moisture content of final product will be 10-12%. Blanching berries for one minute in boiling water before drying gives a lustrous black colour to the dried berries and prevents mouldiness of berries. But prolonged blanching should be avoided; since it can deactivate enzymes responsible for browning process (Nybe, 2001).The dried berries are garbled, graded and packed in polythene lined gunny bags. The dried pepper is cleaned to remove dirt, stones, leaves etc. and the berries are graded according to size and density. Garbling machines remove the extraneous matter and grade pepper according to density. Manual cleaning is also resorted to remove the foreign matter. Good garbled pepper should have a bulk density of 500-600 g /litre. Malabar Garbled, Malabar Ungarbled, Tellicherry Garbled Extra Bold etc. are some of the important grades of black pepper.

White pepper

It is prepared from ripe, red berries. These berries are detached from stalk and packed in gunny bags. Bags are allowed to soak in slow running water for about one week during which the pericarp gets loosened. Then these berries are trampled under feet to remove any adhering pericarp, washed in clean water and spread out in drying floors. The final product will be cream or white in colour with moisture content of 10-12%.

Pepper oil

Pepper oil is obtained by steam distillation of crushed black pepper. The oil yield is in the range of 2.5-3.5% and is colourless to pale green, which is used in perfumery and for flavouring dishes.

Pepper oleoresin

Pepper oleoresin is obtained by solvent extraction of black pepper with organic solvents like acetone, ethanol etc. which yields 10-13% oleoresin. Pepper oleoresin contains the same flavour, odour and pungency as that of the spice.

1.8 Pomegranate

(*Punica granatum* L. F. Punicaceae)

Pomegranate is best suited to regions with long, hot dry summer and cool winters or in areas with continuously warm and dry weather. In humid weather, the quality of fruit is not good. The edible portion of the seed is the aril, the juicy pulp covering the seeds which are red, pink or yellowish white in colour. It is a very rich source of anthocyanin pigment especially, the red and pink types. It also contains sufficient amounts of riboflavin, vitamin C and minerals. Pomegranate is important as a special diet for sick and aged persons (Yadav, 2007).

1.8.1 Post harvest technology of pomegranate

Storage, Grading and Packaging

Fruits are harvested when the rind turns slightly yellow and the fruit gives a metallic sound when knuckled. Fruits become ready for harvesting 5-7 months after flowering. The respiratory behavior and ethylene production clearly indicates the non climacteric nature of ripening of pomegranate fruit. Sorting is done to discard malformed, immature, diseased and cracked fruits. The fruits are graded into different categories namely, A-over 750 g, spotless; B-500-750 g, spotless; C-400-500 g, spotless; D-300-400 g; E-250-300g. After sorting, the fruits are cleaned with muslin cloth or by dipping in ethyl olate to remove dirt and to improve appearance. A and B grade fruits are individually wrapped in tissue paper or plastic liner. CFB cartons of 3 or 5 kg capacity with cushioning material can be used for packaging fruits for marketing and export (Anon, 2003). The average weight (300-450g) of fruit of cv. Ganesh is preferred for export to the Middle East while 250-300g in the UK, the Netherlands and Germany. The fruits have better keeping quality as compared to many tropical fruits due to the thicker rind. The optimum storage conditions for pomegranate are 5^{0}C temperature and 90-95% relative humidity. Waxing of the fruit reduces water loss and shriveling during storage.

Processing

Pomegranate fruit is not extensively processed as it is mostly relished in the fresh form. However, during gluts and when there are numerous cracked fruits, processing becomes a viable option.

Pomegranate juice

Pomegranate juice is a much sought- after product. Pressing fruit quarters in a rack and cloth hydraulic press yields 42% juice where as pressing with a basket press yields 70 % juice with low tannin. The juice, after extraction from the fruits is clarified by heating in a flash pasteurizer followed by cooling, settling for 24 hours, racking up and filtering. The clear juice is then preserved by heat treatment (Bose *et al.*, 1999).

Pomegranate rub

Equal quantities of juice and sugar are mixed and the syrup is concentrated on a slow fire to a thick consistency (70-75^0 Brix).

Pomegranate wine

The whole fruit is pressed without crushing and sugar is added to the juice to make up the brix to 22-23^0 TSS. Potassium metabisulphite is added as preservative. Wine yeast is added as starter to facilitate fermentation. The wine contains 11.9% alcohol, 0.85% total acidity, 0.07% volatile acidity and 11.5^0 Brix. TSS.

Grenadine

Equal parts of pomegranate juice and sugar are mixed and allowed to stand for three days. This mixture is brought to boil and simmered for 3 minutes and finally strained into sterilized bottles and sealed.

Pomegranate rind powder

The rind is cut into small pieces and dried at 60^{0}C for 48 hours. The dried peel is ground and sieved with a 60-mesh sieve and packed. The rind powder yield is about 13% and the powder is rich in carotene (12.52%), polyphenols (16.5%) and minerals (5.35%).

Anardana

Pomegranate arils when sun dried or dehydrated yields "anardana", which is used as an acidulant in Indian curries especially, in North India. Fruits from pomegranate trees growing abundantly in certain pockets of Jammu & Kashmir, Himachal Pradesh and Punjab are collected and used for preparation of anardana. In these regions, the fruits are harvested when they become ripe and brownish red in colour. Seeds along with the pulp are separated by hand from the rind and are sun-dried for about 10-15 days until their colour turns reddish brown. The drying ratio is normally in the range of 5:1. Singh *et al.* (2007) have described a methodology for producing good quality anardana. After separation from the fruits, the arils are subjected to steam blanching for five minutes. The arils are dried in cabinet drier at optimum temperature of 55^{0}C. The prepared anardana can be stored up to six months at room temperature and eight to ten months at low temperature (8±2^{0}C) when packed in LDPE 200 gauge bags and glass bottles. Pruthi and Saxena (1984) proposed Indian Standard specification for anardana which is detailed as under.

Physico- chemical parameters	Limit	%(m/m)
Foreign matter	Max.	5 .0
Moisture	Max	12.0
Acidity	Min.	9. 0
Total ash	Max.	4. 0
Acid-insoluble ash	Max	1. 2
Crude fibre	Max	29.0

Besides juice and anardana, squash, syrup, jam, fruit juice concentrates etc. are prepared from pomegranate fruits. In California, wine is prepared from pomegranate fruits with an alcohol content of 11.9%.

1.9 Bael

(*Aegle marmelos* F. Rutaceae)

Bael is an indigenous fruit crop. It is well known for its medicinal properties due to marmelosin content. Marmelosin is most probably the therapeutically active principle of bael fruit. It has been isolated as a colourless crystalline compound (Dixit and Dutta, 1932). The ripe fruit is tonic, restorative, laxative and good for heart and brain (Chadha, 2002). The ripe fruit is laxative and unripe fruit is prescribed for diarrhoea and dysentery. It has a great demand from Indian systems of medicine, such as Ayurvedic (Kirtikar and Basu, 1935). It is the richest source of riboflavin among fruits.

Fruits become fully mature 8 months after fruit set. At maturity, the rind colour changes from deep green to light green and pulp from light yellow to deep yellow. Fruits are harvested individually along with a portion of stalk and not allowed to drop on the ground. Harvesting by shaking the tree is discouraged as the fruits are likely to develop cracks on impact because of brittle peel.

1.9.1 Post harvest technology bael

Storage, Grading and Packaging

From the respiratory studies, the bael fruit can be classified as a climacteric fruit (Roy, 2001). Fruits harvested at full maturity can be stored for about 15 days, whereas those harvested at ripe stage can be stored only for a week. The shelf life of ripe fruits can be extended to about 3 months at 9°C temperature and 85-90% relative humidity. Marked physiological breakdown was noticed when storage temperature was below 9°C (Roy and Singh, 1979). Ripening of bael can be accelerated by a combination of high temperature and exogenous application of ethylene. Ripe bael could be made available 2-3 months ahead of normal ripening by treating the fruits with ethrel at 1000-1500 ppm and keeping the fruits at 30°C after harvesting them in January. It takes 18-24 days for the fruit to be artificially ripened. The composition of bael, whether ripened artificially or naturally, did not vary much; only the accumulation of sugar was slightly less in artificially ripened fruits compared with the natural ones (Roy and Singh,1981). Fruits are graded according to size. Fruits are mostly packed in gunny bags and sometimes in baskets for transportation and marketing.

Processing

Raw or mature green fruits are ideal for making preserve. Ripe fruits can be processed into RTS, nectar, squash, jam, leather etc. Extraction of pulp from ripe bael is the main hindrance to processing. Conventional methods may not give satisfactory pulp yield because of mucilaginous texture and the pulp may develop off flavour and discolouration due to enzymatic activity. Therefore, addition of water and subsequent heat inactivation of enzyme is done. Satisfactory pulp yield can be obtained by adding equal quantity of water to pulp, followed by adjusting pH to 4.3 with citric acid and heated to 80° C for 1 minute. The loss of peroxidase activity is taken as an indication of loss of activity of all other deteriorative enzymes (Eskin *et al.* 1971). It has been reported that some other products from bael fruit can also be made such as mixture of bael and milk powder, bael fruit flakes and enzyme treated clear bael juice (Singh and Roy, 1984).

1.10 Passion Fruit

(*Passiflora edulis* F. Passifloraceae)

Tropical America is the home of passion fruit. Passion fruit is viny in growth habit and there are two morphological forms of the fruit; purple and yellow types. It is consumed fresh as well as in the processed form. Passion fruit is cherished for its unique flavour and aroma. It excels many other fruit crops in the small quantity of pulp required to produce acceptable processed products. Passion fruit is a fair source of provitamin A and a good source of ascorbic acid, riboflavin, niacin and has a high mineral content. Besides its unique flavour, the high acid content of passion fruit juice is its most remarkable characteristic, which is important in processing and formulation of products. The floral-mixed fruit character of the juice has been attributed to the presence of deep yellow/orange volatile compounds, in particular ethyl butyrate, ethyl hexanoate, hexyl butyrate and hexyl hexanoate (Bose *et al.*, 1999). Hexyl hexanoate is the principal compound responsible for flavour.

1.10.1 Post harvest technology

Storage, Grading and Packaging

Passion fruit is non-climacteric. It begins to lose moisture after a few days of storage at ambient temperature which results in shrivelling of skin and some loss of flavour. Fruits can be stored without much deterioration in quality for about 4 to 5 weeks at 7 to 10 ⁰ C and 85-90 % relative humidity. Waxes and fungicidal dips have been used in Australia to extend shelf life of fresh fruits.

Fresh market acceptance in Australia is related to traditional consumer preference for a blemish-free, well-filled, flavoursome fruit with an attractive purplish black skin. Processors place high demands on pulp recovery, juice content and flavour, but acidity and soluble solids to a lesser degree. Fresh fruit in Queensland must comply with the following standards; full mature, half to full ripe but not over ripe, large (>40 mm in diameter) with at least 35% pulp recovery (Menzel *et al.*, 1988).Fruits are mechanically graded for size and for colour by hand, and then packed in 8 to 10 kg cartons. Smooth, blemish-free, dark purple, well graded fruits attract good prices. The demand is for fruits with a count of 120 to 140 per carton. Tray packs offer good presentation but the cost of such packaging is higher.

Processing

Passion fruit can be cut in halves and the pulp should be scooped out and added to fruit salads, ice-cream or fruit juices. The pulp can be preserved by heat processing or in frozen storage until consumed. Frozen storage has been found to be the best way for retention of flavour. Because of its strong acidic and aromatic flavour characteristics, passion fruit juice is generally diluted (1:10) when consumed as nectar by itself or used as a minor constituent when combined with other fruit juices. Passion fruit has a delightful flavour (Whitfield and Last, 1986), which is used for flavouring drinks or for blending with other juices. The common processed products of passion fruit are ; fruit nectar, *sherbet*, nectar combinations with citrus, pineapple and other juices, jam, jelly, squash, syrup and concentrate.

Passion fruit juice has a higher starch content compared with other fruit juices. If the starch is not removed, thermal processing produces a highly viscous product that is ideal for some uses. However, the starch may be removed in a decanting centrifuge, leaving a juice that can be concentrated to about 50°Brix. Casimir *et al.*, (1981) have discussed the types of evaporation equipment suitable for concentrating passion fruit juice. The important requirement for the equipment is a very short residence time in the evaporator to prevent thermal damage, which will destroy the flavour of the juice.

1.11 Saffron

(*Crocus sativus* L. F. Iridaceae)

Saffron is one of the most expensive spices traded in international markets. The dried trilobed stigma of saffron is the commercial product. Though it originated in South Europe, saffron is also cultivated widely in Jammu and Kashmir. Price of saffron is in the range of Rs. 45,000-50,000/ kg. Flowers are large, blue or lavender in colour with trifid orange coloured stigmas along with style tips, are dried to produce saffron of commerce. It is a very valuable ingredient in medicinal preparations of Ayurvedic and Unani systems of medicine. Saffron contains glycosides, crocin and picrocrocin and also lycopene, b-carotene, g-carotene and zeaxanthin. Crocin is the chief colouring principle in saffron. Chief flavouring component is safronal (Auria *et al.*, 2004)

Plants flower in October –December. Flowers are collected daily, early in the morning after dew disappears. Flowers are cleaned and style and stigma are separated from perianth. These post-harvest operations require skilled labour and are usually done by women. About 147,000 to 166,000 flowers give 5 kg of fresh stigma, which will ultimately yield 1kg of pure dry saffron. About 1, 60,000 flowers are hand picked to produce one kg of good quality of dried saffron (Singh *et al.*, 2008).

1.11.1 Post harvest technology of saffron

Colour, flavour and overall appearance are the factors reckoned in the quality of saffron. Trilobed stigmas are plucked from freshly harvested flowers and dried in sun, to a moisture content of 8-12%. This constitutes first grade, which is known by the name 'Shahi saffron' in trade. Sometimes, whole flowers are dried in sun for 3-5 days, beaten lightly with sticks and then passed through coarse sieves. Material which passes through is then thrown into water; petals that float on water are thrown away and parts which sink are collected and dried. This forms second grade saffron (Mongra saffron). Discarded parts of flowers are again subjected to beating process and process of throwing entire pounded mass in water is repeated. Product which sinks is collected, which is much inferior in value and this constitutes the third grade (Lacha saffron). During bumper harvest whole flowers are dried in sun and stigmas are separated by hand from dried flowers. Final product sold in market is a loosely matted mass of dark, reddish brown flattened stigmas with a characteristic aromatic odour and butter taste. The dried stigmas are packed in cloth bags or butter paper and finally in closed tins or in tight wooden boxes.

1.12 Long Pepper

(*Piper longum* L., F. Piperaceae)

Long pepper is the dried fruit of *Piper longum*, a slender creeping shrub but does not climb on trees as in black pepper. It is an important medicinal plant and is also used in flavoring dishes in many countries, particularly in Asia. It is also known by the name "Thippali" or "Pipali". It is an important ingredient in many herbal formulations. It is used in the treatment of diseases of respiratory tract, pain reliever, antihelminthic; dysentery etc. 'Piperine', an important constituent of long pepper enhances bioavailability. Piperine was shown to enhance the bioavailability of antitubercular drugs. The essential oil of the fruit showed antibacterial and antifungal activities. The major constituents are piperine, piperlongumine, piperlonguminine and methyl-3, 4, 5-trimethoxy cinnamate. The dried spikes are acrid, mildly thermogenic, stomachic, aphrodisiac, carminative, expectorant, tonic, laxative and antiseptic.

Spikes become ready for harvesting six months after planting, when they are green in colour. Spikes can be harvested 60-70 days after their formation. If left without picking, they ripe and their pungency is lost to a great extent. Harvested spikes are sun dried on drying yards for 4 to 5 days until they are perfectly dry. The green spike to dry spike ratio is around 10:1.5. Properly dried spikes will be black in colour. The dried spikes have to be stored in moisture proof containers.

Long pepper is boiled with ginger, mustard oil, butter milk and curd, it forms an ointment used in paralysis. In the Konkan region, the roasted spikes are beaten with honey and used in rheumatism. Long pepper in powdered form is mixed with black pepper and rock salt in 2:3:1 ratio in half teaspoon doses and used in colic (NIIR, 2007). Green long pepper is also used in pickling and for culinary purposes.

1.13 Conclusion

Efficient post harvest management of potential nutraceutical crops helps in reduction of gluts and increases the per capita availability of nutrients. Post harvest technology and agro-processing are vital links between agriculture and industry. They are also generators of gainful employment in the unorganised sector under village and cottage industries. Value addition to nutraceutical crops brings about changes in the physical form of the product, enhances its shelf life and improves its presentation to consumers. Concentrated efforts are needed to increase production and value addition of nutraceutical crops so that the concept of "Green Food" can be materialised. Branded, packaged products of nutraceutical crops have huge potential in the domestic and export markets. In this era of globalisation, development of technology and its adoption at primary production and subsequent post harvest stages will play a major role in meeting domestic requirements and enhances competitiveness in foreign markets.

References

Anonymous.2003. Dalib Vrit (in Marathi) Maharashtra Pomegranate Growers Research Association, Pune.

Auria, M.D., Mauriella, G and Rana, G.L.2004. Volatile organic compounds from saffron. *Flavour and Fragrance J.*19 (1):17-23

Bajpai, P.N.1969.*Hort.Adv.*, 8:15-16

Benny, M., Benny, A and Pranitha P.2006. Quantitative determination of Emblicanin-A, present in standardized amla extracts by HPTLC. *Spice India* 19 (3):22-24.

Bose, T.K., Mitra, S.K., Farooqui, A.A and Sadhu, M.K. (1991). Tropical Horticulture, Naya Prakash, Calcutta.

Casimir, D.J., Kefford, J.F and Whitfield, F.B.1981. Technology and flavour chemistry of passion fruit juices and concentrates. *Advances in Food Research*, 27; 243

Chadha, K.L. (2002) Handbook of Horticulture, Indian Council of Agricultural Research. p. 140

Coursin, W.E.1970. Relationship of nutrition to mental growth and development. *Mod. Med* 38:81-84

Dixit, B.B.K and Dutta, S.1932.*J.Indian Chem.Soc*, 9:271-279.

Eksin, N.A.M., Henderson, H.M and Townsend, R.J.1971. In: *Biochemistry of Food*, Academic Press, London.

Fuhrmann, B., Rosenbaldt, M., Hayek, T., Coleman, R., Aviram, M.2000.Ginger extract consumption reduces plasma cholesterol, inhibits LDL oxidation and attenuates development of atherosclerosis, apoliprotein L-deficient mice. *J nutr.* 130 (5):1124-1131.

Gomez Saji and Khurdiya D.S. 2007. Value addition in aonla through product diversification: *Proceedings of the National Seminar on Nutrition and Life style Diseases.* Thrissur.pp 26-35.

Gupta, G.K and Bopaiah.1986. *Indian Hort.*, 31:15

Jayachandran, B. K., Bai, M.M., Salam M.A and Mathew, K.P.1992. Storage of seed ginger. *Spice India* 5(12):4-8

Joshi, G.D., Nadkarni, H.R and Patil, R.S.1999. Processing of kokum (*Garcinia indica*) and citrus. (In): *Post Harvest Handling and Processing of Horticultural Crops*.Directorate of Extn.Edn.,Dr.Balasaheb Sawant Konkan Krishi Vidyapeeth, Dapoli, Dist.Ratnagiri,pp.20-21.

Kirtikar, K.R and Basu, B.D.1935.In: Indian Medicinal Plants, Vol I (ED.L.M.Basu), Allahabad, India

Kumar, S and Nath, V.1993.*J.Food Sci.Technol*, 30:202-203.

Lele, R.D.2001.Ayurveda and Modern Medicine. Bharatiya Vidya. (In: *Spice India* 19 (3) :

Lewis, Y.S.1973. The importance of selecting the proper variety of a spice for oil and oleoresin extraction. *Proc. Conference on Spices*. T.P.I. London, pp.183-185

Mojzisova, G and Kuchta, M.2001.Dietary flavanoids and risk of coronary heart disease. *Physiol. Res* 50:529-535

Nadkarni, A.K.1927.*Indian Materia Medica*, S.K.Bijur, Bombay

Naik, A.G and Chundawat, B.S.1993. In: *Golden Jubilee Symp. Horticultural Research-Changing Scenario. Bangalore, Hort.Soc.India*, May.24-28, 1993, Abst.No.18.6, p.337

National Institute of Industrial Research.2007. Cultivation and Processing of Selected Medicinal Plants. Asia Pacific Business Press, New Delhi, India. pp 435-439

Nybe, E.V.2001.*Three Decades of Spices Research at KAU*. Kerala Agricultural University, Thrissur, India, 131p.

Nybe, E. A., Mini Raj, N., Peter, K.V.2007.*Spices*. New India Publishing Agency, New Delhi.

Pandey, U.B.2004.Garlic. In: Hand Book of Herbs and Spices. Vol.I. (ed. Peter, K.V.). Wood head Publishing Limited, England, pp.180-194.

Pantastico, Er.B.1975. *Postharvest Physiology, Handling and Utilization of Tropical and Subtropical fruits and Vegetables*, AVI Publishing Co., WestPoint, Conn.

Pruthi, J.S. 2001.Minor spices and Condiments-Crop Management and Post Harvest Technology. Indian Council of Agricultural Research, New Delhi.

Pruthi, J.S and Saxena, A.K. 1984. Studies on anardana (Dried pomegranate seeds); Composition and packaging requirements of anardana *J .Fd Sci. Technol*.21 (5) 296-9.

Rai, S and Hossain, M.1998. Comparative studies of three traditional methods of seed rhizome storage of ginger (*Zingiber officinale* Rosc.) cultivars and their association. *New Zealand J.Crop and Hort. Sci*.27:79-82

Regeena S .2006.Turmeric-the spice that fights cancer. *Spice India* 19 (3):15-16

Sampathu, S.R and Krishnamurthy, N. 1982. Processing and utilization of kokam (*Garcinia indica*).*Indian Cocoa, Arecanut and Spices J*.6 (1):12-13

Roy, Susanta K.2001. In: *Fruits: Tropical and Subtropical*, Vol II (Eds.Bose et.al), Naya Udyog, Calcutta, India.

Roy, Susanta K and Singh, R.N.1979.*Progressive Hort.*11:21-28

Roy, Susanta K and Singh, R.N.1981.*Punjab Hort.J.*21:74-82.

Singh, D.B., Kingsly, A.R.P and Patil, R.T.2007. Anardana-A promising spice. *Spice India* 19 (3):21-22

Singh, R and Kumar, S. (1997). Effect of different storage conditions on the shelf life of aonla cv. Chakaiya. *Haryana J.Hort. Sci.*, 26 (1-12): 12-15.

Singh Rekhi, Singh Purshotam and Singh Parmeet.2008.Scientific cultivation of saffron. *Spice India* 19 (3):10-13.

Singh, R.N and Roy, S.K.1984. *The Bael*, ICAR, New Delhi, pp.21-22.

Sreenivasan, V and Rema, J. 2002. Cultivation practices for sustainable production of black pepper, ginger and turmeric. Spices Production Technology (eds.Korikamath, Faleiro, V.S., Manjunath, B.L. and Ashok Kumar, J.). ICAR Research Complex for Goa, Ela, Old Goa, pp.21-29

Subar, A.F., Krebs-Smith, S.M., Cook, A and Kahle, L.L.1998. Dietary sources of nutrients among US children, 1989 to 1991. *Pediatrics* 102:913-923

Vajragupta,O.,Boonchoong,P.,Watanabe,H.,Tohda,M.,Kummaasud,Nand Sumanaont,Y.2003.Manganese complexes of curcumin and its derivatives-evaluation for the free radical scavenging ability and neuroprotective activity. *Free Radic.Bio.Med.* 35.1632-1644

Whitfield, F.B and Last, J.H.1986.The flavour of passion fruit-a review. *Progress in Essential Oil Research.* Walter de Gruyter, Berlin, Germany

Yadav. P.K.2007. Fruit Production Technology. International Book Distributing Co. Lucknow.p.308

□□□

13

Quality Control of Herbal Plant Material

M. Geetha

CONTENTS

1.1 Introduction 202

1.2 Quality Criteria 203
- 1.2.1 Identity 203
- 1.2.2 Purity 203
- 1.2.3 Content or assay 203
- 1.2.4 Safety 204
- 1.2.5 Microbial limits 204

1.3 Quality Evaluation 204
- 1.3.1 Organoleptic evaluation 204
- 1.3.2 Microscopical evaluation 204
- 1.3.3 Physical evaluation 205
  - 1.3.3.1 Ash values 205
  - 1.3.3.2 Extractive values 206
  - 1.3.3.3 Volatile oil determination 206
  - 1.3.3.4 Physical constants 206
  - 1.3.3.5 Microbial limits 206
  - 1.3.3.6 Pesticide residue 206
  - 1.3.3.7 Heavy metals 207
  - 1.3.3.8 Aflatoxins 207

1.4 Thin Layer Chromatography 207
1.5 High Performance Thin Layer Chromatography 207
1.5.1 Markers 207
1.5.2 Assay or content 208
1.5.3 Gravimetry 208
1.6 Chemical Method 208
1.7 Spectroscopic Methods 208
1.8 Biological Methods 209
1.9 Miscellaneous Methods 209
1.10 DNA Fingerprinting 209
1.11 Hyphenated Techniques 209
1.12 Conclusion 209
References 210

1.1 Introduction

WHO estimated that 80% of world's population in the developing country depends on the herbal drugs. Today in the Western countries people make use of thousands of indigenous plants that are native to America, Europe, Africa and Australia (IHP, 2002). About 500 herbs are employed with in the conventional medicine, although whole plants are rarely used. From ancient times many herbs played pivotal role in providing lead compounds (WHO, 1998).

Herbal market in India is estimated at Rs.7, 000 crores for the current year. Rs. 3,600 crores worth of raw materials derived from medicinal plants are exported (BHP, 1996). Significant growth is likely to takes place by 2012 and this is expected to touch Rs.14, 500 crores. In 1990, sale of OTC phytomedicine in seven European countries were about $2.4 billion (Rao, 2000). 70,000 plant species are used for treating various ailments all over the world. Chinese Pharmacopoeia included 5,700 plant medicines. In Ayurveda 2,000 medicinal plant are considered to have medicinal value (WHO, 1992).

Exporters of medicinal plant material from India face stringent quality norms imposed by the EU through the Traditional Herbal Medicinal Products Directive (THMPD), Food Supplement Directive (FSD) which encourages high quality products. In Germany under Germen medicines Act, Commission E governs herbal product import. In Canada, Health directorate of Canada issues a Drug Identification Number (DIN) for each herb. In Australia, Therapeutic Goods Administration (TGA) formed a committee known as

Complementary Medicines Evaluation committee (CMEC) which is looking after the import, in USA herbal drugs import and sale are monitored by Food and Drug Administration (FDA) and in China by Chinese state FDA.

As per WHO (1998) herbal drugs are defined as plant or plant parts that have been converted into phytopharmaceutical by means of simple processes involving harvesting, drying and storage. Crude products derived from plants which no longer show any organic structure, such as essential oil, fixed oil, resins and gums.

Derived or isolated compounds in the processed state such as extracts or even isolated purified compounds (e.g.: Sennoside- B from *Cassia angustifolia*) will no longer considered as herbal drugs.

Herbal medicines constitute a wide spectrum of raw materials which makes difficulty in defining the identity and potency of plant materials (Johnson, 1940; WHO, 1995). The quality herbal drugs posses difficulty mainly because, the activity of herbal products in many cases due to mixture of many complex secondary metabolites, active principle(s) in most cases unknown, selective analytical methods or reference standard may not be available commercially. Process such as, harvesting, drying storage, transportation may vary from place to place. Environmental factor such as, climate, altitude, rainfalls have huge effect on plants. Genetic factors such as polyploidy, mutation may also show variation. The plant material collected from wild will differ from the cultivated plant products. The quality criteria therefore based on scientific definitions of these raw materials.

1.2 Quality Criteria

Quality of raw material has to be based on pharmacopoeia definition such as, Identity, Purity and Content or Assay.

1.2.1 Identity

Authenticity or proper botanical identity of plant material is the first step in establishing the quality. The authentication can be done by, herbarium specimen and/or with the help of taxonomist.

1.2.2 Purity

It is considered with reference to the other material or adulteration or admixture.

1.2.3 Content or assay

It deals with the amount of active constituents responsible for the biological action. Depending on the type of crude drug the percentage active constituents measured or tits potency is checked.

1.2.4 Safety

Commercially cultivated drugs may show toxic material such as total organic chloride, total organic phosphorous as pesticide residue, heavy metals and in some herbs aflatoxins may be present.

1.2.5 Microbial limits

The amount of pathogenic and pathogenic micro organism is measured and it should be as per the WHO limit.

1.3 Quality Evaluation

To govern the different standards of quality of the plant material, it has to under go a set of evaluation parameters. They are Organoleptic Evaluation, Microscopical Evaluation, Physical Evaluation, Chemical Evaluation and Biological Evaluation.

1.3.1 Organoleptic evaluation

The characters such as colour, odour, taste and external character of a plant are evaluated by using sense organs (Evan *et al.*, 2002; Brain and Turner, 1985). Morphological or gross anatomical characters such as size, shape of leaf, stem, bark, root and rhizome. The margin, base, apex type of venation are the identification feature of leaf. Broken surface character is measure for bark, root and stem. For very small seed the number of grains per gram is used as measurement. From the colour of Chamomile flowers one can assess whether it is collected in proper season, Fennel exhausted using water will be dark coloured and having foul odour. By pressing between filter papers the quality of clove can be found.

1.3.2 Microscopical evaluation

For initial identification, microscopy is the important parameter for any herb. Staining reagent is used for identification and location of various tissues and its contents. Phloroglucinol and HCl used for lignified tissues, tincture of alkana for suberized or cuticularized walls and ruthenium red for gums and mucilage (Johnson, 1940; Kokoshki *et al.*, 1985). Permanent slides of authentic crude drugs or direct comparison with authentic crude drug or description given in the reference books are used as primary standard for the anatomical characters. Histochemical testes for ergastic substance like starch, fixed oil, aleurone grain, mucilage and calcium oxalate can be done in microscopically.

Leaf constants such as, Stomatal number, Stomatal index, Palisade ratio, vein termination and vein islet number are important diagnostic character for the authentification, to find seasonal variation, inter species admixture and adulteration with some other leaf.

Microscopical evaluation for the powered drug is equally important for the presence of adulterant and foreign organic matter. Micrometers are used to authenticate cell dimension, sclerides and trichomes. Quantitative microscopy can be employed using Lycopodium Spore Method for the drugs which fails to produce significant chemical or biological characters. For the quantitative measurement the plant material should possess characteristic features such as, stone cells, calcium oxalate, phloem fiber and starch grains.

1.3.3 Physical evaluation

Foreign Organic Matter (FOM) means the plant part(s) other than mentioned as per definition. In pharmacopoeia for Senna leaf 5% FOM is allowed that means the stalks, fruits and flowers are limited to 5%. Generally pharmacopoeia prescribe 2% limit for FOM. If there is standard is mentioned the 100% pure material should be supplied, for e.g.: *Nuxvomica* Seed. But, in all the case it will be difficult, costly and also it may not pose any threat to the quality.

Insects due to its small size, negligible weight contribute to FOM. For e.g. a common drug room beetle *Stegobium paniceum* weighs between 0.75-2 mg. If we take into consideration of 2% FOM then 27 beetles per gram will be present. That is too much infestation in such case, drug should be free from gross deterioration, insects, moulds, hairs of rodents.

Moisture triggers the enzymatic reaction and microbial growth and increase the weight of the drug. In some country the value of medicinal products was done not by weight but by the content of active constituents. The moisture content may be measured by loss on drying, azotropic distillation and Karl Fischer method.

1.3.3.1 Ash values

Total ash values or Physiological ash

Carbon and organic matter in the drug is converted into as at 450°C. Silicates, calcium oxalate, calcium hydroxides are converted into ash. Total ash value is misleading for the drug which has wide range of ash value for e.g.: Rhubarb has total ash value of 8-40% (Malti *et al.*, 2001).

Acid insoluble ash or Non Physiological ash

This is most reliable than total ash for the drugs like Rhubarb. Total ash is treated with dilute HCl acid to remove inorganic salts to yield mainly silica (Kokate *et al.*, 1985).

Water soluble ash

It is used for the drugs which contain water soluble constituents such as tea and ginger. From the total ash, water insoluble salts are removed and weighed.

1.3.3.2 Extractive values

Biological activity is mainly due to active constituents which has different solubility in different solvents (Osel *et al.*, 1975). The extractive values used as assay methods as per certain pharmacopoeias.

1.3.3.3 Volatile oil determination

As per British Pharmacopoeia Clavenger's apparatus is used for hydro or steam distillation. Volatile oil gets separated at the boiling point and gets collected at graduated receiver. For the drugs which contain volatile oil heavier than water light modified Clavenger's apparatus is used.

1.3.3.4 Physical constants

Melting point, boiling point, freezing point and sublimation point are used for the drugs such as volatile oil, fixed oil, alkaloid, glycoside etc. Refractive index, optical rotations are important parameter for volatile oil and fixed oil.

1.3.3.5 Microbial limits

Parameters	Raw material	Finished product
Staphylococcus aureus / g	Absent	Absent
Salmonella sp./10g	Absent	Absent
Pseudomonas aeruginosa	Absent	Absent
E. coli /10g	10^4/g	10^1 /g
Total aerobic bacteria	10^5 /g	Absent
Total Yeast and Mould	10^3/g	Absent

1.3.3.6 Pesticide residue

Pesticides	Permissible Limits
Organochloro Group Test	1 ppm
Quinolphos*	0.01 ppm
DDE*	1 ppm
Alderin*	0.05 ppm
Dieldrin*	0.05 ppm
DDT*	1.0 ppm
DDD	1.0 ppm
HCH (Hexachloro cyclohexane)	0.30 ppm

Note:* to be tested only when sample fails in organo chloro Group Test

1.3.3.7 Heavy metals

S. No	Heavy Metal content	Permissible Limits
1	Lead (Pb)	10 ppm
2	Cadmium (Cd)	0.30 ppm
3	Arsenic (As)	10 ppm
4	Mercury (Hg)	1 ppm

1.3.3.8 Aflatoxins

S. No	Aflatoxin	Permissible Limits
1	B1	0.5 ppb
2	G1	0.5 ppb
3	B2	0.1 ppb
4	G2	0.1 ppb

1.4 Thin Layer Chromatography

According to the affinity to solvent system the chemical constituents move in support of adsorbent. It is widely used due to simplicity, versatility, specific sensitivity and simple sample preparation. It is the first characteristic finger print (FP) of the herb. By establishing identity of herb with different active constituent, qualitative & quantitative data can be gathered. This method is used for the drugs which contain volatile oil, resins and gums (Kokate *et al.*, 1985). This will be helpful when macro & microscopy fails to identify. Co-TLC with the reference standard is the routine procedure to authenticate the sample. The visualization of the spots of the compounds under study can be done by using UV light or various visualizing agents which develops coloured spots. For steroids, terpenoids 1% vanillin-sulphuric acid and Dragendorff"s reagent for alkaloidal constituents are used.

1.5 High Performance Thin Layer Chromatography

Every step is sophisticated from TLC. Spotting of the solute is done by using injector and drying of spot was achieved by nitrogen gas (Sharma,1997). For the development of chromatogram twin trough chambers are used which reduce amount of solvent. Derivatization achieved by dipping the chromatogram in the visualization reagent. UV detector or diode array detector are mainly employed for detection.

1.5.1 Markers

Marker testing is in no way a substitute for physico-chemical parameters and microscopic characterization. Markers can be of two types, bio Markers and chemical markers.

The secondary metabolite with known pharmacological activity, e.g.:

Ashwagandha (*Withania somnifera)* - Withanolides
Guggulu (*Comifora mukul*) - Guggulusterones
Neem (*Azadirachta indica*) - Azadirachtin or Nimbidine
Haridra (*Curcuma longa*) - Curcumunoids.

Without established activity but chemically identifiable and predominantly available. E.g.:

Bilva (*Aegle marmelos*) - Aegelin
Shatawari (*Asparagus recemosus*) - Satawarine
Aragvadha (*Cassia fistula*) - Fistulin

1.5.2 Assay or content

After preliminary purification or fractionation of the active constituents of the drug is required and for this chromatography is increasingly used. Crude drugs assayed for particular group of constituents for e.g. the total alkaloids in belladonna, the total glycosides of digitalis.

1.5.3 Gravimetry

The constituents are converted into insoluble salts and the residue is weighed. For e.g.: Colchicine in Colchicum, Resin of Podophyllum and Total Balsamic ester of Tolu Balsam.

1.6 Chemical Method

Acid-Base titrimetry is useful in alkaloid base, non-aqueous titration for alkaloid salts such as, Strychnine in Nux vomica, Morphine in opium.

Free alcohol determination in Pepper mint oil and acid value, saponification value and iodine value for fixed oils are used.

1.7 Spectroscopic Methods

UV - Visible Spectroscopy often very fast and reproducible. The product to be analyzed is diluted the absorbance value is measured against blank. The value is generally calculated A % or E%. This is made possible as many characteristic of the plant absorb UV rays at a set wave length which is invariable and characteristic of the plant (Sharma, 1997; Pulok and Mukherjee, 2007).

IR spectroscopy can be used to evaluate the herbs. Even though it conform the plant, it is expensive.

1.8 Biological Methods

Biological assays remain important for screening plant material when there is no other satisfactory method is available. They are time consuming and now a days replaced by chemical and physical assays. Biological assays usually employed for cardio active and anti-tumor drugs.

1.9 Miscellaneous Methods

Radio Immuno assay for Cardinolides, Sennosides, Morphine, Enzyme Immuno assay for Podophyllotoxin, Quassin, Ergot alkaloids

Microbial assay for v.t. B12, Calcium pantothanate, Niacinamide are employed.

1.10 DNA Fingerprinting (based on PCR)

It reveals the specific DNA profile for a particular organism which is as unique as a fingerprint, DNA FP is generally independent of environment, and DNA FP is consistent throughout different parts and developmental stages of the organism.

1.11 Hyphenated Techniques

HPLC-DAD, GC-MS, CE-DAD, HPLC-MS and HPLC-NMR. They provide additional information on spectral characteristics for qualitative analysis and even on-line structural elucidation.

1.12 Conclusion

Even though the active principle is known, selective analytical methods or reference compounds may not be available commercially. Further the source of raw material and the procedures used for processing them are variable. Standardization of herbal drug involves adjusting its preparation to a defined content of a constituent or group of substances with known therapeutic activity by adding excipients. Such modification is desired to provide uniformity in the quality of the product as well as the adjustment of the potency of the medicinal plant. Many international authorities and agencies, including the World Health Organization, European Agency for the Evaluation of Medicinal Products and European Scientific Cooperation of Phytomedicine, US Agency for Health Care Policy and Research, European Pharmacopoeia Commission, Department of Indian System of Medicine have started creating new mechanisms to induce and regulate quality control and standardization of herbal drugs.

In India, Good Manufacturing Practice (GMP) guidelines have been introduced so as to ensure quality control. Medicinal plant boards have been constituted at state and central level to inspire farmers for adopting cultivation of medicinal plants. Herbal gardens are developed to make common man conversant with rich heritage of Indian system of medicine.

References

Anonymous, 2002. Indian Herbal Pharmacopoeia, IDMA, Mumbai, pp. 493-494

BHP, 1996. British Herbal Pharmacopoeia, British Herbal Medicine Association, 1996.

Brain, K.R. and Turner T.D. 1975. The practical Evaluation of Phyto Pharmaceuticals, Wright Scientchnica, Bristol, pp. 36-53.

Dyre, J.R. 1997. Application of Absorption Spectroscopy of Organic compounds, Prentice Hall of India Pvt Ltd., New Delhi.

Evan W.C., 2002. Pharmacognosy, W.B. Saunders (Ed.), London, 513: 4-547.

IHP, 2002. Indian herbal pharmacopoeia, Indian Drug Manufacture's Association, Mumbai.

Johnson, D.A., 1940. Plant micro technique, Mc Grow Hill Book Co., New York and London 1 st Ed., pp. 182-203

Kokate, C.K., Kumar, A., Rao, P.V.P., Rambhau, D. and Rao, Y.M. 1985. *Journal of Research Education and Industrial Medicine*, 5: 25-36.

Kokoshki, C.J. Kokoshki, R.J. and Aim, F.J. Ji Amer. 1985. *Pharmaceutical Association*, 47: 715-728.

Malti, G.C., Hina K.P. and Patel, K.N. 2001. *Ayu international,* 1: 1-7.

Osel, A., Gennaro, A.R. and Martin, A.B. 1975. Remington's Pharmaceutical Sciences, Mack Publishing Company, Pennsylvania, pp. 125-142.

Pulok, K. and Mukherjee, S. 2007. Quality control of herbal drugs, Galaxy printers, pp. 112-517.

Rao E.V. 2000. *Eastern Pharmaceuticals*, XXXI: 35-38.

Sharma, B.K. 1997. Instrumental methods of chemical Analysis, Goel Publishing housed, Meerut, 16th Edision.

WHO, 1992. Expert committee on Specification for Pharmaceutical preparation, pp. 44-52.

WHO, 1995. *Technical report series*, 823: 75-76.

WHO, 1998. Quality Control Methods for Medicinal Plant Materials, Geneva.

□□□

14

Toxicity Evaluation of Herbal Drugs in Animal Models

R. Ilavarasan

CONTENTS

1.1	Introduction	212
1.2	Toxicity Study	212
1.3	OECD Guideline for Toxicity Studies	212
1.4	Conclusion	220
	References	220

1.1 Introduction

Herbal drugs have made a great contribution to maintaining human health. Majority of the world's population in developing countries still relies on herbal drugs. WHO supports the use of herbal drugs and encourage the remedies that have been proven to be safe and effective. A few herbal drugs have withstood scientific testing, but others are used simply for traditional reasons to protect, restore or improve health. No evidence to answer the question of safety and efficacy about most of the herbal drugs now in use. It is still need to be validated scientifically. Toxicity evaluation of herbal drugs is necessary to document the toxicological profile and boost the export of herbal drugs manufacturing in India.

1.2 Toxicity Study

It is essential for any new compound having biological activity and applicable to use in human beings.

1.A cute toxicity study: Adverse effects occurring within a short time following administration of single dose.

2.Repeated dose toxicity study

a.Sub-acute toxicity- 6-8 weeks.

b.Chronic toxicity- Period of 1 year or more.

3.Dermal toxicity test (Acute and repeated dose dermal toxicity studies)

4.Special toxicity studies- Period of 2 years or more

a.Mutagenicity test

b.Carc inogenicity test

c. Reproductive and development toxicity tests

Paracelsus (1493-1541) stated that "All substances are poisons; the right dose differentiates a poison and remedy" this concept is the fundamental principle of toxicology and hazard assessment. WHO- guidelines have given the important criteria to establish the safety profile of the drugs.

Toxicological study results play an important safety assessment for Herbal Drugs, Pharmaceuticals, food additives, pesticides and other chemicals.

1.3 OECD Guideline for Toxicity Studies

420 Acute oral Toxicity fixed dose method

423 Acute oral Toxicity acute toxic class method

425 Acute oral Toxicity Up and Down method

402 Acute dermal Toxicity

404 Acute dermal Irritation/ Corrosion
403 Acute inhalation Toxicity
405 Acute Eye Irritation /Corrosion
406 Skin sensitization
407 28 days repeated oral Toxicity studies in rodents
408 90 days repeated oral Toxicity studies in rodents
409 90 days repeated oral Toxicity studies in non rodents
410 90 days repeated Dermal Toxicity
411 90 days inhalation Toxicity study
412 28/14 days repeated dose inhalation Toxicity study
413 90 days repeated dose inhalation Toxicity study
414 Prenatal Developmental Toxicity study
421 Reproduction /Development toxicity screening test
422 Neurotoxicity study in rodents
451 Carcinogenicity studies
452 Chronic Toxicity studies
453 Combined chronic toxicity/carcinogenic studies

1. OECD 423 (Acute toxic class method)

Principles: Based on a stepwise procedure with the use of a minimum number of animals per step. Absence or presence of compounds related mortality of the animals dosed at step will determine the next step. No further testing is needed. Dosing of the three additional animals with the same dose. Dosing of three additional animals with the next higher or the next lower dose level.

Description of the method

- *Preparation of the animals*: Randomly selected animals are kept in their cages for at least 5 days prior to start of the test to allow of acclimatization of the laboratory condition.
- *Preparation of dose*: Vehicles other than water, the toxic characteristics of vehicle should be known (1% Gum acacia,1% SCMC, Tween 20&80, Honey, Vegetable oils)
- *Procedure*: Three animals in each step.
- *Doses*: 5,50,300,2000 mg/kg b/w

Administration of Doses

- *Rodents*- Volume not to exceed 1 ml/100gm b.w., aqueous solution 2 ml/100g b.w. can be considered.
- *Rats*- Overnight fasted prior to dosing but free assess to drinking water.
- *Mouse*- 3-4 hrs fasted prior to dosing but free assess to drinking water. After administration of drugs, food may be withheld for 3-4 hrs in rats and 2 hours in mice.

Observation

- Observed for 14 days. Closely observed for first four hours after the dosing and than at least once daily thereafter up to 14 days.
- *Parameters*: Mortality, Changes in skin, eyes, mucus, Behavioral pattern changes, Respiratory, Circulatory, ANS, CNS, Observation of tremors, convulsions, salivation, diarrhea, sleep and coma, body weight (at least once in a week), Pathology parameters (gross necropsy and microscopic examination if organs showing any changes).

2. OECD 420 (Acute Oral Toxicity – Fixed dose procedure)

- Fixed doses - 5, 50, 300, 2000, 5000 mg/kg
- *Animal*: Rat of female sex preferable and 10-12 weeks old.
- *Number* : Total of five animals used for each dose level
- *Dose*: single dose, Rat- overnight fasted, Mice- 3-4 hours fasted
- *Time interval between dosing*: Period of 3 or four days between dosing at each dose level is recommended. Onset, duration and severity of toxic sign.
- *Observation*: 14 days. Closely observed for first four hours and then at least daily thereafter up to 14 days. Body weight (Individual weight of animals recorded before administration weekly thereafter). Pathology (Gross pathological changes recorded for each animal).

3. OECD 425 (Acute Oral Toxicity – up and down method)

- This is a sequential test uses a maximum of 5 animals dose2000, 5000 mg/kg.
- Limit test of 2000 mg/kg.
- Dose of animals at the test dose.

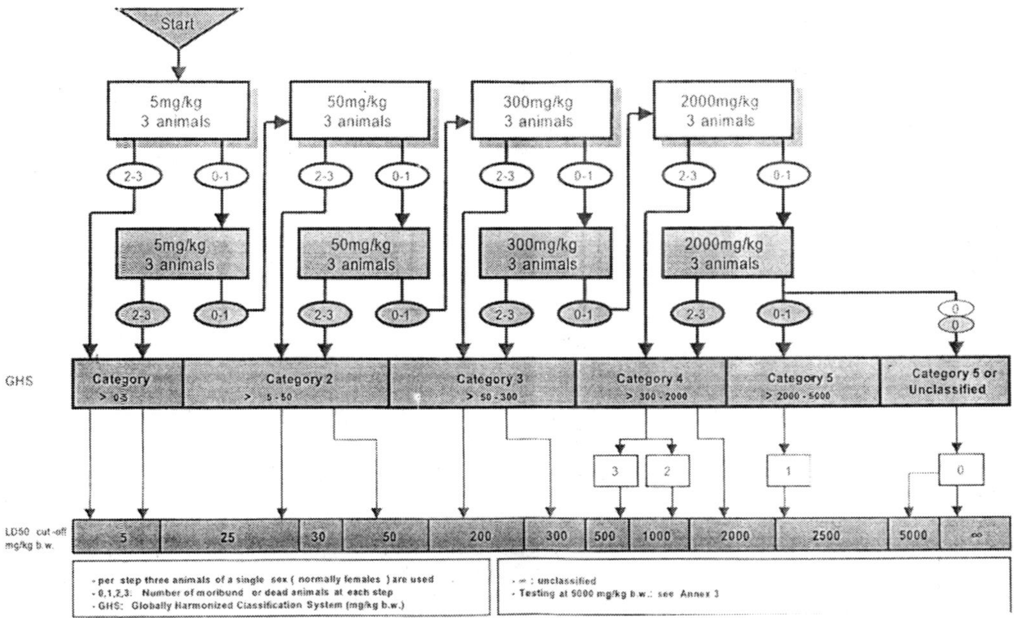

OECD 423

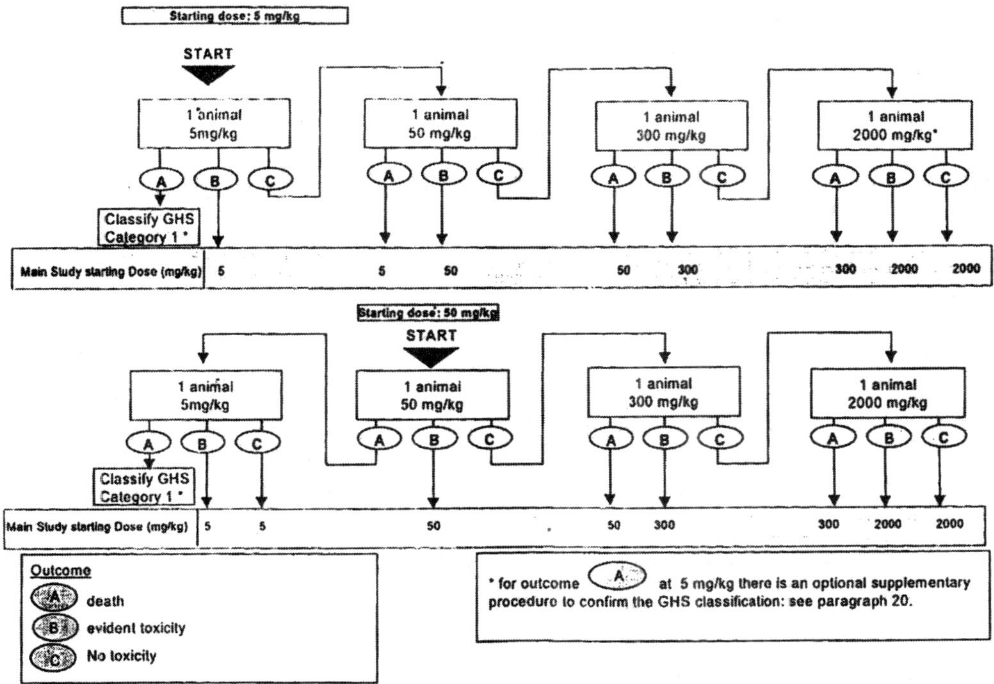

OECD 420 Sighting Study

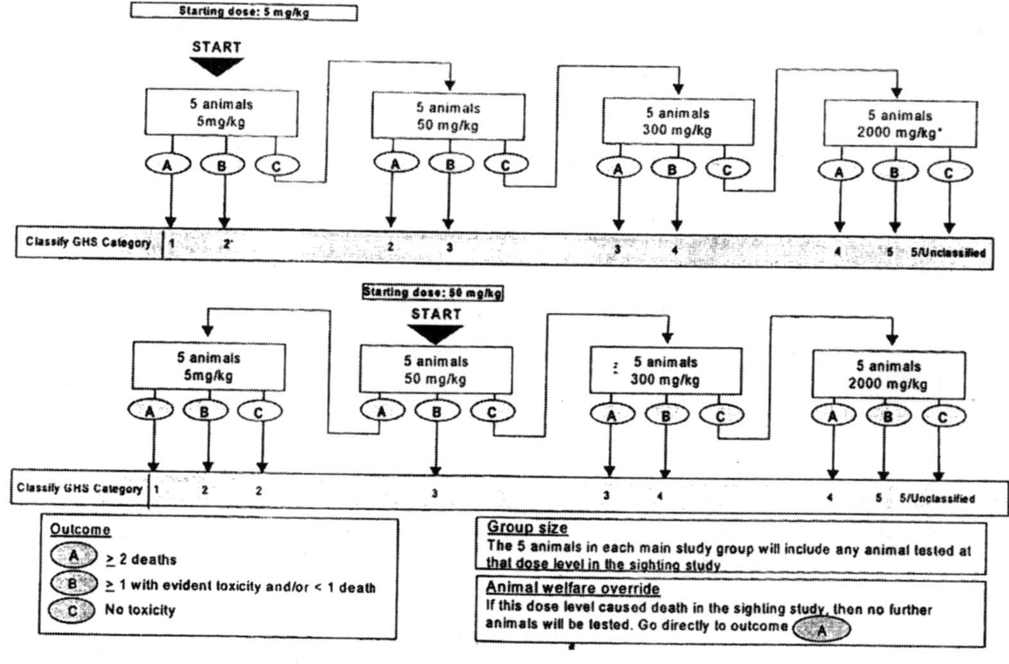

OECD 420 Main Study

- Animal dies, conduct the main test to determine the LD_{50}.
- Animal survives, dose four additional animals sequentially observe 14 days.
- The LD_{50} less than the test dose (2000 mg/kg) when 3 or more animals die.
- LD_{50} is greater than the test dose (2000 mg/kg) when 3 or more animal survive.

Main test

- Single animals are dosed in sequence usually at 48 hour intervals (dosing is determined by the onset, duration and severity of toxic signs). The first animal is dosed below the best preliminary estimate of LD_{50}. If the animal survives. The second animal receives a higher dose. If the first animal dies or appears morbid and the second animal receive a lower dose.
- Selected sequence dose 1.25, 5.5, 17.5, 55, 125, 550, 2000, (or) 1.25, 5.5,17.5,55, 175, 550,1750 and 5000 for specific regulatory needs.

Observation : 14 days

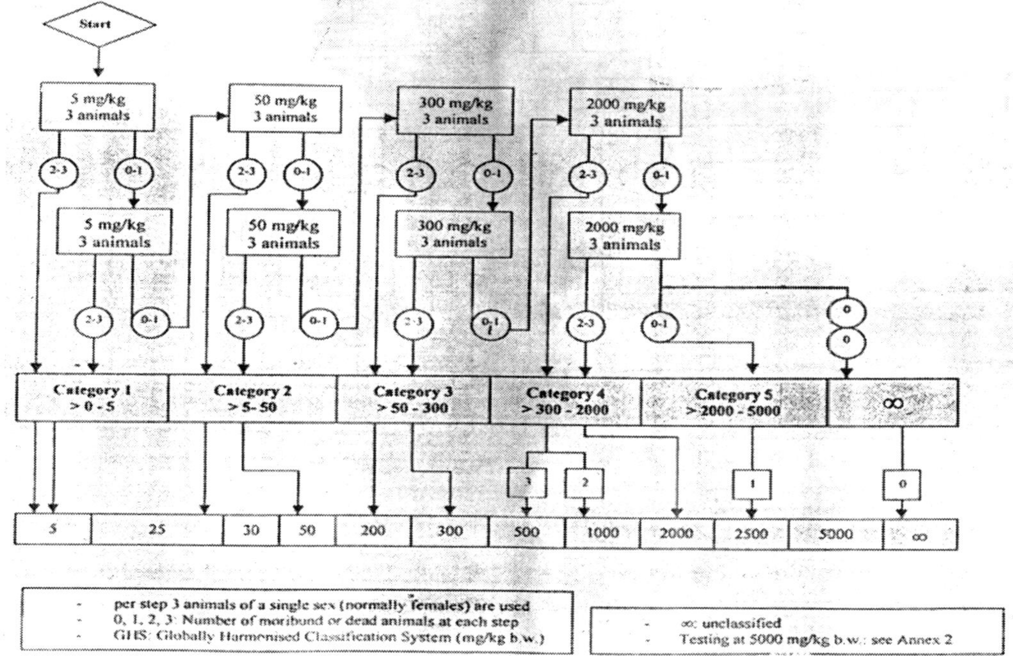

4. OECD-407 (28 day repeated dose oral Toxicity study)

Animal species: Rat or Mice of nine weeks old. Feed with laboratory diet with water and housed individually, small groups in same sex, not more than five/cage.

Housing condition

- *Temperature*: 22^0 C ± 3^{0}C
- *Humidity*: 50-60%
- *Lighting*: 12 hr. light 12 hr. dark

Preparation of doses : Administration by gavages or via the diet or drinking water. Oral administration depends upon the physical/chemical properties of the materials. Vehicles other than water the toxic properties of the vehicle are must be known.

- *Number and sex of animal*: 10 animals (5 female and 5 male) for each dose
- *Dosage*: 1000 mg/kg/day for 28 days daily
- *Volume*: Suspension (1 ml/100 gm b.w.) and aqueous solution (2 ml/100 gm b.w.)

Observation: 28 days

1. General clinical observation at least once a day.
2. Morbidity and mortality at least twice daily
3. Changes in skin, eyes, mucous membrane, secretion and excretion.
4. *ANS activity*- lacrimation, piloerection, pupil size, respiratory pattern, grip strength, motor activity assessment.
5. *Body weight, food consumption, water consumption*: weekly
6. *Hematology parameters*: Haematocrit, Hb, RBC, WBC, DC, platelet, clotting time.
7. *Clinical Biochemistry:* Liver and kidney function test.
8. *Plasma or serum parameters*: Na, K, glucose, cholesterol, urea, creatinine, SGOT, SGPT, total protein, albumin, ALP, Gamma glutamyl transpeptidase.
9. *Urine analysis*: Volume, appearance, specific gravity, pH, protein glucose and blood cells.
10. *Metabolic profiles*: Calcium, phosphate and triglycerides.
11. *Pathology*: Liver, kidney, adrenals, testes, thymus, spleen, brain, stomach, intestine, breast, lungs, urinary bladder, peripheral nerve, bone marrow.

5. OECD 408 (90 day repeated dose oral Toxicity)

Animal species: Rat or Mice of nine weeks old. Feed with laboratory diet with water and housed individually, small groups in same sex, not more than five/cage.

Housing condition

- Temperature: $22^0C \pm 3^0C$
- Humidity: 50-60%
- Lighting: 12 hr. light 12 hr. dark

Preparation of doses : Administration by gavages or via the diet or drinking water. Oral administration depends upon the physical/chemical properties of the materials. Vehicles other than water the toxic properties of the vehicle are must be known.

- *Number and sex of animal*: 10 animals (5 female and 5 male) for each dose

- *Dosage*: 1000 mg/kg/day for 28 days daily
- *Volume*: Suspension (1ml/100gm b.w.) and aqueous solution (2ml/100gm b.w.)

Observation: 90 days

1. General clinical observation at least once a day.
2. Morbidity and mortality at least twice daily.
3. Changes in skin, eyes, mucous membrane, secretion and excretion.
4. *ANS activity*- lacrimation, piloerection, pupil size, respiratory pattern, grip strength, motor activity assessment.
5. *Body weight, food consumption, water consumption*: weekly.
6. *Haematology parameters*: Haematocrit, Hb, RBC, WBC, DC, platelet, clotting time.
7. *Clinical Biochemistry*: Liver and kidney function test.
8. *Plasma or serum parameters*: Na, K, glucose, cholesterol, urea, creatinine, SGOT, SGPT, total protein, albumin, ALP, Gamma glutamyl transpeptidase.
9. *Urine analysis*: Volume, appearance, specific gravity, pH, protein glucose and blood cells.
10. *Metabolic profiles*: Calcium, phosphate and triglycerides.
11. *Pathology*: Liver, kidney, adrenals, testes, thymus, spleen, brain, stomach, intestine, breast, lungs, urinary bladder, peripheral nerve, bone marrow.

6. OECD- 410 (21/28 day Repeated dose dermal Toxicity study)

Principle: The test substances applied daily to the skin in graduated doses to several groups of experimental animals, one dose per group for 21/28 days.

Animals: Adult rat (200-300 gm), Rabbits (2-3 kg), Guinea pigs (350-450 gm). The animals are accommodating at same environmental condition. They all are individually housed in cage. The temperature maintain at 22^0 C (±3^0C) and relative humidity at 30- 70%. 12 hr light and 12 hr dark light cycle provided to animals and they are maintaining on standard laboratory diet + *ad libitum* supply of drinking water.

Number and sex: 10 animals (5 female and 5 male)

Dose level: At least three dose level with control and vehicle control if necessary.

Limit test: One dose level at least 1 gm/kg, b.w. Test drug apply not less than 10% of the body surface area uniformly for 6 hours for a day.

Observation: Sign and toxicity, ANS, CNS, Behavior pattern, Food consumption weekly, Weight variation weekly.

Clinical examination

- *Haematology parameters*: Haemocrit, Hb, RBC, WBC, DC, clotting time, prothrombin time, platelet count
- *Clinical biochemistry*: Liver and Kidney function test, Na, K, Cl, Ca, glucose etc.
- *Pathology*: Gross necropsy, Histopathology of organs showing gross changes.

1.4 Conclusion

The toxicological studies assess the safety of herbal drugs and also determine the relationship between the dose and adverse effect of drugs. The repeated dose toxicity studies in animal models to identify the target organs and / or systems for toxicity and helps the possible toxicity from short-term and long term use.

Literature survey by OECD for OECD guidelines (420, 423, 425 and 407)

References

ASTM (1987). E 1163-87, Standard Test Method for Estimating Acute Oral Toxicity in Rats. American Society for Testing and Materials, Philadelphia Pa, USA.

BMDP Statistics Software, Inc. (1990). BMDP Statistical Software Manual. W.J. Dixon, Chief Ed. 1990 rev. or later. University of California Press, Berkeley, CA, USA

British Toxicology Society Working Party on Toxicity (1984). Special report: a new approach tothe classification of substances and preparations on the basis of their acute toxicity. *Human Toxicol.*, 3, 85-92.

Bruce R.D. (1985). An Up-and-Down Procedure for Acute Toxicity Testing. Fundam. *Appl. Tox.*, 5, 151-157.

Chan P.K and A.W. Hayes (1994) Chapter 16 Acute Toxicity and Eye Irritation In: Principles and Methods of Toxicology. 3rd Edition. A.W. Hayes, Editor. Raven Press, Ltd. New York, USA.

Choi S.C. (1990). Interval estimation of the LD50 based on an up-and-down experiment. *Biometrics* 46, 485-492.

Crofton K.M., Howard J.L., Moser V.C., Gill M.W., Reiter L.W., Tilson H.A.,MacPhail R.C. (1991). Interlaboratory Comparison of Motor Activity Experiments: Implication for Neurotoxicological Assessments. *Neurotoxicol. Teratol.* 13, 599-609.

Diener W., and Schlede E. (1999) Acute Toxicity Class Methods: Alternatives to LD/LC50 Tests. ALTEX 16, 129-134.

Diener W., Mischke U., Kayser D. and Schlede E. (1995). The Biometric Evaluation of the OECD Modified Version of the Acute-Toxic-Class Method (Oral). *Arch. Toxicol.* 69, 729-734.

Diener W., Sichha L., Mischke U., Kayser D. and Schlede E. (1994). The Biometric Evaluation of the Acute-Toxic-Class Method (Oral). *Arch. Toxicol.* 68, 559-610.

Dixon W.J. The Up-and-Down Method for Small Samples (1965). *J. Amer. Statist. Assoc.* 60, 967-978.

Dixon W.J. (1991). Staircase Bioassay: The Up-and-Down Method. *Neurosci. Biobehav. Rev.*, 15, 47-50.

Dixon W.J. (1991). Design and Analysis of Quantal Dose-Response Experiments (with Emphasis on Staircase Designs). Dixon Statistical Associates, Los Angeles CA, USA.

Dixon W.J. and A.M. Mood. (1948). A Method for Obtaining and Analyzing Sensitivity Data. *J. Amer. Statist. Assoc.*, 43, 109-126.

Gad, S.C. (1982). A Neuromuscular Screen for Use in Industrial Toxicology. *J. Toxicol Environ. Health*, 9, 691-704.

Hsi B.P. (1969). The multiple sample up-and-down method in bioassay. *J. Amer. Statist. Assoc.* 64, 147-162.

IPCS (1986). Principles and Methods for the Assessment of Neurotoxicity Ass ociated with Exposure to Chemicals. Environmental Health Criteria Document No. 60

Lipnick, R.L., Cotruvo, J.A., Hill, R.N., Bruce, R.D., Stitzel, K.A., Walker, A.P., Chu, I., Goddard, M., Segal, L., Springer, J.A. and Myers, R.C. (1995). Comparison of the Up-and-Down, Conventional LD50, and Fixed-Dose Acute Toxicity Procedures. Fd. Chem. *Toxicol.* 33, 223-231.

Meyer O.A., Tilson H.A., Byrd W.C., Riley M.T. (1979). A Method for the R outine Assessment of Fore- and Hindlimb Grip Strength of Rats and Mice. Neurobehav. *Toxicol.*, 1, 233-236.

Moser, V.C., McDaniel, K.M., Phillips, P.M. (1991). Rat Strain and Stock Comparisons Using a Functional Observational Battery: Bas eline Values and Effects of Amitraz. *Toxicol. Appl. Pharmacol.*, 108, 267-283.

Noordwijk van A.J. and van Noordwijk J. (1988). An accurate method for estimating an approximate lethal dose with few animals, tested with a Monte Carlo procedure. *Arch. Toxicol.* 61, 333-343.

OECD (1998). Harmonised Integrated Hazard Classification for Human Health and Environmental Effects of Chemical Substances as endorsed by the 28th Joint Meeting of the Chemicals Committee and the Working Party on Chemicals in November 1998, Part 2, p.11 [http://webnet1.oecd.org/oecd/pages/home/displaygeneral/0,3380,EN-documents-521-14-no-24- no-0,FF.html].

OECD (2000) Guidance Document on Acute Oral Toxicity. Environmental Health and Safety Monograph Series on Testing and Assessment No 24.

OECD (2000). Guidance Document on the Recognition, Assessment and Use of Clinical Signs as Humane Endpoints for Experimental Animals Used in Safety Evaluation. Environmental Health and Safety Monograph Series on Testing and Assessment No 19.

OECD (Paris, 1992). Chairman's Report of the Meeting of the ad hoc Working Group of Experts on Systemic Short-term and (Delayed) Neurotoxicity.

Roll R., Höfer-Bosse Th. And Kayser D. (1986). New Perspectives in Acute Toxicity Testing of Chemicals. *Toxicol. Lett., Suppl.* 31, 86.

Roll R., Riebschläger M., Mischke U. and Kayser D. (1989). Neue Wege zur Bestimmung derakuten Toxizität von Chemikalien. Bundesgesund-heitsblatt 32, 336-341.

SAS Institute Inc. (1990). SAS/STAT® User's Guide. Version 6, Fourth Ed. or later. Cary, NC, USA.

Schlede E., Mischke U., Diener W. and Kayser D. (1994). The International Validation Study of the Acute-Toxic-Class Method (Oral). *Arch. Toxicol.* 69, 659-670.

Schlede E., Mischke U., Roll R. and Kayser D. (1992). A National Validation Study of the Acute-Toxic-Class Method - An Alternative to the LD50 Test. *Arch. Toxicol.* 66, 455-470.

Stallard, N. and Whitehead, A. (1995). Reducing numbers in the fixed-dose procedure. *Human Exptl. Toxicol.* 14, 315-323.

Stallard, N., Whitehead, A. and Ridgeway, P. (2000). Statistical evaluation of modifications to the fixed dose procedure (manuscript in preparation).

Tupper, D.E., Wallace, R.B. (1980). Utility of the Neurologic Examination in Rats. *Acta Neurobiol. Exp.,* 40, 999-1003.

Vågerö M. and R. Sundberg. (1999). The distribution of the maximum likelihood estimator in up-and-down experiments for quantal dose-response data. *J. Biopharmaceut. Statist.* 9(3), 499-519.

Van den Heuvel, M.J., Clark, D.G., Fielder, R.J., Koundakjian, P.P., Oliver, G.J.A., Pelling, D., Tomlinson, N.J. and Walker, A.P. (1990). The international validation of a fixed-dose procedure as an alternative to the classical LD50 test. Fd. *Chem. Toxicol.* 28, 469-482.

Van den Heuvel, M.J., Dayan, A.D. and Shillaker, R.O. (1987). Evaluation of the BTS approach to the testing of substances and preparations for their acute toxicity. *Human Toxicol.,* 6, 279-291.

Whitehead, A. and Curnow, R.N. (1992). Statistical evaluation of the fixed-dose procedure. *Fd. Chem. Toxicol.,* 30, 313-324.

❑❑❑

15

Standardization of Marine Herbal Preparations for the Management of Malaria

S. Ravikumar, G. Ramanathan, K. Kathiresan and S. Jacob Inbaneson

CONTENTS

1.1	Introduction	224
1.2	Marine Plants	226
1.3	Description about Mangroves	226
1.4	Protocol for the Standardization of Marine Herbal Preparations	230
	1.4.1 Biological standardization	230
	1.4.2 In vivo antiplasmodial activity	231
	1.4.3 Toxicity studies	232
	1.4.4 Microbiological studies	234
	1.4.5 Organoleptic standardization	234
	1.4.6 Chemical standardization	236
1.5	Conclusion	241
	References	241

1.1 Introduction

Traditional medicines as practiced in India like Ayurveda, Siddha etc., have gained popularity in recent years. The physicians practicing Ayurveda dispensing Kasthaushadi, namely, predominantly plants based drugs, claim that, Ayurveda is one of the proven systems of medicine and have been time tested for ages. But then, the market share of India, with one of the twelve mega biodiversities in the world and the presence about 350 flowering medicinal plants, is just one percentage of the global market in plant based medicine, in spite of the hub of our traditional medicines like Ayurveda, Siddha etc. This is because our plant based medicine are not backed by rigorous scientific studies to test their safety and efficacy. In other words, they are not standardized. First of all, the herbalists or the traditional medical practitioners do not want any standardization. They seem to worry that standardization is nothing but 'chemicalisation' or 'pharmaceuticalisation'. They say that herbs have been active for thousands of years and do not need any standardization. This worries apart what is standardization? Most consumers and even many manufactures think of standardization as a fairly recent phenomenon, brought about by applying modern "state of the art" scientific methods to the production of herbal products. They would certainly be quite surprised to know that, the call for standardization products has been a rallying cry in the industry for at least three hundred years.

The term standardization means different things to different people. For some it is manufacturing a product to contain a specific amount of the herb in a dosage unit. For others, it is consistent manufacturing practice. For still others, it is manufacturing a product to contain certain marker compound(s) to a certain specific level. The industry in general has largely been successful in convincing the consumers that standardization means that a product contains a specified amount of an "active ingredient" what is increasingly done now is to fix a specific plant component(s) identifiable by assay and to standardize extracts to their content. Some believe that this quantitative measure will ensure and efficacious product.

At the heart of controversy surrounding standardization appears to be a great deal of confusion and misunderstanding as to what purpose of standardization is and what the process actually involves. By nature, plant based products may be highly variable in their chemical makeup. There are numerous factors that may affect the ultimate chemical profile of herb and the content of a specific marker, including intrinsic factors such as genetics and extrinsic factor such as growing, harvesting, drying and storage conditions. For example in a breeding study of St. John Wort, it was found that there was significant variation, not only in the marker content throughout the growing season, but also between the same accessories grown at different sites, as well as between different accessories grown at the same site. This natural variation in the chemical make up of herbs presents a considerable challenge, especially for researchers who must use product that are consistent in strength in order to obtain reproducible results. With the current technology, it is not possible to quantify the hundreds of chemical constituents present in herbal material in a timely and cost efficient manner.

The compromise solution to this dilemma is to select a marker compound(s) and then ensure that every batch of the product contains the same amount of that marker compound(s). This approach to ensuring consistency is based upon the assumption that the content of other constituents will vary in proportion to the marker compound. That is if each batch contains the same "Standardized" amount marker, the content if other constituents will also be relatively consistent. How does a manufacture actually produce these standardized products with the same amount of marker in every batch? Strictly speaking, 'standardization' of marker content is achieved my blending different batches. This is superior method for attaining consistency, according to the American Herbal Products Association Guidelines. However in spite of claims that their products are "standardized" many manufactures employ normalization to produce products with the target market content - adjusting the extraction ratio and/or adding fillers to achieve the targeted marker content. This, it is believed is acceptable only with in narrow limits and large adjustments are only appropriate in cased where it has been established that, the marker is responsible for the pharmacological activity. It is further specified that the addition of marker compounds to an extract is not an acceptable method for achieving "standardized" marker content and if such additions are not stated on the label, the resulting product must be considered adulterated. By itself, providing a certain quantity of a particular marker, while physiologically reassuring, in many cases it does not necessarily guarantee superior quality or better results. Several researchers have reported that even in products that contain the specified amount of marker(s), the content of other important constituents and/or therapeutic effectiveness can vary substantially. In Germany, where this practice originated, standardization to a specific content of active or analytical marker is only one of several means of maintaining product quality. In the recent guidance documents published by the APHA, a much broader definition of standardization is delineated: *Standardization refers to the whole body of information and controls necessary to produce a material of reasonable consistency. This is achieved through minimizing the inherent variation of the natural product composition through quality assurance practices applied to medicinal plant growing, extraction and formulation development.*

1.2 Marine Plants

Plants that have been adapted to thrive greater than 0.5% NaCl are called halophytes *eg.* mangroves and seaweeds. Mangroves are woody plants that grow at the interface between land and sea in tropical and sub-tropical latitudes, which are often classified either as excretives and succumbers or excluders and includers. Excretives / excluders have glandular cells capable of secreting excess salts from plant organs *eg. Avicennia germinans.* Succumbers / includers used to accumulate water within large vacuoles to minimize salt toxicity *eg. Salicornia, Suaeda, Sesuvium portulacastrum, Allenrolfea, Arthrocremum* and *Halimione.*

1.3 Description about Mangroves

The word mangrove is usually considered a compound of the Portuguese word 'Mangue' and the English word 'grove'. The corresponding French words are 'manglier' and 'paletuvier' (Macnae, 1968) while the Spanish term is 'mangler'. The Dutch use 'vloedbosschen' for the mangrove community and 'mangrove' for the individual trees. German usage is similar to the English. The word 'mangro' is a common name for *Rhizophora* in Surinam (Chapman, 1975). It is believed that, all these words originated from the Malaysian word, 'manggi-manggi' meaning above the soil.

The use of mangroves falls into two major categories. Firstly, the indirect use of the mangrove ecosystem are in the form of vital ecological function such as control of coastal erosion, protection of coastal land, stabilization of sediment and natural purification of coastal water from pollution. Secondly, the economic benefits are many and varied. Several economically important species are associated with mangroves, that includes crab, shrimp, oyster, lobster and fish. Traditionally, the mangrove have been exploited for fine wood and charcoal and their uses include construction of dwellings, furniture, boats and fishing gear and production of tannins for dying and leather manufacturing. Varieties of traditional products and artifacts from mangrove provide food for mangrove dwellers. The mangrove leaves are useful contributors to the nutrient systems of mangrove environment. It is known that, mangrove leaves contain sufficient amount of minerals, vitamins and amino acids which are essential for the growth and nourishment of marine organisms and livestock. Mangrove foliage plays an important role in the formation of detritus which is utilized by several estuarine and marine detritovorous organisms and mangrove leaves make a superior fodder due to their high salt and iodine content (Bandaranayake, 1998 and Vannucci, 1997). Numerous mangrove plants were used in folklore medicine. Recent works proved that, extracts from mangroves and mangrove associated species exhibited sensitivity against human, animal and plant pathogens (Table 1).

Among the 20 different families, about 80 species of true mangrove tree/ shrubs are recognized, of which 50 – 60 species make a significant contribution to the structure of mangrove forest (Tomlinson, 1986; Bandaranayake, 1998). In India, mangrove flora comprises of 35 species under 16 genera and 13 families. Of these, 31 species (16 genera under 13 families) are present along the East coast; 18 species (10 genera under 7 families) on the West coast; and 31 species (15 genera under 12 families) on the Andaman and Nicobar islands (Kathiresan, 1999)

Table 1 : Literature on the bioactivity of Mangroves and Associates

Year	Author (s)	*Plant species*	Biological activity
1969	Loder and Russell	*Bruguiera sexangula* *Bruguiera exaristata* *Bruguiera paviflora*	Antitumor
1972	Molyneux	*Derris trifoliata*	Toxicity to fish
1974	Ganguly and Sircar	*Rhizophora mucronata* *Sonneratia apetala*	Plant growth hormone
1977	Chou *et al.*	*Xylocarpus granatum* *Xylocarpus moluccensis*	Insect antifeedant
1981	Jongsuvat	*Acanthus ilicifolius*	Anticancer
1981	Rollet	*Languncularia racemosa* *Rhizophora lamarckii* *Derris uliginosa*	Antihypertensive Herbicide toxicity Antiviral Antispasmodic Antihaemorrhage
1982	Balasooriya *et al.*	*Thespesia populnea* *Aegiceras corniculatum*	Antisteroidogenic Cytotoxic Antifertility Antiviral Toxicity to fish
1982	Singh and Dhawan	*Bacopa monniera*	Rat poison
1983	Mc Millan	*Halophila ovalis*	Mosquito larvicidal
1985	Ghos *et al.*	*Suaeda maritima*	Antiviral
1987	Chaudhari and Mahapatra	*Pluchea indica*	Antiinflammatory
1988	Thangam and Kathiresan	*Rhizophora apiculata*	Mosquito larvicidal

Contd...

Table 1. Contd...

1989	Goyal and Rani	*Thespesia populnea*	Antibacterial
1989	Miles *et al.*	*Heritiera littoralis*	Antifeedant Antifungal
1990	Collins *et al.*	*Acanthus ebracteatus*	Analgesic
1990	Kokpol *et al.*	*Acanthus ilicifolius*	Analgesic Antiinflammatory
1991	Ramamurthi *et al.*	*Exoecaria agallocha*	Biotoxicity towards fish and field crab
1992	Premanathan *et al.*	*Acanthus ebracteatus* *Sonneratia apetala* *Sesuvium portulacastrum* *Lumnitzera racemosa*	Antiviral
1993	Padmakumar *et al.*	*Avicennia marina*	Analgesic
1993	Lin *et al.*	*Lumnitzera racemosa*	Antihypertensive
1994	Padmakumar and Ayyakkannu	*Acanthus ebracteatus*	Antiviral
1994	Kapil *et al.*	*Acanthus ilicifolius*	Leishmanicidal
1995	Krishnamoorthy *et al.*	*Exoecaria agallocha*	Larval toxicity to fresh water prawn
1996	Premanathan *et al.*	*Acanthus ebracteatus*	Antiviral
1996	Thongpraditchote *et al.*	*Pluchea indica*	Neumotrasmittans
1997	Devi *et al.*	*Sonneratia caseolaris*	Mosquito larvicidal
1997	Padmakumar and Ayyakkannu	*Rhizophora mucronata* *Exoecaria agallocha* *Avicennia marina* *Avicennia officinalis* *Bruguiera cylindrica* *Salicornia brachiata*	Antiviral
1997	Madhu and Madhu	*Rhizophora mucronata* *Acanthus ilicifolius* *Avicennia officinalis* *Bruguiera cylindrica*	Biotoxicity on fingerlings of fish
1997	Thangam and Kathiresan	*Rhizophora apiculata*	Mosquito larvicidal

Contd...

Table 1. Contd...

1998	Alarcon *et al.*	*Rhizophora mangle*	Antihyperglycaemic
1998	Konishi *et al.*	*Exoecaria agallocha*	Antitumor
1999	Masuda *et al.*	*Exoecaria agallocha*	Antioxidant
1999a	Premanathan *et al.*	*Rhizophora apiculata*	Antiviral
1999b	Premanathan *et al.*	*Rhizophora mucronata*	Anti HIV
1999c	Premanathan *et al.*	*Ceriops decandra* *Exoecaria agallocha*	Antiviral
2000	Ramanathan and Raju	*Rhizophora apiculata* *Exoecaria agallocha* *Avicennia officinalis* *Ceriops decandra* *Aegiceras corniculatum* *Lumnitzera racemosa* *Acanthus ilicifolius* *Suaeda monoica* *Bruguiera cylindrica* *Rhizophora mucronata*	Antibacterial
2001	Prakash	*Rhizophora apiculata* *Exoecaria agallocha* *Avicennia officinalis* *Ceriops decandra* *Aegiceras corniculatum* *Lumnitzera racemosa* *Acanthus ilicifolius* *Suaeda monoica* *Bruguiera cylindrica* *Rhizophora mucronata*	Antibacterial
2001	Chitra	*Rhizophora apiculata* *Exoecaria agallocha* *Avicennia officinalis* *Ceriops decandra* *Aegiceras corniculatum* *Lumnitzera racemosa* *Acanthus ilicifolius* *Suaeda monoica* *Bruguiera cylindrica* *Rhizophora mucronata*	Antibacterial
2004	Babu selvam	*Rhizophora mucronata* *Acanthus ilicifolius* *Avicennia officinalis* *Avicennia marina* *Exoecaria agallocha* *Ceriops decandra* *Lumnitzera racemosa* *Suaeda maritima*	Spermicidal

1.4 Protocol for the Standardization of Marine Herbal Preparations

Standardization and Quality Evaluation of Herbal Drugs

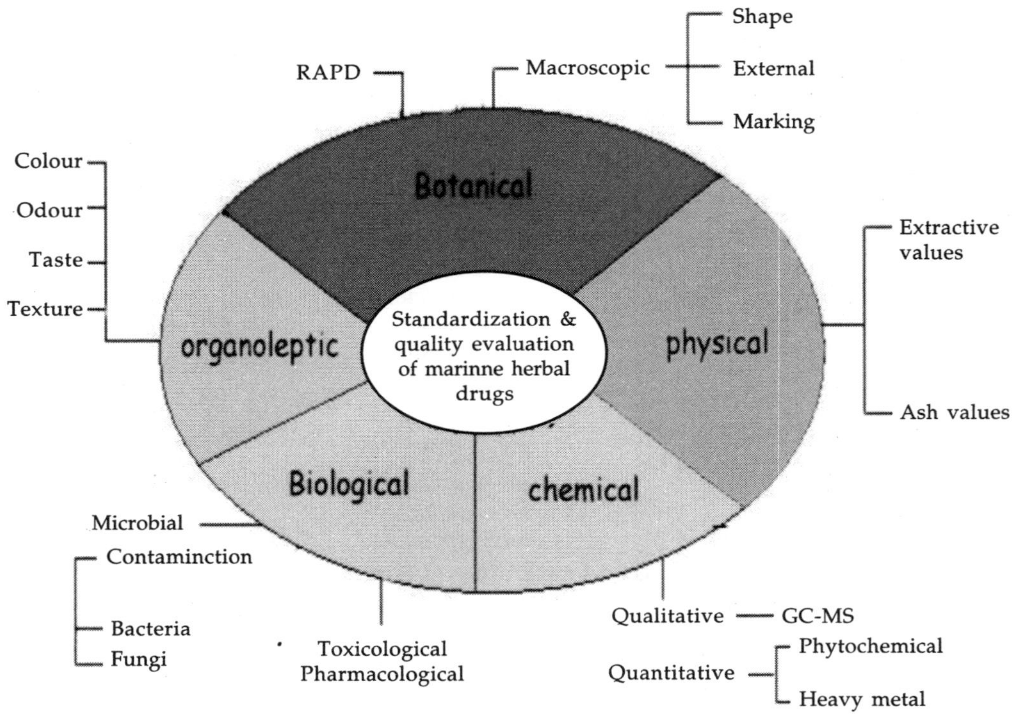

1.4.1 Biological standardization

People who use herbal medicine for self-diagnostic health concerns run potential health risks. These risks range from unknown effects of herbals and their interactions with prescribed drug use, misleading of symptoms of diseases, thereby, losing the opportunity for early diagnosis and treatment (Cupp, 1999). The risk involves people of all ages and in young persons usage of herbs has resulted in fatal consequences. Similarly, older persons experienced serious medical complications when they were admitted to the hospital due to severe interruption and prescribed medications. For instance, some people developed kidney disease after taking Chinese herb prescribed by a sliming clinic (Vickers and Zollman, 1999), due to lack of adequate information on the quality and side effects of the herbal medicine. The physician often doesn't address the potential risks of combining herbs and medications.

In addition to their direct pharmacological effect, herbal products may be contaminated with microorganisms. Many herbal medicines are known to contain toxic reactions, possible mutagenic effects and allergic reactions (Ernst, 1998; Senior, 1998). Herbs are greatly safe if it is used properly but they can be harmful and even fatal if misused (Roan, 2000). The concern is that no safe use has been established, scientific data is scanty, potency and packaging information is inconsistent. While there are many unknown factors associated with the use of herbs due to the absence of standardization, number of herbs becomes unpopular. For example, St. John's wort is used extensively for mild to moderate depression and some preliminary test indicated no ill effects or interactions. Likewise, saw palmetto has shown promising but limited results when used for prostrate problems. Garlic has been shown to have some possible beneficial effects on cholesterol levels; Echinacea for treatment and prevention of common cold and for its calming and sleep inducing effects (O' Hara *et al.*, 1998; Youngkin and Israel, 1996). According to the European pharmacopoeia, these drugs should in general, meet modern hygiene hazards, which aim at low microbial viable counts and the absence of pathogenic microorganisms.

Marine natural products have gained attention from chemist and pharmacologist. A number of institutions around the world have been engaged in active research for the past decades. Morethan 600 secondary metabolites from marine algae (Faulkner, 1986) have been identified to cure variety of diseases. A great variety of drugs have been scientifically useful in clinical treatment of human neoplasm at acceptable levels of toxicity. For instance the marine herbal preparations from marine origin can be carried out by the following ways:

1.4.2 *In vivo* antiplasmodial activity

Crude polyherbal extracts of marine plants could be tested *in vivo* for antiplasmodial activity by 4-day suppressive test against *Plasmodium berghei* (Peters *et al.*, 1975). On day 0 experimental as well as control groups of Swiss albino mice (20-25 g) were inoculated intraperitoneally with 1 x 10^7 (70-80% parasitaemia) *P. berghei* infected blood cells diluted in PBS (pH 7.2). After inoculation, each animal must be dosed orally with 15, 30, 60, 120 and 240 $mg.kg^{-1}$ of polyherbal extracts of *Aegiceras corniculatum* + *Chaetomorpha antennina* in 1ml H_2O. Further doses of extract could be given on day 1, 2, 3 on day 4. Tail blood smear must be taken and stained with Giemsa and the percentage suppression of parasitaemia must be determined using the following formula: Average % suppression of parasitaemia = (Av% parasitaemia in control – Av% parasitaemia in test / Av% parasitaemia in control) x 100. The results must be represented as follows:

In vivo antiplasmodial activity of mangrove extracts against *Plasmodium berghei*

Concentration of extracts (mg.kg^{-1}) (*Aegiceras corniculatum* + *Chaetomorpha antennina*)	**% Suppression of parasitaemia**
15	3.62 ± 0.506
30	6.52 ± 0.425
60	26.08 ± 0.065
120	50.57 ± 0.465
240	42.28 ± 0.229

1.4.3 Toxicity studies

(1) Experimental animal

Adult male albino rats of wistar strain 4-6 weeks old, weighing 150-200g must be housed in large spacious cages should be maintained in controlled environment of temperature, humidity and light/dark cycles. They must fed with standard pelleted diet and water *ad libitum*. The experiment must carried out after obtaining ethical committee approval by the respective institution.

(2) Acute toxicity studies

Rats should kept for overnight fasting and divided in to 4 groups of 10 rats each. The extract under study should administered intraperitoneally in different doses to different groups. The animals then observed for signs of toxicity and mortality for every 2 hours and at the end of 24 hrs, number of animals dead should be counted in each group. Percentage death should be calculated and transferred to probits using probit table. Probits must be then plotted in Y-axis against the log dose of the drug in X-axis. The LD_{50} must be determined by the dose that corresponds to probit 5.

(3) Sub acute toxicity studies

Fifteen albino rats must be taken and divide in to three groups. Group I served as control. Group II can be administered with drug extract of 150 mg.kg^{-1} per day and Group III can be administered with drug extract of 300 mg.kg^{-1} per day. This dose must be given to animal for 28th days. All the animals then sacrificed at 29th day, different visceras were separated, grossly inspected and then subjected to histopathological studies. Histological changes must be observed in liver and kidney adopting paraffin method. Haematoxilin and Eosin stains should be used to stain the slides. Blood must be drawn from the carotid artery and serum should be separated for the analysis of Asparatase amino transferase (AST), Alanine amino transferase (ALT) (Reitman and Frankel, 1957), Alkaline phosphatase (ALP), Acid phosphatase (ACP) (Kind,

1967), Protein (Lowry *et al.*, 1957), Bilurubin (Malloy and Evelyn, 1937), cholesterol (Zak, 1977), urea (Varley, 1976) by standard methods. Moreover, factors like superoxide dismutase (SOD) (Das *et al.*, 2000), Glutathione peroxidase (Rotruck *et al.*, 1973), catalase (Sinha, 1972), Triglycerides (Grossman *et al.*, 1976), Thiobarbituric acid (TBA) (Ohkawa *et al.*, 1979), Very low density lipoprotein (VLDL), Low density lipoprotein (LDL), High density lipoprotein (HDL) (Warnick *et al.*, 1982) must be analyzed in tissue homogenates. Data can be interpreted as follows (Table 2).

Table 2 : Effect of chosen extracts on the serum and tissue homogenate parameters in mice

Parameters	Control	Treatments	
		150 mg.kg^{-1}	300 mg.kg^{-1}
AST (µm.l^{-1})	37.70 ± 1.18	41.20 ± 1.20 (9.25)	45.40 ± 2.30 (16.96)
ALT (µm.l^{-1})	51.08 ± 0.20	63.10 ± 1.55 (19.04)	69.80 ± 4.70 (26.80)
ALP (µm.l^{-1})	15.92 ± 0.72	35.60 ± 1.40 (55.28)	33.60 ± 2.40 (52.60)
ACP (µm.l^{-1})	10.50 ± 0.064	16.00 ± 0.12 (34.37)	15.70 ± 0.90 (33.12)
Bilurubin (µm.l^{-1})			
Direct	0.098 ± 0.02	0.21 ± 0.015 (53.33)	0.18 ± 0.008 (45.55)
Total	0.155 ± 0.03	0.37 ± 0.22 (58.10)	0.23 ± 0.01 (32.60)
Protein (mg.dl^{-1})	1.91 ± 0.06	3.30 ± 0.28 (42.12)	3.60 ± 0.13 (46.90)
Urea (mg.dl^{-1})	22.0 ± 0.23	21.00 ± 1.20 (-4.76)	22.80 ± 1.10 (3.50)
TGL (mg.dl^{-1})	76.12 ± 4.12	82.70 ± 5.60 (7.95)	83.74 ± 4.90 (9.09)
HDL (mg.dl^{-1})	24.50 ± 1.32	34.50 ± 2.10 (28.98)	34.40 ± 2.80 (28.70)
LDL (mg.dl^{-1})	39.12 ± 4.13	42.96 ± 0.58 (8.93)	29.80 ± 1.20 (-37.27)
VLDL (mg.dl^{-1})	15.22 ± 0.82	16.54 ± 1.12 (7.98)	16.74 ± 0.98 (9.08)
Total cholesterol (mg.dl^{-1})	78.84 ± 6.27	94.00 ± 3.80 (16.12)	80.94 ± 4.98 (2.59)
Catalase (mg. liver protein^{-1})	297.07 ± 18.10	325.58 ± 6.81 (8.77)	386.27 ± 7.54 (23.11)
SOD (mg. Liver protein^{-1})	77.91 ± 6.21	68.10 ± 0.75 (-14.40)	88.55 ± 4.65 (12.0)
Glutathione peroxidase (mg. Liver protein^{-1})	0.96 ± 0.030	0.97 ± 0.021 (1.30)	0.91 ± 0.074 (-5.49)
TBA (mg. liver protein^{-1})	1.27± 0.384	1.67 ± 0.11 (23.95)	1.77 ± 0.12 (28.24)

Values are the average of three replicates and found 1 % level significance between treatments and control; Values in parentheses indicates percentage increase/decrease over control.

1.4.4 Microbiological studies

One gram of samples from polyherbal extracts must be stored at refrigerator (4°C) and room temperature (27°C) and serially diluted with sterile water up to 10^{-10} dilutions. From each dilution, 1 ml of sample must be taken and used for the enumeration of total heterotrophic bacteria (Nutrient agar, HiMEDIA, Mumbai), fungi (Rose Bengal agar, HiMEDIA, Mumbai), *Escherichia coli* (Mc Conkey agar, HiMEDIA, Mumbai) and *Salmonella* sp. (Bismuth sulphite agar, HiMEDIA, Mumbai), *Enterobacter* (EMB agar, HiMEDIA, Mumbai) using appropriate culture medium. All the plates must be incubated for 24 hrs, but 72 hrs incubation should be maintained for the growth of fungi. After the incubation period, the colony forming unit (CFU) of bacteria and fungal populations must be observed and tabulated. The numbers of colonies appearing on dilution plates are counted and the number of $CFU.g^{-1}$ of sample must calculated and represented here under (Table 3).

Table 3 : Microbial counts in chosen extract stored at different temperatures

Microorganisms	Number of $CFU.g^{-1}$	
	(27°C)	(4°C)
THB	87	78
Fungi	Nil	Nil
E. coli	Nil	Nil
Salmonella sp.	Nil	Nil
Enterobacter sp.	Nil	Nil

1.4.5 Organoleptic standardization

Man has been using plant materials for the treatment of his ailments since time immemorial due to their safety, efficacy, cultural acceptability and lesser side effects. Most of the herbal drugs used in various countries have not been evaluated scientifically. Standardized drugs of consistent quality are needed for reliable therapeutic use. Thus, the importance of developing test parameters for quality assessments of the single drugs and compound formulations need to be emphasized. The perfect quality control profiles of herbal medicine based on physical, chemical, biological evaluation packed with their stability and bioavailability parameters are need to be established. In order to ensure the acceptance of phytopharmaceutical as an integral part of modern drug therapy, the important method of quality assessment is organoleptic and physical standardization of marine herbal products.

Analysis of physical properties

Physical properties such as total ash, acid insoluble ash, water soluble ash, sulphated ash, extractive values, ethanol and water soluble extractives can be analyzed by following the method of Trease and Evans (1972).

Analysis of organoleptic parameters

The organoleptic parameters *viz.*, colour, odour, taste and consistency must be checked as per the standard procedures.

Organoleptic properties of crude methanolic extract of mangroves

Plant species	Colour	Taste / Flavour	Consistency
Acanthaceae			
Acanthus ilicifolius	Dark brown	Sour, Spicy odour	Pasty
Myrsinaeceae			
Aegiceras corniculatum	Dark brown	Sour, Spicy odour	Pasty
Avicenniaceae			
Avicennia marina	Reddish brown	Bitter, Spicy odour	Pasty, Sticky
Avicennia officinalis	Redish brown	Sour, Spicyodour	Pasty, Sticky
Rhizophoraeceae			
Bruguiera cylindrica	Redish brown	Sour, Spicy odour	Pasty, Sticky
Rhizophora mucronata	Brown	Bitter, Spicy odour	Pasty Sticky
Rhizophora apiculata	Redish brown	Sour, Spicy odour	Pasty
Rhizophora annamalayana	Redish brown	Sour, Spicy odour	Pasty
Ceriops decandra	Redish brown	Sour Spicy odour	Pasty, Sticky
Euphorbiaceae			
Exoecaria agallocha	Dark brown	Sour, Spicy odour	Pasty, Sticky
Combretaeceae			
Luminetzera racemosa	Brown	Sour, Spicy odour	Pasty
Aizoceae			
Sesuvium portulacastrum	Dark brown	Bitter, Spicy odour	Pasty
Chaenopodiaceae			
Suaeda monoica	Dark brown	Bitter, Spicy odour	Paste

Physical properties of crude methanolic extracts of mangroves

Plant species	Total ash (%)	Acid in soluble ash	Water soluble ash	Sulphated ash	Extractive value (%)	Alcohol soluble extractives (%)	Water soluble extractives (%)
Acanthaceae							
Acanthus illicifolius	10 + 4.082				11.4 + 0.327	12.30 + 0.245	23.80 + 0.082
Myrisinaceae							
Aegiceras corniculatum	10 + 4.082	-	+	-	9.1 + 0.082	27.43 + 0.351	16.21 + 0.172
Avicenniaceae							
Avicennia marina	12 + 1.633	-	+	-	12.3 + 0.245	15.44 + 0.359	18.67 + 0.057
Avicennia officinalis	10 + 4.083	-	-	-	15.7 + 0.572	15.67 + 0.057	21.73 + 0.596
Rhizophoraeceae							
Bruguiera cylindrica	10 + 4.083	-	-	-	5.9 + 0.735	21.20 + 0.163	11.34 + 0.278
Rhizophor mucronata	20 + 4.082	-	-	-	12.1 + 0.082	26.82 + 0.016	12.70 + 0.572
Rhizophora apiculata	10 + 4.082	-	-	-	7.4 + 0.327	28.44 + 0.359	12.13 + 0.106
Rhizophora annamalayana	10 + 4.082	-	-	-	5.6 + 0.490	22.05 + 0.041	9.28 + 0.154
Ceriops decandra	10 + 4.082	-	-	-	5.5 + 0.408	17.43 + 0.351	10.17 + 0.057
Euphorbiaceae							
Exoecaria agallocha	20 + 4.082	-	-	-	13.2 + 0.163	11.27 + 0.579	14.00 + 0.408
Combretaceae							
Luminetzera racemosa	12 + 1.633	-	-	-	7.4 + 0.327	16.74 + 0.604	12.20 + 0.163
Aizoceae							
Sesuvium portulacastrum	20 + 4.082	-	-	+	4.2 + 0.163	5.20 + 0.041	6.98 + 0.065
Chenopodiaceae							
Suaeda monoica	10 + 4.082	-	-	-	6.3 + 0.245	9.29 + 0.073	1.64 + 0.033

+ : Soluble ; - : Insoluble ; Values are the average of three replicates

1.4.6 Chemical standardisation

(1) Analysis of minerals

Herbal medicine is still the main stage of about 75-80 % of the world population, mainly in the developing countries due to better cultural acceptability, better compatibility with human body and lesser side effect. However, medicinal plant part should be authentic, free from harmful chemicals *viz.*, heavy metals etc. Presence of heavy metals in the drugs are considered to be either constructive or destructive. The recent revolution on the presence of excess of heavy metal sent a warning signal to the herbal industry. The origin of the metal content could be due to the absorption of metal by plants and also addition of heavy metal in the plant product to get the desired effect. This has also created a debate on the relevance of metal content in the products. In order to understand the complexity or the argument, it is necessary for the analysis of trace element and heavy metals in marine plant extracts as the marine environment is rich in halogens mainly in the form of chloride and bromide salts and other chemical entities such as sulphates (Vasanthi, 2002). Marine plants use these elements in biosynthetic process to produce compound such as halogenated terpenes, acetogenins and alkaloids, which are unique to the marine environment (Fenical, 1975).

About 500 mg of crude methanolic extract of mangroves can taken for heavy metal/ trace element analysis. Necessary precautions must be taken at every step to avoid metallic contamination in any form. Pre cleaned silica crucible should be maintained at 800°C until the weight of the crucible was constant. Extracts from mangrove can be taken in the silica crucible and maintained in a Muffle furnace (TECHNICO, HYBRID, India) at 800°C for 4 hours and further cooled at room temperature by keeping in a dessicator.

The ash obtained must be ground well with 5ml of nitric acid and perchloric acid with the ratio of 4:1 and further concentrated to 2ml by heating. The concentrated ash should filtered through Whatman paper No.1. The filtrate must be made up to 20ml using deionized water and stored in tightly capped plastic tubes. Then the prepared solution will be directly used for the determination of various mineral elements *viz.*, Aluminium (Al), Cadmium (Cd), Cobalt (Co), Chromium (Cr), Copper (Cu), Iron (Fe), Manganese (Mn), Nickel (Ni), Lead (Pb), Zinc (Zn) and Arsenic (As) using an Atomic Absorption Spectrophotometer (GBC 932 AA, AUSTRALIA) by following standard procedure described by Rajendran (2003).

(2) GC-MS analysis

Plants produce a great array of secondary metabolites as a result of metabolic activities. These compounds either alone or in combination is being responsible for specific physiological changes or the therapeutic action in the human body when administered as a medicament or a health supplement. Hence, it will be practical to subject the individual herb for testing of these compounds. One of the best methods of standardizing herbs or herbal formulations based on modern scientific herbal characteristics is chromatography. It helps to find out the presence of such compounds in the system which also helps in regulating herbal sanctity. The unique chemical constituents can serve as a marker in the finished form of herbal drug.

About 10 mg of crude extract will be dissolved separately in 1 ml of methanol. From that, 0.1 µl must be injected in to GC-MS (GC 17A, Japan) with standard specification (column size-0.25 mm x 25 m, Carrier gas-Helium, Column -5% phenyl polysiloxane, flow rate-0.4 m/min, Sample injection temperature - 250ºC, acceleration and reflector temperature - 10ºC/min, initial temperature 70ºC). The compounds showed maximum peak area will be identified by chemical library search (TUTOR.LIB, WILEY 139.LIB).

Detection of bioactive metalobites from *Aegiceras corniculatum* using GC-MS analysis

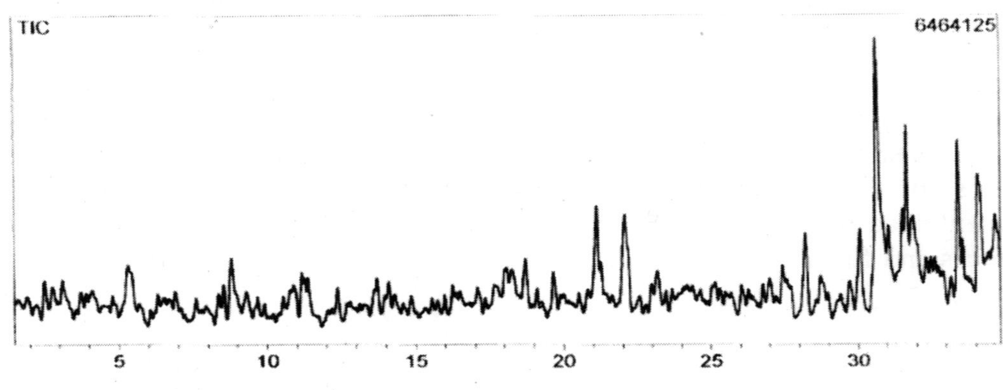

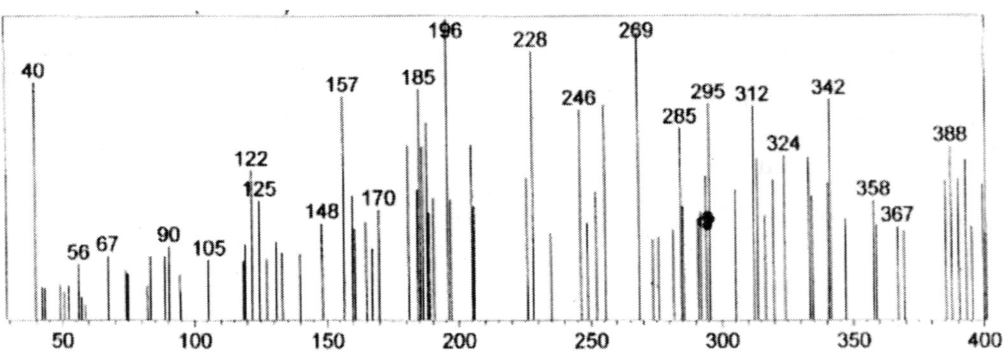

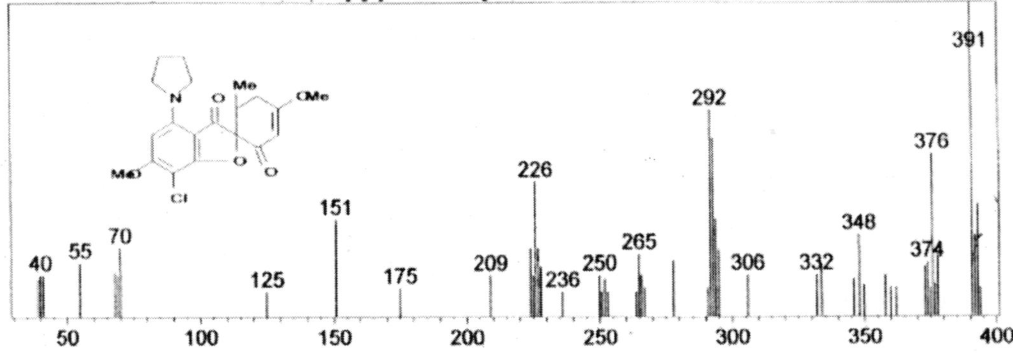

Identified chemical constituents (Spiro (benzofuran-2(3 H), 1'- (3 Cyclohexane)- 2', 3-dione, 7- chloro-4',6) present in the crude extract of *Aegiceras corniculatum*

(3) Phytochemical analysis

Phytochemistry deals with the determination of chemical compounds in plant extracts in order to understand the therapeutic properties of drugs. In recent years, phytochemical investigation of Indian medicinal plant have been progressive steadily and assured extraordinary importance due to the uses of indigenous drugs. Moreover, quality control of plant extracts is problematic due to lack of understanding of which component are important for therapeutic activity. These chemical screening must be performed to allow localization and targeted isolation of new useful type of constituents with potential activities. Studies concerned with the presence of phytochemical constituents in mangrove were exhaustively carried out in plant tissues. But, recently it has been recognized that, phytochemicals present in plant of extract could be helpful for better understanding of the therapeutic activity rather than the plant tissue. The extracts of mangroves can be screened for the presence of phytochemical constituents by following the method of Sofowora (1982) and Kepam (1986).

(4) Molecular Standardisation (RAPD)

Herbal drug technology is used for converting botanical materials to medicine, where standardization and quality control with proper integration of modern scientific technique and traditional knowledge are important. DNA based techniques have utility in the fields of taxonomy, physiology, embryology, genetics etc., which have also been widely used for authentication of plant species of medicinal importance. Especially in case of those species that are frequently substituted or adulterated with other species or varieties that are morphologically or phytochemically indistinguishable. Current focus on chemotype driven finger printing and related techniques require integration with genotype driven molecular techniques so that an optimal characterization of botanical material is possible. For instance, dried fruit samples of *Lycium barbarum* were differentiated from its related species using RAPD markers (Zhang *et al.*, 2001). The RAPD techniques have also been used for determining the component of a Chinese herbal prescription, Yu-Ping-Fengsan (Cheng *et al.*, 1998).

Padmesh *et al.* (1998) reported the genetic variability in *Andrographis paniculata*, a hepatoprotective drug by using RAPD markers. Genetic and geographic variations among mangrove plants have been carried out by several workers. Abeysinghe *et al.* (2000) reported the genetic and geographic variation of the mangrove tree, *Bruguiera* in Srilanka by using RAPD technique. Bartleson *et al.* (2003) reported the genetic diversity of the red mangrove, *Rhizophora* mangle in Southern Florida using the methods of PCR assisted RAPD markers. Hayden *et al.* (2003) reported the genetic variations of two morphologically different seaweeds *viz., Ulva* and *Enteromorpha.* Fresh leaf samples of mangrove plants must be collected and authenticated through morphological characters (Kathiresan, 2002) and their herbarium must be maintained at a museum. The collected samples will be subjected to DNA isolation by CTAB method proposed by Thompson and Henry (1993).

After the extraction of DNA, its quality and purity must be determined measuring the absorbance at 260 nm and 280 nm respectively using UV visible Spectrophotometer (Shimadzu, Japan) and further with agarose gel electrophoresis technique. The amount of DNA can be measured using the following formula, Amount of DNA = OD at 260 nm x 50 (conc. of DNA at OD) x Dilution factor/1000. After that, the DNA will be amplified by using three different primers (possibly OPB-11, OPB-18 and OPB-20) in PCR maintained at 94°C for 3 min followed by 44 cycles of 94°C for 1 min to denature the DNA, 36°C for 1 min to allow primers to anneal, 72°C for 2 minutes to extend the DNA followed by 15 minutes 75°C for final extension.

The concentrations of chemicals taken for PCR mixture are tabulated as follows:

Name of the chemicals	Quantity needed	Volume taken
Buffer	10mM Tris HCl	2.5 ml
$MgCl_2$	2.0 mM	2.5ml
dNTP mix	100mM	0.5ml
Taq. DNA polymerase	0.2 m/ml (50)	0.2ml
Primers	15 picomoles	1.2ml (400 stock)

The name of the primer sequence, primer direction and the GC content are represented as follows:

S. No.	Primer Name	Primer sequence from 5′ to 3′	Primer direction	GC content
1.	OPB – 18	5′ – CCACAGCAGT – 3′	Arbitrary	60%
2.	OPB – 20	5′ – GGACCCTTAC – 3′	Arbitrary	60%
3.	OPB –11	5′ – GTAGACCCGT – 3′	Arbitrary	60%
4.	OPB – 12	5′ – CCTTGACGCA – 3′	Arbitrary	60%

With appropriate addition of loading dye, PCR product can runout on a 1.2% Agarose gel and stained with Ethidium bromide. Picture could be taken on the UV light. The RAPD finger printing generated and can be compared with other plant species belonging to different genera to find out the genetic similarity. Data should be scored manually based on the presence and absence of bands at identical molecular size and designated as + or -. The polymorphic bands were scored as ++. Comparisons can be carried out between samples amplified using the same primer in a pair-wise manner. The similarity index between individuals must be calculated following the method of Nei and Li (1979). Genetic similarity between individuals A and B (S_{AB}) can be calculated using the formula: $S_{AB} = 2N_{AB} / (N_A+N_B)$. Where, N_{AB} is the RAPD bands shared in common between individuals A and B when N_A and N_B are the total number of band scored for individual A and B respectively. Estimation of genetic distance between different samples should be calculated using the Jaccard's distance co-efficient (Sneath and Sokal, 1973). Dendrogram must be produced using group average and further Neighbour Agglomerative unweighted pair-group method using Arithmetic average (UPGMA) clustering (Manly, 1986) by $NTSYS_{PC}$ 2.1. software. The statistical analysis must be done by SHAN to perform sequential, agglomerative, hierarchical and nested clustering methods. The program can find alternative trees where there are ties in the input matrix. Genetic similarity for variation among the chosen halophyte will be calculated by primer-wise comparison of individuals. The average genetic similarity index can be analysed using primer pair contribution based on UPGMA Cluster analysis. The similarity between the halophytes could be analyzed and found the chosen species differ genetically with very low percentage of similarities.

1.5 Conclusion

It is concluded from this study that, the methanolic extracts from marine halophytes both mangroves and seaweeds sourced from southern Indian coast has enormous potential of antiplasmodial activity in combined form with standard organoleptic, physical, biological and chemical parameters and thus it could be used for the further development of antimalarial drug after completing the appropriate clinical trials.

References

Abeysinghe, P.D., Triest, L., Greef, B.D. Koedam, N. and Hettiarachi, S. 2000. Genetic and geographic variation of the mangrove tree *Bruguiera* in Srilanka. *Aquatic Botany*, 67: 131-141.

Bandaranayake, W.M. 1998. Traditional and medicinal uses of mangroves. *Mangroove Salt and Marshy plants*, 2: 133-148.

Bartleson, S.M., Downs, E., Hall, J.F., Johnson, R.C., Lefebvore, D.L., Nielson, J.T., Powell, R.L., Stallings, G.R., Wang, B. and Wang, D. 2003. The genetic diversity of red mangrove. *Rhizophora mangle*, in southern Florida, NSF Research in Ecology, program. Howard Hughes Medical Institute.

Chapman, V.J. 1975. 'Mangrove vegetation'. A. R. Ganther Verlag, Komman ditgesellschaft. F2 94901 Vaduz.

Cheng, K.T., Chen, C.F., Tsay, H.S. and Chou, T.W. 1998. *Planta Medica*, 64: 563-565.

Cupp, M. 1999. Herbal remedies : Adverse effects and drug interactions. *American Family Physician*, 59: 1239-1244.

Das, K., Samanta, L. and Chainy, G.B.N. 2000. A modified spectrophotometric assay of SOD using nitrite formation by superoxide radicals. *Indian Journal of Biochemistry and Biophysics*, 37: 201-204.

Ernst, E. 1998. Harmless Herbs ? A review of the recent literature. *American Journal of Medicine*, 104: 170-178.

Faulkner, D.J. 1986. Marine natural products. *Natural Products Reports*, 3:1.

Fenical, W. 1975. "Halogenation in the Rhodophyta - A review". *Journal of Phycology*, 11: 245-259.

Grossman, S.H., Mollo, E. and Ertingshausen, G. 1976. Simplified, totally enzymatic method for determination of serum triglycerides with a centrifugal analyzer. *Clinical Chemistry*, 22: 1310-1313.

Hayden, H.S., Blomster, J., Maggs, C.A., Silva, P.C., Stanhope, M.J. and Walland, J. 2003. Linnaeus was right all along : *Ulva* and *Enteromorpha* are not distinct genera. *European Journal of Phycology*, 38: 277-294.

Kathiresan, K. 1999. Impact of mangrove biodiversity on associated fishery resource and fishers income. Project Report. WWF. Washington, USA, 108.

Kathiresan, K. 2002. Why are mangrove degrading? *Current Science*, 83: 1246-1249.

Kepam, W. 1986. Qualitative organic analysis (Spectrochemical techniques_. Ed. II. Mc Graw Hill, London, 40 – 58.

Kind, J. 1967a. The hydrolases - acid and alkaline phosphatase, In: Practical Clinical Enzymology (Ed. Van. D.) London, Norstand Company Limited, 191.

Lowry, O.H., Rosebrough, N.J., Farr, A.L. and Randall, R. 1951. Protein measurement with the Folin's phenol reagent. *Journal of Biological Chemistry*, 193: 265-275.

Macnae, W. 1968. A general account of the fauna and flora of mangrove swamps and forest in the Indo-Pacific region. *Advances in Marine Biology*, 6: 73-270.

Malloy, H.T. and Evelyn, K.A. 1937. The determination of bilirubin with the photoelectric colorimeter. *Journal of Biological Chemistry*, 119: 481.

Manly, K. 1986. Multivariate statistical methods. A Primer London: Chapman and Hall, London.

Nei, M. and Li, W.H. 1979. Mathematical model for studying genetic variation in terms of restriction endonucleases. *Proceedings of Academy of Sciences USA*, 74: 5267-5273.

O'Hara, M., Kiefer, D., Farrell, K. and Kemper, K. 1998. A review of 12 commonly used medicinal herbs. *Archieves of Family Medicine*, 7: 523-536.

Ohkawa, H., Ohishi, N. and Yagi, K. 1979. Assay for lipid peroxides in animal tissues by thiobarbituric acid reaction. *Analytical Biochemistry*, 95: 351-318.

Padmesh, P., Sabu, K.K., Seeni, S. and Pushpangadan, P. 1998. The use of RAPD in assessing genetic variability in *Andrographis paniculata*. Nees, A hepato protective drug. Plant Biotechnology Division, Tropical Botanical Garden and Research Institute, Thiruvananthapuram, India.

Peters, W., Porrus, J.H. and Robinson, B.L. 1975. The chemotherapy of rodent malaria XXII : The value of drug resistant strains of P. berghei in screening for blood Schizontocidal activity. *Annals of Tropical Medicinal Parasitology*, 155-171.

Rajendran, P., Muthukrishnan, J. and Gunasekara, P. 2003. Microbes in heavy metal remediation. *Indian Journal of Experimental Biology*, 41: 935-944.

Reitmans and Frankel, 1957. A Colorimetric method for determination of serum Glutamate oxaloacetate and pyruvate Transaminases. *American Journal of Clinical Pathalogy*, 28: 56-63.

Roan, S. 2000. Quality control for herbs. Los. Angeles Times.

Rotruck, J.T., Pope, A.L., Ganther, H.F., Swanson, A.B., Hafeman, D.G. and Heksta, W.G. 1973. Selenui biochemical role as a component of Glutathione peroxidase. *Science*, 179: 585-595.

Senior, K. 1998. Herbal medicine under scrutiny. *Lancet*, 352: 1040.

Sinha, A. K. 1972. *Analytical Biochemistry*, 47: 389.

Sneath, P.H.A. and Sokal, R.R. 1973. Numerical taxonomy. The principles and practise of numerical classification. W. H. Freeman and Company, San Francisco, USA.

Sofowora, A., 1982. Medicinal plants and traditional medicine in Africa. John Wiley and Sons Ltd.

Thompson, D. and Henry, R. 1993. Use of DNA from dry leaves for PCR and RAPD analysis. *Plant Molecular Biology Reports*, 11: 202-206.

Tomlinson, P.B. 1986. 'The Botany of Mangroves', Cambridge University Press, Cambridge, UK, 413.

Trease, G.E. and W.C. Evans. 1972. Pharmacognosy, Ed. Baillere and Tindall, London.

Vannucci, M. 1997. Supporting appropriate mangrove management intercoast network. Special edition, 1: 1-3.

Varley, H. 1976. In: Practical Clinical Biochem., 4th Edition, New York, William Heinmann Publishers Pvt. Ltd., 452.

Vasanthi, A.H. 2002. Biochemical and pharmacological studies of some marine algae of Gulf of Mannar. Ph.D. Thesis, Tamil University, Thanjavur, India.

Vickers, A. and Zollman, Z. 1999. Herbal medicine. *British Medical Journal*, 7216: 1050-1054.

Warnick, G.R., Bendeson, J. and Albers, J.J. 1982. Dextran sulfate Mg^{2+} precipitation procedure for quantitation of high-density lipoprotein cholesterol. *Clinical Chemistry*, 28: 1379-1388.

Youngkin, E. and Israel, D. 1996. A review and critique of common herbal alternative therapies. *The Nurse Practitioner*, 21: 39-49.

Zak, B. 1977. Cholesterol methodologies: A review. *Clinical Chemistry*, 23: 1201.

Zhang, K.Y., Leung, H.W., Yeung, H.W. and Wong, R.N. 2001. *Planta Medica*, 67: 379-381.

□□□

16

In vitro Culture and Studies on Secondary Metabolites in Nothapodytes nimmoniana

Singh, Indramohan and N. Kumaravadivel

CONTENTS

1.1	Introduction	246
1.2	In vitro Culture	246
1.3	Callus Induction	247
1.4	Culture Media and Growth Regulators	247
1.5	Micropropagation	248
1.6	In vitro Rooting and Hardening	248
1.7	Secondary Metabolites	249
1.8	Camptothecin (CPT) and its Derivatives as Important Anticancer Secondary Metabolites	249
	1.8.1 Occurrence	249
	1.8.2 Properties	250
	1.8.3 Pharmacology and therapeutic uses	250
	1.8.4 Mechanism of action of CPT	251
	1.8.5 Biosynthesis of CPT	252
1.9	Secondary Metabolites from N. nimmoniana from Plants and In vitro Culture	252
1.10	Extraction and Characterization of Medicinal Compounds	253
1.11	Analysis of Secondary Metabolites in Extracts of In vitro Culture	254
1.12	Conclusion	255
	References	255

1.1 Introduction

India, being a tropical country, is rich in medicinal plant diversity and is one of the twelve-mega diversity hot spot regions of the world. *Nothapodytes nimmoniana* (Grah) Mabb. (Synonyms, *Nothapodytes foetida* and *Mappia foetida*) belonging to family Icacinaceae, is an important anti-cancer endangered medicinal plant. The Genus *Nothapodytes* consists of more than 16 species and sub-species. Globally this species is distributed in the Indo-Malaysian region and China. Within India it is distributed in the Western Ghats of Maharashtra, Karnataka, Kerala, and Tamil Nadu as well as in the Eastern region i.e. West Bengal and Assam. Because of destructive harvesting, habitat loss, the population of this species has declined 50-80%. Recently the tree has been assigned a vulnerable status. (Padmanabha *et al.*, 2006). It is an excellent source of quinoline alkaloids, camptothecin (CPT) and 9-methoxycamptothecin (9-OMeCPT) which are used clinically as such or after derivatization as anticancer agents for the treatment of solid tumours (Govindachari and Vishvanathan, 1972).

Camptothecin was first reported in the Chinese tree *Camptotheca acuminata* (Wall and Wani, 1996) and later discovered in *Nothapodytes nimmoniana*. The stem, wood, bark, seeds, leaves and various organs of the tree are rich sources of alkaloids. In *C. acuminata*, the camptothecin content was 0.005% of dried plant as against 0.14–0.24% in *N. nimmoniana* (Puri *et al.*, 1999). Recent interest in the possible anti cancer effects has focused the need for the extensive study of the bioactive compounds from *Nothapodytes nimmoniana*.

1.2 *In vitro* Culture

In vitro culture is very important in this endangered medicinal plant. For sterilization of explant a thorough washing of seeds (without seed coat) with soap water followed by tap water and presoaking for 24 hrs in distilled water with 1 mg L^{-1} gibberellic acid before sterilization was found suitable (Ciddi and Shuler, 2000). Different methods for surface sterilization of seeds have been adopted. Sterilization with 70 percent ethanol for 1 minute followed by 20 percent Clorox ($NaClO_4$) for 10 minutes was reported. (Ciddi and Shuler, 2000). A very simple method for seed sterilization was used by Fulzele *et al.* (2002). Decoated seeds were washed in running tap water and surface disinfected in 70 percent ethanol for 2 minutes followed by immersing in 0.1 percent aqueous mercuric chloride solution and finally 4-5 washing with sterile distilled water.

Thengane *et al.* (2003) used 10 percent savlon for disinfection of seeds and 1 percent mercuric chloride for final sterilization followed by 4-5 times rinsing and presoaking in sterile distilled water for 24 hours. Anuradha *et al.* (2006) sterilized the seeds by treating first with Teepol for 15 minutes and then dipped in 70 percent ethanol for 5 minutes followed by repeated washing with distilled water. The seeds were then treated with cetrimide solution and 100 PPM bavistin for 45 minutes followed by a thorough washing with sterile distilled water. Final surface sterilization was carried out with 0.1 percent aqueous mercuric chloride for 15 minutes followed by repeated washing with sterile distilled water.

1.3 Callus Induction

Nothapodytes nimmoniana seeds and seed parts are the main explants used for studies on callus induction. In addition, use of leaves, stem parts and roots have also been reported. Ciddi and Shuler (2000) used excised embryos to obtain callus culture. Callus cultures were also initiated from stem parts of *Nothapodytes foetida*. (Fulzele *et al.*, 2001). Successful callus initiation was achieved with dissected cotyledons from seeds of *N. foetida*. (Thengane *et al.*, 2003). Callus cultures were also initiated from immature zygotic embryos (Fulzele *et al.*, 2003). Sundravelan *et al.* (2004) established callus cultures from leaf explants.

1.4 Culture Media and Growth Regulators

Murashige and Skoog (MS) medium (Murashige and Skoog, 1962) was used for *in vitro* culture studies. (Ciddi and Shuler, 2000 ; Fulzele *et al.*, 2001, 2003; Thengane *et al.*, 2001, 2003; Sundravelan *et al.*, 2004; Anuradha *et al.*, 2006). Different auxins like 2, 4-D (Fulzele *et al.*, 2001; Thengane *et al.*, 2001, 2003; Sundravelan *et al.*, 2004), 2, 4, 5-T (Thengane *et al.*, 2003), NAA (Ciddi and Shuler, 2000; Fulzele *et al.*, 2001; Thengane *et al.*, 2003; Anuradha *et al.*, 2006), Picloram (Ciddi and Shuler, 2000; Sundravelan *et al.*, 2004) in different concentration either alone or in combination with other auxins or cytokinins have been tried for *in vitro* culture studies.

Ciddi and Shuler (2000) found Picloram (2 mg L^{-1}) the most suitable plant growth regulator for callus induction from isolated embryos. Studies showed that in the presence of NAA, maximum concentrations of camptothecin (0.035 mg/ml) and 9-methoxycamptothecin (0.026 mg/ml) were found in the medium. Alkaloid production was reduced in presence of 2, 4-D in the culture medium. (Fulzele *et al.*, 2001). Reports indicated that auxins 2, 4-D and 2, 4, 5- T induced white friable callus while that on NAA was compact and creamy. (Thengane *et al.*, 2003). MS medium supplemented with Picloram (2 mg L^{-1}), N^6- benzyl adenine (BA) (1 mg L^{-1}) and gibberellic acid (GA_3) (1 mg L^{-1}) was found to be suitable for the establishment of callus cultures from leaves of *N. foetida* (Sundravelan *et al.*, 2004).

The combination of different auxins and cytokinins, in different concentrations, also found to be significant for callus initiation. Response of cotyledon explants for callus induction was maximum on combination of 2,4-D (2.26 μM) with BA at a concentration of 2.22μM (Thengane *et al.*, 2003). Callus cultures were also initiated with zygotic embryos cultured on Murashige and Skoog's (MS) medium supplemented with 2, 4-D, 6-BA and Kn with concentrations of 9.05 μM, 4.44 μM and 2.32 μM respectively (Fulzele *et al.*, 2003). Significant percentage of callus was obtained with MS basal medium supplemented with 5.37 μM NAA and 4.44 μM BAP (Anuradha *et al.*, 2006).

1.5 Micropropagation

Most of the reports on *Nothapodytes nimmoniana* contain callus induction and analysis of camptothecin and other secondary metabolites content. There are only a few reports on Micropropagation studies in this particular plant. These reports indicated regeneration of adventitious shoots, somatic embryogenesis and multiple shoot induction from different explants.

Regeneration of adventitious shoots from *Nothapodytes foetida* has been achieved using different seedling explants. (Thengane *et al.*, 2001). Multiple shoots were induced on hypocotyl segments of 20–25-d-old seedlings of *N. foetida* cultured on Murashige and Skoog (MS) medium supplemented with different concentrations and combinations of cytokinins (Rai, 2002). Somatic embryogenesis was established from callus cultures derived from isolated embryos cultured *in vitro* on suitable culture media (Anuradha *et al.*, 2006). MS medium was mostly used successfully as a basal for Micropropagation studies in *N. nimmoniana*. ½ MS and ¼ MS were also tried in some of the experiments (Rai, 2002), but with little or non-significant success percentage.

Ciddi and Shuler (2000) obtained multiple shoots with leaves in a hormonal combination of Kn (1 mg L^{-1}) and NAA (0.2 mg L^{-1}). Regeneration of adventitious shoots has been achieved in Murashige and Skoog's (MS) basal medium supplemented with various concentrations of TDZ. The optimum levels of TDZ concentrations were 0.91-4.45 μM. (Thengane *et al.*, 2001). The highest shoot multiplication was achieved on medium containing TDZ (2.2 μM). (Rai, 2002). MS medium devoid of plant growth regulators favored the development of globular somatic embryos that differentiated further into plantlets (Fulzele *et al.*, 2003). Significant percentage of somatic embryogenesis was found with the isolated embryos inoculated on MS basal supplemented with NAA (10.74 μM) & BAP (4.44 μM) (Anuradha *et al.*, 2006).

1.6 *In vitro* Rooting and Hardening

For *in vitro* rooting of shoots, shoots of 1.5–2 cm in height with two or three leaves derived from both primary and secondary cultures were transferred to half-strength or one-fourth MS medium containing IAA and IBA either individually or in combination. Among the combination of auxins tested, the best rooting response was obtained on ¼ MS medium with IBA (2.4mM) combined with IAA (5.7 mM) after 30–35 d of culture (Rai, 2002). *In vitro* rooted plantlets were first transferred to a growth chamber after 8 wk of culture, before transfer to greenhouse conditions, In growth chamber, the plantlets were transferred to plastic pots (7.5 cm diameter) containing Soil-rite. Plantlets were removed from the growth chamber after 4 wk and then transferred to larger pots (9.5 cm diameter) containing soil: compost (1:1) and kept in the greenhouse (Rai, 2002).

1.7 Secondary Metabolites

Secondary metabolites are compounds belonging to extremely varied chemical groups, such as organic acids, aromatic compounds, terpenoids, steroids, flavonoids, alkaloids, carbonyls, etc. Their function in plants is usually related to metabolism and/or growth regulation, lignification, coloring of plant parts and protection against pathogen attack (Payne *et al.*, 1991). The secondary metabolites are also referred to as "Natural Products". It is believed that more than 100,000 different structures of secondary metabolites may be synthesized by organisms; to a tune of 10^9 tons per year (Bowles and Leyser, 1994).

Some secondary metabolites are considered as metabolic waste products, for example, alkaloids may function as nitrogen waste products. However, a significant portion of the products derived form secondary pathways serve either as *protective agents against various pathogens* (e.g. insects, fungi or bacteria) or *growth regulatory molecules* (e.g. hormone-like substances that stimulate or inhibit cell division and morphogenesis). Due to these physiological functions, secondary metabolites are potential anti-cancer drugs, since either direct cytotoxicity is affected on cancer cells or the course of tumor development is modulated and eventually inhibited. (Kintzios and Barberaki, 2004).

1.8 Camptothecin (CPT) and its Derivatives as Important Anticancer Secondary Metabolites

1.8.1 Occurrence

CPT is a potent antitumor TIA first isolated by Monroe E. Wall and Mansukh C. Wani in 1958 from extracts of *Camptotheca acuminata* , a deciduous tree native to China and Tibet, which has been extensively used in traditional Chinese medicine. In addition, CPT is also produced by a number of other medicinal plants, including *Nothapodytes foetida, Pyrenacantha klaineana, Merrilliodendron megacarpum* (Icacinaceae), *Ophiorrhiza pumila* (Rubiaceae),

Ervatamia heyneana (Apocynaceae) and *Mostuea brunonis* (Gelsemiaceae), species belonging to unrelated orders of angiosperms (Lorence and Nessler, 2004).

Immature leaves serve as major sinks of photosynthates and sites of production for some phytohormones. They are nutrient-rich and tender in physical structure, features that make them attractive to herbivory and pathogens. Many species defend their juvenile developmental stages metabolically. It is likely that the 5–6-fold of the CPT content in young *C. acuminata* leaves compared to mature ones and the peak of CPT production in young seedlings (Lopez-Meyer and Nessler, 1997; Lu and McKnight, 1999), represent a chemical defense mechanism deployed by young leaves and seedlings to deter attacks by herbivores, pathogens, or both.

1.8.2 Properties

CPT is a weak acid and therefore is the lactone ring highly susceptible to ring opening by hydrolysis, forming carboxylate. The open ring form is inactive and it must therefore by closed to inhibit topo I. The closed form is favored in acidic condition, as it is in many cancer cells microenvironment. CPT is transported in to the cell by passive diffusion. Cellular uptake is favored by lipophilicity, which enhances intracellular accumulation. Lipophilicity makes compounds more stable because of improved lactone partitioning into red blood cells and consequently less hydrolysis of the lactone. CPT has affinity for human serum albumin (HSA), especially the carboxylate form of CPT. Because of that, the equilibrium between the lactone ring and the carboxylate form is driven toward the carboxylate. Reduced drug-HSA interactions could result in improved activity.

1.8.3 Pharmacology and therapeutic uses

The CPT derivatives, irinotecan and topotecan, are used throughout the world for the treatment of various cancers, and over a dozen more CPT analogues are currently at various stage of clinical development. In spite of the rapid growth of the market, CPT is still harvested by extraction from bark and seeds of *C. acuminate and N. foetida* (Lorence and Nessler, 2004).

Although the antitumor activity of CPT and its derivatives has been the focus of research for groups in both the industry and academic arenas, CPTs have also been studied as potent inhibitors of replication, transcription, and packing of double stranded DNA-containing adenoviruses, papovaviruses, and herpes viruses, and the single-stranded DNA-containing autonomous parvoviruses (Pantazis *et al.*, 1999). CPT inhibits viral functions by poisoning topo I, the host cell enzyme required for initiation and completion of viral functions. If properly developed, CPTs could prove to be powerful antiviral drugs for several DNA viruses, which are causative agents for a large number of diseases.

In the mid-1990s, CPT was also shown to have promising activity against parasitic trypanosomes and Leishmania (Bodley and Shapiro, 1995). Researchers at the National Cancer Institute screened 2000 diverse compounds for functional inhibition of the hypoxia-inducible factor 1 (HIF-1), a master regulator of the cancer cells' ability to survive under oxygen deprivation. Only four compounds exhibited HIF-1 inhibitory activity, and three of these were CPTs (Rapisarda *et al.*, 2002). Hence, these drugs may have other desirable activities against solid tumors that are independent of topo I poisoning.

1.8.4 Mechanism of action of CPT

Topoisomerases (topo) are ubiquitous enzymes that solve topological problems generated by key nuclear processes such as DNA replication, transcription, recombination, repair, chromatin assembly, and chromosome segregation. There are two types of topoisomerases: topo I and topo II. Topo I catalyzes changes in the linking number of DNA (the number of times one strand of DNA crosses the other) by breaking and resealing phosphodiester bonds (Lorence and Nessler, 2004). The topo I mediated reaction can be divided into four steps: (1) enzyme- DNA binding, (2) DNA single-strand cleavage by reversible trans-esterification, in which the 50 oxygen of a phosphodiester bond is exchanged by the hydroxyl group of tyrosine 723 of human topo I (through which the protein becomes covalently linked to the 30 terminus of the cleaved DNA strand), (3) single-strand passage, and (4) re-ligation of the cleaved DNA strand.

Topo I is the cellular target of CPT and its derivatives (Hsiang *et al.*, 1985). CPT binds only weakly to normal B-DNA under physiological conditions, and it does not bind to topo I alone. CPT induces topo I-linked DNA breaks by preventing DNA re-ligation (Avemann *et al.*, 1988). The efficacy of CPTs as anticancer agents may be related to the fact that drug-mediated cleavage complexes stabilized during S phase are much more toxic than those stabilized during G1 or G2 phases (Horwitz and Horwitz, 1973). Cross-linking studies have suggested the interaction of CPT with both, topo I and the DNA, forming a ternary complex that stabilizes the trans-esterification intermediate (Hertzberg *et al.*, 1990; Pommier *et al.*, 1995). The structure of this ternary complex is of great interest for elucidation of the detailed mechanism of action of CPT and development of new topo I poisons.

Relevant events in determining the cytotoxic potency of CPT and its derivatives are: (a) uptake, (b) lysosomal or mitochondrial sequestration and, (c) nuclear localization and stabilization of the cleavable complex.

1.8.5 Biosynthesis of CPT

Several investigators have recognized plant cell cultures as a tool to unravel the biosynthetic pathway that leads to CPT formation in plants and as a potential system for its production. All TIAs, including CPT, are derived from the common precursor strictosidine, which is the product of a condensation reaction between the indole tryptamine and the terpenoid secologanin catalyzed by the enzyme strictosidine synthase (Kutchan, 1995). Strictosidine is then converted to strictosamide via intermolecular cyclization and this compound is a precursor of CPT as proven by incorporation of the radio labeled precursor.

It has been postulated that CPT could be formed from strictosamide by three trans-formations: (1) oxidation-recyclization of the B- and C-rings, (2) oxidation of the D-ring and removal of the C-21 glucose moiety, and (3) oxidation of ring E (Hutchinson *et al.*, 1979). A presumed precursor of CPT was isolated from the polar fraction of a large-scale extraction of C. acuminata (Carte *et al.*, 1990). 3(S) - Pumiloside, 3(S) - and 3(R)-deoxypumiloside, plausible CPT precursors have been found in O. pumila (Aimi *et al.*, 1989; Kitajima *et al.*, 1997; Yamazaki *et al.*, 2003b).

The precise mechanism of transport and storage of CPT and its precursors remains unclear. One possible mechanism of transport of such an insoluble compound is its conversion into a more water soluble form such as a glucoside, e.g. chaboside (Aimi *et al.*, 1990) and transport to other parts of the plant (Yamazaki *et al.*, 2003b).

1.9 Secondary metabolites from *N. nimmoniana* from plants and *In vitro* culture

Camptothecin and 9-methoxy camptothecin are the major secondary metabolites identified in *Nothapodytes nimmoniana* (Govindachari and Vishvanathan, 1972). There are several reports of extraction of these secondary metabolites both from *in vivo* and *in vitro* grown parts of the plant. Among the various *in vivo* parts of the plant analyzed, the highest CPT content was found in root bark (0.333%) followed by stem bark (0.236%). The other tissues used in analysis were leaves, root wood, stem wood, seeds and root tips (Padmanabha *et al.*, 2006). The earliest report on the production of CPT and 9-methoxy-CPT from callus cultures of *N. foetida* obtained from seeds (Roja and Heble, 1994). The callus was reported to produce 1 µg per g dry weight of 9-methoxy-CPT along with traces of CPT.

Ciddi and Shuler (2000) showed the presence of CPT and 9-methoxy-CPT after four passages in maintenance medium. The content of CPT in callus was found to be 9.5 µg per g dry weight along with traces of 9-methoxy-CPT.

Production of CPT and 9-methoxy-CPT was also achieved with untransformed root cultures grown on MS basal supplemented with NAA (71.36 μM) and BA (8.87 μM). This was the first report to show the production of these secondary metabolites from untransformed root cultures. (Fulzele *et al.*, 2002). Use of dissected cotyledons as explants for successful *in vitro* production of CPT and 9-methoxy-CPT was reported. Media containing 2, 4-D (0.9-13.57 μM) with BA (2.22 μM) produced significant amount of CPT (0.12-1.3%) (Thengane *et al.*, 2003). The first report of somatic embryogenesis from isolated embryo cultures and their potentiality to synthesize CPT and 9-methoxy-CPT was given by Anuradha *et al.* (2006).

1.10 Extraction and Characterization of Medicinal Compounds

Different organic solvents with a variety of methods, like water bath extraction, Soxhlet apparatus, sonication etc. have been used for the extraction of secondary metabolites from the medicinal plants. Various chromatographic techniques, spectrophotometric and NMR studies have been done in order to characterize these medicinal compounds.

Puri *et al.* (2005) used methanol, water, dichloromethane and chloroform as solvents for getting the continuous extract of camptothecin by sonication method. The extract was analyzed by LC-MS. 100 mg of the dried plant material (6-8% moisture contents) was extracted in a centrifuge tube with 2ml of methanol by sonification (2x30 sec). The mixture was mixed with 18 ml water and 20 ml dichloromethane and the material was stirred vigorously for 5 min. on a magnetic stirrer. Centrifugation for 10 min at 2000 rpm yielded two phases. The dichloromethane phase was separated and evaporated to dryness. The residue was dissolved in 1 ml of chloroform. These extracts were dried to give camptothecinoids residues. The residues were reconstituted in chloroform: methanol (3:1) mixture and filtered through 0.2 μm filter paper and analyzed by LC-MS.

Padmanabha *et al.* (2006) reported analysis of camptothecin from ethanolic extract of different plant samples. The samples were dried at 60°C for 96 hours in a hot air oven and then ground. The ground sample was used to prepare ethanolic extract with 61% ethanol. The compound was analyzed by reversed phase HPLC. The HPLC conditions were 254 nm as detector wavelength, 1.6 mL/min flow rate and 10 μM sample loop. The mobile phase was 40% acetonitrile and 60% water with 0.1% tri-fluoroacetic acid (TFA) in an isocratic mode. The standard was prepared using DMSO and methanol in 1:50 (v/v) ratio respectively. The retention time was 3.5 minutes for every five runs.

1.11 Analysis of Secondary Metabolites in Extracts of *In vitro* Culture

Extracts of in vitro culture are abundant in CPT and its derivatives, besides they also contain other unidentified alkaloids.

Ciddi and Schuler (2000) freeze dried and extracted callus cultures of *N. foetida* with 3 vol methanol and kept overnight in deep freezer. After thawing, the methanolic layer separated and the process was repeated three for complete extraction. The combined extract was evaporated to dryness and extracted with chloroform (3×15ml). The chloroform extract evaporated and 2 ml HPLC-grade chloroform was added. The resulting solution was analyzed by TLC and HPLC. Results showed that callus was able to produce CPT but the production was 80 times less than of intact plant and 200 times less than callus cultures of *C. acuminata.*

Fulzele and co-workers (2002) developed untransformed root cultures and dried it in an oven at 55°C for 16 hrs. Dried powered material was extracted with 2 ml methanol by sonication for 30 min at room temperature. The methanolic extract was evaporated to dryness and the concentrate was redissolved in HPLC grade methanol and centrifuged at 9000 g for 5 minutes. The clear supernatant was used for Chromatographic analysis. Results revealed that the concentration of CPT and 9-methoxy-CPT was 12 times less as compared to the normal roots of two year old intact plants raised from somatic embryos.

Fulzele and co-workers (2003) also used callus cultures, somatic embryos, regenerated plantlets and 2 years old plants for the determination of alkaloids. The biomass was dried at 55°C for 16 hrs and ground to a powdered form. The extraction procedure was the same as that for untransformed root cultures.

Callus cultures and white globular embryos of *N. foetida* synthesized trace amounts of alkaloids. Somatic embryos showed the presence of CPT (0.011% dry weight, DW) and 9-methoxy-CPT (0.0028% DW). When well-developed plantlets were grown in MS medium with BA, production of 0.061% DW camptothecin and 0.0053% DW 9-methoxy-CPT was achieved. The plants in paper cups synthesized CPT (0.094% DW) and 9-methoxy-CPT (0.0088% DW). Two-year-old plants were harvested and their stem, roots, and leaves separated and analyzed for the determination of alkaloids. Roots synthesized maximum levels of CPT (0.20% DW) and 9-methoxy-CPT (0.097% DW). Leaves produced the lowest level of CPT and 9-methoxy-CPT as compared with the roots and stem (0.045% and 0.013% DW, respectively). A moderate concentration of alkaloid was produced by stems.

Thengane and co-workers (2003) used one month old fresh callus developed from dissected cotyledons of *N. foetida*. They dried the calli and extracted with 10 ml methanol per g of callus three times at room temperature and evaporated under reduced pressure. The extract was defatted with MeOH and again extracted with 20 ml of ether. The defatted extract was dissolved in chloroform and extracted with 5 ml water to remove polar compounds. The resulting solution was analyzed for secondary metabolites contents.

Anuradha *et al.* (2006) prepared extracts for embryogenic masses, obtained *in vitro*. Dried embryogenic masses were defatted with petroleum ether and filtered. The residual mass was extracted repeatedly with methanol five times and evaporated to dryness. The methanolic extract was dissolved in water and repeatedly extracted with chloroform 4-5 times. The combined chloroform extract was evaporated and analyzed.

TLC was carried out with mobile phase chloroform: ethyl acetate (8:2) (v/v). CPT and 9-methoxy-CPT were observed under UV light at 254 and 365 nm. CPT appeared as blue fluorescent spot and 9-methoxy-CPT as green fluorescent spot. Quantitative HPLC was carried out with solvent system KH_2PO_4:CH_3CN: MeOH (7:2:1). The flow rate was adjusted to 1.2 ml min^{-1}. Results showed that embryogenic masses have the biosynthetic potentiality to produce CPT (0.001 %) and 9-methoxy-CPT (0.0007%). Successive petroleum ether, chloroform and methanol extracts of *Mappia foetida* (Synonym of *Nothapodytes nimmoniana*) leaves and stem were tested for their antibacterial activity. The methanol fractions were found to be most effective against all the tested organisms (Kumar *et al.*, 2002).

1.12 Conclusion

There have been only a few reports of *in vitro* propagation and secondary metabolites production from *Nothapodytes nimmoniana* which also showed anti bacterial and anti viral properties besides more pronounced anti cancer activities. Therefore, more studies are needed to optimize the *in vitro* propagation and correct stage & procedure for extracting the medicinal compounds in order to get the maximum quantity without disrupting the natural occurrence of the plant.

References

Aimi, N.H., Nishimura, H.M., Sakai, S.I. and Haginiwa, J. 1990. Chaboside, first natural glycocamptothecin found from *Ophiorrhiza pumila*. *Tetrahedron Letters*, 31: 5169-5172.

Anuradha, M., Indu, B.K., Rajani, R. and Divakar, K.M. 2006. Somatic embryogenesis in *Nothapodytes foetida*: A potent Antineoplastic Plant. *Plant cell Biotechnology and Molecular Biology*, 7: 77-80.

Avemann, K., Knippers, R., Koller, T. and Sogo, J.M. 1988. Camptothecin, a specific inhibitor of type I DNA topoisomerase, induces DNA breakage at replication forks. *Molecular and Cell Biology*, 8: 3026-3034.

Bowles, D. and Leyser, O. 1994. 'TheBig Green Book' Plant Biotechnology – Centre for exploitation of science and technology (CEST).

Ciddi, V. and Shuler, M.L. 2000. Camptothecin from callus cultures of *Nothapodytes foetida*. *Biotechnology Letters*. 22: 129-132.

Fulzele, P.D., Satdive, R.K. and Pol, B.B. 2002. Untransformed root cultures of *Nothapodytes foetida* and production of camptothecin. *Plant Cell, Tissue and Organ Culture*, 69: 285-288.

Fulzele D.P., Satdive, R.K. and Pol, B.B. 2001. Growth and production of camptothecin by cell suspension cultures of *Nothapodytes foetida*. *Planta Medica*, 67:150-152.

Fulzele, D.P. and Satdive, R.K. 2003. Somatic embryogenesis, Plant regeneration, and the Evaluation of camptothecin content in *Nothapodytes foetida*. *In Vitro Cellular and Development Biology - Plant*, 39: 212-216.

Roja, G. and Rao, P.S. 2000. Anticancer Compounds from Tissue Cultures of Medicinal Plants. *Journal of Herbs, Spices & Medicinal Plants*, 7: 71-102.

Govindachari, T.R. and Vishvanathan, N. 1972. Alkaloids of *Mappia foetida*. *Phytochemistry*, 11: 3529-3530.

Hertzberg, R.P., Busby, R.W., Caranfa, M.J., Holden, K.G., Johnson, R.K., Hecht, S.M. and Kingsbury, W.D. 1990. Irreversible trapping of the DNA-topoisomerase I covalent complex. Affinity labeling of the camptothecin binding site. *Journal of Biological Chemistry*, 265: 19287-19295.

Kintzios, S.E. 2006. Terrestrial plant derived anticancer agents and plant species used in anticancer research. *Critical reviews in Plant Sciences*, 25: 79-113

Krishnamurthy, K.V. 2003. Influence of medium composition on callus induction and camptothecin(s) accumulation in *Nothapodytes foetida*. *Plant Cell, Tissue and Organ Culture*, 72: 247-251.

Kumar, R.N., Vishwanathan, H., Suresh, T. and Mohan, P.S. 2002. Antibacterial activity of *Mappia foetida* leaves and stem, *Fitoterapia*, 73: 734-736.

Lopez-Meyer, M., Nessler, C.L. and McKnight, T.D. 1997. Sites of accumulation of the antitumor alkaloid camptothecin in *Camptotheca acuminata*. *Planta Medica*, 60: 558-560.

Lorence, A. and Nessler, C.L. 2004. Molecules of Interest Camptothecin, over four decades of surprising findings. *Phytochemistry*, 65: 2735-2749.

Murashige, T. and Skoog, F. 1962. A revised medium for rapid growth and bioassay with tobacco tissue cultures. *Physiology of Plants*, 15: 473-497.

Padmanabha, B.V., Chandrashekar, M., Ramesha, B.T., Hombe Gowda, H.C., Gunaga, R.P., Suhas, S., Vasudeva, R., Ganeshaiah, K. N. and Uma Shaanker, R. 2006.

Patterns of accumulation of camptothecin, an anti-cancer alkaloid in Nothapodytes nimmoniana Graham. in the western ghats, India: Implications for identifying high yielding sources of alkaloid. *Current science*, 90: 95-99.

Pommier, Y., Kohlhagen, G., Kohn, K.W., Leteurtre, F., Wani, M.C. and Wall, M.E. 1995. Interaction of an alkylating camptothecin derivative with a DNA base at topoisomerase I-DNA cleavage sites. *Proceedings of National Acadamy of Sciences USA*, 92: 8861-8865.

Puri, S.C., Handa, G., Gupta, R.K., Srivastava, T.N., Somal, P. and Sharma, S.N. 1999. Quantitation of camptothecin in *Nothapodytes foetida*. *Journal of Indian Chemical Society*, 76: 370-371.

Rai, V.R. 2002. Rapid clonnal propagation of *Nothapodytes Foetida* (Weight) sleumer- A threatened medicinal tree. *In Vitro Cell Developmental Biology – Plant*, 38: 347-351.

Sundravelan, R.B., Desireddy and Ciddi, V. 2004. Production of camptothecines from callus cultures of *Nothapodytes foetida* (Weight) Sleumer. *Indian Journal of Biotechnology*, .

Thengane, S.T., Kulkarni, D.K., Shrikande, V.A., Joshi, D.P., Sonawane, K.B. and Krishnamurthy, K.V. 2003. Influence of medium composition on callus induction and camptothecin(s) accumulation in *Nothapodytes foetida*. *Plant Cell Tissue and Organ Culture*, 72: 247-251.

Wall, M.E. and Wani, M.C. 1996. Camptothecin and taxol: from discovery to clinic. *Journal of Etnopharmacology*, 51: 239-254.

Yamazaki, Y., Sudo, H., Yamazaki, M., Aimi, N. and Saito, K. 2003. Camptothecin biosynthetic genes in hairy roots of *Ophiorrhiza pumila*: Cloning, characterization and differential expression in tissues and by stress compounds. *Plant Cell Physiology*, 44: 395-403.

□□□

17

Modern Techniques Involving Tissue Culture for Medicinal Plants

M. Anbazhagan, R. Rajendran and M. Kalpana .

CONTENTS

1.1 Introduction 260
1.2 Surface Sterilization 260
1.2.1 Basal culture media 261
1.2.2 Phytohormones 261
1.2.3 Physical factors 262
1.3 Micropropagation 262
1.4 Callus Culture and Plant Regeneration 262
1.5 Somatic Embryogenesis and Conservation 263
1.6 Protoplast Culture 264
1.7 Rhizogenesis 265
1.8 Acclimatization and Field Establishment 265
1.9 In vitro Development of Synthetic Seeds 266
1.10 In vitro Production of Secondary Metabolites 266
1.11 Applications of In vitro Propagation 267
1.11.1 In vitro mutagenesis 267
1.11.2 Somaclonal variation 267
1.11.3 Cryopreservation 268
1.11.4 Genetic transformation systems for medicinal plants 269
1.11.5 Haploid production 270
1.12 Conclusion 270
References 271

1.1 Introduction

India has a very rich diversity of medicinal plants. Generally forests harbour a large number of medicinal plant species. But deforestation is the main cause of rapid loss of medicinal plant wealth and in this way valuable medicinal plants are under the threat of extinction. Pharmaceutical companies depend largely upon materials procured from naturally occurring stands that are being rapidly depleted. Plant tissue culture is an alternative way of commercial propagation (George and Sherrington, 1984) and is being used widely for the production of a large number of plant species, including many medicinal plants.

Experimental approaches used for propagation of medicinal plants through tissue culture can be divided into three broad categories. The most common approach is to isolate organized meristems like shoot tips or axillary buds and induce them to grow into a complete plant. This system of propagation is commonly referred to as micropropagation. In the second approach, adventitious shoots are initiated on leaf, root, stem segments and floral parts or on callus derived from those organs. The third system of propagation involves induction of somatic embryogenesis in cell and callus cultures. This method is most efficient as large numbers of somatic embryos can be obtained.

The use of plant cell and tissue culture methodology as a means of producing medicinal metabolites has a long history (Verpoorte *et al.*, 2002). Since plant cell and tissue culture emerged as a discipline within plant biology, researchers have endeavored to utilize plant cell biosynthetic capabilities for obtaining useful products and for studying the metabolism (Misawa, 1994). Cultured plant cells synthesize, accumulate and sometimes exude many classes of metabolites. Medicinal compounds are of particular interest and much effort has been devoted to obtaining some of the most active and precious therapeutics. Numerous alkaloids, saponins, cardenolides, anthraquinones, polyphenols and terpenes have been reported from *in vitro* cultures and reviewed several times (Misawa, 1994; Verpoorte *et al.*, 2002; Vanisree and Tsay, 2004).

Biotechnology involving modern tissue culture, cell biology and molecular biology offers the opportunity to develop a new germplasms that are well adapted to changing demands. In this text, a new and a different biotechnological approach have been provided for the micropropagation and regeneration of medicinal plants on onehand and the production of secondary metabolites on the other.

1.2 Surface Sterilization

Explants collected from field-grown plants are usually contaminated by various microorganisms. Besides surface contaminants, there may be some internal infectants that can be expressed even after years in culture due to some endophytic bacteria or fungi. Common sterilizing agents like sodium or calcium hypochlorite (5–10%), ethyl alcohol (50–95%) and mercuric chloride (0.01–0.1%) are used to exclude the surface contaminants by washing in the appropriate solution for 10–25 minutes followed by several rinses in sterile water. Rigorous screening of the stock cultures for bacterial contamination is highly essential.

1.2.1 Basal culture media

Now a days, more than 50 types of basal media formulations are used for the *in vitro* culture of plant tissues (Gamborg *et al.*, 1976; Huang and Murashige, 1977) and the formulation described by Murashige and Skoog (1962) (MS medium) is the most commonly and widely used basal medium. Basically, a nutrient medium consists of all the essential major and minor nutrient elements, vitamins, plant growth regulators and a carbohydrate as a carbon source with other organic substances like tender coconut water, tomato juice, extract of banana fruit juice etc. as optional additives.

The macronutrients N, P, K, Ca, Mg and S are required in all types of plant tissue cultures, but the optimal concentration of each may vary with plant species. Although, plant cells in culture may grow on nitrates alone, the pH of the medium (5.0–6.0) is usually more stable and better results are obtained when the medium contains both nitrate and ammonium ions sources of nitrogen. The essential micronutrients required in trace amounts for whole plants and for tissue in culture include Fe, Mn, Zn, B, Cu, Cl, Mo and Ni. Iron is usually added in a chelated form to avoid precipitation and facilitate absorption. The carbohydrate requirement in culture media is usually met by the incorporation of 2–3% sucrose. In addition to their role as a carbon source, the carbohydrates act as osmotica and help in the maintenance of osmotic potential in the culture medium. Generally plants are synthesized all the vitamins and biocatalysts required for their normal growth and development. But some specific vitamins namely (B1), nicotinic acid (B3), pyridoxin (B6) and myo-inositol are usually required as an additives addition to cell and tissue culture for better growth and development. Although, cultured cells are normally capable of synthesising all of their required aminoacids, addition of L-glutamine, L-cysteine, L-proline and L-tyrosine or an amino acid mixture like caesin hydrolysate to the medium may increase cell growth. Agar is commonly used for preparing semisolid or solid culture media.

1.2.2 Phytohormones

Growth regulators are organic compounds naturally synthesized in higher plants, which influence growth and development. The levels and kinds of plant growth substances added in the culture medium largely determine the success of tissue culture process. Root and shoot initiation and the process of differentiation from unorganized callus tissue are closely regulated by the relative concentrations of auxins and cytokinin in the medium (Ammirato, 1983; Bajaj *et al.*, 1988; Rout and Das, 1997). The gibberellins stimulate growth of organs but generally do not favour organ initiation.

1.2.3 Physical factors

Light, temperature and relative humidity are important parameters in culture incubation. Photosynthetic activity is not very important during initial phases of *in vitro* culture but at later stages the culture materials are induced to become autotrophic to a certain degree. Light is essential for morphogenetic processes like shoot and root initiation and somatic embryogenesis. Both quality and intensity of light as well as the photoperiod are very essential to the success of certain culture experiments (Murashige, 1977). Exposure to light for 12–16 h per day under 35–112 $\mu mol\ m^{-2}s^{-1}$provided by cool, white fluorescent lamps is usually recommended. Blue light promotes shoot formation while, red light induces rooting in many species (Murashige, 1977). The temperature usually employed in the culture incubation room is approximately 25°C. Tropical species usually require higher temperature of 27–30°C.

1.3 Micropropagation

Micropropagation of various plant species, including many medicinal plants, has been described by many authors. From a practical and pharmaceutical point of view, propagation from existing meristems is not technologically difficult and it yields plants that are genetically identical with the donor plants (Hu and Wang, 1983). Micropropagation of medicinal plants has been achieved through rapid proliferation of shoot tips and axillary buds in culture.

1.4 Callus Culture and Plant Regeneration

The induction of callus growth and subsequent differentiation and organogenesis are accomplished by the differential application of growth regulators and the control of conditions in the culture medium. With the stimulus of endogenous growth substances or by addition of exogenous growth regulators to the nutrient medium, cell division, cell growth and tissue differentiation are induced. Several reports are available on the regeneration of various medicinal plants via callus culture (Davidonis and Knorr, 1991; Cacho *et al.*, 1991; Das *et al.*, 1997; Basu and Chand, 1998).

Saxena *et al.* (1997) reported plant regeneration via organogenesis from callus cultures derived from mature leaves, stems, petioles and roots of young seedlings of *Psoralea corylifolia.* The calli differentiated into green nodular structures which developed into dark green shoot buds in the media supplemented with 2.5 mg/IBA and 1.0 mg/l NAA. The addition of 5.0 mg/l adenine sulfate (Ads) in the culture medium resulted in quick growth of shoot buds within 4 weeks of culture. Rout and Das (1997a) described *in vitro* organogenesis of *Zingiber officinale* via callus culture. Shoot regeneration was maximal on media supplemented with 5.0 mg/l BA, 1.0 mg/l IAA and 100 mg/l of adenine sulfate. They also reported that shoot-bud regeneration was highest under continuous illumination.

1.5 Somatic Embryogenesis and Conservation

Plant regeneration via somatic embryogenesis from single cells that can be induced to produce an embryo and then a complete plant has been demonstrated in many medicinal plant species (Wakhlu *et al.*, 1990; Kumar, 1992; Kackar *et al.*, 1993; Zhou *et al.*, 1994; Purohit *et al.*, 1994; Gastaldo *et al.*, 1996; Choi *et al.*, 1997; Das *et al.*, 1999). Arumugam and Bhojwani (1990) noted that the development of somatic embryos from zygotic embryos of *Podophyllum hexandrum* on MS medium containing 2 μM BA and 0.5 μM IAA. Further development of the embryos was achieved when the sucrose concentration increased from 4% to 6%. They also indicated that somatic embryogenesis was completely suppressed if the temperature was raised above 27°C or the cultures were transferred to light. Zhou *et al.* (1994) reported that thidiazuron (TDZ) had a positive effect on the induction of somatic embryogenesis in cells from *Cayratia japonica.* The production of somatic embryos per culture was stable up to the 12th subculture. Subsequently, somatic embryos were germinated on MS basal salts supplemented with 0.01 mg/l NAA and 2% (w/v) sucrose.

Efficient development and germination of somatic embryos are prerequisites for commercial plantlet production. Omission or lowering of growth regulator concentrations in culture media has improved embryo development and germination of many medicinal plants (Arumugam and Bhojwani, 1990; Wakhlu *et al.*, 1990; Kumar, 1992). Germination of the somatic embryos were also achieved on MS medium without the growth regulator (Zhou *et al.*, 1994; Hirai *et al.*, 1997). However, Arumugam and Bhojwani (1990) noted that inclusion of BA (2μM) and GA_3 (2.8μM) in the medium stimulated embryo development of *Podophyllum hexandrum,* although 75% of the embryos did germinate on basal MS medium devoid of growth regulator. Wakhlu *et al.* (1990) reported that the somatic embryos of *Bunium persicum* matured and germinated on the basal MS medium supplemented with 1.0 mg/l kinetin. Further, Kunitake and Mii (1990) reported that 30–40% of somatic embryos of *A. officinalis* germinated after being treated with distilled water for a week; they were subsequently transferred to half-strength MS medium supplemented with 1.0 mg/l IBA, 1.0 mg/l GA_3 and 1% sucrose. Ghosh and Sen (1996) reported that the development of somatic embryos of *A. cooperi* were obtained on medium containing 1.0 mg/l NAA, 1.0 mg/l kinetin, 0.2 mg/l absicsic acid and 2900 mg/l KNO_3. They also reported that the inclusion of absicsic acid (0.2 mg/l) in the culture medium increased the embryo development and germination of plantlets. Somatic embryos also germinated into plantlets on half-strength MS medium supplemented with 2% sucrose and either IAA or IBA (Johri and Aminuddin Pal, 1996). Kunitake *et al.*, (1997) observed that the somatic embryos of *A. officinalis.*

1.6 Protoplast Culture

Recent progress on the isolation, culture and fusion of protoplasts has made a significant contribution to plant improvement (Bajaj *et al.*, 1987). Plants have been regenerated from isolated protoplasts of a number of medicinal plants and somatic hybrids. The use of protoplasts not only enables mass production of plants, but it can also be a rich source of genetic diversity and somaclonal variation. Protoplast cultures have proven particularly useful for increasing the production of alkaloids and other natural products in plants (Bajaj, 1988). Cell lines with high shikonin contents were obtained from protoplast derived callus of *Lithospermum erythrorhizon* (Kumar, 1992; Fujita *et al.*, 1985) and *Asparagus* (Elmer *et al.*, 1989).The induction of disease resistance was also introduced through protoplast fusion between *Nicotiana tobacum* × *N. osophila* (Evans *et al.*, 1981). In addition, the fusion between protoplasts of *Hyoscyamus muticus* × *Nicotiana tobacum* resulted in somatic hybrid plants that showed high growth rates (Jia *et al.*, 1983). Regeneration of plants from protoplasts of different medicinal plants has been reported (Bajaj *et al.*, 1988; Hsu *et al.*, 1990; Gilmour *et al.*, 1989; Kunitake and Mii, 1990; Zafar *et al.*, 1995). The interaction between genotypes and the inclusion or exclusion of plant growth regulators in the protoplast culture medium helped in the plating efficiency and in colony and somatic embryo formation of *A. officinalis.* Conversion to plantlets occurred as rapidly as 1–2 weeks after transfer to embryo maturation medium or 5–6 weeks after initial protoplast culture (May and Sink, 1995). Chen *et al.*, (1997a, b) standardized the protocol on protoplast isolation from shoot tip cultures of *Asparagus.* Subsequently, they achieved plant regeneration from protoplasts. About 80% of the protoplasts showing cell division and developed into colonies of which 25% initiated somatic embryos.

1.7 Rhizogenesis

In vitro induction of roots from growing shoots has been achieved in standard media containing auxin and in media in the absence of auxin depending on plant genotype (Rout *et al.*, 1989). There is marked variation in the rooting potential of different plant species and systematic trials are often needed to define the conditions required for root induction. Moderate to high concentrations of all cytokinins inhibit rooting. Rooting improved in many woody and herbaceous species when the concentration of macro salts was lowered to half or less and the concentration of sucrose was lowered from 2 or 3% to 0.5% (Webb and Street, 1977). Lal *et al.* (1988) examined the effectiveness of various auxins on rooting of *Picrorhiza* microshoots and found that NAA at 1.0 mg/l was superior to IAA. Rooting in 90% of the cultured shoots of *Saussurea lappa* was achieved within 25–30 days on media containing 0.5–1.0 μM NAA (Arora and Bhojwani, 1989). Generally, the microshoots of various medicinal plants rooted on MS medium without the growth regulators (Cristina *et al.*, 1990; Mao *et al.*, 1995; Saxena *et al.*, 1998). But addition of IAA, IBA and NAA to the MS medium produced more rooting (Faria and Illg, 1995; Hossain *et al.*, 1993; Jha and Jha, 1989; Patra *et al.*, 1998).

1.8 Acclimatization and Field Establishment

Acclimatization of micropropagated plants in the greenhouse under field conditions was reviewed by Preece and Sutter (1991). Satheesh and Bhavanandan (1988) reported that when micropropagated plants of *Plumbago rosea* were transferred to pots containing a 1:1 soil and sand mixture under greenhouse conditions, about 60% of the plants survived. Jha and Jha (1989) noted the highest survival of *Cephaelis ipecacuanha* when the plants were maintained for a 4-week period in liquid MS medium and then transferred to greenhouse conditions. Mathur *et al.* (1988) reported that 60% of the encapsulated shoot buds of *Valeriana wallichii* were rooted in pots with vermiculite under greenhouse conditions. Hossain *et al.* (1993) indicated that the rooted plantlets of *A. marmelos* were hardened with 4.9mM IBA for 1–2 weeks and subsequently transferred to pots containing 2:1 garden soil: compost. Mao *et al.* (1995) reported that rooted plantlets of *Clerodendrum colebrookianum* required a minimum of 2 weeks in a mist tunnel, after which they could be transferred successfully into greenhouse conditions. Approximately 60% of the rooted plants of *Centella asiatica* survived in pots containing a 1:1:1 mixture of soil, sand and well-rotted cow dung (Patra *et al.*, 1998). The plants were supplied with MS inorganic salts twice a week before transfer to the greenhouse (Palai *et al.*, 1997). Saxena *et al.* (1997) reported that rooted plantlets of *Psoralea corylifolia* were successfully transferred to a 1:1 mixture of soil and sand. About 95% of the regenerated plants survived in the greenhouse.

1.9 *In vitro* Development of Synthetic Seeds

The protocols of synthetic seed technology for successful micropropagation, storage and transport of plants have been defined for several crops (Ganapathi *et al.*, 1994). Seeds, somatic embryos, shoot-tips and pollens can be encapsulated in different concentrations of sodium alginate for storage and micropropagation. Somatic embryos of celery (*Apium graveolens*) immobilized in alginate were successfully grown into a whole plants (Redenbaugh *et al.*, 1986). Mathur *et al.* (1988) developed a technique for micropropagation of *Valeriana wallichii* using encapsulation of apical and axial shoot buds. Subsequently, Mathur *et al.* (1989) used 4% sodium alginate and 100μM $CaCl_2$ for development of synthetic seeds for micropropagation. The shoot tips of *Plumbago zeylanica* were encapsulated in 3.5% sodium alginate and cultured in various media for germination. Among the different gel matrices tested, MS 0.1 mg/l IBA showed the best response for germination in synthetic media. About 96% of cultures germinated on media containing MS 0.1 mg/l IBA within a week. The developing shoots and roots emerged through the alginate matrix and grew into plantlets in 2 weeks.

1.10 *In vitro* Production of Secondary Metabolites

Plants are the traditional sources of many chemicals used as pharmaceuticals. Most valuable phytochemicals are products of plant secondary metabolism and possess sufficient chemical or structural complexity so that artificial synthesis is difficult or not currently possible. Plant cell culture holds much promise as a method for producing complex secondary metabolites *in vitro* (Barz and Ellis, 1981; Deus and Zenk, 1982; Ravishankar and Venkataraman, 1988a). A breakthrough in cell culture methodology occurred with the successful establishment of cell lines capable of producing high yields of secondary compounds in cell suspension cultures (Zenk, 1978). Constable *et al.* (1981) reported that alkaloid production from single leaf cell lines of *Catharanthus roseus* became variations. Stafford *et al.* (1986) reported that the accumulation of secondary products in plant cell cultures depends on the composition of the culture medium, including the types and amounts of plant growth regulators, mineral salts and carbon sources used and on environmental conditions, including temperature, light and gas composition during culture. Kargi and Rosenberg (1987) observed that immobilization of plant cells improved the production of secondary metabolites. The production of secondary metabolites in plant cell suspension cultures was reviewed by Rokem and Goldberg (1985). Panda *et al.* (1989) also provided an extensive review on the role of plant cell reactors on the production of secondary metabolites. Nigra *et al.* (1987) reported the production of solasodine from calli of *Solanum eleagnifolium* and Pyrrolizidine alkaloids were isolated from root cultures of *Senecio vulgaris, S. vernalis* and *S. erucifolius* (Toppel *et al.*, 1987). Scragg *et al.* (1990) reported that quinoline alkaloids were produced in significant quantities from globular cell suspension cultures of *Cinchona ledgeriana* on B_5 medium supplemented with 0.1 mg/l NAA. Significant amounts of sanguinarine were produced in cell suspension cultures of *Papaver somniferum* using bioreactors (Park *et al.*, 1990). Ravishankar and Grewal (1991) reported that the influence of media constituents and nutrient stress influenced the production of diosgenin from callus cultures of *Dioscorea deltoidea.* Pradel *et al.* (1997) reported that the biosynthesis of Cardenolides was maximal in the hairy root cultures of *Digitalis lanata* compared to leaf. Production of tropane alkaloids was high in differentiated roots of *Hyoscyamus muticus* cultured on MS medium containing 0.05 mg/l NAA and 0.5 mg/l IBA. Srividya *et al.* (1998) reported that the production of azadirachtin and nimbin was higher in cultured shoots and roots of *Azadirachta indica* compared to field grown plants.

1.11 Applications of *In vitro* Propagation

1.11.1 *In vitro* mutagenesis

Most of the available genetic variation used in breeding programs has occurred naturally and exists in germplasm collections of new and old cultivars, land race and genotypes. This variation through crosses is recombined to produce new and desired genes combinations (Maluszynski *et al.*, 1995). Then existing germplasm fails to provide the desired recombinants and it is necessary to resort to other resources of variation. Since spontaneous mutations occur with extremely low frequency, mutation induction techniques provide tools for the rapid creation and increase in variability in medicinal plants. The impact of mutation techniques on crop improvement has already been evaluated (Broertjes and Van Harten, 1988; Micke, 1999). *In vitro* culture methods has facilitated the use of mutation techniques for improvement of both seed and vegetatively propagated plants. In many vegetatively propagated crops mutation induction in combination with *in vitro* culture techniques may be the only effective method for plant improvement (Jain, 2002). Up to now, the mutation studies were helped to induce high production of metabolites of commercialized medicinal plants, similar to those obtained by spontaneous mutation. The combination of micropropagation and induced mutations can develop and multiply elite mutants in a short period of time in most of the medicinal plants.

1.11.2 Somaclonal variation

Somaclonal variation involves all forms of variation among regenerated plants derived from tissue culture (Larkin and Scowcroft, 1981; Jain *et al.*,

1998a; Jain and De Klerk, 1998), such as: (i) physical and morphological changes in undifferentiated callus; (ii) differences in the ability to organize and form organs *in vitro*; (iii) changes manifested among differentiated plants and (iv) chromosomal changes. Somaclonal variation has been reviewed at length (Skirvin, 1978; Scowcroft and Larkin, 1988) and has proven useful in plant improvement (Skirvin *et al.*, 1993; Jain *et al.*, 1998a, b; Jain and De Klerk, 1998), and could be of much interest to the medicinal plant breeders.

The selected somaclones should be genetically stable in seed and vegetatively propagated crops for routine induction of genetic variability through tissue culture and this aspect should be thoroughly checked before using them in regular crop improvement programs. Somaclonal variation is unpredictable in nature and can be both heritable (genetic) and non-heritable (epigenetic) in regenerated plants. DNA methylation causes genetic instability in somaclones, which probably comes from epigenetic changes (Jain, 2001). Since somaclonal variation can broaden the genetic variation in number of crop plants, a broader range of plant characteristics can be altered, including plant height, yield, no. of flower/plant, early flowering, resistance to diseases, insects and pests and salt. The reduction, and even the total loss of regeneration ability, is a general phenomenon observed during undifferentiated cell culture.

1.11.3 Cryopreservation

Cryopreservation of tissue and cells has been utilized for long term storage of elite genetic material. In the mid 1900s, the science of cryobiology improved rapidly with the discovery of the beneficial effects of cryoprotectant substances that are added to cell freezing solutions. Prior to freezing, the cells must be treated with a cryoprotectant solution such as glycerol, dimethyl sulfoxide (DMSO) or ethylene glycol. These substances protect the cells and their membranes from damage during the freezing process. After the cells have been exposed to the freezing medium containing the cryoprotectant, they must be dehydrated so that the water inside the cells will not form ice crystals that can also damage the cell / tissues. The cells / tissues are dehydrated and cooled slowly prior to plunging them into liquid nitrogen. Sakai (1960) first reported the survival of plant tissues after exposure to ultra low temperature -196°C and the significance of using DMSO as a cryoprotectant (Quatrano, 1968). Preservation of cultured plant tissues in liquid nitrogen is efficient and appropriate method of germplasm conservation. This technique will greatly help in the *in vitro* conservation of germplasm on long-term basis and that will lead to the establishment of cryo-storage bank.

1.11.4 Genetic transformation systems for medicinal plants

Direct manipulation of DNA sequences to alter gene expression in medicinal plants is an area that is ripe for expansion. Although the primary target for trait manipulation in medicinal plants is the content of active compounds, for development as crops, basic agronomic characters related to uniformity, stability, growth and development, and resistance to biotic and abiotic stresses, must also be improved (Bajaj, 1998; Dubey and Guerra, 2002; Bernath, 2002). The application of biotechnological approaches to medicinal plants does not start from a low baseline. There is already considerable interest in manipulating plant biosynthetic pathways to produce drug precursors, food components or pesticides; for example, through trichomes, specialized surface organs that manufacture, store and exude secondary metabolites (Stevenson, 2004). In *Mentha spp* (mints), biosynthetic pathways have been engineered to modify essential oil production in the trichomes and to enhance the resistance of the plant to fungal infection and abiotic stresses (Veronese *et al.* 2001). There is a long history of experimental and commercial production of high-value phytochemicals by tissue culture, an *in vitro* system for growing plant organs, explants, tissues, cells or protoplasts. Genetic transformation of cultures using bacterial vectors to transfer genes into the cultured plant DNA has been widely employed to improve product output in such systems. For example, hairy root cultures, transformed by infecting cells with the bacterium *Agrobacterium rhizogenes*, often sustain stable and high productivity in hormone-free culture conditions (Sevon and Oksman-Caldentey, 2002). An efficient transgene delivery system is thus already in place for several important medicinal plants, including *P. somniferum* (Park and Facchini, 2000), *Artemisia spp.* (wormwoods) (Chen *et al.*, 1999), members of the family Solanaceae (Moyano *et al.*, 199) and *Taraxacum platycarpum* (Lee *et al.*, 2004). Transformation systems based on *Agrobacterium tumefaciens* are well established for Taxus (yew) (Han *et al.*, 1994), Echinacea (Wang and To, 2004), Scrophularia (figwort) (Park *et al.*, 2003), Digitalis (foxglove) (Sales *et al.*, 2003), Thalictrum (meadow rues) (Samanani *et al.*, 2002) and Artemisia (Chen *et al.*,2000). Problems of regenerating whole plants from cultures have been overcome for many plants, but some important species remain recalcitrant, notably *G. biloba* (Balz *et al.*, 1999). As well as being essential elements of a system for rational engineering of the genetic makeup of medicinal plants, tissue culture and regeneration could make important contributions to other aspects of plant breeding (Rout *et al.*, 2000). Tissue culture often promotes genetic disturbances, which result in somaclonal variation, greatly extending the range of useful variation available to the breeder (Rout *et al.* 2000; Kukreja and Dhawan, 2000; Sevon and Oksman-Caldentey, 2002). Tissue culture can also contribute to the conservation of valuable biodiversity (Purohit *et al.*, 1994; Augustine and D'Souza, 1997) and somatic embryogenesis, the micropropagation of whole plants from cultured cells, has a potentially significant part to play in establishing breeding material from wild populations (Kintzios and Michaelakis,1999) and in mass-producing material for selection or engineering (Prakash *et al.,* 2002).

1.11.5 Haploid production

Haploid plants are produced from the male or female gametes of the plant, most commonly the immature pollen or microspores, via either intact anther culture or isolated microspore culture. Ovules also are a possible alternative source for haploid production in plants, particularly in species where microspores or pollen are unresponsive to manipulation *in vitro.* Gynogenesis has, therefore, been used for obtaining haploid plants through culture of unfertilized ovules *in vitro* and production of gynogenic haploids has been reported in several crop species. As they arise without fertilization, these haploid plants have half the genetic complement of a diploid plant. After chemical doubling of their chromosome complement they become 'doubled haploid' (DH) and instantly homozygous. The development of homozygosity in a single generation using DH technology accelerates breeding by three to four years compared to conventional breeding methods. DH's are produced in many of the major food crops (rice, barley, wheat, canola etc.) to accelerate breeding, however the medicinal plants have proven recalcitrant to attempts to develop haploid plants. The *in vitro* production of haploids is extremely valuable in plant breeding and genetics, and with haploids, the establishment of homozygous lines of new varieties is possible in a short period of time, resulting in an important reduction in the number of selection generation.

1.12 Conclusion

Medicinal plants are produced mainly for their medicinal value, thus the propagation and improvement of quality attributes such as active compounds, for development as crops, basic agronomic characters related to uniformity, stability, growth and development and resistance to biotic and abiotic stresses, must also be improved and the creation of novel variation are important economic goals for medicinal plant culturists. Successful *in vitro* propagation of medicinal plants is now being used for commercialization. Many commercial laboratories and national institutes worldwide use *in vitro* culture system for rapid plant multiplication, germplasm conservation, elimination of pathogens, genetic manipulations and for secondary metabolite production (O'Riordain, 1999). The great potential of micropropagation for large-scale plant multiplication can be tapped by cutting down the cost of production per plant by applying low-cost tissue culture, which is to adopt practices and proper use of equipment and resources to reduce the unit cost of micropropagule and plant production without compromising the quality. Bioreactor technology may cut down the cost of plant production provided proper precautions are taken to prevent contamination. Somatic embryogenesis facilitates cryopreservation, synthetic seed development, mutations and genetic transformation. Plant transformation methods and enhanced gene silencing technology can effectively be used to evaluate and authenticate newly discovered endogenous genes to characterize their function in plants as well as to genetically manipulate trait quality and productivity (Dandekar, 2003).

Recent progress in tissue culture involving modern techniques has opened a new era and possibilities for the overall progress of medicinal plants like large scale production of plantlets, somatic embryo development, synthetic seed production and their conservation, production of secondary metabolites, *in vitro* mutagenesis, haploid production, *in vitro* flowering and genetic variability.

References

Ammirato, PV. Embryogenesis. In: Evans DA, Sharp WR, Ammirato PV, Yamada J, editors. 1983. Handbook of Plant Cell Culture, Vol. 1. New York: MacMillan,

Arora, R. Bhojwani, SS. 1989. *In vitro* propagation and low temperature storage of *Saussurea lappa* C.B. Clarke: an endangered medicinal plant. *Plant Cell Reports*,8: 44–7.

Arumugam, N. Bhojwani, SS. 1990. Somatic embryogenesis in tissue cultures of *Podophyllum hexandrum*. *Can J Bot*, 68: 487–91.

Augustine, A.C. and D'Souza, L. 1997. Regeneration of an anticarcinogenic herb, *Curculigo orchioides* (Gaertn.). *In Vitro Cell. Dev. Biol. Plant*, 33: 111–113.

Bajaj, Y.P.S. 1998. Biotechnology for the improvement of medicinal plants. *Acta Hort*. 457: 37–45.

Bajaj, YPS. Furmanowa, M. Olszowska, O. 1988.Biotechnology of the micropropagation of medicinal and aromaticplants. In: Bajaj YPS, editor. *Biotechnology in Agriculture and Forestry*, Vol. 4. New York: Springer Verlag, pp. 60–103.

Bajaj, YPS. Gosal, SS. Gill, MS. 1987. Somatic hybridization: a biotechnological approach to plant breeding. In: Ganguli PK, Mandal AK, editors. *Advances in Plant Breeding*. New Delhi: Oxford, IBA,

Bajaj, YPS.1988. Regeneration of plants from frozen (-196°C) protoplasts of *Atropa belladonna, Datura innoxia* and *Nicotiana tabacum*. *Ind J Exp Biol*, 26: 289–92.

Balz, J.P. *et al*. 1999. Production of ginkgolides and bilobalide by *Ginkgo biloba* plants and tissue cultures. *Planta Med*, 65: 620–626.

Barz, W. Ellis, B. 1981. Potential of plant cell cultures for pharmaceutical production. In: Beal JL, Reinhard E, editors. *Natural Products as Medicinal Agents.* Stuttgart: Hippokrates, pp. 471–507.

Basu, P. Chand, S. 1998. Tropane alkaloids from callus cultures, differentiated roots and shoots of Hyoscyamus muticus L. *Plant Biochem Biotech,* 7: 39–42.

Bernath, J. 2002. Strategies and recent achievements in selection of medicinal and aromatic plants. *Acta Hor.* 576: 115–128.

Broertjes, C. Van Harten AM, editors. 1978. Application of Mutation Breeding Methods on the Improvement of vegetatively propagated crops. *The Netherlands*: Elsevier Science Publ.; p. 325.

Broertjes, C. Van Harten AM, editors. 1988. Applied mutation breeding for vegetatively propagated crops. *Developments in crop science The Netherlands*: Elsevier Publ., p:197–204.

Cacho, M. Moran, M Herrera, MT. Fernandez-Tarrago, J. Corchete, MP. 1991. Morphogenesis in leaf, hypocotyl and root explants of Digitalis thapsi L. cultured in vitro. *Plant Cell Tissue and Organ Culture,* 25: 117–23.

Chen, D.H. *et al.* 1999. Ri-mediated transformation of Artemisia annua with a recombinant farnesyl diphosphate synthase gene for artemisinin production. *Plant Cell Tissue Organ Cult.* 57: 57–162.

Chen, D.H. *et al.* 2000. Expression of a chimeric farnesyl diphosphate synthase gene in *Artemisia annua* L. transgenic plants via *Agrobacterium tumefaciens*-mediated transformation. *Plant Sci,* 155: 179–185.

Chen, GY. Conner, AJ. Christey, MC. Fautrier, AG. Field, RJ. 1997a. Culture and regeneration of protoplasts from shoots of asparagus cultures. *Int J Plant Sci,* 158: 543–55.

Chen, GY. Conner, AJ. Christey, MC. Fautrier, AG. Field, RJ. 1997b. Protoplast isolation from shoots of asparagus cultures.*Int J Plant Sci,* 158: 537–42.

Choi, YE. Kim, JW. Soh, WY. 1997. Somatic embryogenesis and plant regeneration from suspension cultures of Acanthopanax koreanum Nakai. *Plant Cell Reports,* 17: 84–8.

Constable, F. Rambold, S. Chatson, KB. Kurz, WGW. Kutney, JP. 1981. Alkaloid production in *Catharanthus roseus* (L.) G. Don. VI. Variation in alkaloid spectra of cell lines derived from one single leaf. *Plant Cell Reports,* 1: 3–7.

Cristina, M. Dos Santos, F. Esquibel, MA. Dos Santos, VP. 1990. *In vitro* propagation of the alkaloid producing plant *Datura insignis* Barb. Rodr. *Plant Cell Tissue and Organ Culture,* 21: 75–8.

Dandekar, AM. 2003. Techniques for manipulating quality and productivity traits in horticultural crops. *Acta Hortic,* 625: 293–305.

Das, P. Palai, SK. Patra, A. Samantaray, S. Rout, GR. 1999. *In vitro* somatic embryogenesis in *Typhonium trilobatum* Schott. *Plant Growth Regulation,* 27: 193–7.

Das, P. Patra, A. Das, AB. Palai, SK. Bose, T. 1997. *In vitro* regeneration of *Typhonium trilobatum* Schott., a medicinal plant, through callus culture. *Orissa J Hort*, 25(2): 57–61.

Davidonis, G. Knorr, D.1991. Callus formation and shoot regeneration in *Vanilla planifolia. Food Biotechnol*, 5: 59–66.

Deus, B. Zenk, MH. 1982. Exploitation of plant cells for the production of natural compounds. *Biotechnol Bioengineer*, 24: 1965–74.

Dubey, T. and Guerra, D.J. 2002. Use of biotechnology for growing medicinal plants. *Recent Progress in Medicinal Plants*, 5: 47–61.

Elmer, WH. Ball, T. Volokita, M Stephens, CT. Sink, KC. 1989. Plant regeneration from callus derived protoplasts of asparagus.*J Am Soc Hortic Sci*, 114: 1019–24.

Evans, DA. Flick, CE. Jensen, RA.1981. Incorporation of disease resistance into sexually incompartble somatic hybrids of the genus *Nicotiana. Science*, 213: 907–9.

Faria, RT. Illg, RD. 1995. Micropropagation of *Zingiber spectabile* Griff. *Scientia Horticulturae*, 62: 135–7.

Fujita, Y. Takahashi, S. Yamada, Y. 1985. Selection of cell lines with high productivity of shikonin derivatives by protoplast culture of *Lithospermum erythrorhizon. Agric Biol Chem*, 49: 1755–9.

Gamborg ,OL. Murashige,T. Thorpe, TA.,Vasil, IK. 1976. Plant tissue culture media. *In Vitro*, 12: 473–8.

Ganapathi, TR. Bapat, VA. Rao, PS. 1994. *In vitro* development of encapsulated shoot tips of cardamum. *Biotechnology Technique*, 8: 234–9.

Gastaldo, P. Caviglia, AM. Carli, S. Profumo, P. 1996. Somatic embryogenesis and esculin formation in calli and embryoids from bark explants of *Aesculus hippocastanum* L. *Plant Sci*, 119: 157–62.

George, EF. Sherrington, PD. 1984. In: Plant Propagation by Tissue Culture. Exegetics Ltd., Eversley, England, pp.39–71.

Ghosh, B. Sen, S. 1996. Suspension culture, somatic embryogenesis and stable regeneration in *Asparagus cooperi* Baker.*Cytobios*, 87: 189–200.

Gilmour, DM. Davey, MR. Cocking, EC. 1989. Production of somatic hybrid tissues following chemical and electrical fusion of protoplasts from albino cell suspensions of *Medicago sativa* and *M. borealis*. *Plant Cell Reports*, 8: 29–32.

Han, K.H. *et al.* 1994. Genetic transformation of mature Taxus: an approach to genetically control the in vitro production of the anticancer drug, taxol. *Plant Sci.* 95: 187–196.

Hirai, G. Kasai, N. Harada, T. 1997. Somatic embryogenesis in mature zygotic embryo culture of Glehnia littoralis. *Plant Cell Tissue and Organ Culture*, 48: 175–80.

Hossain, M. Karim, MR. Islam, R. Joarder, OI. 1993. Plant regeneration from nucellar tissues of *Aegle marmelos* through organogenesis. *Plant Cell Tissue and Organ Culture*, 34: 199–203.

Hsu, JY. Yeh, CC. Yang, TP. Lin, WC. Tsay, HS. 1990. Initiation of cell suspension cultures and plant regeneration from protoplasts of asparagus. *Acta Hortic*, 271: 135–43.

Hu, CY. Wang, PJ. 1983. Meristem, shoot-tip and bud culture. In: Evans DA, Wang WR, Ammirato PV, Yamada Y, editors. *Handbook of Plant Cell Culture*, Vol. 1. New York: MacMillan, pp. 177–277.

Huang, L. Murashige, T. 1977. Plant tissue culture media: major constituents; their preparation and some applications.*Tissue Culture Assoc Manual*, 3: 539–48.

Jain, SM. Brar, DS. Ahloowalia, BS, editors. 1998a. Somaclonal variation and induced mutations in crop improvement. Dordrecht, *The Netherlands*:Kluwer Acad. Publ. p. 603.

Jain, SM. De Klerk, GJ. 1998. Somaclonal variation in breeding and propagation of ornamental crops. *Plant Tissue Cult Biotechnol*, 4(2): 63–75.

Jain, SM. 2001. Tissue culture-derived variation in crop improvement. *Euphytica*, 118: 153–66.

Jain, SM. 2002. Feeding the world with induced mutations and biotechnology. Proc. Int. Nuclear Conference 2002 – Global trends and Perspectives. Seminar 1: *Agriculture and bioscience*. Bangi, Malaysia: MINT, p: 1–14.

Jain, SM. Ahloowalia, BS. Veilleux, RE. 1998b. Somaclonal variation in crop plants. In: Jain SM, Brar DS, Ahloowalia, editors. Somaclonal variation and induced mutations in crop improvement. Dordrecht, The Netherlands: Kluwer Acad. Publ.p. 203–18.

Jha, S. Jha, TB. 1989. Micropropagation of *Cephaelis ipecacuanha* Rich. *Plant Cell Reports*, 8: 437–9.

Jha, S. Sen, S. 1985. Regeneration and rapid multiplication of *Bowiea volubilis* Harv. in tissue culture. *Plant Cell Reports*, 4: 12–4.

Jia, J. Potrykus, I. Lazal, GB. Saul, M.1983. *Hyoscyamus- Nicotiana* fusion hybrid selected via auxotroph concentration and verified by species specific DNA hybridization. In: Protoplasts, 1983, Poster Proceeding, Biekhauser,Basel Boston Stuttgart, pp. 110–1.

Johri, JK. Aminuddin Pal, A. 1996. Regeneration of betelvine (*Piper betle* L.) through somatic embryogenesis. *Ind J ExptBiol*, 34: 83–5.

Kackar, A. Bhat, SR. Chandel, KPS. Malik, SK. 1993. Plant regeneration via somatic embryogenesis in ginger. *Plant Cell Tissue and Organ Culture*, 32: 289–92.

Kargi, F. Rosenberg, MZ. 1987. Plant cell bioreactors: Present status and future trends. *Biotechnology Progress*, 3(1): 1–8.

Kintzios, S. and Michaelakis, A. 1999. Induction of somatic embryogenesis and *in vitro* flowering from inflorescences of chamomile(*Chamomilla recutita* L.). *Plant Cell Rep*, 8: 684–690.

Kukreja, A.K. and Dhawan, O.P. 2000. Somaclonal breeding for improvement of herb and essential oil yield in mints: a case study. *J. Med. Aromatic Plant Sci*, 22: 431–442.

Kumar, A. 1992. Somatic embryogenesis and high frequency plantlet regeneration in callus cultures of *Thevetia peruviana*. *Plant Cell Tissue and Organ Culture*, 51: 47–50.

Kunitake, H. Mii, M. 1990. Somatic embryogenesis and plant regeneration from protoplasts of asparagus (*Asparagus officinales* L.). *Plant Cell Reports*, 8: 706–10.

Kunitake, H. Nakashima, T. Mori, K. Tanake, M. 1997. Normalization of Asparagus somatic embryogenesis using a maltose- containing medium. *J Plant Physiol*, 150: 458–60.

Lal, N. Ahuja, PS. Kukreja, AK. Pandey, B. 1988. Clonal propagation of *Picrorhiza kurroa* Royle ex Benth. By shoot tip culture. *Plant Cell Reports*, 7: 202–5.

Larkin, PJ. Scowcroft, WR. 1981. Somaclonal variation – a novel source of variability from cell cultures for plant improvement. *Theor Appl Genet*, 60: 197–214.

Lee, M.H. *et al.* 2004. *Agrobacterium rhizogenes* -mediated transformation of *Taraxacum platycarpum* and changes of morphological characters. *Plant Cell Rep*. 22: 822–827.

Maluszynski, M. Ahloowalia, BS. Sigurbjornsson, B. 1995. Application of *in vitro* and in vivo mutation techniques for crop improvement. *Euphytica*, 85:303–15.

Mao, AH. Wetten, A. Fay, M. Caligari, PDS. 1995. *In vitro* propagation of *Clerodendrum colebrookianum* Walp; a potential natural anti-hypertension medicinal plant. *Plant Cell Reports*, 14: 493–6.

Mathur, J. Ahuja, PS. Lal, N. Mathur, AK. 1989. Propagation of *Valeriana wallichii* D.C. using encapsulated apical and axial shoot buds. *Plant Sci*, 60: 111–6.

Mathur, J. Ahuja, PS. Mathur, AK. Kukreja, AK. Shah, NC. 1988. *In vitro* propagation of *Valeriana wallichii*. D.C. *Plant Med*, 54: 82–3.

May, RA. Sink, KC. 1995. Genotype and auxin influence direct somatic embryogenesis from protoplasts derived from embryogenic cell suspensions of *Asparagus officinalis* L. *Plant Sci*, 108: 71–84.

Micke A. Mutation in plant breeding. In: Siddiqui BA, Khan S, editors. 1999. Breeding in crop plants – mutations and *in vitro* mutation breeding. New Delhi, India: Kalyani Publ.; p. 1–19.

Misawa M. 1994. Plant tissue culture: an alternative for production of useful metabolite. FAO Agricultural Services Bulletin No. 108. Roma, Italy: Food and Agriculture Organization of the United Nations.

Moyano, E. *et al.* 1999. Effect of Agrobacterium rhizogenes T-DNA on alkaloid production in Solanaceae plants. *Phytochemistry*, 52:1287–1292.

Murashige, T. 1977. Manipulation of organ culture in plant tissue cultures. *Botanical Bull Acad Sinica*, 18: 1–24.

Murashige, T. Skoog, F. 1962. A revised medium for rapid growth and bioassays with tobacco tissue cultures. *PhysiolPlant*, 15: 473–97.

Nigra, HM. Caso, OH. Giulietti, AM. 1987. Production of solasodine by calli from different parts of *Solanum eleagnifolium* cav. plants. *Plant Cell Reports*, 6: 135–7.

O'Riordain, F. 1999. Directory of European plant tissue culture laboratories, 1996–97. COST Action, vol. 822. Brussels: Commission of the European Communities.

Palai, SK. Rout, GR. Das, P. 1997. Micropropagation of ginger *(Zingiber officinale* Rosc.): interaction of growth regulators and culture conditions. Proc Biotechnology of Spices, *Medicinal and Aromatic Plants* Kerala, India. pp. 20–4.

Panda, AK. Mishra, S. Bisaria, VS. Bhojwani, SS. 1989. Plant cell reactors a perspective. *Enzyme Microb Technol*, 11: 386–97.

Park, JM. Giles, KI. Songstad, DD. 1990. Industrial production of Sanguinarine from the cell suspension culture of *Papaver sominiferum*. Proc. *APBiotechnology. Kyungiu Publ.*, Korea. pp. 184–7.

Park, S.U. and Facchini, P.J. 2000. *Agrobacterium rhizogenes* mediated transformation of *Opium poppy, Papaver somniferum* L., and *California poppy,* Eschscholzia californica Cham., root cultures. *J. Exp. Bot.* 51:1005–1016

Park, S.U. *et al.* 2003. Genetic transformation of the figwort, *Scrophularia buergeriana* Miq., an Oriental medicinal plant. *Plant Cell Rep.* 21: 1194–1198.

Patra, A. Rai, B. Rout, GR. Das, P. 1998. Successful plant regeneration from callus cultures of *Centella asiatica* (Linn.) Urban. *Plant Growth Regulation,* 24: 13–6.

Pradel, H. Dumkelehmann, U. Diettrich, B. Luckner, M. 1997. Hairy root cultures of Digitalis lanata. Secondary metabolism and plant regeneration. *J Plant Physiol,* 151: 209–15.

Prakash, L. *et al.* 2002. Production of four commercially cultivated Echinacea species by different methods of *in vitro* regeneration. *J. Hort. Sci. Biotech,* 77: 158–163.

Preece, JE. Sutter, EG. 1991. Acclimatization of micropropagated plans to the greenhouse and field. In: Debergh PC, Zimmerman RH, editors. Micropropagation. The Netherlands: *Kluwer Academic Publ.,* pp. 71–93.

Purohit, S.D. *et al.* 1994. Micropropagation of safed musli (*Chlorophytum borivilianum*), a rare Indian medicinal herb. *Plant Cell Tissue Organ Cult* ,3: 93–96.

Purohit, SD. Dave, A. Kukda, G. 1994. Somatic embryogenesis and plantlet regeneration in'Safed Musli' (*Chlorophytum borivilianum*). *Ind J Plant Genet Resources,* 7(1): 65–71.

Quatrano, RS. 1968. Freeze preservation of cultured flax cells utilizing DMSO. *Plant Physiol,* 43: 2057–61.

Ravishankar, GA. Grewal, S. 1991. Development of media for growth of *Dioscorea deltoidea* cells and *in vitro* diogenin production: influence of media constituents and nutrient stress. *Biotechnology Lett,* 13(2): 125–30.

Ravishankar, GA. Venkataraman, LV. 1988a.Plant cell culture for high value metabolites. *Science Reporter,* 7: 379–84.

Redenbaugh, K. Viss, P. Slade, D. Fujii, J. 1986. Scale-up: Artificial seeds. In: Somers DA, Gengenbach BG, Biesboer DD, Hackett WP, Green CE, editors. Proc VI Int Cong *Plant Tissue and Culture.* Minneapolis, MN: Univ Minnesota, p. 132.

Rokem, JS. Golgberg, I. 1985. Secondary metabolites from plant cell suspension cultures: methods for yield improvement. In: Mizrahi A, vanWezel AL, editors. *Advances in Biotechnological Process,* Vol. 4. New York: Alan R. Liss, pp. 241–74.

Rout, G.R. *et al.* 2000. *In vitro* manipulation and propagation of medicinal plants. *Biotechnol. Adv,* 18: 91–120.

Rout, GR. Das, P. 1997a. *In vitro* organogenesis in ginger (*Zingiber officinale* Rosc.). *J Herbs Spices Medicinal Plants,* 4(4): 41–51.

Rout, GR. Debata, BK. Das, P 1989. *In vitro* mass-scale propagation of Rosa hybrida cv. landora. *Curr Sci* , 58(15): 876–8.

Rout, GR. Saxena, C. Samantaray, S. Das, P. 1999. Rapid clonal propagation of *Plumbago zeylanica* Linn. *Plant Growth Regulation,* 28: 1–4.

Sakai, A. 1960. Survival of the twigs of woody plants at -196 °C. *Nature,* 185: 392–4.

Sales, E. *et al.* 2003. *Agrobacterium tumefaciens* -mediated genetic transformation of the cardenolide-producing plant Digitalis minor L. *Planta Med.* 69: 143–147.

Samanani, N. *et al.* 2002. *In vitro* regeneration and genetic transformation of the berberine-producing plant, *Thalictrum flavum* ssp. *glaucum. Physiol. Plant,*116: 79–86.

Satheesh, KK. Bhavanandan, KV. 1988. Micropropagation of *Plumbago rosea* Linn. *Plant Cell Tissue and Organ Culture,* 15: 275–8.

Saxena, C. Palai, SK. Samantaray, S. Rout, GR. Das, P. 1997. Plant regeneration from callus cultures of *Psoralea corylifolia* Linn. Plant Growth Regulation, 22: 13–7.

Saxena, C. Rout, GR. Das, P. 1998. Micropropagation of *Psoralea corylifolia* Linn. *J Med Aromatic Plant Sci,* 20: 15–8.

Scowcroft, WR. Larkin, PJ. 1988. Somaclonal variation. In: Bock G, March L, editors. *Applications of plant cell and tissue culture.* CIBA Foundation Symp.Chichester: Wiley; p. 21–35.

Scragg, AH. Ashton, S. York, A. Bond, P. Stepan-Sarkissian, G. Grey, D. 1990. Growth of *Catharanthus roseus* suspensions for maximum biomass and alkaloid accumulation. *Enzyme Microb Technol,* 12: 292–8.

Sevon, N. and Oksman-Caldentey, K.M.2002. *Agrobacterium rhizogenes*-mediated transformation: root cultures as a source of alkaloids. *Planta Med,* 68: 859–868.

Sevon, N. and Oksman-Caldentey, K.M. 2002. *Agrobacterium rhizogenes*-mediated transformation: root cultures as a source of alkaloids. *Planta Med.* 68: 859–868.

Skirvin, RM. Norton, M. MCPheeters KD. 1993. Somaclonal variation: has it proved useful for plant improvement. *Acta Hortic* , 336: 333–40.

Skirvin, RM. 1978.Natural and induced variation in tissue culture. *Euphytica,* 27:241–66.

Srividya, N. Sridevi, BP. Satyanarayana, P. 1998. Azadirachtin and nimbin content in in vitro cultured shoots and roots of *Azadirachta indica* A. Juss. *Ind J Plant Physiol,* 3(2): 129–34.

Stafford, A. Morris, P. Fowler, MW. 1986. Plant cell biotechnology: a perspective. *Enzyme Microb Technol*, 8: 578–87.

Stevenson, R. 2004. Chemical harvest. *Chemistry & Industry 6th Sept* pp. 16–18

Su, WW. Hwang, WI. Kirn, SY. Sagawa, Y. 1997. Induction of somatic embryogenesis in *Azadirachta indica. Plant Cell Tissue and Organ Culture*, 50(2): 91–5.

Teshima, D. Ikeda, K. Satake, M. Aoyama, T. Shimomura, K. 1988. Production of emetic alkaloid by *in vitro* culture of *Cephaelis ipecacuanha* A. Richard. *Plant Cell Reports*, 7: 278–80.

Toppel, G. Witte, L. Riebesehl, B. Borstel, KV. Hartmann, T. 1987.Alkaloid patterns and biosyntic capacity of root cultures from some pyrrolizidine alkaloid producing Senecio species. *Plant Cell Reports*, 6: 466–9.

Vanisree M, Tsay HS. 2004. Plant cell cultures-an alternative and efficient source for the production of biologically important secondary metabolites. *Int J Appl Sci* Eng 2:29–48.

Veronese, P. *et al.* 2001. Bioengineering mint crop improvement. *Plant Cell Tissue Organ Cult.* 64:133–144.

Verpoorte R, Contin A, Memelink J. 2002. Biotechnology for the production of plant secondary metabolites. *Phytochem Rev*,1:13–25.

Wakhlu, AK. Nagari, S. Barna, KS. 1990. Somatic embryogenesis and plant regeneration from callus cultures of *Bunium persicum Boiss. Plant Cell Reports*, 9: 137–8.

Wang, H.M. and To, K.Y. 2004. *Agrobacterium* -mediated transformation in the high-value medicinal plant *Echinacea purpurea. Plant Sci.* 66:1087–1096.

Webb, KJ. Street, HE. 1977. Morphogenesis *in vitro* of Pinus and Picea. *Acta Hortic*, 78: 259–69.

Zafar, Y. Nenz, E. Damiani, F. Pupilli, F Arcioni, S. 1995 Plant regeneration from explant and protoplast derived calluses of *Medicago littoralis. Plant Cell Tissue and Organ Culture*, 41: 41–8.

Zenk, MH. 1978. The impact of plant tissue culture on industry and agriculture. In: Thorpe TA, editor. *Frontiers of Plant Tissue Culture*, Proc Fourth Int Cong Plant Tissue Culture. Calgary: Canada, pp, 1–13.

Zhou, J. Ma Hui, Guo F, Luo X. 1994. Effect of thidiazuron on somatic embryogenesis of *Cayratia japonica. Plant Cell Tissue and Organ Culture*, 36: 73–9.

□□□

18

DNA Barcoding of Indian Medicinal Plants - A Case Study

S. Dhivya, S. Rajeev Kumar, V. S. Ramachandran and R. Sathishkumar

CONTENTS

1. 1 Introduction 280
1.1.1 Characteristic features of DNA barcode 280
1.1.2 Animal DNA barcoding 281
1.1.3 Advantages of cytochrome oxidase 1 as DNA [barcode] 281
1.2 Plant DNA Barcoding 282
1.2.1 The proposed plant DNA barcodes 282
1.3. DNA Barcoding of Traditional Medicinal Plants 283
1.3.1 The plant family apocynaceae 283
1.3.2 The matK gene 284
1.3.3 Major steps involved in plant DNA barcoding 285
1.3.3.1 Samples for analysis 285
1.3.3.2 Genomic DNA isolation 285
1.3.3.3 Primer design and PCR amplification 285
1.3.3.4 Sequence retrieval and alignment 286
1.3.3.5 Steps involved in multiple sequence alignments 287
1.3.3.6 Phylogenetic analysis 288
1.4 Conclusion 288
Aknowledgement 289
References 289

1.1 Introduction

DNA Barcoding is a tool for identifying species, using variations in a segment of short DNA sequence. The identification is solely based on the significant variation between the species. The use of nucleotide sequence variations to investigate evolutionary relationship is not a new concept as rRNA, several molecular markers and Mitochondrial DNA have been successfully used in molecular systematic for decades. Now, DNA Barcoding has become a standardized approach for identifying animals and plants. Recent advancements in molecular biology and computational biology, have led to the development of barcodes to not only help recognize quickly and cheaply the known species and to retrieve information about them, but also can speed upto the discovery of the millions of species. In long run DNA Barcodes can be used to manage earth's immense biodiversity. In 2003, Paul D.N. Hebert from the University of Guelph, Ontario, Canada had initiated the use of DNA barcodes and compiled a datasets which provide a new master key for identifying plant species.

The potential uses of DNA Barcodes are,

- *Identification from any fragments of Biological material*: as DNA barcoding helps in distinguishing them at the Species level.
- *Forensics*: DNA barcoding can be employed in forensics just like molecular markers.
- *Authentication of herbal medicines / foodstuffs*: DNA barcoding helps in identifying duplicates from originals.
- *Biosecurity and trade in controlled species*: In case of illegal import and export of economically valuable biological materals, DNA barcodes aid to promote legal trade.
- *Inventory and ecological surveys*: DNA barcodes will help in creating digital herbariums with both molecular and other conventional datasets.

1.1.1 Characteristic features of DNA barcode

The essential features of DNA barcode regions are:

- It should possess high interspecific, low intraspecific sequence divergence.
- It should undergo standard amplification with universal primers.
- The amplicon should be easy to sequence.
- Amplicon should be short so that one sequencing reaction will cover the entire length.
- The Barcode regions should have few insertions/deletions.

For DNA barcoding, construction of comprehensive, consistent reference library of DNA sequences and development of cost effective technologies for species identification are essential. The goal is that anyone, anywhere, anytime be able to identify quickly and accurately the species of a specimen, whatever its condition. Results so far suggest that a mitochondrial DNA barcodes will enable identification of most animal species. For plants, mitochondrial genes do not vary sufficiently to distinguish among closely related species. Hence many barcode genes have been proposed and are under study.

1.1.2 Animal DNA Barcodes

Mitochondrial gene Cytochrome Oxidase 1 (COX1) is successful for animals as DNA barcode. Recovering mitochondrial DNA sequences are easier than nuclear sequences, especially from small or partially degraded samples. Sequence differences among closely related species are 5 to 10 fold higher in mitochondrial than that of nuclear genes. Thus, shorter segments of mitochondrial DNA can distinguish the species which is also cost effective. Intraspecific variation in mitochondrial DNA is low in most animal species. This may reflect rapid loss of ancestral ploymorphisms due to maternal inheritance or selective sweeps following the emergence of advantageous mutations. Regardless of the cause, small intraspecific and large interspecific differences signal distinct genetic boundaries between most species, enabling precise identification with a mitochondrial barcode. Mitochondrial genes rarely contain introns, which are non-coding sequences interspersed between the coding regions of a gene. The following limitations have been encountered in DNA barcoding, as there are groups with little sequence diversity, resolution of recently diverged species, hybrids and nuclear pseudogenes.

1.1.3 Advantages of *Cytochrome Oxidase 1* as DNA [Barcode]

The COX1 barcoding gene is 648 base pairs in length, easy to sequence in one step with current technology and thus it is cheaper. Other advantages include,

- Easy to recover from diverse taxa, using a limited set of primers
- Easily aligned for sequence comparisons
- Effective in distinguishing among closely related species from a variety of invertebrate and vertebrate taxa

COX1 barcode sequence shows significant differences within species. For example, among 260 species of North American birds, differences between closely related species averaged 18 times higher than differences within species.

Comparison of COX1 barcode sequence shows, it differ from one another by only one or two base pairs out of 648, while we differ from chimpanzees at about 60 locations and of gorillas about 70 locations.

1.2 Plant DNA Barcoding

Plant taxonomist employ conventionally, morphological characters for discriminating taxa at several ranks namely family, genus, species, etc. Different types of evidence are utilized in plant classification, such as morphological, chemical and molecular features. Skilled professionals are required to interpret the morphological characters and navigate lengthy and complicated taxonomical keys to discriminate the plant species. This process has become expensive and time consuming (Newmaster *et al.*, 2009).

Several molecular tools have been applied for detection and identification of plant samples, these are designed to function primarily at the species or population levels of the taxonomic hierarchy. Some of the common molecular markers include RFLP, AFLP, RAPD, ISSR etc., (Leroy et al., 2000 and Weising *et al.*, 2005). These methods are not character based and hence they are not conducive for routine plant identification in conjunction with a high-throughput and bioinformatics system. The above molecular marker techniques are also not optimal for the taxonomically wide plant identification because of problems associated with high relatedness within taxon variability and lack of confident assignment of orthology. DNA sequences have potential to overcome these hurdles (Blaxter, 2004). Some DNA regions are sufficiently conserved to permit primer sets for PCR amplification and contain sufficient informative sequence variation to discriminate up to the species level. For plants, many genes have been proposed as DNA Barcodes till now.

1.2.1 The proposed plant DNA barcodes

The chloroplast trnL (UAA) intron as a DNA barcode showed to be potential gene candidate (Taberlet *et al.*, 2007). Barcodes such as ITS and trnH-psbA served as good starting point for large groups of angiosperms. These DNA segments have been proved to be successful for barcoding 8000 species of flowering plants of Costa Rica (Kress *et al.*, 2005). Chloroplast gene rbcL is found to be a core locus in a multigene approach for barcoding of land plants, which has showed some drawbacks due to its uneven and variable distribution. Many taxa were misidentified and showed only 0.79% to 13.89% of interspecies distance (Newmaster *et al.*, 2006). Plastid gene maturaseK (matK) has been examined and evaluated as potential locus for addressing particular evolutionary queries of the monocot families of Poaceae, Joinvillaceae, Cyperaceae and Smilaceae (Khidir *et al.*, 1999). The gene matK has also been identified as molecular marker for examining genetic diversity and phylogeny of Thai Boesenbergia of Zingiberaceae (Jiranan *et al.*, 2006). The plastid gene matK and trnK intron reveals evolutionary significance for the family Piperales (Wanke *et al.*, 2007). The chloroplast psbA-trnH intergenic spacer has been tested as potential DNA barcoding regions for species identification of cycadales (Sass *et al.*, 2007). The coding (rpoB, rpoC1, rbcL, matK and 23SrDNA) and non coding (trnH-psbA, atpF-atpH and psbK-psb1) regions of plastid has been studied as multilocus barcodes for plant species (Fazekas *et al.*, 2008). The plastid gene matK has been proposed as a barcode for the family Zingiberaceae. (Dhivya *et al.*, 2008). The gene trnH-psbA has been shown to distinguish ten species of Polygynaceae and also able to recognize eight other species of Polygynaceae (Jingyuan *et al.*, 2009). The psbA-trnH intergenic spacer region has been studied as a barcode, to distinguish and differentiate various *Dendrobium* species (Hui *et al.*, 2009).

Due to joint effort of some botanical institutes, DNA Barcode database system for entire floras with a repository called the Barcode of life Data System (BOLD) www.boldsystems.org has been initiated (Ratnasingham and Hebert, 2007). BOLD is an informatics workbench aiding acquisition, storage, analysis and publication of DNA barcodes. There are also automated identification tool for plants like Herbarium Information Management System (HIMS), Laboratory Information Management System (LIMS) and AIT (Automated Identification technology). AIT is high throughput system which saves time and cost but requires high volumes of plant sample for analysis (Newmaster *et al.*, 2009).

1.3 DNA Barcoding of Traditional Medicinal Plants

Traditional medicines are groomed for generations by practitioners. Throughout the history of mankind, many infectious diseases have been treated with herbals. The traditional medicine practice and usage have increased for the treatment of many infectious diseases recently. A number of scientific investigations have highlighted the medicinal importance of the many plant families like Asteraceae, Liliaceae, Apocynaceae, Solanaceae, Caesalpinaceae, Rutaceae, Piperaceae, Sapotaceae, Zingiberaceae etc. Medicinal plants play a vital role in the development of new drugs and hence it's of great commercial importance (Perumalsamy *et al.*, 1998). We have chosen Apocynaceae family for DNA barcoding study because of its immense medicinal importance and also due to its wide presence in South India.

1.3.1 The plant family-apocynaceae

Apocynaceae is the family of flowering plants containing 5 subfamilies, 17 tribes, 411 genera and 4650 species. The characteristics feature of this family is production of milky latex. Some plant species belonging to this family is believed to contain medicinal properties which are being used traditionally. Some of the Apocynaceae members and their medicinal importance are listed below :

PLANT NAME	MEDICINAL IMPORTANCE
Alstonia scholaris	Hypotensive, anticancerous
Catharanthus roseus	Anticancerous
Nerium oleander	Diuretic
Plumeria rubra	Anti-inflammatory, diuretic
Rauvolfia serpentine	Hypnotic, sedative
Holarrhena pubescens	Antidysenteric
Tabernaemontana coronaria	Anti-inflammatory
Thevetia peruviana	Cardiotonic, purgative
Wrightia tinctoria	Antidiarrhoeal

In the study, matK was selected as a DNA barcode candidate for Apocyanaceae, after extensive screening of different potential barcode candidates using *in silico* methods.

1.3.2 The *matK* Gene

The *matK* gene of chloroplast codes for *maturaseK*, which is involved in Group II intron splicing. The gene spans 1500 base pairs, which is located within the intron of the *trnK*. The two exons of the *trnK* gene that flank the *matK* will be lost, leaving the gene intact in the event of splicing (Wolfe *et al.*, 1992). The gene contains high substitution rates within species and has been used in many systematic and evolutionary studies (Johnson *et al.*, 1995). A homology search indicates that the carboxyl terminus are structurally related to portions of maturase like polypeptide and this might be involved in splicing of Group II introns. The *matK* gene has been used effectively in phylogenetic studies in the families Saxifragaceae and Polemoniaceae (Steele *et al.*, 1994), Poaceae (Liang *et al.*, 1996), Orchidaceae tribe Vandeae (Jarrell *et al.*, 1995) and Myrtaceae (Gadek *et al.*, 1996). Among the available chloroplast genes, *matK* has got high substitution rates, which makes it ideal to suit the needs of a perfect DNA barcode.

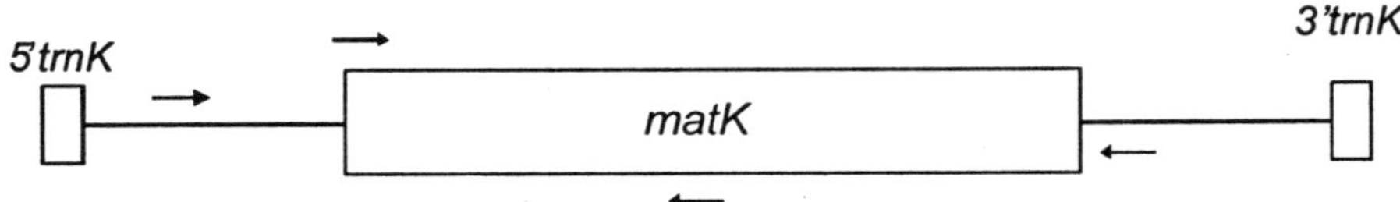

Fig. 1 : The Linear Structure of a Complex Gene *trnK-matK*

1.3.3 Major Steps Involved in Plant DNA Barcoding

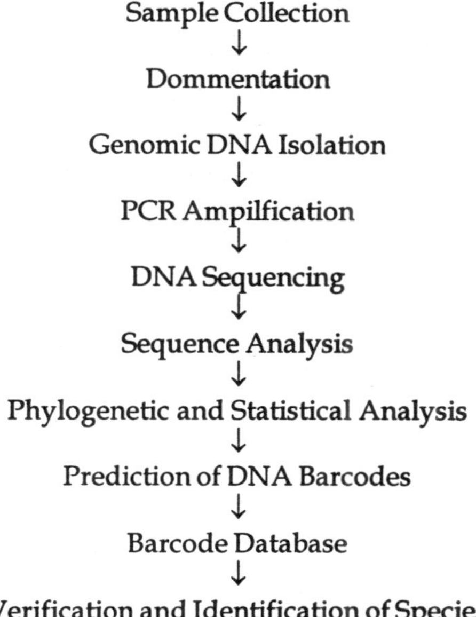

1.3.3.1 Samples for analysis

The plant samples for this study were collected from different places like, Western Ghat regions, Institute of Forest Genetics and Tree Breeding campus, Coimbatore, Tamil Nadu and Indian Centre for Medical Heritage Herbarium, Kanjikode, Kerala which included *Allamanda blanchetii, Adenium obesum, Alstonia scholaris, Carissa carandas, Holarrhena pubescens, Plumeria alba, Rauvolfia tetraphylla, Rauvolfia serpentine, Thevetia peruviana and Vularis solanacea*

1.3.3.2 Genomic DNA Isolation

Genomic DNA isolation for Apocynaceae plants was carried out by modified CTAB Cetyl Trimethyl (Ammonium Bromide) method (Khanuja *et al.*, 1999), which is designed specifically for the plants producing large amounts of secondary metabolites and oil.

1.3.3.3 Universal primers for *matK* and PCR amplification

Universal primers specific for *mat*K gene was designed. The genomic DNA was used as template for PCR. Universal primers of *mat*K were used for amplification with the following PCR condition- initial denaturation at 94°C for five minutes followed by 25 cycles of denaturation at 94°C for one minute, annealing at 56°C for 1 minute, extension at 72°C for one minute and final extension at 72°C for 10 minute. The PCR product was then separated in 0.8% agarose gel.

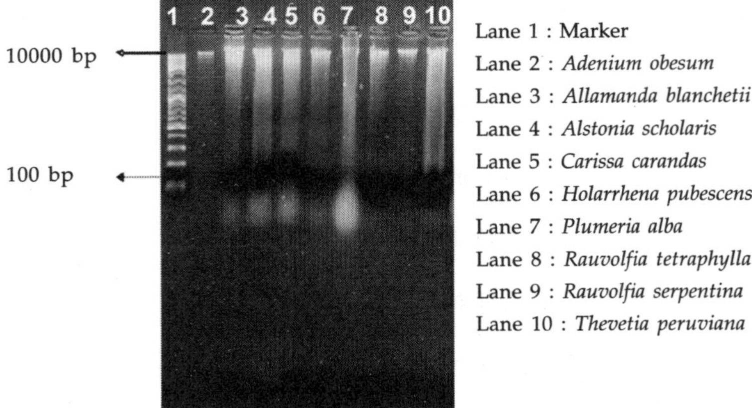

Fig. 2 : Genomic DNA from the plant species of Apocynaceae.

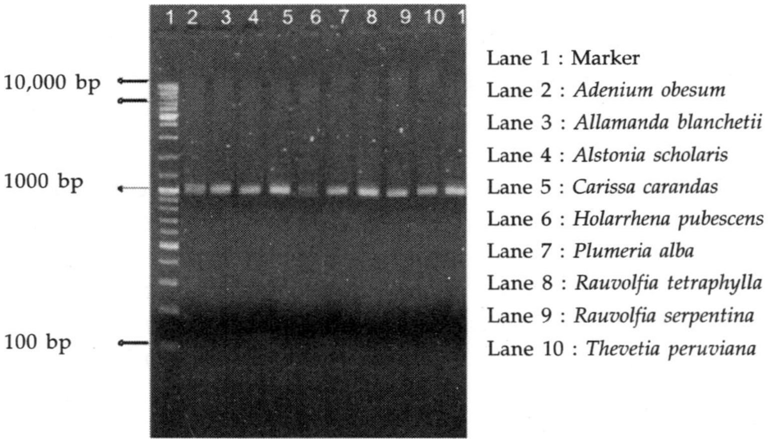

Fig. 3 : PCR Amplification of matK the plant species of Apocynaceae.

1.3.3.4 Sequence retrieval and alignment

Sequences of *matK* gene were collected from Taxonomy database and National Center for Biotechnology Information (NCBI), which provides various information regarding nucleotide sequences, protein sequences, 3D structures, EST's and journals. The DNA sequences of subfamily, tribes and related subspecies available were collected for the comparative analysis.

Sequence alignment was performed using multiple sequence alignment tool ClustalX. Alignment quality analysis was performed and low scoring segments have been highlighted. Input sequences were in different file formats like FASTA (Pearson), Clustal, EMBL/SWISSPROT and NBRF/PIR.

1.3.3.5 Steps involved in multiple sequence alignments

Pairwise alignment

All sequences were compared to each other and dendrogram was constructed.

Alignment parameters

- For every pair of sequence alignment, distance was calculated and same was used to construct the phylogenetic tree. Dynamic programming and Wilbur and Lipman methods were used for calculating the distance.
- Dynamic programming is slow and accurate where as Wilbur and Lipman programming is extremely fast and approximate.

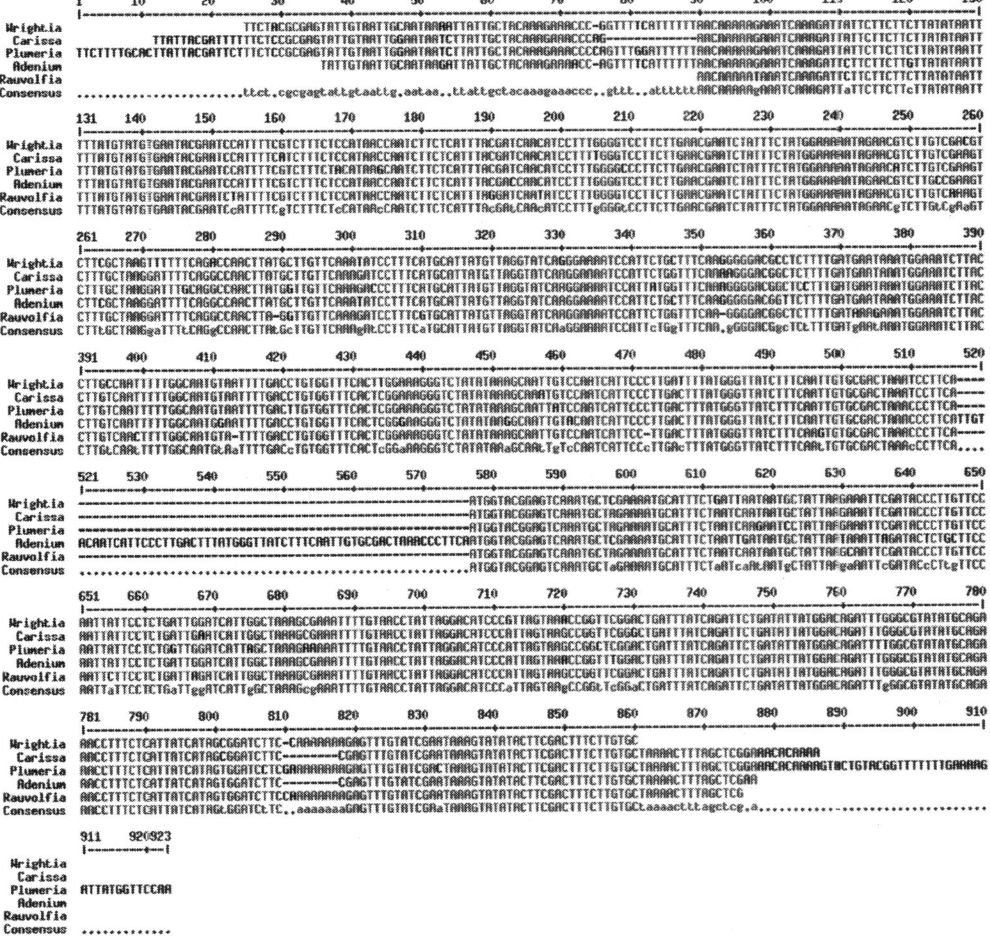

Fig. 4: Multiple sequence alignment for the genera of apocynaceae.
(The blue coloured region indicates the variable sites and red shows the conserved region)

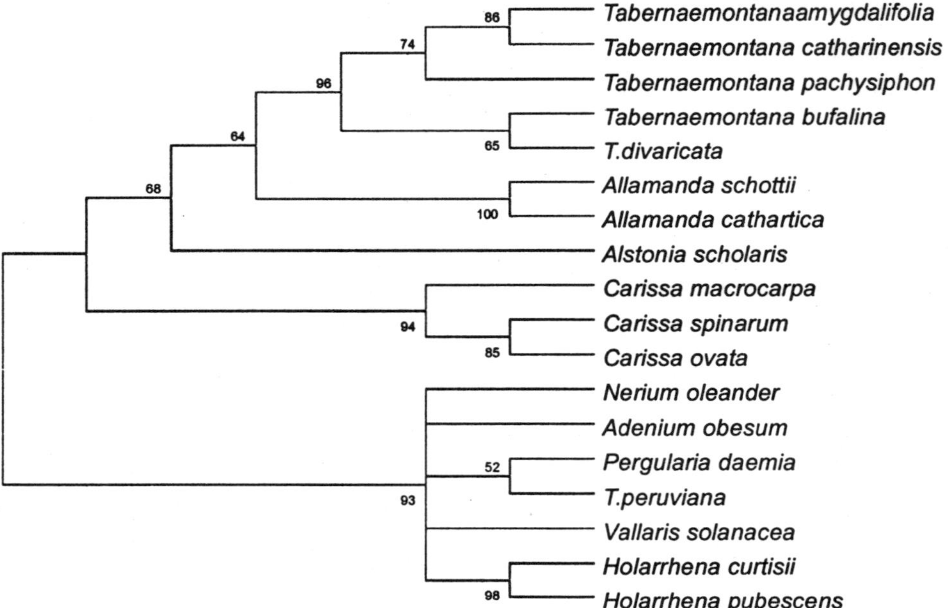

Fig. 5 : Phylogenic tree represents the evolutionary relationship between the plant species of Apocynaceae.

1.3.3.6 Phylogenetic analysis

Molecular Evolutionary Genetic Algorithm (MEGA) is an integrated tool for automatic and manual sequence alignment, to construct phylogenetic tree, to estimate the rates of molecular evolution and to test evolutionary hypothesis. In this study, phylogenetic analysis was performed for Apocynaceae subspecies, related tribes and genera. Steps involved in the phylogenetic analysis included,

- Aligning sequences
- Estimating evolutionary distances
- Building trees
- Testing tree reliability
- Marking genes and domains
- Testing and grouping of sequences
- Constructing tree from the distance data

1.4 Conclusion

Phylogenetic analysis complements and often outperforms the similarity searches, identifying variants, patterns and transition/transversion rate, when addressing sequence identity, especially with the reference database. Portable software MEGA framework for qualified identification of nucleotide sequences of Apocynaceae family provided inter and intra species relationships. From this analysis it is clear that *matK* gene differentiates at species level. The overall gene distance show 0.041 and was calculated by Kimura 2 parameter. The variation indicates 10.8 at genus level and 2% at species level. These results suggest that plant DNA barcoding can be effectively used for other family members too in future.

Acknowledgement

The project is funded by University Grants Commission (UGC), New Delhi. Mrs. S. Dhivya, thank UGC, New Delhi for the award of the Research Fellowship in Science for Meritorious Students (RFSMS) under UGC-SAP grant to pursue her PhD programme.

References

Blaxter M.L. (2004). The promise of a DNA taxonomy. *Philosophical Transaction of the Royal Society of London Series B Biological Science.* **359**,669-679.

Dhivya S., Rajeev Kumar S., and Sathishkumar R. (2008). Phylogenetic analysis of chloroplast *mat*K gene from Zingiberaceae for plant DNA barcoding. *Bioinformation,* 3(1), 24-27.

Fazekas A., J, Burgess K.,S, Kesanakurti P.,R, Graham S.,W, Newmaster S.,G, Husband B.,C, Percy D.,M, Hajibabaei M and Barrett S.,C. (2008) Multiple multilocus DNA barcodes from the plastid genome discriminate plant species equally well. *PLos One* (3)7, e2802.

Gadek P.A., Wilson P.G., and Quinn C.J. (1996). Phylogenetic reconstruction in Myrtaceae using *mat*K, with particular reference to the position of Psiloxylon and Heteropyxis. *Aust. Syst. Bot.* 9, 283-290.

Jarrell D.C, and Ciegg M.T. (1995). Systematic implications of the chloroplast - encoded *mat*K gene in the tribe Vandeae (Orchidaceae), *Am. J. Bot*. 82: 137.

Jingyuan S., Hui Y., Ying L., Xiwen L.,Chang L., Jianping H., Caixiang X and Shilin C.(2009).Authentication of the family Polygonaceae in Chinese pharmacopoeia by DNA barcoding technique.*Journal of Ethnopharmacology*. 124(3): 434-439.

Jiranan T., Chatchai N., Sudsangvan C and Thaya J. Sirawut K. (2006).Genetic Variation and Species Identification of Thai Boesenbergia, (Zingiberaceae) Analyzed by Chloroplast DNA. Polymorphism. *Journal of Biochemistry and Molecular Biology,* 39(4): 361-370.

Johnson L.A, and Soltis D.E, (1995). Phylogenetic inference in Saxifragaceae sensu stricto and Gilia (Polemonoiaceae) using *mat*K sequences. *Ann. Mo. Bot. Gard,* 82 : 149-175.

Khidir W.H., Lawrence A., A and Hongping L. (1999). Phylogeny of poaceae Inferred from *matK* sequences. Ann. Mo.Bot. Gard, 86,835- 851.

Kress W.J., Wurdack K.J., Zimmer E.A., Weigt L.A. and Janzen D.H. (2005) Use of DNA barcodes to identify flowering plants. *Proc Natl Acad Sci USA*, 102(23):8369-8374.

Leroy X.J., Leon K and Branchard M. (2000). Plant genomic instability detected by micro-satellite primers .*Electronic Journal of Biotechnology*, 3(2): 140-148.

Liang H., and Hilu K.W. (1996). The matK gene sequence variation and application in gross systematics. *Am. J. Bot*, 84(6): 830-839.

Newmaster S.G., Fazekas A.J. and Ragupathy S. (2006). DNA barcoding in the land plants: evaluation of *rbcL* in a multigene tiered approach. *Can. J. Bot.*, 84, 335–341.

Newmaster S.G., Ragupathy S and Janovec J.(2009). A botanical renaissance : state-of-the art DNA barcoding facilitates an Automated identification Technology system for plants.*Int.J.Computer Applications in Technology*. 35

Ratnasingham S and Hebert D.N Paul. (2007). Barcodding Bold: The Barcode of Life Data System.(www.barcodinglife.org). *Molecular Ecology notes.*

Sass C., Little D.P., Stevenson D.W. and Specht C.D. (2007). DNA barcoding in the cycadales: testing the potential of proposed barcoding markers for species identification of cycads. *PLos One* 2(11), e1154.

Steele K.P., and Vilgalys R. (1994) Phylogenetic analysis of Polemoniaceae using nucleotide sequences of the plastid gene *mat*K. *Syst. Bot* 19,126-142.

Taberlet P., Coissac E., Pompanon F., Gielly L., Miquel C., Valentini A., Vermat T., Corthier G., Brochmann C. and Willerslev E. (2006). Power and limitations of the chloroplast trnL (UAA) intron for plant DNA barcoding. *Nucleic acid research*, 35, *e*14.

Wanke S., Jaramillo M.A., Borsch T., Samain M.S., Dietmar Q and Neinhuis.C (2007) Evolution of Piperales-matKgene and trnK intron sequence data reveal lineage specific resolution contrast. *Molecular Phylogenetics and Evolution*, 42(2): 477-497.

Weising K., Nybirm H., Wolff K. and Kahl G. (2005) DNA Fingerprinting in plants. Principles, methods and Applications *CRC Press, Boca Raton, FL.*

Wolfe K. H., Morden C. W. and Palmer J. D. (1992) Function and evolution of a minimal plastid genome from a nonphotosynthetic parasitic plants. *Proceedings of the National Academy of Sciences, USA* 89(22): 10648-10652.

YaO. H., Song, J.Y., Ma X.Y., Liu. C., Li. Y., Xu. H.X., Iean J.P., Daun L.S., and Chen S.L., (2009). Identification of Dendrobiuno Species by a Candidate DNA barcode sequence: *The Chloroplast psbA-trn4 intergenic region. Plasta med.* 75(6): 667-669.

❑❑❑

19

Metabolic Engineering for Therapeutics

M. Muthukumar, S. Seal, P. N. Sivalingam
R. Chandra, V. Satheesh and S. Anandhan

CONTENTS

1.1 Introduction 292
1.1.1 Definitions 292
1.1.2 Basis and current status of metabolic engineering 293
1.2 Metabolic Engineering in Microorganisms 294
1.2.1 Metabolic engineering in bacteria 294
1.2.2. Metabolic engineering in fungi and yeasts 296
1.3. Metabolic Engineering in Plants 298
1.3.1 Plants as bioreactors for therapeutics 298
1.3.2 Transgenic plants for therapeutics and biopharmaceuticals 299
1.4 Metabolic Engineering in Animal Cell Culture Systems 299
1.5 Inverse Metabolic Engineering 302
1.6 Conclusion 302
References 302

1.1 Introduction

Metabolic engineering is the science that combines systematic analysis of metabolic and other pathways with molecular biological techniques to improve cellular properties by designing and implementing rational genetic modifications (Koffas *et al.*, 1999). This multidisciplinary field therefore, draws principles from chemical engineering, computational sciences, biochemistry and molecular Biology. With the recent advances developed in analytical methods and mathematical tools, it is further gaining momentum in a way of redirecting metabolic fluxes for industrial and medical purposes. It is clear that metabolic engineering deals with the measurement of metabolic fluxes and elucidation of their control as determinants of metabolic function and cell physiology. It is also well suited as a framework for the analysis of genome-wide differential gene expression data, in combination with data on protein content and in vivo metabolic fluxes. Thus, a novel aspect of metabolic engineering is that it departs from the traditional reductionist paradigm of cellular metabolism, taking instead a holistic view.

In essence, metabolic engineering involves the application of engineering principles of design and analysis to the metabolic pathways in order to achieve a particular goal. This goal may be to increase process productivity, as in the case of production of antibiotics, biosynthetic precursors or polymers, or to extend metabolic capability by the addition of extrinsic activities for chemical production or degradation. It has applications in bioprocess (e.g. methanol production) and pharmaceutical industries (e.g. antibiotic production). The insights of the integrated view of metabolism generated by metabolic engineering will have profound implications in biotechnological applications, as well as in devising rational strategies for target selection for screening candidate drugs or designing gene therapies. In this chapter, we have concentrated on the basic principles and concepts related to metabolic engineering in general and their applications mainly on therapeutic production in detail.

1.1.1 Definitions

Metabolic engineering has been defined in different ways at different points of time by several eminent metabolic engineers. The most commonly adopted definitions and accepted world over are presented below.

1. Metabolic engineering has been defined as the targeted and purposeful alteration of metabolic pathways in an organism in order to better understand and utilize cellular pathways for chemical transformation, energy transduction, and supramolecular assembly (Lessard, 1996).

2. Neilsen (1998) has defined metabolic engineering as the purposeful modification of intermediary metabolism using recombinant DNA techniques.

3. Stephanopoulous (1999) has referred to it as the direct improvement of cellular properties through the modification of specific biochemical reactions or introduction of new ones, with the use of recombinant DNA technology.

1.1.2 Basis and current status of metabolic engineering

The total chemical reactions that occur within a living organism constitute the metabolism and the various compounds that take part in or formed by these reactions are called metabolites. Both the synthesis (anabolism) and breakdown (catabolism) of most of the compounds occurs by a number of reaction steps, and the reaction sequences pertaining to those compounds is termed as metabolic pathway. Manipulations can be made in the cellular metabolic pathways which will lead to enhanced production of any desired product. Here comes the role of metabolic engineering. Metabolic engineering focuses on altering the enzymes and other components of the metabolic pathways of the cell. In other words, it is genetically engineering a cell to produce a specific enzyme in higher levels or to replace one enzyme by another with more desirable properties.. A typical metabolic engineering approach focuses on a particular metabolic intermediate or product of a particular metabolic pathway such as starch, vitamins, carotenoids, amino acids, antibiotics etc. There are several ways to metabolically engineer an organism and the most common ways are:

1. Blocking a metabolic flux (re-channel).
2. Channelling a metabolic flux into new cell compartments.
3. Inducing a metabolic flux.
4. Introducing a new metabolic pathway into the organism.

These methods of metabolic engineering are explained in detail subsequently. Through metabolic engineering several approaches have been used for the modification of the host cell to achieve the desired goal. These include: the removal of the enzyme or its inhibition to eliminate a competitive pathway (Gasson *et al.*, 1996; Shimada *et al.*, 1998) or a toxic byproduct (Aristidou *et al.*, 1994; Cameron *et al.*, 1998), the amplification of a gene or group of genes to improve the synthesis of existing products (Shimada *et al.*, 1998; Lu and Liao; 1997), the expression of an heterologous enzyme(s) to extend the substrate range (Panke *et al.*, 1998; Sprenger, 1996), to produce novel product (Chopra and Vageeshbabu, 1996; Hershberger, 1996; Misawa and Shimada, 1998), to provide pathways for the degradation of toxic compounds (Chen and Wilson, 1997; Keasling *et al.*, 1998; Xu *et al.*, 1996), or to design a more environmentally resistant plant (Smirnoff, 1998). Alternatively, the deregulation of existing enzymes might be necessary to overcome the existing control mechanisms of a rigid node. A combination of modifications may be needed to achieve the goal. Additional manipulation of the central metabolic pathway may also be required to generate the precursor, cofactors, and energy needed to sustain the modified pathway (Stephanopoulos and Sinskey, 1993, Farmer and Liao, 1996). The manipulation of central metabolism has been exemplified in the production of aromatic compounds in *E. coli* growing in glucose (Berry, 1996). The significance and strategies of metabolic engineering in microorganisms, plants and animals differ in the methodology and purpose. They are dealt in different heads in detail.

1.2 Metabolic Engineering in Microorganisms

Metabolic engineering in microorganisms especially bacteria and yeast for the production of therapeutically important compounds and biopharmaceuticals is commercially attempted by several industries because in bacteria and yeast, it is comparatively easier to introduce a new metabolic pathway and the induction of a metabolic flux. A detailed list of therapeutic compounds produced through metabolic engineering in different microorganism is listed in the Table 1. Successful metabolic engineering in bacteria and yeast could be due to their following properties:

1. Higher yields of metabolites in comparison to plants.
2. Ability to use simple and inexpensive carbon sources (e.g. glycerol).
3. Fast growing organism (in bioreactors).
4. Host preferred codon usage.
5. C omparatively easier downstream processing for purification of the desired product.

1.2.1 Metabolic engineering in bacteria

Metabolic engineering in bacteria has been employed by several industries for production of industrial enzymes, proteins, amino acids, organic molecules and antibiotics (Martin *et al.*, 2003). There are three strategies adopted in general for metabolic engineering in bacteria. They include:

(i) Inserting new pathways in microorganisms with the aim of producing novel metabolites, e.g., production of polyketides by Streptomyces spp.

(ii) production of heterologous peptides, e.g., production of human insulin, erythropoietin, and tPA.

(iii) improvement of both new and existing processes, e.g., production of antibiotics and industrial enzymes.

Despite the prospect of obtaining major improvement through metabolic engineering, this approach is, however, not expected to completely replace the classical approach to strain improvement - random mutagenesis followed by screening in industrial biotechnology. Identification of the optimal genetic changes for improvement of a given process requires analysis of the underlying mechanisms, at best, at the molecular level. To reveal these mechanisms a number of different techniques may be applied, viz., detailed physiological studies, metabolic flux analysis (MFA), metabolic control analysis (MCA), thermodynamic analysis of pathways, and kinetic modeling. Of all these methods MFA and MCA are most successful mathematical tools for analysis which are discussed briefly.

a) *Metabolic flux analysis (MFA)*: Metabolic flux analysis (MFA) is based on a known biochemistry framework. A linearly independent metabolic matrix is constructed based on the law of mass conservation and on the Pseudo-steady state hypothesis (PSSH) on the intracellular metabolites. The formulation resulted in a set of linear equations that can be expressed as a stoichiometric matrix A of dimension m by n with vectors for net accumulation, r (mx1), and metabolic flux, u (nx1). No kinetic data is required. Dynamic flux-balance analysis can be obtained from the same derivation as shown in the following Equation with b as the transport term.

$$r = \frac{dX}{dt} = A.v - b$$

Typically, the system that results is under-determined system where m>n. However, under certain conditions, some pathways are inoperative and can be neglected. The system may become completely determined or over-determined and can be solved along with the measurements of external or internal fluxes (Delgado and Liao, 1997; Chen *et al.*, 1997).

b) *Metabolic Control Analysis (MCA):* Metabolic Control Theory was independently developed by Kacser and Burns (Kacser and Burns, 1973) and Heinrich and Rapoport (1997) to identify the kinetic constraints in a biochemical network. In 1992, a more structural approach to MCT has been developed by Fell. Based upon the kinetic data, the control coefficients can be calculated. This approach is used to determine the rate-limiting reaction in a network. However, a single rate-limiting step may not exist and several steps may share the control of the metabolic network. Three commonly used normalized sensitivity measures, which quantify how the control of steady state fluxes and concentrations is distributed between different reactions in a metabolic network, have been defined as:

1. Flux Control Coefficient (FCC)
2. Concentration Control Coefficient (CCC)
3. Elasticity

The flux control coefficient is a measure of how a change in the concentration of the enzyme E affects the steady-state flux through that particular enzyme. That is, it is the measure of the degree of control exerted by enzyme E on this steady-state flux. The concentration control coefficient is a measure of the extent of control exerted by the enzyme E on the steady-state concentration X, while the elasticity is a measure of the response of the reaction rate upon changes in the concentration X.

The extension of MCT is applicable under large changes in enzyme activities and effector concentration (Small and Kacser, 1993a, 1993b). This extension would allow for experimental calculations of the deviation index from MFA data. A top-down approach for the determination of control coefficients was developed by Brown *et al.* (1990), which allows for the determination of control coefficients based on relative flux alone and avoids the need for values of enzyme activities. In addition, more information is obtained from a single application. But, there are certain limitations to this method as it is invalid if more than one kinetic link exist between the two ends of a pathway (i.e., more than one enzyme catalyzes the same reaction). Moreover, a small overall flux control coefficient (FCC) does not necessarily mean the individual FCCs are small. However, it can be used in certain limited situations (Brown *et al.*, 1990).

1.2.2 Metabolic engineering in fungi and yeasts

Metabolic engineering has been successfully reported in fungi and yeasts. Secretory recombinant glycoproteins are of more therapeutic value as the carbohydrate moiety and glycosylation process helps to determine their pharmacokinetic properties. Recombinant therapeutic glycoproteins have been produced by engineering of N-glycosylation pathway of glycans in yeast (Roy *et al.*, 2000). Another example is integrating the transcriptional and metabolite profiles in Aspergillus terreus to direct the production of lovastatin. Lovastatin is a potent hydroxymethylglutaryl coenzyme A (HMG-CoA) reductase inhibitor to reduce the serum cholesterol levels. Geodin is derived from anthraquinone emodin, an intermediary compound in the biosynthesis of natural products. These are polyketide-derived secondary metabolites whose production are correlated and engineering of lovE gene cluster under FLO11 fungal promoter by gene replacement of geodin resulted in lovastatin producing Aspergillus terreus (Askenazi *et al.*, 2003).

Table 1 : Recent developments of therapeutics produced through metabolic engineering in microorganisms

Metabolically engineered microorganism	Engineering strategy	Target therapeutic	Reference
Streptomyces clabuligerus	Engineering kinetic model of biosynthetic pathway by incorporation of lysine-epsilon aminotransferase (lat) gene	Cephalosporin C	Khetan *et al.*, 1996
Escherichia coli	Overproduction of amino acid precursors at the bottom of pathway.	β-lactam antibiotics	Thykaer and Nielsen, 2003
Streptomyces clavuligerus	Gene replacement technology of lat gene by cvm1 gene	Clavulanic acid	Paradkar *et al.*, 2001
Streptomyces spp.	Heterologous expression of novobiocin, clorobiocin and coumermycin genes	Amino coumarin antibiotics	Heidi *et al.*, 2009
Streptomyces venezuelae	Heterologous expression of noviose biosynthetic genes (novW and novU and novS)	Hybrid antibiotic noviosylate 10-deoxymethy-nolide/ narbonolide	Pageni *et al.*, 2008
Methylomonas spp. strain 16A	Site specific integration of carotenoid gene cluster under control of promoter for chloramphenicol resistance	Carotenoids	Sharpe *et al.*, 2007
Streptomyces coelicolor strain M512	Inhibition of methyl transferase gene (cloP) in the clorobiocin's biosynthetic gene cluster	Anticaumarin antibiotics	Freitag *et al.*, 2005
Pichia ciferrii (Yeast)	Over expression of tetraacetylphytosphingosine (TAPS) gene	Tetraacetyl phyto-sphingosine	Jung-Hoon Bae *et al.*, 2003
Aeromicrobium erythreum (actinomycete)	Reverse engineering of mutB and cobA genes by tagged mutagenesis in methyl melonyl coA pathway	Erythromycin	Reeves *et al.*, 2004
Lysobacter lactangenus	Engineering the cephabacin biosynthetic pathway	Cephabacins, one of the β-lactam antibiotics	Sohn *et al.*, 2001

1.3. Metabolic Engineering in Plants

Next to bacteria and yeasts, plants present themselves as a potent organism to be genetically modified for being metabolically engineered for production of therapeutics. Plants have considerable potential for the production of biopharmaceutical proteins and peptides because they are easily transformed and provide a cheap source of protein. The production of pro-vitamin-A in rice is the most successful demonstration of metabolic engineering in plants for production of therapeutics. The following are reasons that may be attributed for possibility of metabolic engineering in the plant systems:

a. Plants are the most prolific factories for small molecules (bioreactors).

b. More than one lakh metabolites have been identified in plant systems.

c. Plants are the richest source of secondary metabolites (5,000 25,000 types per plant).

The flow of metabolites through metabolic and biosynthetic pathways is regulated to respond to development and environmental conditions in plant systems. The distribution of metabolites (partitioning) controls the flux of carbon compounds towards the synthesis of sugars, starch and oil in storage organs, and may change to meet developmental states, environmental constraints or in the activation of the diverse array of defence responses when a plant is attacked by pathogens. Metabolic fluxes can be manipulated, and an understanding of the control of partitioning between metabolic pathways is required for metabolic engineering- this knowledge can be used to increase the production or yield of commercially important metabolites and products. Strategies to modify partitioning include:

(i) Changing levels of signaling metabolites and hormones.

(ii) Removal of end products to change reaction equilibria.

(iii) Expression of heterologous enzymes to bypass endogenous regulatory mechanisms.

(iv) Expression of transcriptional regulators that control pathways.

(v) Switching off gene expression by sense/antisense suppression.

1.3.1 Plants as bioreactors for therapeutics

A bioreactor is a device in which a substrate of low value is utilized by living cells or enzyme to generate products of higher value. Plant are used as bioreactors for production of biomolecules like carbohydrates, lipids, proteins and edible vaccines because of the three major reasons such as:

1. *Post Translational Modifications*: Microbes do not provide a good cellular environment for post-translational modifications of eukaryotic proteins, which are often recovered from bacteria as insoluble aggregates. Solubilization of these aggregates and refolding of the polypeptides to obtain native proteins involves considerable costs. The cellular environment of plants, on the other hand, is as good as that of animals for post translational modification and folding of the proteins so that animal proteins recovered from plants are usually at least as good as the native animal proteins.

2. *Storage Costs*: Microbial biomass has to be immediately processed and the purified proteins have to be ordinarily stored under cold storage. But in the case of plant seeds, they could be stored under ambient conditions from which the proteins may be isolated as per need. This would greatly reduce the storage costs of the products.

3. *Ethical Considerations*: Biochemical production from transgenic animals raises several public and ethical concerns, while no such problem is associated with plants.

1.3.2 Transgenic plants for therapeutics and biopharmaceuticals

Several biotechnology companies are now actively developing, field testing, and patenting plant expression systems, while clinical trials are proceeding on the first biopharmaceuticals derived from them. One transgenic plant-derived biopharmaceutical, hirudin, is now being commercially produced in Canada for the first time. Product purification is potentially an expensive process, and various methods are currently being developed to overcome this problem, including oleosin-fusion technology (Fig.1, Goddijn and Janpen, 1995), which allows extraction with oil bodies. In some cases, delivery of a biopharmaceutical product by direct ingestion of the modified plant potentially removes the need for purification. Such biopharmaceuticals and edible vaccines can be stored and distributed as seeds, tubers, or fruits, making immunization programs in developing countries cheaper and potentially easier to administer. Some of the most expensive biopharmaceuticals of restricted availability, such as glucocerebrosidase, could become much cheaper and more plentiful through production in transgenic plants (Giddings *et al.*, 2000). The list of therapeutics and biopharmaceuticals that are attempted to be expressed in plants by metabolic engineering is presented in the Table 2.

1.4 Metabolic Engineering in Animal Cell Culture Systems

Metabolic engineering is playing a central role in the development of more efficient methodologies for producing high-value biopharmaceuticals in mammalian, insect cells, and other eukaryotes. The majority of human biopharmaceuticals are secreted glycoproteins produced by mammalian cell culture. These glycoprotein products include monoclonal antibodies such as Rituxan, Hercepten, and Remicade, erythropoietin (EPO), tissue plasminogen activator (tPA), Factor VIII, Protein C, and many other proteins in clinical trials. As glycoproteins, these biotherapeutics typically include oligosaccharide (carbohydrate) chains attached to the protein at specific amino acid residues. The number, type, and location of the carbohydrate attachments on the protein can affect key properties of commercial biopharmaceuticals including clearance rate, immunogenicity, biological specific activity, solubility and stability against proteolysis.

Humans will typically accept only those biotherapeutics that have particular types of carbohydrate attachments and will often reject glycoproteins that include non-mammalian oligosaccharide attachments. As a result, mammalian cells Chinese Hamster Ovary (CHO), Baby Hamster Kidney (BHK), and Human Embryonic Kidney-293 (HEK-293) are used for the production of the vast majority of these glycoprotein therapeutics because of their capacity to generate glycoforms and perform other post-translational processing patterns that are accepted by human patients. Unfortunately, production of biotherapeutics in mammalian cells can be expensive due to the need to grow these cells in costly cell culture environments and because mammalian cells often produce the proteins in low yields.

Table 2 : Transgenic plants altered for producing biotherapeutics/biopharmaceuticals

Transgenic Plant	Method	Target therapeutic	Potential use
Potato	AMT	Human serum albumin	Recombinant protein for curing liver cirrhosis
Tobacco	AMT	Gluocerebrosidase	Enzyme replacement therapy for curing Gaucher's disease (recessive inherited lysosomal storage disorder)
Rice	PB	Human Antitrypsin	Therapeutic potential against cystic fibrosis, liver disease and hemorrhage
Maize	PB	Human Aprotinin	Trypsin inhibitor for transplantation surgery
Oilseed	AMT	Human enkephalins	Wound repair and control of cell proliferation
Tobacco	AMT	Human erythropoietin	Anaemia

AMT- Agrobacterium mediated transformation, PB – Particle bombardment
Source: Giddings *et al.*, 2000; Goldstein and Thomas, 2004.

Two different metabolic engineering strategies are being implemented to improve production of high quality biotherapeutics. One strategy to increase the efficiency of producing expensive biotherapeutics is to enhance the capabilities and capacity of the mammalian cell hosts through metabolic engineering. Metabolic engineering of mammalian metabolism, cell cycle, apoptosis, and glycosylation can increase cell densities and yields and improve the quality of products from cell culture. Two specific examples of mammalian cell metabolic engineering will be presented:

1. Elimination of a glycosylation processing pathway to improve antibody product quality
2. Inhibition of cell death (apoptosis) pathways to increase monoclonal antibody product yields. Glycosylation pathways and other post-translational processes can also be manipulated to obtain higher quality products from these mammalian cell hosts.

An alternative metabolic engineering strategy is to alter the physiology of non-mammalian hosts in order to allow these organisms to produce biotherapeutics accepted by humans. These alternative hosts currently lack the genetic and metabolic content for generating high-quality mammalian forms. However, these cells can often generate higher product yields than mammalian cells and metabolic pathways are being engineered so that these organisms can generate the more desirable higher value glycoproteins. Some examples of therapeutics that are expressed in mammalian cells are listed in (Table 3). Increasing biotherapeutic production efficiencies in mammalian, insect cells and other eukaryotes will lower the costs to develop valuable biopharmaceuticals and ultimately lower health care costs for patients receiving these life-saving drugs (Wurm, 2004).

Table 3 : Metabolic engineering in animal cell culture system for therapeutic use

Animal cell culture system	Metabolic engineering strategy	Therapeutic use	Reference
Hamster ovary cells (mammalian expression host)	Control of endogenous gene expression with antisense DNA and gene targeting	Production of recombinant glycoprotein therapeutics	Warner, 1999
Liver cells of mice	Blocking fatty acid metabolism	Decreased regulatory signal malonyl-CoA and Reduction in obesity	Anonymous, 2009

1.5 Inverse Metabolic Engineering

The classical approach of metabolic engineering, as discussed above, requires detailed knowledge of the enzyme kinetics, the system network, and intermediate pools involved, and on such bases, a genetic manipulation is proposed for some presumed benefits. In contrast, the concept of inverse metabolic engineering is first to identify the desired phenotype, then to determine environmental or genetic conditions that confer this phenotype, and finally to alter the phenotype of the selected host by genetic manipulation (Bailey et al., 1996, Delgado and Liao, 1997). As in the case of expression of the oxygen binding protein, VHb, in *E. coli*, the observed phenotype of high heme cofactor levels in an obligate aerobe *Vitreoscilla* under oxygen limitation suggested that synthesis of the haemoglobin could improve growth of other organisms under similar limitations (Bailey *et al.*, 1996).

1.6 Conclusion

Metabolic engineering in microorganisms has helped in therapeutics production especially antibiotics by minimizing production cost. But, plants are now gaining widespread acceptance for the large scale production of therapeutics rather than microorganisms. The first plant derived recombinant pharmaceutical proteins are reaching final stages of clinical evaluation and many more are in the development pipeline. Though the plant derived pharmaceuticals are lower in yield and inconsistent in quality, attempts are made to overcome these limitations. On the other hand, metabolic engineering of animal cell expression hosts are gaining momentum for the purpose of enhancing the production of recombinant glycoproteins of pharmaceutical importance, targeted drug delivery and gene therapy. Thus, metabolic engineering is the youngest of the modern sciences, still in its infancy, has not only benefited the pharmaceutical industry but also opened new vistas in functional genomics and drug discovery.

References

Annonymous. 2009. Metabolic pathway engineered in mice to prevent diet-induced obesity. Biotechdaily, Online Bioresearch News.

Aristidou, A.A., San, K.Y., Bennett, G.N. 1994. Modification of central metabolic pathway in *Escherichia coli* to reduce acetate accumulation by heterologous expression of the *Bacillus subtilis* acetolactate synthase gene. *Biotechnology & Bioengineering*, 44: 944-951.

Askenazi, M., Driggers, E.M., Holtzman, D.A., Norman, T.C., Iverson, S., Zimmer, D.P., Mary-Ellen Boers, Blomquist, P.R., Martinez, E.J., Monreal, A.W., Feibelman, T.P. Mayorga, M.E., Maxon, M.E., Sykes, K., Tobin, J.V., Cordero, E., Salama, S.R. Trueheart,

J., Royer, J.C., K.T. Madden. 2003. Intergrating transcriptional and metabolic profiles to direct the engineering of lovastatin-producing fungal strains. *Nature Biotechnology*: 1-7.

Bailey, J. E., Shurlati, A., Hatzimanikatis, V., Lee, K., Renner, W.A., Tsai, P.E. 1996. Inverse metabolic engineering a strategy for directed genetic engineering of useful phenotypes. *Biotechnology & Bioengineering* 52:109-121.

Berry, A. 1996. Improving production of aromatic compounds in *Escherichia coli* by metabolic engineering. *Trends in Biotechnology* 14:219-259.

Brown, G.C., Hafner, R.P., Brand, M.D. 1990. A 'top-down' approach to the determination of control coefficients in metabolic control theory. *European Journal of Biochemistry*, 188:321-325.

Cameron, D.C., Altaras, N.E., Hoffman, M.L., Shaw, A.J. 1998. Metabolic engineering of propanediol pathways. *Advances in Biochemical Engineering*, 14: 116-125.

Chen, R., Hatzimanikatis, V., Yap, W., Postma, P.W., Bailey, J.E. 1997. Metabolic consequences of Phosphotransferase (PTS) mutation in a phenylalanine-producing recombinant *Escherichia coli*. *Biotechnology Progress* 13:768-775.

Chen, S.L., Wilson, D.B. 1997. Genetic engineering of bacteria and their potential for Hg^{2+} bioremediation. *Biodegradation* 8:97-103.

Chopra, V.L., Vageeshbabu, H.S. 1996. Metabolic engineering of plant lipids. *Journal of Plant Biochemistry & Biotechnology*, 5:63-68.

Delgado, J., Liao, J.C. 1997. Inverse flux analysis for reduction of acetate excretion in *Escherichia coli*. *Biotechnology Progress*, 13:361-367.

Farmer, W.R., Liao, J.C. 1996. Progress in metabolic engineering. *Current Opinion in Biotechnology* 7:198-204.

Fell, D.A. 1992. Metabolic control analysis: a survey of its theoretical and experimental development. *Biochemical Journal* 286:313-330.

Freitag, A., Rapp, H., Heide, L., Shu-Ming Li. 2005. Metabolic engineering of aminocaumarins: inactivation of the methyltransferase gene *cloP* and generation of new clorobiocin derivatives in a heterologous host. Chembiochem: *A European journal of chemical biology*, 6(8): 1411-8.

Gasson, M. J., Benson, K., Swindell, S., Griffin, H. 1996. Metabolic engineering of the *Lactococcus lactis* diacetyl pathway. *Lait*, 76: 33-44.

Giddings, G., Allison, G., Brooks, D., Carter, A. 2000. Transgenic plants as factories for biopharmaceuticals. *Nature Biotechnology*, 18:1151 – 1155.

Goddijn, O.J.M., Janpen. 1995. Plants as bioreactors. Trends in biotechnology, 13:79-387

Goldstein, D.A., Thomas, J.A. 2004. Biopharmaceuticals derived from genetically modified plants. *Q.J. Med.*, 97: 705-716.

Heide, L. 2009. Genetic engineering of antibiotic biosynthesis for the generation of new aminocoumarins. Biotechnology advances, 1873-99.

Heinrich, R., Rapoport, S.M. 1997. Metabolic regulation and mathematical models. In Progress in Biophysics and Molecular Biology, Vol. 32, Butler, J. A. V., Noble, D., Ed., Pergamon Press: Oxford, UK, 1977, pp 1-82.

Hershberger, C.L. 1996. Metabolic engineering of polyketide biosynthesis. *Current Opinions in Biotechnology* 7: 560-2.

Jung-Hoon Bae, Jung-Hoon Sohn, Chang-Seo Park, Joon-Shick Rhee and Eui-Sung Choi. (2003). Intergrative transformation system for the metabolic engineering of the sphingoid base-producing yeast, *Pichia ciferrii*. *Applied and Environmental Microbiology*, 69(2):812-9.

Kacser, H., Burns, J.A. 1973. The control of flux. Symposium of the Society of Experimental Biology 27:65-104.

Keasling, J.D., Vandien, S.J., Pramanik, J. 1998. Engineering polyphosphate metabolism in *Escherichia coli* - implications for bioremediation of inorganic contaminants. *Biotechnology & Bioengineering* 58:231-239.

Khetan, A., Malmberg, L.H., Sherman, D.H., W.S. Hu. 1996. Metabolic engineering of cephalosporin biosynthesis in *Streptomyces clavuligerus*. *Annals of the New York Academy of Sciences*, 782:17-24.

Koffas, M., Roberge, C., Lee, K., Stephanopoulos, G. 1999. Metabolic engineering. *Annual Review of Biomedical Engineering*, 1: 535 -557.

Lessard, P. 1996. Metabolic engineering, the concept coalesces. Nature Biotechnology 14: 1654-1655.

Lu, J. L., Liao, J.C. 1997. Metabolic engineering and control analysis for production of aromatics-role of transaldolase. *Biotechnology & Bioengineering*, 53: 132-137.

Martin, V.J.J., Pitera, D.J., Withers, S.T., Newman, J.D., J.D. Keasling. 2003. Engineering a mevalonate pathway in *E. coli* for production of terpenoids. *Nature Biotechnology*, 21: 796 – 802.

Misawa N., Shimada H. 1998. Metabolic engineering for the production of carotenoids in non-carotenogenic bacteria and yeasts. *Journal of Biotechnology* 59:169-181.

Nielsen, J. 1998. Metabolic engineering: Techniques for analysis of targets for genetic manipulations. *Biotechnology and Bioengineering*, 58:125-132.

Pageni, B., Tae-Jin Oh, Lee, H., J. Sohng. 2008. Metabolic engineering of noviose: heterologous expression of novWUS and generation of a new hybrid antibiotic, noviosylated 10-deoxymethylnolide/narbonolide, from *Streptomyces venezuelae* YJ003-OTBP1. Biotechnology letters.

Panke, S., Sanchezromero, J. M. 1998. Engineering of quasi-natural *Pseudomonas putida* strains for toluene metabolism through an ortho-cleavage degradation pathway. *Applied and Environmental Microbiology*, 64: 748-751.

Paradkar, A.S., Mosher, R.H., Anders, C., Griffin, A., Griffin, J., Hughes, C., Greaves, P., Barton, B., S.E. Jensen. 2001. Applications of gene replacement technology to

Streptomyces clavuligerus strain development for clavulanic acid production. *Applied and Environmental Microbiology*, 67(5):2292-7.

Reeves, A.R., Cernota, W.H., Brikun, I.A., Wesley, R.K., J.M. Weber. 2004. Engineering precursor flow for increased erythromycin production in *Aeromicrobium erythreum*. *Metabolic engineering*, 6(4):300-12.

Roy, S.K., Chiba, Y., Jigami. Y. 2000. Production of therapeutic glycoproteins through the engineering of glycosylation pathway in yeast. Biotechnol. *Bioprocess Eng*. 5: 219-226.

Sharpe, P.L., Dicosimo, D., Bosak, M.D., Knoke, K., Tao, L., Cheng, Q., Rick, W. Y. 2007. Use of transposon promoter probe vectors in the metabolic engineering of the obligate methanotroph *Methylomonas sp*. Strain 16a for enhanced C40 carotenoid synthesis. *Applied and environmental microbiology*, 73(6):1721-8.

Shimada, H., Kondo, K., Fraser, P. D., Miura, Y., Saito, T., Misawa, N. 1998. Increased carotenoid production by the food yeast *Candida utilis* through metabolic engineering of the isoprenoid pathway. *Applied Microbiology and Biotechnology*, 64: 2676-2680.

Small, J.R., Kacser, H. 1993a. Responses of metabolic systems to large changes in enzyme activities and effectors. 1 The linear treatment of unbranched chains. *European Journal of Biochemistry* 213:613-624.

Small, J.R., Kacser, H. 1993b. Responses of metabolic systems to large changes in enzyme activities and effectors. 2 The linear treatment of branched pathways and metabolite concentrations. Assessment of the general non-linear case. *European Journal of Biochemistry* 213:625-640.

Smirnoff, N. 1998. Plant resistance to environmental stress. *Current Opinion in Biotechnology* 9:214-219.

Sohn, Y.S, Nam, D.H., D.D. Ryu. 2001. Biosynthetic pathway of cephabacins in *Lysobacter lactamgenus*: molecular and biochemical characterization of the upstream region of the gene clusters for engineering of novel antibiotics. *Metabolic engineering*, 3(4):380-92.

Sprenger, G. A. 1996. Carbohydrate metabolism in *Zymomonas mobilis* - a catabolic highway with some scenic routes. *FEMS Microbiology Letters*, 145: 301-307.

Stephanopoulos, G. 1999. Metabolic fluxes and metabolic engineering. *Metabolic Engineering* 1:1-11.

Stephanopoulos, G., Sinskey, A. J. 1993. Metabolic engineering methodologies and future prospects. *Trends in Biotechnology* 11:392-396.

Thykaer, J., J. Nielsen. 2003. Metabolic engineering of beta-lactam production. Metabolic *Engineering*, 5(1):56-69.

Warner, T.G. 1999. Enhancing therapeutic glycoprotein production in Chinese hamster ovary cells by metabolic engineering endogenous gene control with antisense DNA and gene targeting. *Glycobiology*, 9(9): 841-850.

Wurm, F.M. 2004. Production of recombinant protein therapeutics in cultivated mammalian cells. *Nature Biotechnology*, 22(11): 1393-1398.

Xu, B.W., Wild. J.R., Kenerley, C.M. 1996. Enhanced expression of a bacterial gene for pesticide degradation in a common soil fungus. *Journal of Fermentation & Bioengineering* 81:473-481.

□□□

20

The Role of High Performance Liquid Chromatography in Herbal Research

J. Dharuman

CONTENTS

1.1 Introduction 308

1.2 Types of High Performance Liquid Chromatography 308

1.2.1 Qualitative HPLC 308

1.2.2 Quantitative HPLC 312

1.2.3 Preparative HPLC 313

1.3 Conclusion 314

References 314

1.1 Introduction

In the year 1906 Russian botanist Tswett invented column chromatography for the purpose of investigation of a plant pigment then in the year 1931 Kuhn and Lederer renewed the interest of use of column chromatography (Apers *et al.,* 2003). This technique was redeveloped by Tiselius in 1940 which could be used to determine the number of components in a mixture. The technique of high performance liquid chromatography was developed in the late 1960s and early 1970s from a knowledge of the theoretical principles that already had been established from column chromatography.

High performance liquid chromatography has taken a pivotal growth and gained prominence over the other analytical instruments in the separation and analysis of compounds of all origin. Chromatography is essential to herbal and phytochemistry research and is the key to obtaining pure compounds for structure elucidation for pharmacological testing or for development into therapeuticals. It also plays a fundamental role as an analytical technique for quality control and standardisation of herbal products and phytotherapeuticals.

1.2 Types of High Performance Liquid Chromatography

The technique of High Performance Liquid Chromatography can be categorized into 3 types according to the domains where it is applied mainly (Beckett and Stanlake, 1998),

- Qualitative HPLC (Finger printing)
- Quantitative HPLC (Using reference standard or markers)
- Preparative HPLC (Separation of mixtures into its individual components)

1.2.1 Qualitative HPLC

The use of chromatographic techniques in the field of herbal research is picking up fast nowadays, because of the safety and efficacy of herbal drugs and nutraceuticals. Qualitative HPLC technique is used in various aspects of herbal drug standardization.

The modern day's research focuses on establishing scientific evidence for the efficacy of herbs commonly prescribed by medicine practitioners for the treatment of diseases like diabetes, hypertension, asthma, cancer and infectious diseases (Calixto, 2000). The areas of study in which the usage of High Performance Liquid Chromatography (HPLC) and related chromatographic techniques includes,

- Chemical characterization of phytochemicals extracted from herbs.
- Study of the effect of herbal extracts on mammalian cells.

- Quantitative assessment of the cell viability and apoptosis (programmed cell death) for both cancer and nonmalignant cells in the presence and absence of herbal extracts.
- Study of protein profile of mammalian cells which have been treated with herbal drug formulations.

World Health Organization (WHO) has laid guidelines for the standardization of herbal formulations in which the use of HPLC and allied technique is evident (Singh *et al.*, 2007). The prime domains in which chromatographic techniques are mainly used includes,

- Quality control of crude drug materials, plant preparations and finished products takes major utility of HPLC and HPTLC techniques.
- Stability assessment and shelf life.
- Safety assessment.
- Standardisation of bioactive extracts on the basis of active principles or major compounds.
- Qualitative and semi quantitative information about the main active constituents present in the crude drug as chemical markers.
- The quality of the drug can also be assessed on the basis of the chromatographic fingerprint.
- Determination of heavy metals - viz., cadmium, lead, arsenic, etc.
- Pesticide residue – **WHO** and **FAO (Food and Agricultural Organization)** set limits of pesticides, which are usually present in the herbs. These pesticides are mixed with the herbs during the time of cultivation. Mainly pesticides like DDT, BHC, toxaphene, aldrin cause serious side-effects in human beings if the crude drugs are mixed with these agents. The limits of presence of these pesticides can be measured mainly by HPLC and HPLC with Mass detection techniques.
- Microbial contamination – usually medicinal plants containing bacteria and molds are coming from soil and atmosphere. Analysis of the limits of *E. coli* and molds can be achieved by chromatographic techniques.
- The substance known as afflatoxins will produce serious side-effects if consumed along with the crude drugs. The presence or quantization of afflotoxins can be very well established by HPLC and LC/MS techniques.
- Chemical and instrumental analyses are routinely used for analyzing synthetic drugs to confirm its authenticity. In the case of herbal drugs, however the scene is different especially for polyherbal formulations,

there is no chemical or analytical methods available. In the case of herbal drugs, the quality of raw materials and products can be furnished by Chromatographic analysis viz., HPLC and LC/MS, etc.

Modern herbal ayurvedic monographs

- In the modern herbal ayurvedic monographs the standardization parameters are discussed in a comprehensive way. According to the modern ayurvedic monograph the quality control protocols includes the evaluation of foreign matter and foreign minerals which could be possible with the use of HPLC technique.
- The analytical tool mainly used for the identification of herbal drug material is chromatographic finger prints.
- Slimming activity of herbal medicines in human.
- Determination of toxic activity of herbal remedies in human body.
- Therapeutic drug monitoring (blood level concentration) of herbal constituents in human body

The modern day's research exploits the use of HPLC and allied techniques to study the most challenging tasks. Interestingly the incorporation of an herbal extract with antioxidant properties resulted in more pronounced cell death in cancer cells compared with non-malignant cells which was studied with the help of HPLC technique.

The presence of aflatoxins can be determined by chromatographic methods using standard aflatoxins B_1, B_2, G_1, G_2 mixtures. Aflatoxin is a product of the microbial strain *Aspergillus flavus*.

The identification and quantification of pesticides and insecticides present in the crude drugs can be easily studied by HPLC technique.

Chromatographic fingerprinting as a means of quality control

The method of Chromatographic fingerprinting has proved the rationality of the basic concepts of traditional systems of medicines. The increasing utility of herbal medicines has necessitated the need of having a tool for better understanding of the rationality of the traditional systems of medicines.

Different steps of operation of the Method of Fingerprinting

While analyzing the traditional medicines we need to have a tool, which will use the terminology of the traditional philosophy and will be able to understand by modern generations. In the present method of Chromatographic finger printing the High performance Liquid Chromatography is used (Hildebert Wagner, 1999). In the conventional UV – Visible detector monitoring system, the samples at one or two selected wave lengths has a limitation of not being able to present the qualitative and quantitative information about some molecules present in the sample because of poor absorptivity in the selected wavelengths. The photodiode array detector will be scanning each of the ingredients eluted from the column from 200-800 nm, covering the entire range of UV visible electro magnetic radiation. This over comes the disadvantage of single wavelength analysis.

All molecules present in a sample are separated over the HPLC column and arranged in a specific order of polarity. In the present method of fingerprinting out of the four windows of data display of a PDA detector the contour and 3-D chromatograms were used as data manipulators. The contour chromatogram will be displaying the retention time on X-axis and UV spectrum on Y-axis. The 3-D chromatogram will have the quantitative values of the same parameters. In the 3-D chromatograms the X-axis of retention time shows the polarity of the molecules from high polar to non-polar. The Y-axis shows the UV spectrum of each molecule from 200-800nm, based on the conjugative properties of the molecules. After arraying all molecules in a specific order of polarity they were correlated with the traditional properties.

There are different stages in which finger printing is done. In the first step the sample is extracted into solvent and subjected for Chromatographic fingerprint analysis. Secondly the chromatogram is developed using the method and the contour and 3- D chromatograms are generated. The third step in the fingerprint is interpretation related to different efficacies of the molecules. The fingerprint has been divided into different therapeutic zones qualitatively and quantitatively based on which the interpretations will be made. Based on the retention time, polarity, spectrum and the total pattern of the constituents will be interpreted towards the efficacy of the each of the molecules and medicine as a whole.

The modern time analysis uses speculative tools for the fingerprinting, like the HERBOPRINT ™ which is useful in herbal analysis back dropped by HPLC data (Lazarowych and Pekos, 1998). This helps to standardize the medicines based on the traditional principles and properties as mentioned in the traditional texts.

When analyzing drug molecules the chromatographer needs to know some basic properties like functional groups, solubility and UV Visible spectrum of the analyte to develop a chromatographic method for analysis. Thus interpretation of the fingerprint having molecules at different retention times and their UV spectrum properties along with polarity is providing information about the chemical and therapeutic efficacy of different herbal drug molecules of any origin, whether modern or traditional.

Over the past decade, herbal manufacturers have increasingly focused on the issue of quality, and standardization in particular, as the key feature for generic herbal products (Lloyd, 1997). Marketing campaigns have emphasized the use of modern scientific techniques to ensure that products contain specified amounts of marker compounds, asserting that this process of standardization guarantees identity, safety, efficacy and stability.

The evaluation of a product in its entirety, so-called "fingerprinting" can be determined by appropriate methods, which may include HPLC-UV(DAD), HPLC-ELSD, HPLC-MS, or a combination of these techniques. The usefulness of fingerprinting is especially evident in the quality control of blackcohosh (*Actaea racemosa* L.), which has been reportedly adulterated with other *Actaea*species, mostly imported from China. The majority of the triterpene glycosides of *A. racemosa* occur in several other *Actaea* species.

1.2.2 Quantitative HPLC

HPLC technique for the quantization of constituents of herbal product is considered to be the meticulous analytical technique because of its rapidity, accuracy and sensitivity. Constituents of a herbal product are quantified using reference standard or marker compounds. The main advantage of HPLC for quantization is simultaneous estimation of constituents. The extent of estimation is so robust that 23 amino acids of an plant extract can be seperatd and quantified within a time span of 20 minutes. Three types of quantization are possible with HPLC,

- Single standard method
- Multi standard method
- Standard addition method

Single standard method

This method uses a known concentration of reference standard or marker. Concentration of a component is calculated using the peak area generated for the reference standard (marker)compound and the plant constituent.

Multi standard method

Here a linearly increasing concentration (within Beer's Law limit) of standard solutions are used and a calibration curve (standard curve) is prepared by plotting the peak area against concentration. Concentration of component in a extract can be measured using graphical method or statistically by Regression analysis.

Standard addition method

This method is used when the component of an extract is presented in a lowest concentration, that the detector fails to sense the presence. A known concentration of preanalysed reference standard is added to the extract. The herbal component is estimated using the combined response produced by the sample and standard.

1.2.3 Preparative HPLC

Chromatography is the lynchpin of phytochemistry and is the key to obtaining pure compounds for structure elucidation, for pharmacological testing or for development into therapeuticals (Sapna Shrikumar *et al.*, 0000). It also plays a fundamental role as an analytical technique for quality control and standardization of phytotherapeuticals.

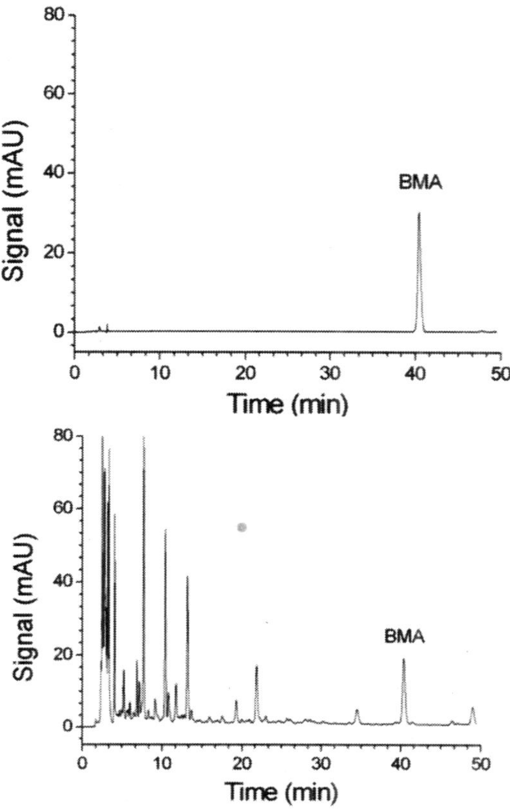

Ref: ***Ying Xie, Hua Zhou, Yuen Fan Wong, Zhongqiu Liu, Hongxi Xu, Zhihong Jiang, and Liang Liu,*** An optimized high-performance liquid chromatography (HPLC) method for benzoylmesaconine determination in *Radix Aconiti Lateralis Preparata* (*Fuzi*, aconite roots) and its products. Chinese Medicine. **3(6)**.2008.

The herbal samples will be extracted into different aqueous and non-aqueous solvents. The extracts will be subjected to preparative chromatographic separation on a large scale and pure compounds will be isolated and the single molecules will be analyzed using different analytical instruments for structure establishment. Then the possible activity is studied by modeling and *invitro –in vivo* studies. The clinical drug discovery process starts now.

1.3 Conclusion

The subject of herbal drug standardization is massively wide and deep. For the purpose of research work on standardization of herbal formulations and nutraceuticals a profound use of Chromatographic methods are used.

India can emerge as the major country and play the lead role in production of standardized, therapeutically effective herbal formulations. India needs to explore the medicinally important plants. This can be achieved only if the herbal products are evaluated and analyzed using sophisticated modern techniques of standardization such as HPLC and LC/MS methods.

References

Apers.S, Van Meenen.E, Pieters.L. Vlietinck.A, 2003. Quality control of liquid herbal drug preparations, *Journal of Pharmaceutical and Biomedical Analysis*, 33, (4),529-537

Beckett. A.H, and Stanlake, 1998. Practical pharmaceutical chemistry, 4th Edition, 85.

Calixto. J.B, 2000. Efficacy, safety, quality control, marketing and regulatory guidelines for herbal medicines (phytotherapeutic agents), *Brazilian Journal of Medical and Biological Research*, 33(2) 179-189.

D. P. Singh, R. Govindarajan, A. K. S. Rawat, 2007. High-performance liquid chromatography as a tool for the chemical standardisation of Triphala -an Ayurvedic formulation. *Phytochemical Analysis*,19(2),164-168.

Hildebert Wagner, 1999. Phytomedicine Research in Germany, *Environmental Health Perspectives*,107(10)779-781, Brogan & Partners.

Lazarowych .N. J, Pekos.P, 1998 . Use of fingerprinting and marker compounds for identification and standardization of botanical drugs, Strategies for applying pharmaceutical HPLC analysis to herbal products. *Drug information journal*, 32, (2),363-392

Lloyd R. Snyder, Joseph L. Glajch, and Joseph J. Kirkland, 1997. Practical HPLC Method Development, 2nd Edition.222.

Sapna Shrikumar, M. Uma Maheswari, A. Suganthi, T.K. Ravi. WHO guidelines for herbal drug standardisation, Pharmainfo.net.

□□□

Medicinal and Neutraceutical Properties of Some less Explored Allium Species of Uttarakhand Himalayas : A Potential Future Crop

Hemant Kr. Pandey, M.S.M. Rawat and Zakwan Ahmed

CONTENTS

1.1	Introduction	316
1.2	Allium Chemistry	319
1.3	Medicinal Properties of Allium Species	320
1.4	Common Medicinal Uses of Allium Species	322
1.5	Evaluation of Neutraceutical Properties of Some Less Explored Allium Species	323
1.6	Nutritional Quality Screening of Allium Species	324
1.7	Macro and Micro Elemental Composition	325
1.8	Amino Acid Contents of Allium Species	326
1.9	Conclusion	327
	References	328

1.1 Introduction

Since time immemorial Himalayas are famous for its rich medicinal and aromatic plant biodiversity. The varying soils, topography, occurrence of different climatic and micro-climatic zones of the region provide very congenial condition to grow a number of therapeutically important medicinal plant species. The Uttarakhand Himalayas have a rich heritage of medicinal and aromatic flora, several herbs possess odoriferous principles and are used as raw material for phyto-chemical, pharmaceuticals, perfumery, food, condiments and flavours based industries (Anonymous, 1980; Chopra *et al.*, 1956). The unscientific, over and irregular exploitation of medicinal plants from its natural habitat has resulted to very fast depletion, as well as extinction of some therapeutically important medicinal species. So development of agro-technology of medicinal plants and their commercial scale cultivation is need of the hour.

The total geographical area of Indian Himalayas is about 5, 94,427 sq. km., which is about 18% of the total area of the country. The length of Indian Himalayas is about 2,400 km and width is around 240 – 320 km.

The major geographic divisions of Himalayas are:

(i) Eastern Himalaya (Assam, Darjeeling, Sikkim Himalayas)

(ii) Central Himalaya (Nepal Himalayas)

(iii) Western Himalaya (Kashmir, Himachal Pradesh, Uttarakhand Himalayas)

India, with its diverse agro-climatic conditions and regional topography, has been considered as the treasure house or botanical garden of plant genetic resources. Hence, India is recognized as one of the world's top 12 mega diversity nations. Our herbal wealth constitutes more than 8,000 species and accounts for around 50 % of all higher flowering plant species of the world. However, available information shows that 1,800 species are used in classical Indian systems of medicines. Ayurveda uses 1,200, Siddha - 900, Unani - 700, Amchi - 600, Tibetan - 450 (Suresh *et al.*, 2006). Although there are around 8,000 medicinal plant species used by different communities in India across different ecosystems, only around 10% of them (880 species) are in active trade. Among these, around 48 species are exported in the form of raw drugs and extracts, while around 42 species are imported. (Vijayalatha, 2004; Singh, 2005).

Allium is an important genus of family Amarylliadaceae, has 700 species. Confined chiefly to the northern temperate and alpine zones of the world, i.e. Europe, northern, middle, south-east Asia, north-west America and the Altai mountains (Buijsen, 1990). Hooker (1892), in 'Flora of British India' had recorded 27 *Allium* species; Blatter (1927-29) in his work 'Beautiful Flowers of Kashmir' reported 12 species and Stearn (1947) reported 32 *Allium* species from India. In addition to the above major contributions, Collet (1921) had described four species of this genus from Shimla, and Rau (1975) mentioned eleven species growing wild in Western Himalaya. According to Dhar and Kachroo (1983), 19 species of *Allium* are found in Kashmir Himalaya, 16 of these reach the alpine region and six are endemic to this region. These six species are *Allium clarkei* HK. (Kashmir), *A. loratum* Baker (Ladakh), *A. consanguineum* Kunth, *A. stracheyi* Baker, *A. thomsonii* Baker and *A. przewalskianum* Regel (Pakistan and Kashmir).

Genetic resource conservation is concerned with the range of variation within each crop plus those wild species which may be of value either now or in future. Scientific interest in the collection of wild *Allium* texa has been based on taxonomy and cytology. In relation to wild *Allium* species the IBPGR have recently initiated collecting programs in Kashmir.

In India the north - western Himalaya and adjoining areas are rich in species of *Allium* as compared to the rest of these ranges. Although, more than 40 of *Allium* have been reported from Himalayan region by various worker, out of which 30 species have been reported from temperate areas of northern Himalaya. Occasionally a few of them are cultivated locally (Negi and Pant, 1990; 1992; Negi and Gaur, 1991). Some wild *Allium* species of Uttarakhand Himalayan region are:

1. *Allium carolinianum* DC
2. *Allium chinense* G. Don.
3. *Allium humile* Kunth
4. *Allium consanguinenum* Kunth
5. *Allium stracheyi* Baker
6. *Allium wallichii*
7. *A llium przewalskianum*
8. *Allium semonovii* Regal
9. *Allium victorialis* L.
10. *Allium cernuum* Roth.
11. *Allium griffithianum* Boiss
12. *Allium clarkei* H. K.
13. *Allium loratum*
14. *Allium proliferum*
15. *Allium auriculatum* Kunth
16. *Allium rubellum* M. Bieb.

In Uttarakhand Himalaya *Allium* is one of the important genus. It is widely distributed in the temperate and alpine regions (Negi, 1986; 1987; Negi *et al.*, 1985; 1991). The wild *Allium* species have been utilized as spice and condiments, locally called Jamboo, Doona, Dunn, Dhun, Pharan etc. *Allium* species are aromatic green used in pulverized state, primarily for seasoning or garnishing food and beverages, characterized by pungency, strong odour, sweet and bitter taste. For condiments point of view, bulbs, leaves and flowers are harvested and used in dried form. Therefore, there is tremendous pressure on these herbs. Due to over-exploitation their rate of depletion is very fast.

Although cultivation of *Allium* species is not much popularized in the Himalayan region, *Allium cepa* L. and *Allium sativum* L. are grown on a limited scale in this region. In addition to that *A. proliferum*, commonly called as Faran, *A. ampeloprasum, A. tuberosum* and *A. ascalonicum* are also being grown in small quantity for their own consumption, mostly in the kitchen gardens.

All wild species of *Allium* found in this region are consumed by the locals whenever, these are found in abundance. The plant parts like leaves, bulbs, scales, flowers, seeds etc. are collected for consumption. Mostly these parts are used as condiments for seasoning various dishes either in fresh or dried form. In some cases the above ground parts are collected, crushed and made into small cakes which are dried in the sun. These cakes and other dried parts of wild *Allium* species are used during off season. Their use is more prevalent in the places which are either not well connected with the main market or remain cut off from rest of the country during some particular seasons, like long winter in Laddakh.

Allium species are important in the sense that all part of the plants are edible. It is a bulbous herb with characteristic odour. Leaves variable, fistular, flat, filliform or liner. Flowers white, yellow, red, rosy, purple, golden yellow and blue, in capital umbel. On the basis of inflorescence characteristics, *Allium* is now placed in the commonly known Daffodil family Amaryllidaceae.

The natural season of wild *Allium* species is from April to November. The plant parts such as leaves, flowers and bulbs are collected from their wild sources and supplied by the local inhabitants of high altitude areas. But in recent years dried leaves of *Allium auriculatum* Kunth, *Allium humile* Kunth, *Allium rubellum* M. Bieb have been in more demand (Negi *et al.*, 1993). Such valuable and useful plants need to be scientifically evaluated to search promising ones for large scale cultivation under standardized agro-techniques.

Over exploitation of the wild species accompanied by a lot of biotic interference has led to considerable shrinking of the populations of various species at different places. The wild *Allium* species are not safe as the shepherds collect them for their folks, while taking their herds to high altitude for grazing.

Since there is no check on these activities, it is leading to erosion of the germplasm of *Allium* species from this region. There is every possibility that some of these important species might get totally wiped -out in the near future.

1.2 *Allium* Chemistry

The distinctive flavour characteristics and purported therapeutic benefits of tissues of plants from the genus *Allium* are conferred by assay of organosulfur compounds generated from non-protein sulfur amino-acid precursors, the S-alk(en)yl L-cystein sulfoxide(ACSO)[1]also called as alliin(Lancaster and Boland, 1990). The profile and quantities of these ACSO are responsible for the nature and intensity of flavours derived from fresh and processed *Allium* tissues. For example the methyl-(MCSO), 1-Propenyl-(1-PECSO) and 2-propenyl-(allyl,2-PECSO) cystein sulfoxide are responsible for yielding character impact compounds of freshly comminuted cabbage leaves, onion bulbs and garlic cloves respectively (Freeman *et al.*, 1976 a, b). On crushing the bulb, the enzyme allinase breaks down alliin to produce allicin, of which the principal ingredient is the odoriferous diallyl disulfide (DADS). Later, allicin rearranges to di allyl disulfide (DADS) and thiosulfinate.

Enzymatic Conversion of Alliin into Allicin

$$\underset{\text{Alliin}}{2C_3H_5\text{-}\overset{\overset{O}{\uparrow}}{S}\text{-}CH_2\text{-}CH(NH_2)COOH} \xrightarrow{\text{Allinase}} \underset{\text{Allicin}}{C_3H_5\text{-}\overset{\overset{O}{\uparrow}}{S}\text{-}S\text{-}C_3H_5} + \underset{\text{Ammonia}}{2NH_3} + \underset{\text{Pyruvic acid}}{2CH_3COCOOH}$$

$$\underset{\text{Allicin}}{2\ C_3H_5\text{-}\overset{\overset{O}{\uparrow}}{S}\text{-}\ S\text{-}\ C_3H_5} \longrightarrow \underset{\text{DADS}}{C_3H_5\text{-}S\text{-}\ S\text{-}\ C_3H_5} + \underset{\text{Thiosulfonate of DADS}}{C_3H_5\text{-}S\text{-}\ \overset{\overset{O}{\uparrow}}{S}\text{-}\ C_3H_5}$$

A distinguishing feature of *Allium* species is the metabolic net-work of sulfur compounds. One class of sulfur compounds, the S-alk(en)yl cysteine sulfoxides, gives rise to the numerous volatile sulfur compounds like allicin (diallyl thiosulfinate),which have caused *Allium* species to be valued as herbal medicines and as food. Intact *Allium* cells have no odour, but when cells are disrupted the enzyme allinase hydrolyzes the S-alk(en)yl cysteine sulfoxide to produce pyruvate, ammonia and many volatile sulfur compounds associated with flavour and odour. The non protein sulfur amino acids collectively referred as S-alk(en)yl cysteine sulfoxide. There are four common ACSO which occur naturally:

$$\underset{\displaystyle \text{S-Methyl-L-cysteine sulfoxide}}{\overset{\displaystyle O \atop \uparrow}{CH_3\text{-}S\text{-}CH_2\text{-}CH(NH_2)COOH}}$$

$$\underset{\displaystyle \text{S-Propyl-L-cysteine sulfoxide}}{\overset{\displaystyle O \atop \uparrow}{CH_3\text{-}CH_2\text{-}CH_2\text{-}S\text{-}CH_2\text{-}CH(NH_2)COOH}}$$

$$\underset{\displaystyle \text{S-(1-propenyl)-L-cysteine sulfoxide}}{\overset{\displaystyle O \atop \uparrow}{CH_3\text{-}CH{=}CH\text{-}S\text{-}CH_2\text{-}CH(NH_2)COOH}}$$

$$\underset{\displaystyle \text{S-(2-propenyl)-L-cysteine sulfoxide}}{\overset{\displaystyle O \atop \uparrow}{CH_3\text{-}CH{=}CH\text{-}S\text{-}CH_2\text{-}H(NH_2)COOH}}$$

1.3 Medicinal Properties of *Allium* species

Stajner *et al.* (2003) investigated anti-oxidant properties of bulbs, leaves and stalks of leek (*Allium sphaerocephalon* L.) and examined anti-oxidant enzyme superoxide dismutase, catalase, peroxidase, glutathion peroxidase, the quantities of malonyldialdehyde, superoxide and hydroxyl radicals, reduced glutathione, flavonoids, chlorophyll, carotenoids and vitamin C contents were examined. The extracts of all plant organs exhibited anti-oxidant activity. The highest anti-oxidant activity was observed in the leaves.

Huang *et al.* (2004) examined the anti-oxidant and anti-glycative effects of six organosulfur compounds diallyl sulfide, diallyl disulfide, S-allyl cystein, S-ethyl cystein, S-methyl cystein and S-propyl cystein. The results suggested that the use of these organo-sulfur agents could protect partially oxidized & glycated LDL or plasma against further oxidative and glycative deterioration, which might benefit patients with diabetic related vascular diseases.

Higuchi *et al.* (2003) evaluated anti-oxidant activity of sulfur containing compounds in *Allium* species for human low density lipoprotein (LDL) oxidation *in-vitro*. He elucidated that alk(en)yl substitutes (Me, Pr and allyl) and number of sulfur atoms in the compounds are important for oxidative activity.

Lea (1996) explained that *Allium* species are rich source of organosulfur compounds. These compounds inhibit the incident and growth of cancer, epidemiological studies have suggested that consumption of garlic including diallyl sulfides can inhibit induction of cancer in experimental animals.

Ejaz *et al.* (2003) revealed that *Allium sativum* extract is useful in cancer chemoprevention; it is helpful in prevention and treatment of different types of cancer. The main constituent allicin penetrates very rapidly into different compartments of cell and is completely metabolized in the liver. The studies provided compelling evidence that garlic and its organo-allyl sulfur compounds are effective inhibitors of tumor growth.

Pinto and Rivlin (2001) studied that *Allium* derivatives have significant anti-proliferative action on human cancers. Both hormone responsive and hormone unresponsive cell lines respond to these derivatives. The effects shown by *Allium* derivatives include induction of apoptosis, regulation of cell cycle progression and modification of pathways of signal transduction, these derivatives appear to regulate nuclear factors involved in immune function and inflammation as well as in cellular proliferation. He proved that *Allium* derivatives inhibit proliferation of human prostate cancer cell line (LN Cap) and the human breast cancer cell line (MCF-7).

Milner (1996) explained that *Allium* derivatives may involve in inhibiting the formation of DNA adducts with carcinogens. The *Allium* constituents have been shown to inhibit the covalent binding of the carcinogen 7, 12-dimethylbenz [a] anthracene to DNA, an intracellular event that correlates with decreased mutagenesis and carcinogenesis. Singh *et al.* (1996), studied that *Allium* derivatives suppress the growth of oncogene -transformed tumors in mice. Li *et al.* (1995), studied the effect of garlic derivatives on human breast cell lines, MCF-7 and MCF-7^{ras} which retain many features of human breast cancer clinically. In cell culture, *Allium sativum* extracts significantly inhibit anchorage - dependent growth of these cells. Caldes and Precott (1969), experimentally proved that aqueous extract of shallot (*Allium ascalonicum*) bulbs had been found to have significant activity against P388 leukemia in mice.

In the year 2004, Kao identified allergenic proteins in *Allium sativum* and investigated their importance in allergies to other *Allium* species (leek, shallot and onions). A major allergen, alliinlyase was identified by mass spectrometry. Garlic allinase showed strong cross-reactivity with alliinlyase from other *Allium* species. The wide distribution of alliinlyase in *Allium* suggested that it may be a new cross-reactive allergen.

Sovova and Sova (2004) studied pharmaceutical significance of *Allium sativum* L. and its hypolipidemic effects on *in-vitro* and *in-vivo* models. Harris *et al.* (2001) reviewed antimicrobial properties of *Allium species* and reported that not only it is antimicrobial, antiviral, antifungal and antiprotozoal, but it also has beneficial effects on the cardiovascular and immune systems. Benkeblia (2004) studied antimicrobial activity of essential oil extracts of various *Allium* species. The high inhibitory activity was observed in essential oil extracts of garlic. He also reported that where seasoning is desired, essential oil extracts of onions and garlic can be used as natural antimicrobial additives for incorporating in various food products.

Ankari and Mirelman (1999) experimented antimicrobial properties of allicin extracted from *Allium sativum.* Allicin in its pure form was found to exhibit (i) antibacterial activity against a wide range of gram negative and gram positive bacteria, including multi drug resistant entero-toxicogenic strains of *Escherichia roll.* (ii) Antifungal activity, particularly against *Candida albicans;* (iii) antiparasitic activity, including some major human intestinal protozoan parasites such as *Entamoeba histolytica* and *Giardia lamblia* and antiviral activity. Liakopoulou and Sinakes (1992), isolated a low molecular weight peptide from *Allium porrum* having strong inhibitory activity on platelet aggregation induced by collagen or ADP.

Current studies indicate that to have an effective garlic product you must deliver a minimum equivalence of four gram of fresh garlic, or a "total allicin potential" (TAPO) of at least 4000 microgram per day.

1.4 Common Medicinal Uses of *Allium* Species

Physiological Effects

- Antibiotic
- Antifungal
- *Anthelmintic*
- Antiviral
- Antioxidant
- Carminative
- Diuretic
- Expectorant
- Lipid lowering effect
- Inhibition of platelet aggregation
- Lowers blood pressure
- General immuno- stimulation

Traditional Uses

Internally

- Indigestion
- Treatment of parasites
- Poison antidote
- Eye diseases
- Menstrual abnormalities
- Laxative
- Edemic conditions
- Infectious diseases
- Childbirth aid
- Cough and cold
- Jaundice
- Asthmatic conditions
- Hemorrhoids
- Kidney stones
- Dysentery
- Hysterical/ Spastic conditions
- Abnormal cramping
- Aphrodisiac

Externally

- Skin Rashes
- Abscesses
- Scabies
- Dentistry
- Infected wounds
- Lice and dandruff control

Modern Uses

- Cardiovascular conditions
- High cholesterol
- High blood pressure
- Atherosclerosis
- Cough / Bronchitis
- Fever and cold
- Yeast infection

Dosage range

Equivalent of 4 g fresh garlic bulbs per day or 4000 micrograms of total allicin potential.

1.5 Evaluation of Nutraceutical Properties of Some Less Explored *Allium* Species

For detailed evaluation of nutraceutical properties following four economically important and less explored *Allium* species have been taken for screening of bio-chemical traits, mineral composition and quantitative estimation of amino acids.

1. *Allium auriculatum* Kunth.
2. *Al lium ampeloprasum* L.
3. *Allium ascalonicum* L.
4. *Allium rubellum* M. Bieb.

The *Allium* species namely *Allium auriculatum* Kunth, *Allium ampeloprasum, Allium ascalonicum* and *Allium rubellum* have been collected form Auli Joshimath (Chamoli) and Harshil (Uttarkashi) regions. The plants were identified at B. S. I., Dehradun. The plant samples were collected in the months of April-May, at the stage of full maturity of plants for evaluation their nutraceuticals properties.

To evaluate the nutraceutical properties the aerial and underground parts of *Allium* species were screened for moisture, total minerals, crude fat, thiamin, riboflavin and sulfur by AOAC methods (AOAC, 1985; 1990), total carbohydrates by Phenol sulphuric acid method (Dubois *et al.*, 1956), protein by Snell and Snell, 1995, dietary fiber according to Maynnard (1970), vitamin C by Ranganna, 1976 method, phenolics as per Malik and Singh1980 and tannins by Schanderi, 1970. The chlorophyll a & b were estimated by Withan *et al.*, 1971 and carotenoids as per the Dietmar *et al.*, 2002 method.

The screening of macro and micro elements were carried out by Flame Photometer and Atomic absorption spectrophotometer as suggested by Issac (1980). Quantitative screening of amino acids was carried out by water's quaternary gradient HPLC, PICO-TAG method (Bidlingmeyer *et al.*, 1984) an integrated technique for pre column derivatization of sample with phenyisothiocynate (PITC). Identification of the amino acids in the samples was carried out by comparison with the retention times of the standards.

1.6 Nutritional Quality Screening of *Allium* Species

The *Allium* sp. taken in the study were screened for different nutraceutical traits viz. moisture, total carbohydrates, minerals, protein, crude fat, dietary fiber, vitamin B_1, vitamin B_2, vitamin C, α-carotene, β-carotene, leutin, phenolics, tannins, chlorophyll a and b contents to evaluate nutritional quality of leaves as well as bulbs of plants individually. Among these *Allium* species, moisture content in leaves samples varied from 85. 10% -89.90% and in bulb samples it ranged 63.82% -85.59%. In the leaves of *Allium ascalonicum* maximum moisture (89.80%) was found. While, in bulb samples maximum moisture (85.59%) was observed in *Allium ampeloprasum.*

Total carbohydrate contents ranged from 66.81-75.31% and 75.55-85. 57% in leaves and bulb samples respectively. Maximum carbohydrate contents (75.31%) were recorded in *Allium rubellum* leaves followed by (71.85%) in *A. ampeloprasum.* While, in bulbs of *A. ampeloprasum* maximum carbohydrates (85.57%) were observed. Minerals varied from 6.92 -11.60% in leaves samples and 2.41-3.46% in bulb samples. In the leaves and bulbs of *A. ascalonicum* highest minerals were found (11.60%) and (3.46%) respectively. Protein content ranged from 0.63-5.12% in leaves samples and 1.40-3.72% in bulbs samples. Maximum protein content (5.12%) and (3.32%) were recorded in leaves and bulbs sample of *A. ampeloprasum.*

Crude fat content varied from 2.34-3.53% in leaves and 1.52-2.37% in bulb samples. Highest fat content (3.53%) was observed in leaves of *A. auriculatum,* while, in bulbs of *A. ascalonicum* maximum fat content (2.37%) was found. Dietary fiber varied from 8.73-14.57% in leaves and 5.72-14.16% in bulbs sample. The *A. auriculatum* leaves and bulb samples recorded maximum dietary fiber content (14.57%) and (14.16%) respectively. Vitamin B_1 was found highest (0.73 mg $100g^{-1}$) and (0.31 mg $100^{-1}g$) in leaves and bulbs samples of *A. ampeloprasum.* Highest concentration of vitamin B_2 (0.84 mg $100g^{-1}$) and vitamin C (45.04 mg 100^{-1} g) contents were recorded in *A. rubellum* leaves. In the bulb sample of *A. ascalonicum* maximum vitamin B_2 (0. 21mg) and vitamin C (24.00 mg) were recorded. α- Carotene content ranged from 1.70-26.01 mg $100g^{-1}$ in leaves and 0.25-1.36 mg $100g^{-1}$ in bulb samples. *A. auriculatum* exhibited maximum α-carotene in leaves (26.01 mg) as well as in bulbs (1.10 mg). β-Carotene varied from 104.62-230.65 mg 100g^{-1} in leave samples and 1.64-9.39 mg 100g^{-1} in bulb samples. Highest quantity of β-carotene in leaves (230.65 mg) and in bulbs (9.39 mg) was found in *A. ascalonicum*, followed by *A. auriculatum* (120.05 mg) and (3.89 mg) respectively.

Phenolics ranged from 0.92-1.048% and 0.38-0.68% in leaves and bulb samples, respectively. In *A. ascalonicum* maximum (1.048%) phenolics in leaves as well as in bulbs were recorded. Tannins varied from 0.94 -1.62% in leaves and 0.36-0.82% in bulb samples. Maximum quantity of tannins (1.62%) was recorded in leaves of *A. auriculatum* followed by (0.82%) in the bulbs of *A. ascalonicum*. Chlorophyll a and b contents ranged from 3.88-4.83 mg and 13.23-18.51 mg 100g^{-1} respectively. Highest chlorophyll a (4. 83 mg) was found in *A. ampeloprasum* followed by (4.68 mg) in *A. rubellum*. Whereas, maximum quantity of chlorophyll b (18.51 mg) was exhibited in the leaves of *A. ascalonicum* followed by (18.47mg) in to *A. ampeloprasum*. Leutin content varied from 0.88 -11.95 mg and 0.11-0.22 mg 100 gm^{-1} in leaves and bulbs of *Allium* samples respectively. Maximum leutin quantity (11.95 mg) was observed in the leaves of *A. ampeloprasum* followed by (3.63 mg) in *A. rubellum*. While the bulbs of *A. ascalonicum* recorded highest concentration (0.22 mg) of leutin.

1.7 Macro and Micro Elemental Composition

The leaves and bulbs of *Allium* species were also screened for macro and micro elements viz. Na, K, Ca, Li, P, S, SiO_2, Fe, Cu, Zn, Mn and Co. In *Allium* species sodium ranged from 11.01-269.48 mg and 21.00-91.65 mg 100g^{-1} in leaves and bulb samples respectively. Highest concentration of Na in leaves (269.48 mg) and in bulbs (91.65 mg) was recorded in *A. ampeloprasum*. Potassium varied from 1010.80–1808.32 mg and 628.29–1225.99 mg 100g^{-1} in leaves and bulbs samples respectively. Maximum quantity of K (1808.32 mg) was recorded in the leaves of *A. auriculatum* and (1225.99 mg) in the bulbs of *A. rubellum*. Calcium ranged from 467.19–697.06 mg in leaves and 144.10–356.82 mg 100g^{-1} in bulb samples.

A. ascalonicum leaves exhibited maximum quantity (697.06 mg) of Ca, where as, in bulbs highest amount (356.82 mg) of Ca was found in *A. ampeloprasum*. The concentration of lithium ranged 5.10-7.60 mg and 3.04–4.57 mg 100 g^{-1} in leaves and bulb samples. *A. auriculatum* leaves, as well as bulbs recorded highest quantity of lithium (7.60 mg) and (4.57 mg) respectively. Phosphorus varied from 165.00-235.50 mg and 150.83-212.83 mg 100 g^{-1} in leaves and bulb samples. Maximum phosphorus (235.50 mg) was recorded in the leaves of *A. ampeloprasum* and (212.83 mg) in bulbs of *A. rubellum*. Sulfur content ranged from 86.00-228.50 mg in leaves and 62.33-

143.17 mg 100 g^{-1} in bulb samples of *Allium* species. *A. auriculatum* exhibited maximum sulfur (228.50 mg) in leaves followed by (188.00 mg) in *A. ascalonicum.* While, in the bulbs of *A. ascalonicum* highest quantity (143.17 mg) of sulfur was found, second highest quantity (96.77 mg) was recorded in *A. auriculatum.* Silica contents varied from 2.16-4.40% in leaves and 0.42-1.41% in bulb samples. Maximum silica was found in the leaves and bulbs of *A. ascalonicum* 4.40% and 1.41% respectively.

Range of micro-element iron was found 7.98-20.11 mg and 1.37-29.12 mg 100 g^{-1} in leaves and bulb samples respectively. *A. auriculatum* leaves exhibited highest concentration (20.11 mg) of iron. While, in *A. rubellum* bulbs maximum quantity (29.12 mg) of iron was recorded, followed by (19.97 mg) A. *auriculatum.* Copper content varied from 0.98-1.85 mg in leaves and 1.20-2.75 mg 100 g^{-1} in bulb samples. Maximum copper content (1.85 mg) was observed in the leaves of *A. ampeloprasum.* Where as, *A. rubellum* bulb samples exhibited highest quantity (2.75 mg) of copper followed by (1.86 mg) in *A. ampeloprasum.* Zinc ranged from 1.71-4.23 mg and 3.18–6.40 mg 100g^{-1} in leaves and bulb samples respectively. *A. rubellum* leaves recorded maximum amount (4.23 mg) of zinc followed by (4.10%) *A. auriculatum.* While, in bulb sample highest zinc (6.40 mg) was found in *A. auriculatum.* Manganese content varied from 2.63-3.91 mg in leaves and 0.70-1.55 mg 100 g^{-1} in bulb samples. *A. ascalonicum* leaves exhibited highest manganese (3.91 mg) followed by *A. ampeloprasum* (3.40 mg). In bulb samples maximum quantity of manganese (1.55 mg) was found in *A. ampeloprasum.* Cobalt content ranged from 0.00-0.94 mg in leaves and 0.00-1.11 mg 100 g^{-1} in bulb samples. *A. ampeloprasum* exhibited highest cobalt content (0.94 mg) in leaves as well as (1.11 mg) in bulb samples respectively.

1.8 Amino Acid Contents of *Allium* species

The quantitative estimation of amino acid contents was carried out by HPLC. Total seventeen amino acids were detected and the separation of the amino acids in the samples was reasonably resolved. All essential amino acids i.e. methionine, leucine, lysine, phenylalanine, arginine, isoleucine, threonine and valine and nine non-essential amino acids were present in these *Allium* species. The total amino acid content in *A. auriculatum* were 35.36 μ mol/g sample, *A. ampeloprasum* 6.077 μ mol/g, *A. ascalonicum* 16.626 μ mol/g and in *A. rubellum* total 5.24 μ mol. amino acids/g were recorded in dry leaves of the plants. Out of total amino acid content present in these species 16.455, 1.825, 6.996 and 1.523 μ mol amino acids/g were made up of essential amino acids in *A. auriculatum, A. ampeloprasum, A. ascalonicum* and *A. rubellum,* respectively. Schuphan and Schwerdtfeger in the year 1972, reported that arginine and glutamic acid were found in abundant quantity than other amino acids in *Allium cepa.*

The ratio of essential amino acids to total amino acid is 1 : 2.14, i.e. almost half (46.67%)of the total amino acid in *A. auriculatum* consist of essential amino acids. The results also indicated that the *A. auriculatum* is rich in alanine, valine, leucine, proline, isoleucine, arginine, glutamic acid, tyrosine and threonine. In *A. ampeloprasum* the ratio of essential amino acids to total amino acid is 1 : 3.35 i.e. almost one third of total amino acids of plant consist of essential amino acids. This species was found rich in alanine, proline, aspartic acid, valine and glutamic acid. The ratio of essential amino acids to total amino acid in *A. ascalonicum* was found 1:2. 38, slightly lower than half of the total amino acid consists of essential amino acids in the plant. This Allium species was found rich in alanine, valine, proline, arginine, leucine and aspartic acid. In *A. rubellum* the ratio of essential amino acids to total amino acid was found 1 : 3.44. This *Allium* species was rich in alanine, cystein, proline and valine.

It is evident from study that *A. ampeloprasum* contains maximum quantity of carbohydrates, crude protein, Vitamin $B_{1,}$ leutin and minerals like sodium, calcium, phosphorus, copper and manganese and *A. ascalonicum* was found rich in minerals, crude fat, Vitamin B_2, β-carotene, phenolics, tannins and minerals viz calcium and manganese. Hence it can be concluded that *A. ascalonicum* and A. *ampeloprasum* are nutritionally superior to other *Allium species*. While, amino acid contents were found more in *A. auriculatum* and *A. ascalonicum*.

1.9 Conclusion

Thus, the study concludes that the aerial and underground parts of these *Allium* species have excellent nutraceutical properties, since the presence of ample quantity of nutrients, antioxidants, minerals and amino acid contents. The edible parts of these plants contain appreciable amount of carbohydrates, minerals, fat, vitamins, protein and dietary fiber. These plants could also serve as good source of mineral elements, especially elements like potassium, calcium, phosphorus, sulfur, iron, zinc, copper etc. Due to presence of substantial amount of nutrients these *Allium* species can contribute significantly to meet out the nutritional requirement of growing population and should be recommended for more and more use. In future, these plant species could be used not only as condiments for flavouring different food stuffs but also as green vegetables due to their demonstrated good nutritional qualities, although further exploration for bioavailability is still required. Keeping in view, the nutritional richness of these less explored *Allium* species, it should be popularized more and more among people. Therefore, development of standard agronomical practices and their large-scale commercial cultivation is the need of hour.

References

AOAC. 1985. Official methods of analysis of the association of official analytical chemists. 11th edn. Washington. 138.

AOAC. 1990. Official methods of analysis of the association of official analytical chemists. 11th edn. Washington.

Ankari, S. and Mirelman, D. 1999. Antimicrobial properties of allicin from garlic (Review). *Microbes and Infection*, 1: 125-129.

Anonymous. 1980. Nutritive value of Indian foods. Indian Council of Medical Research, New Delhi-4.

Benkeblia, N. 2004. Antimicrobial activity of essential oil extracts of various onion (*Allium cepa)* and garlic *(Allium sativum). Lebensmittle –Wissenschaft and Technologie*, 37: 263-268.

Bhujsen, J. R. M. 1990. Taxonomic survey of *Allium* species cultivated in South-East Asia. The Netherlands.

Bidlingmeyer, B. A., Cohen, S. A. and Tarvin, T. L. 1984. Rapid analysis of amino acids using precolumn derivatization. *Journal of Chromatography*, 93-104.

Caldes, G. and Precott, B. 1969. Lecture given in National institute of allergy and infections diseases, National Inst. of Health Bethesda, Maryland, 20014, USA.

Chopra, R. N., Nayar, S. L. and Chopra, I. C. 1956. Glossary of Indian medicinal plants. CSIR, New Delhi.

Dietmar E., Breithaupt and Bamedi, A. 2002. *Journal of Agriculture and Food Chemistry*, 50: 7175-81.

Dubois, M., Galles, K. A., Hamilton, J. K., Rebers, P. A. and Smith, F. 1956. *Anal Chemistry*, 26: 350.

Ejaz, S., Woong, L. C. and Ejaz, A. 2003. Extract of garlic (*A. sativum*) in cancer chemoprevention. *Eksperi. Onkolo.* 25: 93-97.

Freeman, G. G. and Whenham, R. J. 1976a. Thiopranal s-oxide, alk(en)yl thiosulphinates and thiosulphonates: simulation of flavour components of *Allium* species. *Phytochemistry*, 15: 187-190.

Freeman, G. G. and Whenham, R. J. 1976b. Synthetic s-alk (en)yl-L-cysteine sulphoxides – alliinase fission products: simulation of flavour components of Allied species. *Phytochemistry*, 15: 521-523.

Harris, J. C., Cottrell, S. L., Plummer, S. and Lloyd, D. 2001. Antimicrobial properties of *Allium sativum* (Review). *Applied Microbiology and Biotechnology*, 57: 282-286.

Higuchi, O., Tateshita, K. and Nishimura, H. 2003. Antioxidative activity of sulfur containing comps. in *Allium* spp. for human low-density lipo-protein (LDL) oxidation *in- vitro. Journal of Agriculture and Food Chemistry*, 51, 7208-14.

Hooker, J. D. 1892. Flora of British India (Reprinted Bishen Singh and Mahendra Pal Singh, Dehradun, 1973). 6: 337-345.

Huang, Chien-Ning, Horng, Joeu-Shyan Yin and Mei-Chin 2004. Antioxidative and antiglycative effects of six organo sulfur comp. *Journal of Agriculture and Food Chemistry*, 52: 3674-78.

Jakubowski, H. 2003. On the health benefits of *Allium* sp. *Nutrition*, 19: 167-168.

Kao, Shao-Hsuan, Hsu, Ching-Hsian, Su, Song, Nan, Hor and Wei- Ting 2004. Identification and immunologic characterization of an allergen, alliinlyase from garlic (*Allium sativum*). *Journal of Allergy and Clinical Immunology*, 113: 161-168.

Lancaster, J. E. and Boland, M. J. 1990. Flavor Biochemistry. In onions and Allied crops, Vol. 3, Biochemistry, Food science, and minor crops, Brewster, J. L. Rabinowitch, H. D. Eds. CRC press: Boca Raton, F. L., 33 -72.

Lea, M. A. 1996. Organosulfur compounds and cancer. *Advances in Experimental and Medical Biology*, 401:147-154.

Li, G. (1995). Antiproliferative effects of garlic constituents on cultured human breast cancer cells. *Oncology Research*, 2: 787-89.

Liakopoulou-Kyriakides, M. and Sinakes, S. 1992. A low molecular weight peptide from *Allium porrum* with inhibitory activity on platted aggregation. *In-vitro Biochemistry International*, 28: 373-78.

Malik, C. P. and Singh, M. B. 1980. In: Plant enzymology and histo-enzymology, Kalyani Publishers, New Delhi. 286.

Maynard, A. J. (Ed.) (1970) Methods in food analysis, Academic Press, New York. 176.

Milner, J. A. 1996. Garlic: its anti-carcinogenic and anti-tumorigenic properties. *Nutrition Research*, 54: 582-586.

Negi, K. S. 1987. A contribution to the flora of Khatling Glacier in the Garhwal Himalaya (District-Tehri) U. P., *Journal of Botany and Natural Historical Society*, 84: 585 - 598.

Negi, K. S. and Gaur, R. D. 1991. Little known endemic wild *Allium* species in the Uttar Pradesh hills. *Mountain Research and Development*, 11: 162-164.

Negi, K. S. and Pant, K. C. 1992. Notes on economic plants. *Economic Botany*, 46: 112-116.

Negi, K. S., Pant, K. C. and Koppar, M. N. 1993. Addition to the *Allium* species of the flora of Chamoli, Uttar Pradesh Himalaya. *Journal of Botany and Natural Historical Society*, 90: 543-547.

Negi, K. S., Pant, K. C., Kopper, M. N. and Thomas, T. A. 1991. Wild relatives of genus *Allium* L. in Himalayas. *Indian Journal of Genetic Resources*, 4: 73-74.

Negi, K. S. 1986. Edible wild plants of Garhwal Himalayas: An ethnological survey. *Ph. D. Thesis*, Garhwal University, Srinagar.

Negi, K. S., Tewari, J. K. and Gaur, R. D. 1985. A contribution to the flora of Dodital – a high altitude lake in the Himalayas (District -Uttarkashi), U. P., *Journal of Botany and Natural Historical Society*, 82: 258-280.

Peach K. and Tracy M. B. 1956. Modern method of plant analysis, Vol. I. Springer-Verlag, Heidelberg.

Pinto, J. T. and Rivlin, R. S. 2001. Anti-proliferative effects of *Allium* derivatives from garlic. *Journal of Nutrition*, 131: 1058s - 1060s.

Ranganna, S. 1976. Handbook of analysis of quality control for fruits and vegetable products, IInd ed. Tata McGraw Hill Pub. Co. Ltd., New Delhi. p. 1-545.

Rau, M. A. 1975. High altitude flowering plants of west Himalayas. BSI, Howrah.

Schanderal, S. H. 1970. Method in food analysis. Academic Press New York. 709.

Singh, S. 2005. Medicinal plants- A natural gift to mankind. *Agriculture Today*, 3: 58-60.

Singh, S. V. 1996. A novel anti-carcinogenic activity of an organosulfide from garlic, inhibition of H-ras oncogene transformed tumor growth *in-vivo* by diallyldisulfide is associated with inhibition of p21H-ras processing. *Biochemistry and Biophysics Research Communications*, 225: 660-665.

Snell and Snell, 1995. *Colorimetric method of analysis*, 11: 813-816.

Sovova, M. and Sova, P. 2004. Pharmaceutical significance of *Allium sativum* L. 5. hypo-lipidemic effect *in-vitro* and *in-vivo*. *Ceska a Slovenska Farmacie*. 53: 117-123.

Stajner, D., Milic-Demarino, M. and Canadanovic-Brunet, J. 2003. Screening for antioxidant properties of leek, *Allium sphaerocephalon* L. *Journal of Herbs, Spices and Medicinal Plants*, 10: 75-82.

Suresh, J., Hemlatha, P. and Selvi, R. 2006. Biodiversity and conservation of medicinal plants, *Biodiversity: Life to our mother earth* Ed. By Dr. S. John William, Loyola College, G. M. Offset Press, Chennai. 231-256.

Vijayalatha, S. J. 2004. An ornamental garden with medicinal plants an indirect approach for conservation of medicinal plants. *Indian Journal of Areca nut, Spice and Medicinal Plants*, 6: 98-107.

Witham, F. H., Blaydes, D. F. and Devlin, R. M. 1971. Experiments in plant physiology, Van Nostrand, New York. 245.

❏❏❏

22

Natural Antioxidant Food Supplements and Human Health

N. Anusuya and S. Manian

CONTENTS

1.1 Introduction 332

1.2 Free Radicals 332

1.3 Biological Roles of Free Radicals 333

1.4 Adverse Effect of Free Radicals 333

1.5 Antioxidants 333

1.5.1 Types of antioxidants 334

1.5.2 Foods containing antioxidative nutraceuticals 335

1.5.3 Antioxidants in medicinal plants 337

1.5.4 Antioxidants as food supplements 339

1.6 Conclusion 340

References 340

1.1 Introduction

Nature has provided an excellent storehouse of remedies to cure all the ailments of mankind. In prehistoric ages, the primitive man combined instinct with his indulgence and started learning lessons from the nature in the art of healing. His observations and dependence on the nature coupled with serendipitous use of plant products led to the initial knowledge about economic and curative properties of plants. He explored his natural surroundings and tried many plants found around him to his afflictions. By trial and error, he succeeded in probing the efficacy of various drugs in controlling particular diseases. With advancement of science and great understanding of chemistry, man was able to produce synthetically or extract a great variety of pharmaceuticals that were previously unknown to the scientific world or unidentified. Inspite of tremendous development in the field of allopathy, medicinal plants and their derivatives still remain as one of the major sources of drugs in modern and traditional systems throughout the world.

Today's global movement towards a more natural lifestyle has brought about resurgence of interest in herbs. It is increasingly being realized that many of today's diseases are due to the "oxidative stress" that results from an imbalance between formation and neutralization of pro-oxidants. Based on the growing interest in free radical biology and the lack of effective therapies for most chronic disease, the usefulness of antioxidants in protection against these diseases is recognized. There are several Indian medicinal plants known for their beneficial therapeutic effects which also might have antioxidant properties. Natural product research continues to explore Indian traditional medicine to develop new novel drugs.

1.2 Free Radicals

Free radicals are defined as molecules having an unpaired electron in their outer orbit. They are generally unstable, highly reactive and energized molecules. Examples of free radicals are superoxide anion ($\bullet O_2^-$), hydroxyl radical ($\bullet$OH), alkoxyl radical (RO$\bullet$), peroxyl radical (ROO$\bullet$) and hydroperoxyl radical ($HO_2^\bullet$). Nitric oxide (NO$\bullet$) and nitrogen dioxide ($NO_2\bullet$) are the two free radicals of nitrogen. Oxygen and nitrogen free radicals can be converted to other non-radical reactive species such as hydrogen peroxide (H_2O_2) hydrochlorous acid (HOCl), hypobromous acid (HOBr) and peroxy nitrite ($ONOO^-$). Reactive oxygen species (ROS), reactive nitrogen species (RNS) and reactive chlorine species are normally produced as part of the defense system and as the by products of cellular metabolic process, utilizing oxygen (Halliwell *et al.*, 1995; Simon *et al.*, 2000). Apart from aerobic respiration, stimulated polymorphonuclear leukocytes and macrophages and peroxisomes appear to be the main endogenous sources of most of the oxidants produced by cells (Halliwell, 1994). Exogenous sources of free radicals include tobacco smoke, inonizing radiation, organic solvents and pesticides (Pepas, 1996). Even too much of fat in the diet and too much exercise will lead to oxidative damage.

1.3 Biological Roles of Free Radicals

Free radicals may play an important role in the origin of life and biological evolution, implicating their beneficial effects on the organism. Benign functions of free radicals are the activation of nuclear transcription factors, gene expression, and a defense mechanism to target tumor cells and microbial infections (Simon *et al.,* 2000). Superoxide anion may serve as a cell growth regulator (Halliwell, 1997). Singlet oxygen can attack various pathogens and induce physiological inflammatory response (Stief, 2003). Nitric oxide is one of the most widespread signaling molecules and participates in every cellular and organ function in the body. It acts as neurotransmitter and an important mediator of the immune response (Fang *et al.,* 2002).

1.4 Adverse Effect of Free Radicals

Despite its necessity in aerobic energy generation, an excess of ROS over the physiologically relevant level has been known to cause oxidative defects in cellular macromolecules, leading to loss of normal cell function (Ames, 1983). ROS causes lipid and protein oxidation, DNA strand break and base modification and modulation of gene expression. Their role is implicated in many degenerative or many pathological process such as aging (Burns et al., 2001), cancer, coronary heart disease, Alzheimer's disease (Ames, 1983; Gey, 1990; Dias et al., 1997), neurodegenerative disorder, atherosclerosis, cataract, inflammation (Aruoma, 1998), diabetes, asthma, liver disorders, autoimmune disease, muscular dystrophy and Parkinson's dementia. Free radicals can also affect food quality, reducing its nutritional content and promoting food deterioration.

1.5 Antioxidants

There is an expanding quest surrounding the use of antioxidative molecules because they have the capacity to quench free radicals thereby protecting cells and tissues from oxidative damage. In the biological system, the antioxidants are defined as 'any substance that when present at low concentration compared to those of an oxidizable substrate significantly delays or prevents oxidation of that substrate'. Antioxidants act as a major defense against radical mediated toxicity by protecting against the damage caused by free radicals. They are often described as 'mopping up' from radicals meaning they neutralize the electrical charge and prevent the free radicals taking electrons from other molecules.

The human body has developed different enzymatic and non-enzymatic protection mechanism to combat free radicals. However, the innate defenses may not be enough for severe and/or continued oxidative stress. Hence, certain amounts of exogenous antioxidant are constantly required to maintain an adequate level of antioxidant to balance the ROS in the human body.

1.5.1 Types of antioxidants

i) Enzymatic and non-enzymatic antioxidants

The removal of free radical is achieved through enzymatic and non-enzymatic antioxidant reactions. They are found in extracellular as well as intracellular environment. Some important enzymatic and non-enzymatic physiological antioxidants are presented in Table 1.

Table 1 : Enzymatic and non-enzymatic antioxidants

Antioxidant	Location	Functions/ properties
Enzymatic antioxidants		
i) Superoxide dismutase	Mitochondria & cytosol	Removal of superoxide anion ($\bullet O_2^-$)
ii) Catalase	Mitochondria & cytosol	Removal of H_2O_2
iii) Gultathione peroxidase	Mitochondria & cytosol	Removal of H_2O_2 & organic hydroperoxide
iv) Gultathione reductase	Mitochondria & cytosol	Removal of hydroperoxide
Non-enzymatic antioxidants		
i) Cartenoids	Lipid soluble antioxidant in membrane tissue	Removal of ROS
ii) Bilirubin	Product of tissue metabolism in blood	Extracellular antioxidants
iii) Gultathione	Non-protein thiol in cell	Cellular antioxidant defense
iv) α-lipoic acid	Endogenous thiol	Serve as substrate for glutathione, recycling Vitamin C
v) Vitamin C	Aqueous phase of cell	Free radical scavenger, recycling Vitamin E
vi) Vitamin E	Cell	Chain breaking antioxidants
vii) Uric acid	Product of purine metabolism	Scavenging of OH radical

ii) Dietary antioxidants

Dietary antioxidant molecules can be classified into synthetic and natural antioxidants. In general, synthetic antioxidants are compounds with phenolic structures of various degrees of alkyl substitution, whereas natural antioxidants can be phenolic compounds (tocopherols, flavonoids, and phenolic acids), nitrogen compounds (alkaloids, chlorophyll derivatives, amino acids, and amines), or carotenoids as well as ascorbic acid.

Synthetic antioxidants such as butylated hydroxylanisole (BHA), butylated hydroxytoluene (BHT) and tertiary butylated hydroquinone (TBHQ) are very effective and are used for industrial processing. However their usage is discouraged because of their side effects and toxic properties to human health (Anagnostopoulou et al., 2006). For this reason and because of the growing consumer preferences for natural products, there is an increasing interest in the investigation of naturally occurring antioxidants from plants.

Natural antioxidants such as phenolic compounds and vitamins are found in various plant products such as fruits, leaves, seeds and oils. Among the natural compounds polyphenols constitute one of the major groups of herbal compounds acting as radical scavengers and antioxidants. Therefore, these substances have been proposed as health-promoting natural products. Polyphenols are generally divided into hydrolysable tannins and phenylpropanoids such as lignins, flavonoids and condensed tannins. The largest and best studied polyphenols are flavonoids. At present, about 4000 flavonoid compounds are known, containing a number of phenolic hydroxyl groups that confer strong antioxidant activity.

1.5.2 Foods containing antioxidative nutraceuticals

Nutraceuticals or functional foods are any food or food ingredients that may provide beneficial health effects beyond the traditional nutrients they contain. Nutraceuticals are also known as medicinal food, nutritional supplements and dietary supplements. Nutraceuticals can be grouped in different ways, depending on the food sources, mode of action and chemical structures. Fruits, vegetables, spices, herbs and beverages such as tea and wine are typical foods containing various antioxidative neutraceuticals. The best way to ensure adequate intake of antioxidant is to eat a variety of fruits and vegetables through a diet consisting of 5 to 8 servings of fruits and vegetables per day.

i) Fruits

Fruits like berries (blue berry, bilberry/wild blueberry, strawberry, black berry, cherry, raspberry, crowberry), pomegranate, grape, apple, plum, pineapple, kiwi fruit and citrus fruits are rich sources of antioxidants. Berries contain large amounts of phenolic compounds such as hydroxylated benzoic acids and cinnamic acid, and flavonoids including anthocyanins, proanthocyanins, flavonols and catechins. Grapes contain polyphenolic compounds such as caftaric acid, catechin and anthocyanins. Citrus fruits have polyphenols such as hydrocinnamic acids that include coumaric acid, caffeic acid and ferulic acid, limnonoids and naringin. Even citrus peels and seeds have high antioxidant activity. Dried fruits are good source of antioxidants by weight/serving size as the water has been removed making the ratio of antioxidants higher. Typical dried fruits are apricots, prunes, dates, raisins, fig and peaches.

ii) Vegetables

Tomato, sweet potato, ginger, garlic, onion, red beets, red cabbage, chili, kale, artichoke, cauliflower, Brussels sprouts, broccoli, spinach and all green leafy vegetables are typical vegetables containing nutraceuticals including vitamins and phenolic compounds such as flavonoids, anthocyanins and hydroxyl cinnamic acid derivatives.

iii) Legumes

Legumes are excellent source of nutraceutical constituents such as vitamins, minerals, fibre, phytic acid and polyphenols as flavonoids, isoflavones, lignans, phyto-oestrogens and tannins. Soy bean, broad bean, pinto bean, black bean, red beans and red kidney bean are some of antioxidant rich legumes.

iv) Nuts and cereals

Nuts are rich source of antioxidants. Typical nuts are almonds, cashewnut, peanut, flax seed, buckwheat, nettle, pecans and walnuts. Cereals contain a wide range of phenolic compounds such as ferulic, caffeic, *p*-hydroxybenzoic, protocatechuic, *p*-coumaric, vanillic, and syringic acids. Cereals rich in antioxidants are barley, rye, millet, oats and corn.

v) Spices

Spices are rich in antioxidant properties but the serving size is too small to be the top-contributors of antioxidants. Typical spices high in antioxidants are cloves, cinnamon, turmeric, cumin, basil, curry powder, pepper, mustered seeds, oregano, sage, rosemary and thyme.

vi) Beverages

Tea contains large amounts of flavonoids including catechin, epicatechin, epigallocatechin, quercetin, epicatechingallate and epigallocatechingallate. Wine contains large amounts of phenolic compounds including coumaric, cinnamic, caffeic, gentistic, ferulic and vanillic acids.

1.5.3 Antioxidants in medicinal plants

Plants are susceptible to damage caused by active oxygen and thus develop numerous antioxidant defense systems resulting in the synthesis of numerous potent antioxidants. Many aromatic and medicinal plants are rich in antioxidative compounds (Table 2). Several medicinal plants which are known for their beneficial therapeutic effects also might have strong antioxidant property. Some of the common herbal antioxidants are *Aloe vera, Ginkgo biloba, Catharanthus roseus, Piper beetle, Centella asiatica, Glycyrrhiza glabra, Phyllanthus niruri, Phyllanthus emblica, Ocimum sanctum, Withania somnifera, Camellia sinensis, Santalum album, Terminalia chebula, Ficus bengalensis, Murraya koenigii, Coccinia indica, Hemidesmus indicus, Rosemarinus officinalis, Scutellaria baicalensis, Curcuma longa, Sida cordifolia* and *Andrographis paniculata.*

Table 2 : Antioxidant neutraceuticals in medicinal plants

Plant	Chemical constituents	Pharmacological action
Aerva lanata L. (Amaranthaceae)	Kampferol 3-rhamnoside, kampferol 3-rhamnogalactose, canthin 6-one, â-carboline, steroids, tannins.	Hepatoprotective, nephroprotective, antioxidant
Allium sativum L. (Liliaceae)	Ascorbic acid, β-sitosterol, caffeic acid, kampferol, allicin, diallyl catechols, protocatechuic acid, ajoene, allyl propyl sulfide.	Antioxidant, anticancer, lower blood sugar
Aloe barbadensis L. (Liliaceae)	Acetylated mannans, polymannans, anthroquinone C-glycosides, anthrones, anthroquinones, lectins	Antidiabetic, wound healing, anticancer, anti-inflammatory, antioxidant, immunostimulant
Andrographis paniculata Nees. (Acanthaceae)	Andrographolide	Antioxidant, anti-inflammatory, antiulcer, antipyretic
Aristolochia bracteolate L. (Aristalocheaceae)	Alkoloid, sapponins, glycosides, steroids, tannins, phenolic compounds, flavonoid glycosides.	Wound healing, antioxidant, anti-inflammatory
Bacopa monnieri L. (Schrophulariaceae)	Bacoside, hersaponin, herpestine, brahmine, flavonoids	Hepatoprotective, antioxidant, antidiabetic
Camellia sinensis (L.) Kuntze (Theaceae)	Catechin, epicatechin, epigallocatechin, epigallocatechingallate, caffeine, theaine, theobromine, chlorgenic acid.	Anticancer, antioxidant, reduce LDL cholesterol, reduce cardiovascular disease
Cassia auriculata L. (Cesalpinaceae)	Tannins, di-(2-ethyl) hexyl phthalate.	Antioxidant, nephroprotective, antibacaterial, antihyperglycemic

Contd...

Table 2. contd...

Catharanthus roseus L. (Apocyanaceae)	Vincristine, vinblastine.	Anticancer, antioxidant, antihyperglycemic, antidiabetic
Centella asiatica L. (Apiaceae)	Vallarine, asiaticoside, oxy-asiaticoside, β-sitosterol, essential oil, fatty oil, tannin, ascorbic acid, pectic acid.	Anticancer, antiulcer, antioxidant
Coccinia indica L. (Cucurbitaceae)	Glucokenin, alkaloids, fatty acids.	Antioxidant, antidiabetic, hyperglycemic activity
Coriandrum sativum L. (Apiaceae)	Quercetin 3- glucuronide, isoquercetin, rutin, β-carotene	Antioxidant, antidiabetic, antimutagenic
Daucus carrota L. (Apiaceae)	α-Tocopherol, ascorbic acid, camphene, eugenol, γ-terpinene	Antioxidant, used against arthritis, edema, rheumatism
Enicostemma littorale Blume (Gentianaceae)	Phenols, tannin, flavonoids, glycosides, anthroquinones, sterols, gentianine, swertiamarin	Antidiabetic, anti-inflammatory, antioxidant, hypoglycemic, hypolipidemic, anticancer
Eugenia caryophylla Thunb. (Myrtaceae)	Eugenol, isoeugenol, ascorbic acid, β-carotene, β-sitosterol, caryophyllene-oxide	Antiarthritis, antioxidant
Ficus benghalensis L. (Moraceae)	Pelargonidin 3-o-alpha rhamnoside, leucopelargonin	Antioxidant, hepatoprotective, antidiabetic
Glycyrrhiza glabra L. (Fabaceae)	Glycyrrhizin, 18-beta-glycyrrhenetic acid, liquiritigenin, glaberidin, 5.8-dihydroxy-flavone-7-o-beta-D-glucuronide, glychinoides, 5-hydroxy 8-methyl-flavone-7-O-beta D-glucuronide	Antioxidant, Antiinflammatory, antiulcer, hypocholesterolaemic, anxiolytic, antimicrobial
Mentha piperata L. (Lamiaceae)	Menthol, limonene	Anticancer, antioxidant
Myristica fragrans Houtt. (Myristicaceae)	Camphene, eugenol, isoeugenol, lauric acid, palmitic acid, γ-terpene, terpenol	Antiarthritis, antioxidant
Phyllanthus amarus L. (Euphorbiaceae)	Phyllanthin, hypophyllanthin, tannin, polyphenol, astragallin, catechin, epicatechin	Anticancer, radioprotective, anti-inflammatory, antidiabetic, antioxidant
Phyllanthus emblica L. (Euphorbiaceae)	Ascorbic acid, gallic acid, ellagic acid, tannin, flavonoids, kampferol, quercetin, phyllantine, β-sitosterol	Hepatoprotective, antioxidant, anti-inflammatory, antiviral
Phyllanthus niruri L. (Euphorbiaceae)	Phyllanthin, hypophyllanthin, lignans, flavonoids, glycosides,	Hepatoprotective, antioxidant hypolipidemic
Ocimum sanctum L. (Lamiaceae)	Eugenol, carvacol, caryophyllene, tannin, ascorbic acid, β-carotene, β-sitosterol	Antiproliferative, antioxidant, antiarthritis
Olea europaea L. (Oleaceae)	Kampferol, α-tocopherol, β-carotene, apigenin, luteolin	Immune modulatory, antioxidant, maintain sugar and cholesterol level.
Origanum vulgare L. (Lamiaceae)	Thymol, rosemarinic acid	Immune booster, antioxidant

Contd...

Table 2. contd...

Piper beetle L. (Piperaceae)	Essential volatile oil containing betel-phenol, tannins, terpene, sesqueterpene, chavicol	Antidiabetic, antioxidant, antibacterial, antifungal
Piper nigrum L. (Piperaceae)	Piperine, piperidine, piperitine, balsamic, chavicin, ascorbic acid, β-carotene, lauric acid, myristic acid, palmitic acid	Anticancer, antibacterial, antimutagenic, antioxidant
Rosemarinus officinalis L. (Lamiaceae)	Rosmanol, epirosemanol, rosemadial, carnosol, methyl carnosate, carnosic acid, genkwanin, eugenol cirsimaritin, β-sitosterol	Antioxidant, antidiabetic, hepatoprotective,
Santalum album L. (Santalaceae)	Phenolic compounds, eugenol, palmitic acid, β-sitosterol	Antioxidant, prevents premature aging
Scutellaria barbata Don. (Lamiaceae)	Gallic acid	Anticancer, antioxidant
Scutellaria baicalensis Georgi (Lamiaceae)	Baicalein, oroxylin, wogonin, baicalin, wogonoside, 2',5,6', 7-tetrahydroxy flavone	Hepatoprotective, anticancer, antioxidant
Sida cordifolia L. (Malvaceae)	Fumaric acid, sitoindoside	Adaptogenic, antioxidant, immunomodulant, hepatoprotective
Solanum nigrum L. (Solanaceae)	Solanine, solanidine, saponine	Antitumor, antioxidant, antiulcer
Terminalia chebula Retz. (Combretaceae)	Chebulinic acid, tannic acid, anthraquinone, gallic acid	Antidiabetic, antioxidant, chemomodulatory
Thymus vulgaris L. (Lamiaceae)	4-Terpenol, caffeic acid, β-carotene, carvacrol, camphene, lycopene, γ-terpinene, myrcene, palmitic acid	Antiarthritis, antioxidant, enhances blood circulation, effective against sprain, muscular pain
Withania somnifera L. (Solanaceae)	withaferin A, viscosalactione	Anticancer, prevents premature aging, immune booster, antioxidant
Zingiber officinale Roscoe (Zingiberaceae)	6-Gingerol, ascorbic acid, lauric acid, myristic acid, palmitic acid	Anticancer, immunomodulatory,

1.5.4 Antioxidants as food supplements

'Healthy foods' containing active scavenger of free radicals are popular nowadays. Crude extracts of fruits, herbs, vegetables, cereals, and other plant materials rich in phenolics are increasingly of interest in the food industry because they retard oxidative degradation of lipids and thereby improve the quality and nutritional value of food. The importance of the antioxidant phytoconstituents in the maintenance of health and protection from chronic diseases is also raising interest among scientists, food manufacturers, and consumers as the trend of the future is moving towards functional food with specific health effects. Food supplements that provide phytonutrients are sometimes a better choice than even some fruits and vegetables. The multivitamin antioxidant supplements energize the body and boost the mental power. Antioxidant supplements can be taken as tablets and capsules. Several antioxidant supplement products are available in every market, from mainstream supermarket to health food stores and even over the internet.

1.6 Conclusion

There are many foods in our diet that are considered rich in antioxidants. The main thing a person needs to know is which food to include in his diet on a daily basis. If people have difficulty in determining which antioxidant they need, supplements can provide the recommended amounts to stay in perfect health. So the antioxidant functional foods should be popularized so that people can plan their diets for their healthy life.

References

Ames, B.N., 1983. Dietary carcinogenesis and anticarcinogenesis, oxygen radicals and degenerative diseases. *Science*, 221: 1256-1264.

Aruoma, O.I., 1998. Free radicals, oxidative stress and antioxidants in human health and disease. *Journal of The American Oil Chemists Society*, 75: 199-212.

Anagnostopoulou, M. A., Kefalas, P., Papageorgiou, V. P., Assimepoulou, A. N. and Boskou, D. 2006. Radical scavenging activity of various extracts and fractions of sweet orange peel (*Citrus sinensis*). *Food Chemistry*, 94: 19-25.

Burns, J., Gardner, P.T., Matthews, D., Duthie, G.G., Lean, M.E. and Crozier, A., 2001. Extraction of phenolics and changes in antioxidant activity of red wines during vinification. *Journal of Agricultural and Food Chemistry*, 49: 5797-5808.

Diaz, M.N., Frei, B., Vita, J.A. and Keaney, J.F., 1997. Antioxidants and antherosclerotic heart disease. *The New England Journal of Medicine*, 337: 408-416.

Gey, K.F., 1990. The antioxidant hypothesis of cardiovascular diseases, epidermiology and mechanism. *Biochemical Society Transactions*, 18: 1041-1045.

Govindarajan, R., Vijayakumar, M. and Pushpangadan, P. 2005. Antioxidant approach to disease management and the role of "Rasayana" herbs of Ayurveda. *Journal of Ethnopharmacology*, 99: 165-178.

Halliwell, B., 1994. Free radicals, antioxidants and human diseases: curiosity, causes or consequences. *Lancet*, 344: 721-724.

Halliwell, B., Murcia, M.A., Chirico, S. and Aruoma, O.I., 1995. Free radicals and antioxidants in food and *in vivo*: What they do and how they work. *Critical Reviews in Food Science and Nutrition*, 35: 7-20.

Pepas, A.M., 1996. Determinants of antioxidant status in human. *Lipids*, 31: 77-82.

Simon, H.U., Haj-Yehia, A. and Levi-Schaffer, F., 2000. Role of reactive oxygen species (ROS) in the apoptosis induction. *Apoptosis*, 5: 415-418.

□□□

23

Cancer and Diet

Sucheta Tripathy

CONTENTS

1.1 Introduction 342

1.2 Conventional Cancer Treatment 342

1.3 Statistics of Cancer Mortality in USA from 1975-2005 343

1.4 Diet Program during Cancer Treatment 344

1.5 What Causes Cancer in the First Place? 344

1.6 How Nutrition Can Help? 346

- 1.6.1 Antioxidants 346
- 1.6.2 Food 348
- 1.6.3 Fats 350
- 1.6.4 Cancer feeds on sugar 350
- 1.6.5 Juice fasting 351
- 1.6.6 Exercise and healthy living 351

1.7 Cancer Alternate Medicine (CAM) 352

1.8 Conclusion 359

Reference 359

1.1 Introduction

While the main course in cancer therapy has always been the conventional chemotherapy, it has been a long debated subject whether diet can help cure cancer and help alleviate side effects of cancer treatments. Many well known cancer research labs follow various types of approaches to subdue cancer growth. Most of them lie with targeting only cancer cells, while reducing toxicity to normal, healthy cells. While, this has been partly successful, but still a large number of patients suffer from unpleasant side effects of cancer treatments that involves surgery, radiotherapy and chemotherapy. Life style changes through diet and exercise provides a promising way to reduce the unpleasant side effects of cancer treatment as well as to eliminate this disease against all odds. The focus of this chapter is to present some of the dietary options available for cancer patients that help alleviate symptoms as well as treat cancer.

The purpose of this section is not to criticize traditional cancer treatments but to make it blatantly obvious that our current cancer treatment options are inadequate and incomplete and that we need to examine some options like nutrition and life style changes in order to be well. An impending question that needs to be addressed urgently is: "Does nutrition reduce the effectiveness of chemotherapy?" There are two parts to this debate.

1) Does nutrition interfere with chemotherapy? "No."
2) Is chemotherapy effective? "Sometimes."

1.2 Conventional Cancer Treatment

Currently, in addition to radiation, surgery and chemotherapy, other treatment options are also available. Often, they are specific for a particular type of cancer.

Brachytherapy: Internal Radiation therapy, where the radiation source looks like a seed, a ribbon or wire and is put inside the body.

Angiogenesis Inhibitors therapy: Agents that are known to stop angiogenesis (new blood vessel formation in tumors) are injected into affected areas. This is administered in colorectal cancer and non-small cell lung cancer.

Biological Therapy: Biological therapy includes usage of ones own immune system against cancer. This involves interferon, interleukins, colony-stimulating factors, monoclonal antibodies, vaccines, gene therapy, and nonspecific immunomodulating agents.

Bone marrow transplantation: Bone marrow transplantation (BMT) and peripheral blood stem cell transplantation (PBSCT) are procedures that restore stem cells that were destroyed by high doses of chemotherapy and/or radiation therapy.

Hyperthermia: Hyperthermia is a type of cancer treatment in which body tissue is exposed to high temperatures (up to 113°F) to damage and kill cancer cells.

Laser Therapy: Laser light is a light of such high intensity and narrow beam that it can be used to do precise surgery to remove cancer or precancerous growths or to relieve symptoms of cancer. It is used most often to treat cancers on the surface of the body or the lining of internal organs.

Photodynamic Therapy: Photodynamic therapy (PDT) combines a drug (called a photosensitizer or photosensitizing agent) with a specific type of light to kill cancer cells.

Targeted Cancer Therapy: Targeted cancer therapies use drugs that block the growth and spread of cancer by interfering with specific molecules involved in carcinogenesis and tumor growth.

1.3 Statistics of Cancer Mortality in USA from 1975-2005

Although dramatic changes have taken place in terms of cancer therapy and approaches, there is only a marginal decline in cancer death since 1975 [fig-1] .

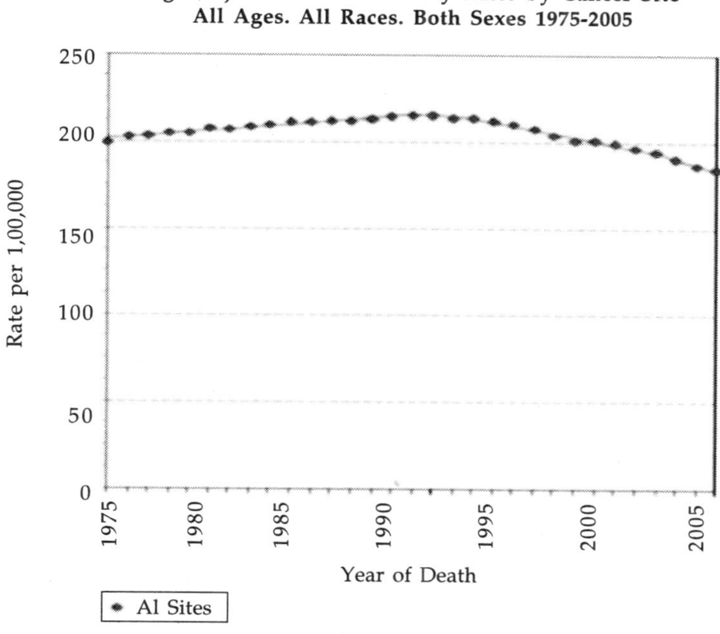

Fig. 1 : Combined Cancer mortality rate from 1975-2005 [*Source :* NCI]

Saturated

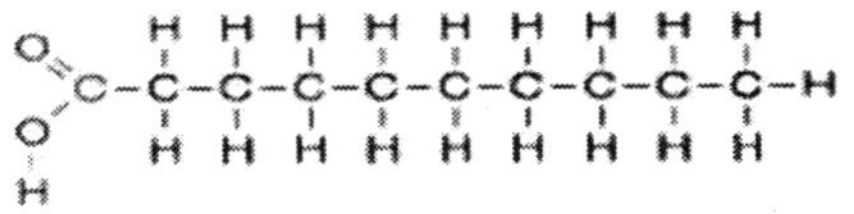

Unsaturated

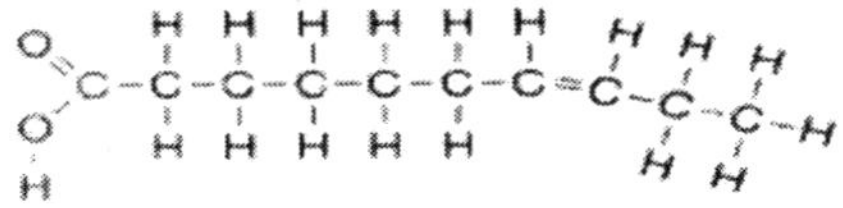

Fig. 2 : Structure of Saturated unsaturated fatty acid

1.4 Diet Program during Cancer Treatment

The extensive research published in the last decade alone proves that what you eat can have a profound effect on your protection against cancer. But mostly what doctor's answer is hardly surprising. It's a little-known fact that nutrition is barely taught in med schools, where the solution to most problems is a drug. And doctors don't trust patients to make lifestyle changes. Many doctors believe that people don't want to make a life style change for the sake of cancer. For them it is just take a pill and forget about it.

Although Cancer Alternative Treatment is gaining momentum, still research is severely lagging behind in this area. A simple Cancer Alternate Medicine (CAM) search in pubmed will just yield about 12,649 (As of January, 2009) odd articles while a cancer therapy search will yield almost 1092783 articles!! This indicates the volume of conventional research against CAM research going on in this area.

In order to understand how alternative treatment as well as CAM can help, it is essential to get deeper into this topic.

1.5 What Causes Cancer in the First Place?

Let's try and understand the cytogenetics of cancer. In cancer, the normally functioning genes get mutated leading to abnormal function which leads to abnormal cell growth and proliferation. This often starts inside a tissue or organ, which later spreads to various other parts of the body through a process called as 'metastasis'. Metastasis is an enormously complex process that remains to be a major problem in the management of cancer. The fact that cancer patients might develop metastasis after years or even decades from diagnosis of the primary tumor makes the metastatic process even more complex (Hunter et al., 2008). The cells acquire the status of immortality, and often help themselves with newer blood vessels through a process called as 'angiogenesis'.

The human body at that point lacks the ability to kill these cells through a process called as 'apoptosis'. Human body has 60 trillion cells that possess the 'blueprint' of life called as the DNA molecules. This long spiral staircase like molecule is constantly under attack at the rate of 1000 to 10000 hits per day (Bruce Ames, PhD, University California Berkeley, unpublished data). Geneticists have estimated that each DNA molecule in our cell has information that can be written down in 500,000 pages!! Imagine if trillions of times these 500,000 pages need to be copied, what is the chance of getting copy error and this copy error most often leads to cancer.

Once cancer happens, it starts weakening the host by installing its own abnormal biochemistry. The following changes take place in the body:

- Changes pH (more towards acidic environment).
- Creation of anaerobic pockets of tissue that resists radiation therapy.
- Blunting the immune system.
- Elevating metabolism and caloric needs, while simultaneously lowering appetite.
- Ejecting toxic by-products that create weakness, apathy, pain and depression in the host.
- Siphoning nutrients out of the blood stream like a parasite.

Cancer happens because the body lacks the ability to kill the mistake cells. These are the outcomes of the immune system malfunctioning. Immunity is compromised because of poor nutrition, stress, sedentary life style, toxic burden etc.

Of the 5 million registered chemicals in the world, mankind comes in contact with 70,000, of which at least 20,000 are known carcinogens. Most common chemical carcinogens that we are exposed to on a day to day basis is pesticides, fungicides, paints, UV radiation, artificial food colorants, hormones, antibiotics in animal products, food preservatives etc. just to name a few.Toxins not only cause DNA breakage, that can trigger cancer, but also subdue immune system, which then allows cancer to become the 'fox in the chicken coop' with no controlling force.

Early Research indicated that once cancer has been up regulated, then no matter of detoxification is going to help. However, newer evidence says otherwise. Cancer growth can be both slowed and even reversed under the right conditions. Research has shown that, in the early stages of cancer development, nutrients alone can reverse 'pre-malignant' cancers [2]. Cancer is reversible, if the toxins that caused it in the first place is detoxified.

1.6 How Nutrition Can Help?

The human body is built from, repaired by and fuelled by substances found in the diet. In the most literal sense 'we are what we eat... and think and breathe and do'. Nutrition therapy merely tries to re-establish 'metabolic balance' in the cancer patients (Singh, 1991).

Nutrition can make medical therapy more of a selective toxin against the cancer and less damaging to the patient. Nutritional therapy is an essential part of comprehensive cancer treatment because:

- Malnutrition kills 40% or more cancer patients.
- Nutrition protects the patient against the toxic effects of chemo and radiation without reducing the effectiveness in killing cancer.
- Nutrients support the immune system in killing cancer.
- Sugar feeds cancer, which makes it crucial to stay away from them.
- Nutrients can become 'biological response modifiers' in changing the way the body works to re-regulate the cancer cells.

Following are some alternative ways to deal with cancer with or without traditional therapy:

1. Antioxidants.
2. Food.
3. Fat.
4. Avoiding Sugar in any form.
5. Juice Fasting.
6. Exercise and healthy life style.
7. CAM.

1.6.1 Antioxidants

By definition anti oxidants are chemicals that act as electron donors and prevents oxidation. Antioxidants protect the body from oxidative damage induced by free radicals and reactive oxygen species by

- suppressing their formation;
- acting as scavengers; and
- acting as their substrate.

Antioxidants are a diverse group of chemicals that can be naturally found in vegetables, fruits and plants in general. Antioxidants are also synthesized in our body. Examples of dietary antioxidants are vitamin E, C, A, phenolic acids, selenium, chlorophyll and cholorphyll derivatives, carotenoids, flavonoids, glutathione, coenzyme Q 10, melatonin, and lycopene. Synthetic dietary antioxidants include butylated hydroxyanisole (BHA) and butylated hydroxytoluene (BHT). Chemicals with antioxidant activity produced in human body include uric acid, high density lipo-protein (HDL), and amino acids such as arginine.

Antioixdants are essential for proper function of the immune system. This is partly because immune cells produce free-radicals for normal defense functions. If the level of free radicals in the immune cells surpasses beyond the normal level, they negatively affect the immune system. On the other hand, antioxidants act as scavengers of the free radicals in cells and therefore promote immunity. Imbalance between free radicals and antioxidants in cells due to deficiency in single or multiple antioxidants has been reported to result in weakness of immunity function.

Antioxidants boost immune system also by playing other important roles such as in cellular metabolism, signal transduction, gene activation, and transcription. High density lipo-protein (HDL) cholesterol has inhibitory effect against oxidation of low density lipoprotein, which is implicated with development of atherosclerosis and cardiovascular diseases. High density lipo-protein (HDL) also prevents endothelial cell dysfunction and activation, maintain the integrity of the endothelial cell lining, and reduce the deleterious effect of oxidized low density lipo-protein (LDL) cholesterol and the levels of thrombogenic factors.

A large number of fruits and vegetables are natural anti-oxidants. A variety of fruits, vegetables, nuts, whole grain cereals and legumes, tea, red wine, and herbs (e.g., rosemary extract) are rich sources of natural antioxidants. Food high in antioxidants includes fruits such as apple, grape, grapefruit, pomegranate, crane berry, black berry and blueberry. Vegetables are also rich antioxidant sources. Red bean, pinto bean, kidney bean, carrot, tomato, garlic and russet potato are among the vegetables high in antioxidant content and categorized among the best antioxidant sources. Other important sources of antioxidants are green tea, black tea, herbal tea, spices, red wine, ginger, and garlic.

Listed is some of the anti-oxidants vis a vis their effects on chemotherapy treatments:

- *Vitamin K*: In one study, patients with mouth cancer who were pretreated with injections of vitamin K prior to radiation therapy doubled their odds (20% vs. 39%) for 5 year survival and disease free status (Krishnamoorthi *et al.*, 1971). In cultured leukemia cells, vitamins K and C added to the chemotherapy drugs of 5FU (5 fluorouracil) and leucoverin provided a 300% improvement over 5FU by itself (Waxman *et al.*, 1982).

- *Vitamin C*: Lung cancer patients who were provided antioxidant nutrients prior to, during, and after radiation and chemotherapy had enhanced tumor destruction and significant longer life span (Jaakkola, 1992).
- *Fish Oil*: A special fat in fish (Eicosapentaenoic acid, EPA) improves tumor kill in hyperthermia and chemotherapy by altering cancer cell membranes for increased vulnerability (Burns et al., 1990). EPA is also known to increase ability of adriamycin to kill cultured leukemia cells (Guffy *et al.*, 1984).
- *Vitamin A and Beta Carotene*: There is a synergistic benefit of using vitamin A with carotenoids in patients who are being treated with chemo, radiation and surgery for common malignancies (Santamaria *et al.*, 1990).

There are also a large number of literatures available for benefits of anti oxidants like Vitamin E, Niacin, Selenium, Carnitine, Quercetin, Ginseng etc.

1.6.2 Food

Let food be thy medicine, thy medicine shall be thy food. - Hippocrates

It is said that we are what we eat. In systemic conditions like cancer it is very important to stay well nourished. Nothing can be more nourishing than whole food:

Whole food contains more than just essential nutrients. There are minor dietary constituents, conditionally essential nutrients, phytochemicals, enzymes, pH stabilizing factors, substances required for a healthy colony of bacteria in the gut, and much more. No pill can substitute for the thousands of useful anti-cancer agents in Nature's trove of delicious, fresh, unprocessed food. While eating, one needs to consider the following points:

a) If a food will not rot or sprout, then throw it out.
b) Avoid human altered food That is packed, preserved food.
c)Eat food in as close to their natural state as possible.

It is a well known fact that taking 1000 mg vitamin C can reduce the risk of stomach cancer, but it is also shown that drinking a glass of fresh orange juice that has merely 37 mg of Vitamin C is twice as effective. So, natural whole some food has a natural symphony of ingredients that is absorbed very well in the body and works much better than extracted and concentrated products.

Top most food in chart for cancer patients are:

- *Vegetables*: Colorful vegetables like Beets, carrots, spinach, tomatoes, squash, capsicum, etc. are very rich in anti oxidants and anti- cancer elements. The colorful pigments in raw vegetables carry 20,000 bioflavonoids and 800 carotenoids. These 'phytochemicals' provide extraordinary protection to our body in addition to providing much desired fibers.

  The cabbage family of vegetables like cabbage, broccoli, cauliflower, Brussels sprout is unique in having a strong anti-cancer element called as indole-3-carbinol and sulforafane. These are very strongly anti cancerous as well as strong detoxifying elements. Scientists at the Johns Hopkins University found that lab animals fed with cruciferous vegetables when exposed to carcinogens had 90% reduction in cancer occurrence (You *et al.*, 1989).

  Alium family vegetables like onion and garlic contain bioflavonoids like quercetin that can actually revert a cancer cell back to a healthy cell. Other anti cancerous compounds in this family of vegetables are allicin, S-allyl cysteine, selenium etc. Chinese researchers have found that intake of onion and garlic reduces the risk of stomach cancer by half (Ansher, 1986).

- *Cold Water Fish*: Salmon, haddock, halibut, bass, cod, mackerel, sardine, tuna and other fish from deep ocean are very good for having large quantities of EPA and DHA (Docosahexaenoic Acid). These fish oil slows down spreading of cancer.

- *Legumes*: Soybeans, kidney beans, pinto beans and many others constitute this group of plant food. They are rich in protease inhibitors, inositol hexaphosphate (IP-6), genistein etc. Genestein is known for shutting down 'angiogenesis'. Researchers have also found that isoflavones and phytoestrogens present in soybean have strong anti cancerous properties (Messina et al., 1991).

- *Colorful Berries*: Raspberries, strawberries, blue berries, cherries, black berries contain a powerful anti-cancerous element called as 'ellagic acid', which causes cancer cells to self destruct through 'apoptosis'.

- *Yogurt*: Yogurt is formed when lactobacillus ferments milk sugar. Since 80% of the immune system lies in the gastro intestinal tract, yogurt acts as an excellent source for fortifying intestinal bacteria and resuming immune function.

- *Green Tea*: Green Tea is known to have catechins that are highly anti cancerous.

- *Healthy Seasoning*: Natural seasoners like mustard, cumin, fenugreek, black cumin, nutmeg, cinnamon, garlic, onion, ginger, sage, rosemary, thyme are excellent resources to provide taste as well as vitality to the food. These can often replace fat, salt and sugar.
- *Clean Water*: Clean water stabilizes acid/base balance and provides healthy flow of nutrients into cells and helps detoxify the body.

1.6.3 Fats

It may sound little bit strange when we talk about fat and good health!! Remember all our 60 trillion cells are surrounded by a membrane made up of fat. The cell membranes play diverse roles starting from gate keeping to cellular transport, cell packing and cell- cell communication. Cell membranes of mammals have a higher composition of polyunsaturated fat (DHA, omega-3 fatty acid) and a lower composition of monounsaturated fat than do reptiles. Higher polyunsaturated membrane content gives greater membrane fluidity (and functionality), commensurate with the higher metabolic rate of the warm-blooded species.

Dietary fats can be divided into good, bad and ugly. Good ones are flax seed oil, olive oil, fish oil, evening primrose oil, rice bran oil, mustard oil, hemp oil etc. Bad fats could be any other vegetable oil, saturated fat etc. Ugly are the deep fried fat that has already been oxidized.

In an experiment, cancer patients were given fish oil and vitamin E against a control group. Omega-3-fatty acid in fish almost always increased survival of patients (Gogos *et al.,* 1998). It is common knowledge that saturated fats are rigid, while unsaturated fats are flexible in their structure. They can bend and maintain more fluid like status [fig-2]. So, what happens when the cell is exclusively made up of saturated fats? The cell will be like a rigid structure with very little permeability. Cell as an entity needs rapid rate of inflow and outflow of substances. If this balance is limited due to membrane rigidity, the cells will seriously start malfunctioning (Pierce, 2002).

1.6.4 Cancer feeds on sugar

Yes literally so...

Cancer cells demonstrate a 3 to 5 fold increase in glucose uptake compared to healthy cells (Demetrakopoulos, 1982). At the famous Sloan Kettering Center hospitals in New York, researchers noticed tumor cells sucking up radio labeled vitamin C molecules like sponges. This is because glucose, the favorite substrate of cancer cells is nearly identical in structure with vitamin C molecules.

Glycolysis was first discovered by famous Nobel Laureate Otto Warburg in 1931. He discovered that cancer cells tend to metabolize anaerobically than most healthy cells, producing lactic acid as a bye product. Lactic acid makes the tissue acidic and thus favors further tumor growth.

Thus, it is very important to avoid refined sugar during cancer treatment and beyond. No doubt reducing sugar will literally starve the cancer cells.

1.6.5 Juice fasting

"One quarter of what you eat keeps you alive;
the other three-quarters keep your doctor alive."

One of the reasons fasting merits attention is that anything else is so often ineffective or downright dangerous. One of the really amazing books of our time is Medical Nemesis, by Ivan Illich (Bantam, 1977). Dr. Illich shows, in exhaustive detail, that medical care has become literally sickening. One in five hospital patients is there because a physician's mistake put him or her there. And with that point, the book is just warming up.

Ever since George Washington's doctors killed him with their prescriptions (No kidding: Medical and Physical Journal, London, 1800. Volume 3, page 409), people have been seeking LESS HARMFUL treatments than conventional medicine has offered. A common "scientific" argument against fasting is that it is always unsafe. Truly the pot is calling the kettle black when drug-based medicine criticizes the safety of the therapy that all animals naturally use, namely fasting. First thing a sick animal does is go off its feed.

1.6.6 Exercise and healthy living

As it is said prevention is better than cure, it is essential to kill the cause of cancer than to deal with it later. As no one suffering from headache - is due to deficiency of aspirin, it's important to find the cause of headache than to stuff yourself with aspirin.

Recent studies have proved that physical activity in form of at least 30 minutes exercise a day reverses gene expression pattern in cancer patients. In colon cancer patients, it is seen that physical activity may protect against aberrant methylation by repressing the BHMT2 gene and thus contribute to a decreased risk of developing cancer (Buehlmeyer et al., 2008). It is also identified that vascular endothelial growth factor (VEGF), angiopoietin-2 (ANG-2) and calciumin dependent phospholipase a2 (iPL-A2) have markedly reduced transcript levels in the mucosa of active rats (Bellizzi et al., 2009).

Another study with cancer survivors having at least 150 minutes of exercise had great quality of life compared to those not doing exercise. This group also had lesser chance of cancer recurrence (Fair and Montgomery, 2009).

Lets look at what exercise does to our body. We have already known that cancer cells prefer anaerobic conditions, which is lack of oxygen transport to our body parts. Exercise, on the contrary makes blood flow more efficient and most of the tissue get oxygenated, thus less occurrence of cancer.

In addition to aerobic exercises, it is also essential to do at least 3 times stress busting yoga, pranayama and meditation to calm the mind. Deep abdominal breathing against shallow breathing helps oxygenating better. Deep breathing is also known to reduce stress and calming mind.

Life style changes also include reduction in toxic intake. In modern civilization more often than not walls are painted with heavy metallic paints. Fruits and vegetables are often sprayed with color and pesticides that increases our toxic load. Applying cosmetics of any kind that is not natural also adds to the toxic burden. So, the rule of thumb is 'avoid anything that is artificial and that can be done away with'. Go natural as much as possible and revert to ancient practices. Avoid artificial drinks, packed food, refined wheat, refined sugar etc. As much possible pre-soak vegetables in warm water before finally washing and using them. Use mild shampoos and soaps for this purpose.

1.7 Cancer Alternate Medicine (CAM)

Become your own Health Detective! For example, go to Pubmed at http://www.ncbi.nlm.nih.gov/entrez/query.fcgi and search on 'beta-glucan cancer'. You will find exciting information such as 'Beta-glucan inhibits the genotoxicity of cyclophosphamide, adriamycin and cisplatin' which describes how beta-glucan pretreatment approximately halves the damage done by chemotherapy drugs. Why not ask your doctor about it if you have decided to receive chemotherapy? Ralph W. Moss, Ph.D., a respected cancer industry analyst, states: "In more than thirty years of studying and chronicling developments in the field of cancer therapy I have seen many useful alternative treatments at first mercilessly vilified and driven underground, only to resurface years later when science eventually confirms that the active principle of such a treatment really does have some recognizable, quantifiable effect against cancer cells."

For example, hyperthermia or heat therapy-once branded as a "worthless remedy" and "quackery" by the ACS in 1967, was removed years later from the unproven methods list. Today, hyperthermia has been hailed by some oncologists as the fifth modality in cancer treatment after surgery, radiation, drugs, and immunotherapy [NCI web site]. This is a long debated topic. Many authors before describing this topic always have a disclaimer section that says 'Don't replace traditional cancer treatment with CAM alone, because it may not work as well it worked with someone else'. I would like to state here that many people who revert to CAM , is because they have already taken traditional treatment that did not work with them. So, they actually have nothing to loose. I have come across a large number of people through mailing lists that have almost cured their cancer with CAM. So, please be discreet while choosing this option.

Following are some of the CAM treatments prescribed by famous physicians and scientists.

- *Dr Johana Budwigs diet plan*: Dr. Johana Budwig was 6 times Nobel Prize nominee in Biochemistry. She worked extensively with terminal cancer patients and has formulated a diet plan that was very popular during 1950's. She treated seriously ill terminal cancer patients with flax seed oil and cottage cheese. After few months of treatment, the patients revived and their cells were once again healthy with missing phosphatids and some of them even got cured (Budwig, 1956).

  Her diet plan includes a daily dosage of freshly ground flax seed with freshly made organic cottage cheese. The diet can be supplemented with freshly squeezed fruit juice without any sugar. The diet has completely forbidden the use of sugar, meat, butter, preserved food, animal fat etc. Fruits and vegetable can be used raw or vegetables can be steamed. Green tea can be consumed.

- *Dr. Mark Gerson Therapy*: Dr. Greson therapy comprises of only freshly squeezed fruit and vegetable juice. He has eliminated sodium from the diet plan. Instead he suggests potassium. Potassium can be found in the outer parts of vegetables. He also suggested liver and body detoxification.

- *Dr. Moerman's Anti-Cancer Diet*: Dr. Moerman was a country doctor in Holland, who did cancer research therapy for more than 55 years!! Like many doctors, he considered that 'Cancer was caused by malfunctioning of the immune system that manifested itself outwardly on the body's weakest organ as tumor'. So, his approach was to enhance immune system. He identified 8 key nutrients that act as immune builders. Vitamin A, Vitamin B complex, Vitamin C, VitaminE, citric acid, Iodine, Iron, sulphur. His diet plan mostly consists of vegetables, fruits, some whole grains, butter, egg yolk etc. He had completely forbidden the use of salt in any form.

- *Dr. George Malkmus*: This diet was developed by Dr. George Malkums. He cured his own cancer with barley green juice and carrot juice!! He mostly consumed 85% raw vegetables a day and 15% cooked food. More on this diet can be found from http://www.hacres.com.

- *Macrobiotic Diet*: This diet plan has achieved a great success over a long period of time. This is also referred to as celebrity diet. Most of the very popular celebrities vow by this diet plan. Macrobiotics was developed by Japanese physician, Sagen Ishizuka (1850-1910). He cured himself of cancer by abandoning the refined diet of affluent Japan and reverting back to the unpurified Japanese diet of brown rice, soybeans, fish, miso soup, sea vegetables and other oriental food. The macrobiotic diet is a mainly vegetarian diet consisting of 50% whole cereal grains, 20 to 30% locally grown vegetables, small amounts of soups, beans and sea vegetables, white meat, fish and fruit in limited amounts. Potatoes, sweet potatoes, tomatoes, eggplants, peppers, asparagus, spinach, beets, zucchini, avocadoes, mayonnaise, tea, coffee and red meat are to be completely avoided.

  Rudolf Breuss Diet Plan: The Breuss-Cure, which originated in Germany, lasts for 6 weeks. A maximum of 500 ml of freshly pressed vegetable juices is used, mainly beet root with some carrot, celery and radish. In addition, herbal tea and onion broths are recommended. The treatment is claimed to have cured thousands of patients. The book, "The Breuss Cancer Cure", details the Breuss diet. It has sold over 900,000 Copies, been translated into five languages, and claims to have led to over 45,000 testimonials from cured sufferers. Both the book and the diet are still actively being used. The Breuss diet is based on a 42 -day fast, but the definition of "fast" used in the Breuss diet actually includes certain types of foods, such as raw fruits and vegetable juices, all taken in liquid form. The theory is that cancer cells can only live on the protein of solid food. Therefore, if you drink nothing but vegetable juice and tea for 42 days the cancerous cells die while the normal cells continue to thrive. "Breuss vegetable juice consists of 55% red beet root, 20% carrots, 20% celery root, 3% raw potato, 2% radishes. The potato is optional except for liver cancer where it plays an important part." His book actually talks about multiple diseases. The diet contains virtually zero glucose and other sugars. Researchers also believe it works because cancer cells are very inefficient at processing glucose and other sugars and that the formula literally starves the cancer cells to death by depriving them of glucose and other sugars. Normal cells can survive on much less glucose and other sugars because they are much more efficient at processing these items. Filtered green tea and filtered Essiac tea have been natural enhancements to the Breuss diet.

- *Turmeric*: Imagine a natural substance so smart it can tell the difference between a cancer cell and a normal cell; so powerful it can stop chemicals in their tracks; and so strong it can enable DNA to walk away from lethal doses of radiation virtually unscathed. Curcumin has powers against cancer so beneficial that drug companies are rushing to make drug versions. Curcumin is all this and more. Curcuma longa is a ginger-like plant that grows in tropical regions. The roots contain a bright yellow substance (turmeric) that contains curcumin and other curcuminoids. Turmeric is used in traditional Indian cooking since time immemorial. Turmeric has been used in Ayurvedic and Chinese medicine for centuries. But it is only within the past few years that the extraordinary actions of curcumin against cancer have been scientifically documented. Among its many benefits, curcumin has at least a dozen separate ways of interfering with cancer." Turmeric (tumeric) has long been revered as the foundation of herbal programs for health. In India's system of Ayurvedic medicine, turmeric has been recognized as a key balancing and detoxifying herb. In Indonesia, Japan, and China people embrace turmeric for its powerful yet safe liver detoxification and in the western medical and herbal traditions, turmeric is considered by many to be one of the most important healing herbs. Extensive trials have been conducted to ascertain its value as an anticancer drug.

Turmeric launches a multiple attack on cancerous cells. Scientists at M. D. Anderson, Texas wrote in January 2003: "Extensive research over the last 50 years has indicated (curcumin) can both prevent and treat cancer. The anticancer potential of curcumin stems from its ability to suppress proliferation of a wide variety of tumor cells, down-regulate transcription factors NF-kappa B, AP-1 and Egr-1; down-regulate the expression of COX2, LOX, NOS, MMP-9, uPA, TNF, chemokines, cell surface adhesion molecules and cyclin D1; down-regulate growth factor receptors (such as EGFR and HER2); and inhibit the activity of c- Jun N-terminal kinase, protein tyrosine kinases and protein serine/threonine kinases."In the latest of a series of reports, the M. D. Anderson says: "Curcumin can suppress tumor initiation, promotion, and metastasis (Aggarwal et al., 2003).

- *Essiac Tea*: Essiac is an herbal cancer treatment developed by a Canadian nurse, Renée Caisse (1888-1978). (Essiac is Caisse spelled backwards.) Ms. Caisse claimed that the formula had been given to her in 1922 by a patient whose breast cancer had been cured by a traditional Native American healer in Ontario, Canada. Thousands of patients have since been treated with this herbal mixture, most of them at Caisse's own Bracebridge Clinic in Ontario. While this clinic was shut down in 1942, the controversy over Essiac simmered for years. Charles Brusch, MD - President John Kennedy's physician is said to have declared that Essiac 'cures cancer'. Essiac cannot be freely marketed in either the US or

Canada. However, a company in Ontario is allowed to provide Essiac to Canadian patients under a special arrangement with health officials there. A number of companies now sell competing "original" Essiac in the form of a tea, but the authenticity of some of these formulas are open to question. Since Essiac is now a very well known treatment, it is important to point out that while Caisse did provide the herbs for oral use, most of her greatest success would seem to have involved the injectible form of the herbs. They would obviously be more potent and fast-acting if administered in this way. Caisse actually felt quite strongly that this method of delivery was the only way to assure that the body could resist malignancy. The four core ingredients of Essiac are: Burdock root (*Arctium lappa*), Indian rhubarb (*Rheum officinale*), Sheep Sorrel (*Rumex acetosella*), Slippery elm (*Ulmus fulva*). All these four herbs normalize body systems by purifying the blood, promoting cell repair, and adding effective assimilation and elimination. When combined, their separate beneficial effects are synergistically enhanced (Foldeak and Dombradi, 1966; Dombradi and Foldeak, 1966; Morita *et al.*, 1984; Rhoads *et al.*, 1984).

- *Noni Juice*: Like grape juice, Noni juice contains a whole slew of cancer fighting nutrients. It kills cancer cells (the anthraquinone damnacanthal and the trace element selenium), it stops the spread of cancer (beta sitosterol, noni-ppt and limonene), it stimulates the white blood cells and other parts of the immune system (polysaccharides) and takes part in a process that enlarges cell membranes so they can better absorb nutrients (proxeronine aids in creating xeronine). And this is only a partial list. Noni juice is reputed to work quickly. It is regarded by many as a very effective cancer treatment. There were noticeable cases where terminal cancer patients were completely cured of their cancer in as little as 10-12 days, but this not necessarily typical. Frequently people wonder why this is not very effective against all patients. Consider this:" It is discovered that liquid Noni must be pasteurized (heat processed) before it can be shipped from Hawaii (where it naturally grows), and that sugar and juices are added to make it more palatable. Its taste and smell are so terrible that even researchers refused to drink it, thus the addition of sugar. Noni must be taken on an empty stomach because stomach acid destroys its properties. When sugar is added to Noni, the digestive process stimulated by the sugar destroys its properties. In other words, most of the Noni available is worthless" (Wong, 2004).

Recently, researchers found that the main reason Noni juice provides so many benefits is that it stimulates the production of Nitric Oxide in the body. The 1998 Nobel Prize for Medicine was awarded to three researchers for the discovery of Nitric Oxide. They found it to be a signaling molecule involved in controlling the circulation of blood regulating activities of the brain, lungs, liver, kidneys, stomach, and other organs. In addition, they found that it effected a "seemingly limitless" range of functions in the body. They found that Nitric Oxide reduces tumor growth, and increases the immune response against the radical replication of cells (Hornick *et al.*, 2003)

- *Laetrile or Nitriloside or Vitamin B 17 or Amygdalin*: Laetrile is found primarily in apricot kernels and comprises about 2-3% of the kernel. It is also available in the kernels of other fruits, such as plums, cherries, peaches, nectarines, pomegranates and apples. All fruit seeds have a healthy form of organic cyanide in them, from apple seed to apricot seed (not 'inorganic' cyanide which is deadly) (Diamond *et al.*, 1997).

  Nitriloside was "rediscovered" in 1920 by a California physician, Ernest Krebs while experimenting with flavorings for bootleg whisky. His son, Dr. Ernest Krebs, Jr. claimed to have purified it and coined the name 'laetrile' in 1952. Krebs' studies showed that when a human or animal system ingests sufficient amount of laetrile (or in its natural form, hydrocyanic acid), this substance becomes selectively toxic to cancer cells. How does laetrile work? One good description of this process is that laetrile is a parcel that contains poisons. When the parcel is unwrapped the poisons are released. Amygdalin is not digested in the stomach by hydrochloric acid, but passes into the small intestine where it is acted on by enzymes that split it into various compounds, which are then absorbed. The specific theoretical function of laetrile is its effect on cancer cells. Normal cells have an enzyme, rhodanase that inactivates the cyanide molecule of the laetrile compound. Cancer cells do not possess this enzyme. In fact, they have an enzyme, beta-glucosidase that releases the cyanide, which then poisons the cancer cells. So, normal cells do not have the power to unwrap the parcel. Only cancer cells have that power. In other words, laetrile is a substance that can be separated (by enzymes, in the presence of water) into glucose, benzaldehyde, and hydrocyanic acid. The last two substances are each, individually, a poison but together they work synergistically (i.e., they are more powerful when combined than separately). The enzyme that unwraps this package is beta-glucoronidase. This enzyme appears in great quantities in and around cancer cells - but not normal cells. The German doctor, Hans Nieper, argued that a synthetic version of laetrile, mandelonitrile, might be even more effective.

- *Wheat Grass Juice*: Fresh wheat grass is a potent source of many vitamins, minerals and plant enzymes. Wheat grass also contains Amygdalin/

Laetrile, although other sources, such as apricot seeds are more potent. This therapy consists primarily of detoxification and consuming a wheat-grass drink several times each day. Ann Wigmore pioneered and promoted wheat-grass juice after curing her own cancer with it in combination with an organic vegetarian diet. If oxygen is considered to be a bullet to kill cancer cells, then wheatgrass could be thought of as a shotgun blast at treating cancer. The number of ways it deals with cancer is reported to be numerous. Firstly, it contains chlorophyll (it is often referred to as "liquid chlorophyll"), which has almost the same molecular structure as hemoglobin.

Chlorophyll increases hemoglobin production, meaning more oxygen gets to the cancer cells. Selenium and laetrile are also contained in wheatgrass, and both have anti-cancer properties. Chlorophyll and selenium also help build the immune system. Furthermore, wheatgrass is one of the most alkaline foods known. It also contains the hormone abscisic acid (also called: dormin), the antioxidant enzyme. Recent studies indicate "Human Colon adenocarcinoma (Caco) cells, stimulated with 12-O-tetradecanoylphorbol-13-acetate, showed activation of the 20S proteasome activities at short incubation times and an increase in intracellular oxidative proteins. Cell treatment with wheat sprout extract led to proteasome inhibition in unstimulated cells and attenuated the effects mediated by TPA. Finally, exposure to the extract affected the expression levels of pro-apoptotic proteins" (Amici, 2008).

- *Maitake Mushroom*: Maitake means 'dancing mushroom' in Japanese. Maitake D-Fraction, extracted from maitake mushroom, has been reported to exert its antitumor effect in tumor-bearing mice by enhancing the immune system through activation of macrophages, T cells, and natural killer (NK) cells. In a previous study, the combination of immunotherapy with the maitake D-Fraction and chemotherapy suggested that the D-Fraction may have the potential to decrease the size of lung, liver and breast tumors in cancer patients (Kodama *et al.*, 2002). In the present study, maitake D Fraction was administered to cancer patients without anticancer drugs, and at the same time NK cell activity was monitored to investigate whether the activity is closely related with disease progression. Maitake D-Fraction hindered metastatic progress, lessened the expression of tumor markers, and increased NK cell activity in all patients examined. Thus maitake D-Fraction appears to repress cancer progression and primarily exerts its effect through stimulation of NK activity. In addition, it is concluded that measurement of NK cell activity may be a useful clinical parameter in monitoring disease progression during and following immunotherapy with maitake DFraction (Kodama *et al.*, 2003).

- *Shiitake - Lentinus edodes*: A vast amount of research into shiitake's medicinal properties has been undertaken and shows that it has the ability to fight tumors and viruses and enhance the immune system. Studies reveal that lenitan; an active component from Shitake mushroom shows significant antitumor property. In addition, "primed" lymphocytes, when given passively to immune deficient mice, were able to retard the development of tumors in these mice (Ngai, 2003).

1.8 Conclusion

Columbia University School of Public Health reports that 95% of cancer caused by diet and environment. Our body is introduced to thousands of new toxic chemical compounds in the last 50 years, but is not yet evolved to handle those in this short period of time. While traditional cancer research is growing phenomenally, the survival rate has not increased significantly. Nature provides cure to many diseases including cancer in form of phytochemicals that are excellent alternatives to traditional therapy. Instead of getting overtly dependent on traditional medicines, one can resort to life style and diet changes as an extremely important alternative for increasing survival.

References

A Cancer Therapy: Results of Fifty Cases and the Cure of Advanced Cancer by Dr. Max Gerson.

Aggarwal, B.B. *et al.*, 2003. *Anticancer Research*, 23: 363-98.

Amici, M. 2008. *Biochimie*, 90: 790-801.

Ansher, S.S. 1986. Federation of Chemistry and Toxicology, 24: 405.

Bellizzi, K.M. *et al.*, 2009. *Journal of Clinical Oncology*, [Epub ahead of print].

Budwig, J. 1956. Cytostatic or cytodynamic control of cancer? *Hippokrates*, 27: 605-12.

Buehlmeyer, K. *et al.*, 2007. *Annals of Anatomy*, 190:71-80.

Burns, C.P. *et al.*, 1990. *Nutrition Reviews*, 48: 233.

California: Future Medicine Publishing, Inc., pp. 812-813.

Demetrakopoulos, G.E. 1982. *Cancer Research*, 42: 7565.

Diamond, W.J. *et al.* 1997. An alternative medicine definitive guide to cancer. Tiburon,

Dombradi, C. and Foldeak, S. 1966. Screening report on the antitumor activity of purified Arctium lappa extracts. *Tumori*, 52: 173.

Dr. Moerman's Anti-Cancer Diet - Holland's Revolutionary Nutritional Program for Combating Cancer by Ruth Jochems (with a Forward by Linus Pauling).

Fair, A.M. and Montgomery, K. 2009. Energy balance, physical activity, and cancer risk. *Methods in Molecular Biology*, 472: 57-88.

Foldeak, S. and Dombradi, G. 1964. Tumor-growth inhibiting substances of plant origin. I. Isolation of the active principle of Arctium lappa. *Acta Physical Chemistry,* 10: 91-93.

Gogos, *et al.,* 1998. *Cancer,* 82: 395.

Guffy, M.M. *et al.,* 1984. *Cancer Research,* 44: 1863.

Hornick, C.A. *et al.,* 2003. *Angiogenesis,* 6: 143-149.

Hunter, K.W. *et al.,* 2008. *Breast Cancer Research,* 10: S2.

Jaakkola, K. 1992. *Anticancer Research,* 12: 599.

Kodama, N. *et al.,* 2002. *Alternative Medicines Reviews,* 7: 236-239.

Kodama, N. *et al.,* 2003. *Journal of Medicinal Foods,* 6: 371-377.

Krishanamurthi, S. *et al.,* 1971. *Radiology,* 99: 409.

Messina, M. *et al.,* 1991. *Journal of National Cancer Research Institute,* 83: 541.

Morita, K. *et al.,* 1984. A desmutagenic factor isolated from burdock (Arctium lappa Linne). *Mutatation Research,* 129: 25-31.

Ngai, P.H. 2003. Life Science, 73:3363-3374.

Pierce, S.K. 2002. Lipid rafts and B-cell activation. *Nature Reviews in Immunology,* 2: 96-105.

Rhoads, P. *et al.,* 1984. Anticholinergic poisonings associated with commercial burdock root tea. *Journal of Toxicology,* 22: 581-584.

Santamaria, L. *et al.,* 1990. Nutrients and Cancer Prevention, p.299, Prasad, Humana Press.

Singh, V.N., 1991. *American Journal of Clinical Nutrition,* 53: 386S.

The Breuss Cancer Cure: Advice for the Prevention and Natural Treatment of Cancer, Leukemia and Other Seemingly Incurable Diseases by Rudolf Breuss.

The Cancer Prevention Diet: Michio Kushi's Macrobiotic Blueprint for the Prevention and Relief of Disease by Michio Kushi, Alex Jack.

The New Cancer Survivors: Living with Grace, Fighting with Spirit by Natalie Davis Spingarn.

Vitamin B-17- forbidden weapon against cancer: The fight for Laetrile by Michael L Culbert. The Wheatgrass Book by Ann Wigmore.

Walters, R. "Essiac" Options: The Alternative Cancer Therapy Book, 110.

Waxman, S. *et al.,* 1982. *European Journal of Cancer Clinical Oncology,* 18: 685.

Wong, D.K. 2004. *Hawaii Medicinal Journal,* 63: 182-184.

You, W.C. *et al.,* 1989. *Journal of the National Cancer Institute,* 81: 162.

❑❑❑

24

Plant Antifungal Protein as Therapeutic Agents

Modhumita Dasgupta

CONTENTS

1.1	Introduction	362
1.2	Classes of Antifungal Proteins from Plants	362
1.3	Antifungal Proteins as Therapeutic Agents	363
	1.3.1 Antifungal/ antimicrobial activity	363
	1.3.2. Plant defensivs	363
	1.3.3 Mode of action	364
	1.3.4 Other antifungal proteins	364
1.4	Anti-HIV Peptides	365
	1.4.1 Ribosome – inactivating proteins (RIPs)	365
	1.4.2 Mode of action	366
	1.4.3 Lectins	367
	1.4.4 Mode of action	367
	1.4.5 Other proteins with anti HIV activity	367
1.5	Anti-Tumor Activity	368
	1.5.1 Lectins	368
	1.5.2 Mistletoe lectins	368
	1.5.3 Other lectins	369
1.6	Ribosome – Inactivating Protein (RIPs)	369
1.7	Other Proteins with Anticancer Activity	370
1.8	Other Therapeutic Roles	370
1.9	Conclusion	370
	References	371

1.1 Introduction

Fungi are an extremely diverse group of organisms, with about 2,50,000 species widely distributed in essentially every ecosystem. Muller and Loeffler (1976) estimated that the weight of fungi on Earth exceed that of humans. They are proficient at colonizing and use plants, humans and animals as substrates. During the last two decades, the incidence of human fungal infections, especially involving immunocompromised patients, has dramatically increased (Cox and Perfect, 1993; Fox, 1993). The majority of fungal exposures and infections are self limiting in intact animal hosts. However, in patients with compromised immune systems, infections even by fungal organisms with low virulence can be life threatening as reported in leukemia patients, where 50% fatalities was accounted to systemic fungal infections (Levitz, 1992; Pfaller and Wenzel, 1992). Nosocomial bloodstream infections have a similar fatality rate (Lyman and Walsh, 1992). Encouragingly, naturally occurring antifungal proteins and peptides, as well as synthetic derivatives have the potential of providing important clinical leads. This review will focus on the use antifungal proteins derived from plants as therapeutic agents.

1.2 Classes of Antifungal Proteins from Plants

Plants require a broad range of defence mechanisms to effectively combat invasion by microbial pathogens and possess both preformed and inducible mechanisms to resist pathogen entry. Some of the responses are constitutive and pathogen non-specific, but a majority of them are induced after recognition of the pathogen. Elicitors produced and released by the pathogen induce defence response, comprising the reinforcement of cell walls, production of phytoalexins and synthesis of defence-related or pathogenesis - related (PR) proteins. PR proteins in plants have been defined as proteins of a host that are induced only in response to attack by pathogens or by a related event (van Loon *et al.*, 1994). The recognized PR proteins have been extensively reviewed and currently comprise 17 families of induced proteins. These include four families of chitinases, one each of 1,3-glucanases, proteinase inhibitors, one specific peroxidase, PR-1 family with unknown biochemical properties, the thaumatin-like PR-5 family, the birch allergen Betv1-related PR-10 family, defensins, lipid-transfer proteins (LTPs), thionins and other proteins including 2S storage albumins and ribosome inactivating proteins (RIPs). Important groups of antimicrobial proteins not induced by pathogen attack and hence not included under PR proteins are lectins and cysteine-rich peptides (Broekaert *et al.*, 2000).

1.3 Antifungal Proteins as Therapeutic Agents

1.3.1 Antifungal/antimicrobial activity

The need for safe and effective antifungal agents increases in parallel with the expanding number of immunocompromised patients at risk for invasive fungal infections. Filamentous fungi belonging to the *Aspergillus* genus, especially *A. fumigatus* and *A. flavus* and *Candida albicans* are important causative agents of invasive mycoses (Mavor *et al.*, 2005). *Candida* bloodstream infections have steadily increased since the 1980s and account for 8–15% of all bloodstream infections (Garbino *et al.*, 2002; Hobson, 2003). Since no fungal vaccines are currently licensed, the only clinical recourse to combat invasive mycoses is the use of antifungal therapeutics (antimycotics). The emergence of fungal pathogens resistant to current therapies further compounds the dearth of antifungal agents. Currently available antifungal compounds act on targets also found in mammalian cells (Debono and Gordee 1994), which may result in toxicity or an adverse drug interaction. It is therefore imperative to discover antifungal compounds that are not toxic to mammalian cells and the past decade has witnessed a dramatic growth in knowledge of natural peptides.

Antifungal peptides are classified by their mode of action. The first group acts by lysis, which occurs via several mechanisms (Shai, 1995). Generally, these include amphipathic peptides, which bind only to the membrane surface and disrupt the membrane structure without traversing the membrane or traverse membranes and interact specifically with certain molecules and peptides, which aggregate in a selective manner, forming aqueous pores of variable sizes, allowing passage of ions or other solutes. The second peptide group interferes with cell wall synthesis or the biosynthesis of essential cellular components such as glucan or chitin (Debono and Gordee, 1994).

1.3.2 Plant defensivs

Several groups of plant antifungal proteins have revealed toxicity to human fungal pathogens and hence provide valuable leads as therapeutic agents. Plant defensivs exhibit antifungal activity against a broad range of phytopathogenic fungi and can also inhibit human fungal pathogens (Thevissen *et al.*, 2004).

Plant defensins, HsAFP1 from coral bells (*Heuchera sanguinea*) (Osborn *et al.*, 1995), RsAFP2 from radish (*Raphanus sativus*) (Terras *et al.*, 1992) and DmAMP1 from dahlia (*Dahlia merckii*) (Osborn *et al.*, 1995) inhibited the growth of *Candida* spp. They interacted with fungus-specific high-affinity plasma membrane binding sites and induced membrane permeabilization (Thevissen *et al.*, 2007). Hs-AFP1 and Rs-AFP2 bound to mannan, chitin, cholesterol, ergosterol, galactocerebrosides and sphingomyelin (De Lucca, 1998). Similarly, the pea defensin Psd1 (Almeida *et al.*, 2000) affected the normal progression of the cell cycle and regulated the interkinetic nuclear migration in retinal neuroblasts (Lobo *et al.*, 2007). MsDef1, a broad-spectrum antifungal defensin from alfalfa (*Medicago sativa*) seed, previously referred to as AlfAFP (Gao *et al.*, 2000) inhibited hyphal elongation of the fungal pathogen *F. graminearum*, causing a hyperbranching phenotype It was demonstrated that MsDef1 blocked the L-type Ca^{2+} channel in mammalian cells (Spelbrink *et al.*, 2004). In susceptible fungi also MsDef1 targeted the Ca^{2+} channels, which are ubiquitous signaling molecule regulating cAMP levels (Iida, *et al.*, 1990), bud formation (Davis, 1995) and hyphal elongation (Jackson and Heath, 1993). Another highly basic defensin, Ib-AMP3 derived from *Impatiens balsamina* reduced the viability of germinated conidia of *A. flavus* and *F. moniliforme* with very high affinity for chitin (Tailor *et al.*, 1997; De Lucca, 1998).

1.3.3 Mode of action

Two groups of defensins were originally described including the morphogenic plant defensins causing reduced hyphal elongation with a concomitant increase in hyphal branching and non-morphogenic plant defensins causing reduced hyphal elongation with no marked morphological distortions (Osborn *et al.*, 1995). Recently it was demonstrated that plant defensins did not directly interact with the plasma membrane phospholipids. (Aerts *et al.*, 2008). Further, in contrast to mammalian and insect defensins, plant defensins does not induce ion permeable pores in artificial membranes composed of phospholipids, nor change their electrical properties.

1.3.4 Other antifungal proteins

Some plants produce lipid transfer proteins, a family of homologous peptides having eight disulfide-linked cysteines, causing toxicity to fungal hyphae. Onion seeds (*Allium cepa*) produce the lipid transfer peptide, ACE-AMP1, which inhibited *F. oxysporum* (Cammue *et al.*, 1995). Uncharacterized proteins showing toxicity to the aflatoxin producing *Aspergillus flavus* was reported from the seeds of *Sorghum bicolor* (Ghosh and Ulaganathan, 1996). Zeamatin, an antifungal peptide purified from the seeds of maize showed membrane - permealizing activity (Roberts and Selitrennikoff, 1990) and was found to cause hyphal rupture in *C. albicans* and *Neurospora crassa*. It was reported to be identical to bifunctional α-amylase and trypsin inhibitor (Richardson *et al.*, 1987; O' Rourke *et al.*, 2001). It had potent antifungal activity *in vitro* against a number of human and plant pathogens (Trudel *et al.*, 1998; Selitrennikoff *et al.*, 2000). Recently, zeamatin in synergy with nikkomycin Z and fluconazole was shown to be effective *in vivo* in systemic murine candidosis model (Selitrennikoff *et al.*, 2000). Peptides in the zeamatin family are also present in *Avena sativa, Sorghum bicolor,* and *Triticum aestivum* seeds (Vigers *et al.*, 1991). A flax seed antifungal peptide similar to zeamatin, in synergy with nikkomycin Z, inhibited *C. albicans* (Bormeyer *et al.*, 1992). Members of the family Rhamnaceae and other plant families produce a basic cyclopeptides with antifungal activity (Gournelis *et al.*, 1997). Frangufoline, amphibine H, rugosanines A and B, and nummularines B, K, R, and S showed significant activity against *A. niger* but not *C. albicans* in zonal inhibition studies (Panday and Devi, 1990).

1.4 Anti-HIV Peptides

A large number of chemical entities, immunogens and topical medications, either synthetic or natural, have been brought into focus to combat AIDS epidemic. A large number of natural products isolated from various plant species, have the potential for development as future drugs for cure of AIDS and could be helpful in replacement of combination therapy presently prevalent for treatment of HIV patients.

1.4.1 Ribosome – inactivating proteins (RIPs)

Many plants contain proteins that are capable of inactivating ribosomes and accordingly are called ribosome-inactivating proteins or RIPs. These typical plant proteins received a lot of attention in biological and biomedical research because of their unique biological activities toward animal and human cells. RIPs are RNA N-glycosidases that inactivate ribosomes through a site-specific deadenylation of the large ribosomal RNA (Endo *et al.*, 1987; Endo and Tsurugi, 1987). Later it became evident that RIPs were also capable of inactivating many nonribosomal nucleic acid substrates and hence were considered polynucleotide:adenosine glycosidases (Barbieri *et al.*, 1997).

Two major classes were distinguished under RIPs: holo-RIPs and chimero-RIPs. Holo-RIPs consist exclusively of a single RNA N-glycosidase domain. Most holo-RIPs consist of a single, intact polypeptide of 30 kDa and are usually referred to as type 1 RIPs. Chimero-RIPs consists of one or more promoters with an N-glycosidase domain linked to a structurally different and functionally unrelated domain. Most chimero-RIPs are the so-called type 2 RIPs (Peumans *et al.*, 2001). Many of the plants from which RIPs are isolated are used in traditional Chinese medicine and may account for some of the reported clinical efficacies of these plants (Au *et al.*, 2000). Ricin, a protein isolated from the seeds of the castor bean plant (*Ricinus communis*) is a lectin and a member of a group of ribosome-inactivating proteins that block protein synthesis in eukaryotic ribosomes (Barbieri *et al.*, 1993). It is a 66 KDa globular protein and the toxic form is a heterodimer consisting of a 32-kDa A chain connected to the 32-kDa B chain through a single disulfide bond (Youle *et al.*, 1976). Ricin was recognized as one of the most exquisitely toxic plant-derived RIPs identified to date (Stirpe, 2004).

Abrin (ABR) belong to the type II ribosome inactivating proteins that are heterodimeric glycoproteins containing a toxophoric A-chain with protein synthesis inhibitory activity and a lectin B-chain binding to D-galactose moieties on the cell membrane (Lin *et al.*, 1970). In addition to its ability to inhibit protein synthesis, recent studies have shown that abrin induce apoptosis (Geier *et al.*, 1996; Komatsu *et al.*, 1998; Keppler-Hafkemeyer *et al.*, 1998). The specificity of the Abrin A chain (ABRA) to a 30-kDa thiol specific antioxidant protein-1 (AOP-1) in mitochondria was demonstrated by Shih et al. (2001). Further, it was reported that ABRA attenuated AOP-1 antioxidant activity in a dose-dependent manner and increased the intracellular level of reactive oxygen species (ROS) to trigger apoptosis.

Trichosanthin (TCS) is a type 1 ribosome-inactivating protein (RIP) isolated from root tuber of *Trichosanthes kirilowii* (Yeung *et al.*, 1985; 1987; Jin *et al.*, 1981). It is the principal active ingredient in the Chinese herb medicine Tian Hua Fen (Wang *et al.*, 1986). This single chain polypeptide depurinates A-4324 of the 28S rRNA and inhibits protein synthesis (Zhang *et al.*, 1992). The discovery that trichosanthin is a potent inhibitor against HIV-l replication in both acutely infected T-lymphoblastoid cells and chronically infected macrophages (McGrath, *et al.*, 1989) has stimulated considerable interest. Under normal enzymatic digestion conditions, trichosanthin also cleaves the supercoiled double-stranded DNA to produce nicked circular and linear DNA (Li *et al.*, 1991).

Clinical trial of TCS on AIDS patients had been conducted, and one of the major drawbacks was anaphylaxis (Byers *et al.*, 1990; 1994). Recently, several TCS derivatives were designed with less antigenicity while retaining most of the activity (Chan *et al.*, 1999; He *et al.*, 1999 a; b). These derivatives are therefore potential antiviral agent. Results suggested that the antiviral action may be partly mediated through enhanced apoptosis of infected cells (Wang *et al.*, 2005).

1.4.2 Mode of action

Trichosanthin elicit a dose-dependent inhibition of cell-free HIV-1 infection and replication. Viral-associated reverse transcriptase activity in HIV-1-infected H9 cells was inhibited in conjunction with the suppression of syncytium formation in the CD4-positive, syncytium-sensitive, Leu3a-sensitive T-cell line CEM-SS and viral core protein p24 expression (Lee-Huang *et al.*, 1995). Further, Trichosanthin also inhibited syncytium formation between infected H9 cells and uninfected Sup-T1 cells (Au *et al.*, 2000).

MAP30 isolated from the seeds and fruits of the bitter melon *Momordica charantia*; GAP31 from seeds of *Gelonium multiflorum* and DAP30 and DAP32 from carnation leaves (*Dianthus caryophyllus*) (De Clercq, 2000) were reported to inhibit HIV infection at noncytotoxic concentrations (Chow *et al.*, 1990). MAP30 also acted as DNA glycosylase/apurinic (AP) lyases, thus explaining the apparent ability of MAP30 to inhibit HIV-1 integrase and irreversibly relax supercoiled DNA. MAP30 has the potential to develop as anti- HIV agent (Wang *et al.*, 1999).

1.4.3 Lectins

Plant lectins represent a well-defined class of anti-HIV (microbicidal) drugs with a novel profile different from other existing anti-HIV drugs. A number of mannose-specific agglutinins from *Galanthus nivalis*, *Hippeastrum* hybrid, *Narcissus pseudonarcissus*, *Listera ovata*, *Cymbidium* hybrid, *Epipactis helleborine* and the N-acetylglucosamine specific lectin from *Urtica dioica* were found to inhibit HIV-1 and HIV-2 infection at similar concentrations as dextran sulfate (IC50: 0.2-0.6 mg/mL) or even lower (IC50: 0.04-0.08 mg/mL). Compared to sulfated polysaccharides, plant lectins also exhibit activity against various enveloped viruses other than HIV including HSV-1, HSV-2, CMV, RSV and influenza virus (Balzarini, *et al.*, 1992). These lectins inhibited virus entry (Balzarini *et al.*, 2004). Jacalin, a multimeric plant lectin was demonstrated to interact with the lymphocyte cell-surface molecule CD4, a known receptor for the HIV-1 and blocked HIV-1 infection of $CD4^+$ lymphoblastoid cells (Corbeau *et al.*, 1994).

1.4.4 Mode of action

Plant lectins would primarily interfere with the virus-cell fusion process. The mode of action is yet to be ascertained.

1.4.5 Other proteins with anti HIV activity

Small macrocyclic peptides, cycloviolins isolated from tropical plant *Leonia cymosa* and circulins, a group of cyclic peptides isolated from *Chassalia parvifolia*, exhibited anti-HIV activity (Gustafson, *et al.*, 1994; Hallock, *et al.*, 2000). Palicourein, a 37 amino acid cyclic polypeptide, isolated from organic extract of the tropical tree *Palicourea condensata* inhibited the *in vitro* cytopathic effects of HIV-1RF infection of CEM-SS cells (Bokesch, *et al.*, 2001). Saporin and luffin also exhibited anti- HIV integrase activity (Au *et al.*, 2000). *Myrianthus holstii* lectin (MHL), a 9284 Da, cysteine-rich protein isolated from aqueous extract showed anti-HIV activity with an EC50 value of 150 nM (Charan *et al.*, 2000).

1.5 Anti-Tumor Activity

1.5.1 Lectins

Several lectins were found to possess anticancer properties *in vitro* and *in vivo*. They are used as therapeutic agents, preferentially binding to cancer cell membranes or their receptors, causing cytotoxicity, apoptosis and inhibition of tumor growth. These compounds can become internalized into cells, causing cancer cell agglutination and/or aggregation. They affect the immune system by altering the production of various interleukins or by activating certain protein kinases. Lectins can bind to ribosomes and inhibit protein synthesis. They also modify the cell cycle by inducing non-apoptotic G1-phase accumulation mechanisms, G2/M phase cell cycle arrest, apoptosis and can activate the caspase cascade. They can also down regulate telomerase activity and inhibit angiogenesis (De Mejía and Prisecaru, 2005).

1.5.2 Mistletoe lectins

Extracts of *Viscum album* (mistletoe) are widely used in complementary cancer therapies in Europe. The aqueous extract of mistletoe has been used in conventional cancer therapy for decades (Bussing *et al.*, 1996). These lectins were identified as the main active principle of mistletoe extracts. They were shown to exhibit cytotoxic effects as well as immunomodulatory activities (Hajto *et al.*, 1989). There are three similar lectins in mistletoe plants with regard to antigenity and chemical structure (Dietrich *et al.*, 1992). The most important and highly investigated lectin is the galactoside-specific VAA-I with biological functions of cytotoxicity, RIP activity, induction of apoptosis, selective binding, release of cytokines and stimulation of natural killer (NK) cells. Similar activity was also demonstrated in the recombinant VAA (rVAA) (Zinke *et al.*, 1996; Hostanska *et al.*, 1999; Hajto *et al.*, 1998). In cultures of human peripheral mononuclear cells, VAA-I could stimulate cytokine production as well as apoptosis in approximately the same concentration as *in vivo* (Bussing *et al.*, 1996).

Similar lectins have been purified from European mistletoe (EM) (Bussing *et al.*, 1996) which enhanced efficiency of NK cells (Mueller *et al.*, 1990). Korean mistletoe (KML-C), a subspecies of European mistletoe, was also shown to have cytotoxic effect against tumor cells (Yoon *et al.*, 2003; Khil *et al.*, 2007). It was reported that KML-C had an immunomodulating activity to enhance the host defense system against tumors and its prophylactic and therapeutic effect on tumor metastasis was associated with the activation of NK cells and macrophages.

Another lectin purified from Chinese mistletoe was referred as active Chinese mistletoe lectin-55 (ACML-55) and demonstrated cytotoxicity, apoptosis and modulation of the cytokine network on human γδ T cell (Gong *et al.*, 2007). ACML-55 was reported to enhance colon cancer immune surveillance through regulating both innate and adaptive immune responses, suggesting its use in complementary therapy for treating colon cancer (Ma *et al.*, 2008). The basic biological effect of mistletoe lectins supports a long-term therapeutic modulation of the natural immune system which is associated with a protective effect in combination with toxic modalities of various therapies (Hajto *et al.*, 2005).

1.5.3 Other lectins

Several lectins present in beans and other edible plant products, have shown to possess anti-tumor activities (Pryme *et al.*, 1996). The roots of *Sophora flavescens* is used in Chinese traditional medicine for thousands of years. Recently, a mannose-binding lectin (designated as SFL) was isolated from the roots with molecular weight of 32 kDa. It agglutinated both rabbit and human erythrocytes (Deng *et al.*, 2000) and caused death of HeLa cells by both direct cytotoxicity and apoptosis in a typical caspase-dependent manner (Liu *et al.*, 2008)

Cytotoxic effects of a lectin prepared from garlic (*Allium sativum*) bulbs strongly reduced the growth and DNA synthesis of human tumor cells in a time and dose-dependent manner and induced apoptosis in the cells at a low concentration (Karasaki *et al.*, 2001).

1.6 Ribosome – Inactivating Protein (RIPs)

Abrin and Ricin isolated from *Abrus precatorius* and *Ricinus communis* respectively are categorized under the RIPs and their antitumor activity was demonstrated decades earlier against Ehrlich ascites tumour (Lin *et al.*, 1970; Walker *et al.*, 1984) and human tumor xenografts in athymic mice (Fodstad *et al.*, 1979). They are peculiar proteins, each having two peptide chains with different functions. One chain, the 'effectomer', is an enzyme capable of inactivating eucaryotic ribosomes. The other chain, the 'huptomer', is a lectin which binds the toxin to cells and facilitates the entry of the effectomer into the cytoplasm (Olsnes and Pihl, 1978).

Further, abrin-derived peptide (ABP) was evaluated for its *in vivo* therapeutic effectiveness in Dalton's lymphoma (DL) mice model. Analysis of the growth inhibitory mechanism in DL cells revealed nuclear fragmentation and condensation with the appearance of sub-G0/G1 peak indicative of apoptosis and was mediated by the reduction in the ratio of Bcl-2 and Bax protein expression and activation of caspase-3 through the release of cytochrome C in DL cells (Bhutia *et al.*, 2008).

The anticancer activity of ricin is well documented and their selectivity at killing tumor cells at low concentrations was reported by Zou and Zhan (2005). Efforts are in progress to use ricin as immunotoxin, where in they are either chemically or genetically modified to reduce severe side effects or nonspecific binding. Anti-B4- blocked ricin, is currently being evaluated in a phase III clinical trial for the treatment of non-Hodgkin's lymphoma (Lam *et al.,* 2004).

Saporin, a protein isolated from the seeds of *Saponaria officinalis* with N-glycosidase activity and categorized as ribosome-inactivating (Stirpe *et al.,* 1983) was reported as toxic on human breast cells (Gasperi Campani *et al.,* 1991).

1.7 Other Proteins with Anticancer Activity

An anticancer peptide was purified from the *Meretrix meretrix* with molecular weight of 31.4 KDa. The peptide inhibited the proliferation of BGC-823 cells up to 60%, destroyed the skeletal structures of the cells and enhanced the activity of enzymes like SOD and alkaline phosphatase (Leng *et al.,* 2005). A novel antitumor protein, coded as TBWSP31, was isolated from tartary buckwheat (*Fagopyrum tataricum*). The protein with the molecular weight of 57 kDa demonstrated antitumor activity against human mammary cancer cell Bcap37 (Guo *et al.,* 2007)

1.8 Other Therapeutic Roles

The PR-5 group of antifungal proteins is structurally related to the sweet tasting protein thaumatin. Receptors which specifically bind to PR-5 proteins were found to be homologous to mammalian adiponectin receptors (Narasimhan *et al.,* 2005) and hence can act as functional homologues of adiponectin and control adiponectin response in mammals. Adiponectin (also called 30-kDa adipocyte complement-related protein - Acrp30) is an antidiabetic and antiatherosclerotic protein hormone in mammals that conditions sensing of energy status, fatty acid oxidation and glucose transport upon interaction with adiponectin receptors. Osmotin, a 26kDa protein isolated from tobacco cells adapted under osmotic stress (Singh *et al.,* 1987) and other related proteins in the PR-5 family with structural homology to lectin-like β barrel domain, when suitably formulated in a pharmaceutically acceptable carrier, can be used as therapeutic agents for the treatment of a wide variety of mammalian disorders like diabetes, arteriosclerosis, and heart disease, in which adiponectin receptor-mediated pathways are implicated.

1.9 Conclusion

Till date, plant sciences have contributed to drug development through the discovery of bioactive compounds that are used to treat infections or other diseases, such as the anti-inflammatory aspirin or the anticancer agent taxol. Possibly, plant proteins are the next hallmark towards contribution of plants to medical sciences. Antifungal proteins and polypeptides have been isolated from diverse groups of organisms including plants, fungi, bacteria, insects and animals. The mechanisms of action of these proteins are as varied as their sources and include degradation of fungal cell wall polymer, membrane channel and pore formation, damage to cellular ribosomes, inhibition of DNA synthesis and inhibition of cell cycle. The mode of action of many proteins remains unknown and is a subject of active research. The range of fungi inhibited by antifungal proteins are extremely broad with humans pathogens being sensitive at micromolar levels, making them excellent leads as antifungal agents in clinical therapies. The other significant roles of antifungal proteins like lectins and RIPs are their potential agents as anti HIV and anti tumor compounds which can provide leads to develop novel drugs to combat these serious maladies.

References

Aerts, A.M., Franaois, I.E.J.A., Cammue, B.P.A. and Thevissen, K. 2008. The mode of antifungal action of plant, insect and human defensins. *Cell and Molecular Life Science*, DOI 10.1007/s00018-008-8035-0

Almeida, M.S., Cabral, K.M., Zingali, R.B. and Kurtenbach, E. 2000. Characterization of two novel defense peptides from pea (*Pisum sativum*) seeds. *Archieves in Biochemistry and Biophysics*, 378: 278-286.

Au, T.K., Collins, R.A., Lam, T.L., Ng, T.B., Fong, W.P. and Wan, D.C.C. 2000. The plant ribosome inactivating proteins luffin and saporin are potent inhibitors of HIV-1 integrase. *FEBS Letters*, 471: 169-172.

Balzarini, J., Neyts, J., Schols, D., Hosoya, M., Van Damme, E., Peumans, W. and De Clercq, E. 1992. The mannose-specific plant lectins from Cymbidium hybrid and *Epipactis helleborine* and the (N-acetylglucosamine) n-specific plant lectin from *Urtica dioica* are potent and selective inhibitors of human immunodeficiency virus and cytomegalovirus replication *in vitro*. *Antiviral Research*, 18: 191-207.

Balzarini, J., Laethem, K.V., Hatse, S., Vermeire, K., Clercq, E.D., Peumans, W., Van Damme, E., Van Damme, A.M., Bohlmstedt, A. and Schols, D. 2004. Profile of Resistance of Human Immunodeficiency Virus to Mannose-Specific Plant Lectins. *Journal of Virology*, 78: 10617-10627.

Barbieri, L., Battelli, M. and Stirpe, F. 1993. Ribosomes-inactivating proteins from plants. *Biochemistry and Biophysics Acta*, 1154: 237–282.

Barbieri, L., Valbonesi, P., Bonora, E., Gorini, P., Bolognesi, A., and Stirpe, F. 1997. Polynucleotide: adenosine glycosidase activity of ribosome-inactivating proteins: effect on DNA, RNA and poly(A). *Nucleic Acids Research*, 25: 518-522.

Bhutia, S.K., Mallick, S.K., Maiti, S. and Maiti, T.K. 2008. Inhibitory effect of *Abrus* abrin-derived peptide fraction against Dalton's lymphoma ascites model. *Phytomedicine,* doi:10.1016/j.phymed.2008.07.001.

Bokesch, H.R., Pannell, L.K. and Cochran, P.K. 2001. A novel anti-HIV macrocyclic peptide from *Palicourea condensata. Journal of Natural Products,* 64: 249-250.

Borgmeyer, J.R, Smith, C.E. and Khai Hutnk, Q. 1992. Isolation and characterization of a 25 kDa antifungal protein from flax seeds. *Biochemistry and Biophysics Research Communications,* 187: 480-487

Broekaert, W.F., Terras, F.R.G. and Cammue, B.P.A. 2000. Induced and preformed antimicrobial proteins. *In*: Slusarenko AJ, Fraser RSS, van Loon LC, eds. Mechanisms of resistance in plant diseases. Dordrecht: Kluwer 371 – 477.

Bussing, A., Suzart, K., Bergmann, J., Pfuller, U., Schietzel, M. and Schweizer, K. 1996. Induction of apoptosis in human lymphocytes treated with *Viscum album* L. is mediated by the mistletoe lectins. *Cancer Letters,* 99: 59-72.

Byers, V.S., Levin, A.S., Waites, L.A., Starrett, B.A., Mayer, R.A., Clegg, J.A., Price, M.R., Robins, R.A., Delaney, M. and Baldwin, R.W. 1990. Phase I/II study of trichosanthin treatment of HIV disease. *AIDS,* 4: 1189–1196.

Byers, V.S., Levin, A.S., Malvino, A., Waites, L., Robins, R.A. and Baldwin, R.W. 1994. A phase II study of e?ect of addition of trichosanthin to zidovudine in patients with HIV disease and failing antiretroviral agents. *AIDS Research Human Retroviruses,* 10: 413-420.

Cammue, B.P.A., Thevissen, K., Hendricks, M., Eggermont, K., Goderis, I.J., Proost, P., Van Damme, J., Osborn, R.W., Guerbette, F., Kader, J.C. and Broekaert, W.F. 1995. A potent antimicrobial protein from onion seeds showing sequence homology to plant lipid transfer protein. *Plant Pathology,* 109: 445-455.

Chan, W.L., Shaw, P.C., Li, X.B., Xu, Q.F., He, X.H. and Tam S.C. 1999. Lowering of trichosanthin immunogenicity by site specific coupling to dextran, *Biochemistry Pharmacology,* 57: 927-934.

Charan, R.D., Munro, H.G. and R. O'Keefe, B. 2000. Isolation and characterisation of *Myrianthus holstii* lectin, a potent HIV-1 inhibitory protein from the plant *Myrianthus holstii. Journal of Natural Products,* 63: 1170-1174.

Chow, T.P., Feldman, R.A., Lovett, M. and Piatak, M. 1990. Isolation and DNA sequence of a gene encoding alpha-trichosanthin, a type I ribosome-inactivating protein. *Journal of Biological Chemistry,* 265: 8670-8674.

Corbeau, P., Haran, M, Binz, H. and Devaux, C. 1994. Jacalin, a lectin with anti-HIV-1 properties, and HIV-1 gp120 envelope protein interact with distinct regions of the CD4 molecule. *Molecular Immunology,* 31: 569-75.

Cox, G. and Perfect, J. 1993. Fungal infections. *Current Opinions in Infectious Diseases,* 6: 422-426.

Davis, T.N. 1995. Calcium in *Saccharomyces cerevisiae.* Adv. Second Messenger. *Phosphoprotein Research,* 30: 339-358.

Debono, M. and Gordee, R.S. 1994. Antibiotics that inhibit fungal cell wall development. *Annual Reviews in Microbiology*, 48: 471-497.

De Clercq, E. 2000. Current lead natural products for the chemotherapy of human immunodeficiency virus (HIV) infection. *Medical Research Reviews*, 20: 323-349.

De Lucca, A.J. 1998. Fungicidal and binding properties of three plant peptides. *Mycopathologia*, 144: 87-91.

De Mejía, E. and Prisecaru, V. 2005. Lectins as Bioactive Plant Proteins: A Potential in Cancer Treatment. *Critical Reviews in Food Science and Nutrition*, 45: 425-445.

De Samblanx, G.W., Goderis, I.J., Thevissen, K., Raemaekers, R., Fant, F., Borremans, F., Acland, D.P., Osborn, R.W., Patel, S. and Broekaert, W.F. 1997. Mutational analysis of a plant defensin from radish (*Raphanus sativus* L.) reveals two adjacent sites important for antifungal activity. *Journal of Biological Chemistry*, 272: 1171-1179.

Dietrich, J.B., Ribereau-Gayon, G., Jung, M.L., Franz, H., Beck, J.P. and Anton, R. 1992. Identity of the N-terminal sequences of the three A chains of mistletoe (*Viscum album* L.) lectins: homology with ricin-like plant toxins and single chain ribosome-inhibiting proteins. *Anticancer Drug*, 3: 507-511.

Endo, Y., Mitsui, K., Motizuki, M., and Tsurugi, K. 1987. The mechanism of action of ricin and related toxic lectins on eukaryotic ribosomes. The site and the characteristics of the modification in 28S ribosomal RNA caused by the toxins. *Journal of Biological Chemistry*, 262: 5908-5912.

Endo, Y. and Tsurugi, K. 1987. RNA N-glycosidase activity of ricin A-chain. Mechanism of action of the toxic lectin ricin on eukaryotic ribosomes. *Journal of Biological Chemistry*, 262: 8128-8130.

Fodstad, O., Johannessen, J. V., Schjerven, L, and Pihl, A. 1979. Toxicity of abrin and ricin in mice and dogs. *Journal of Toxicology and Environmental Health*, 5: 1073-1084.

Fox, J.L. 1993. Fungal infection rates are increasing. *ASM News*, 59: 515-518.

Gao, A.G., Hakimi, S.M., Mittanck, C.A., Wu, Y., Woerner, B.M., Stark, D.M., Shah, D.M., Liang, J. and Rommens, C.M. 2000. Fungal pathogen protection in potato by expression of a plant defensin peptide. *Nature Biotechnology*, 18: 1307-1310.

Garbino, J., Kolarova, L., Rohner, P., Lew, D., Pichna, P. and Pittet, D. 2002. Secular trends of candidemia over 12 years in adult patients at a tertiary care hospital. *Medicine (Baltimore)* 81: 425-433.

Gasperi Campani, A., Zoli, W., Volpi, A., Roncuzzi, L. and Amadori, D. 1991. Inhibition of growth of breast cancer cells *in vitro* by the ribosome-inactivating protein saporin 6. *Anticancer Research*, 11: 1007-1011.

Geier, A., Bar-shalom, I., Beery, R., Haimsohn, M., Hemi, R., Malik, Z., Lunenfeld, B., and Karasik, A. 1996. Induction of apoptosis in MDA-231 cells by protein synthesis inhibitors is suppressed by multiple agents. *Cancer Investigation*, 14: 435-444.

Ghosh, M. and Ulaganathan, K. 1996. Mature seeds of Sorghum contain proteins toxic to Aflatoxin producing *Aspergillus flavus. Journal of. Stored Products Research,* 32: 339-343.

Gong, F., Ma, Y., Ma, A., Yu, Q., Zhang, J., Nie, H., Chen, X., Shen, B., Li, N. and Zhang, D. 2007. A lectin from Chinese mistletoe increases gamma delta T cell-mediated cytotoxicity through induction of caspase-dependent apoptosis. *Acta Biochimistry and Biophysics Sin. (Shanghai)* 39: 445-452

Gournelis, D.C., Laskaris, G.G. and Verpoorte, R. 1997. Cyclopeptide alkaloids. *Natural Products Report,* 14: 75-82.

Guo, X., Zhu, K., Zhang, H. and Yao, H. 2007. Purification and characterization of the antitumor protein from Chinese tartary buckwheat (*Fagopyrum tataricum* Gaertn.) water-soluble extracts. *Journal of Agricultural and Food Chemistry,* 55: 6958-6961.

Gustafson, K.R. Sowder, R.C., Henderson, L.E., Parsons, I.C., Kashman, Y., Cardellina, J.H., McMahon, J.B., Buckheit Jr., R.W., Pannell, L.K., Boyd, M.R. 1994. Circulins A and B. Novel human immunodeficiency virus (HIV)-inhibitory macrocyclic peptides from the tropical tree *Chassalia parvifolia. Journal of American Chemical Society,* 116: 9337-9338.

Hajto, T., Hostanska, K., Gabius and H.J. 1989. Modulatory potency of the galactosidespecific lectin from mistletoe extract (Iscador) on the host defense system in vivo in rabbits and patients. *Cancer Research,* 49: 4803-4808.

Hajto, T., Hostanska, K. and Weber, K. 1998. Effect of a recombinant lectin, *Viscum album* agglutinin (rVAA) on secretion of interleukin-12 in cultured human peripheral blood mononuclear cells and on NK cell-mediated cytotoxicity of rat splenocytes *in vitro* and in *in vivo. Nature Immunology,* 16: 34-46.

Hajto, T., Hostanska, K., Berki, T., Palinkas, L., Boldizsar, F. and Nemeth, P. 2005. Oncopharmacological Perspectives of a Plant Lectin (*Viscum album* Agglutinin-I): Overview of Recent Results from In vitro Experiments and *In vivo* Animal Models, and Their Possible Relevance for Clinical Applications. *ECAM,* 2: 59-67.

Hallock, Y.F. Sowder, R.C., Pannell, L.K., Hughes, C.B., Johnson, D.G., Gulakowski, R., Cardellina, J.H. and Boyd, M.R. 2000. Cycloviolins A–D, anti-HIV macrocyclic peptides from *Leonia cymosa. Journal of Organic Chemistry,* 65: 124-128.

He, X.H., Shaw, P.C., Xu, L.H. and Tam, S.C. 1999a. Site-directed polyethylene glycol modification of trichosanthin: effects on its biological activities, pharmacokinetics, and antigenicity, *Life Science,* 64: 1163-1175.

He, X.H., Shaw, P.C. and Tam, S.C. 1999b. Reducing the immunogenicity and improving the *in vivo* activity of trichosanthin by site-directed pegylation, *Life Science,* 65: 355-368.

Hobson, R.P. 2003. The global epidemiology of invasive Candida infections is the tide turning? *Journal of Hospital Infections,* 55: 159-168.

Hostanska, K., Hajto, T. and Fischer, J. 1999. Selective modulation of phosphatitylserine expression on various subpopulations of human peripheral blood lymphocytes by a plant lectin, *Viscum album* agglutinin VAA-I and its recombinant form (rVAA) *in vitro. Cancer Detection and Prevention,* 23: 511-523.

Iida, H., Sakaguchi, S., Yagawa, Y. and Anraku, Y. 1990. Cell cycle control by Ca^{2+} in *Saccharomyces cerevisiae*. *Journal of Biological Chemistry*, 265: 21216–21222.

Jackson, S.L. and Heath, I.B. 1993. Roles of calcium ion in hyphal tip growth. *Microbiology Reviews*, 57: 367-382.

Jiratchariyakul, W., Wiwat, C., Vongsakul, M., Somanabandhu, A., Leelamanit, W., Fujii, I., Suwannaroj, N. and Ebizuka, Y. 2001. HIV inhibitor from Thai bitter gourd. *Planta Medica*, 67: 350-353.

Karasaki, Y., Tsukamoto, S., Mizusaki, K., Sugiura, T. and Gotoh, S. 2001. A garlic lectin exerted an antitumor activity and induced apoptosis in human tumor cells. *Food research International*, 34: 7-13.

Keppler-Hafkemeyer, A., Brinkmann, U. and Pastan, I. 1998. Role of caspases in immunotoxin-induced apoptosis of cancer cells. *Biochemistry*, 37: 16934-16942.

Khil, L.Y., Kim, W., Lyu, S., Park, W.B., Yoon, J.W. and Jun, H.S. 2007. Mechanisms involved in Korean mistletoe lectin-induced apoptosis of cancer cells. *World Journal of Gastroenterology*, 13: 2811-2818.

Komatsu, N., Oda, T., and Muramatsu, T. 1998. Involvement of both caspase-like proteases and serine proteases in apoptotic cell death induced by ricin, modeccin, diphtheria toxin, and pseudomonas toxin. *Journal of Biochemistry*, 124: 1038-1044.

Lam, L., Lam, C. and Cao, Y. 2004. Immunotoxins - a new class of anticancer drugs. *Drugs of the Future*, 29: 609-612.

Lee-Huang, S., Kung, H.F., Huang, P.L., Li, B.Q., Huang, P., Huang, H.I. and Chen, H.C. 1991. A new class of anti-HIV agents: GAP31, DAPs 30 and 32. *FEBS Letters*, 291: 139-144.

Lee-Huang, S., Huang, P.L., Chen, H.C., Huang, P.L., Bourinbaiar, A., Huang, H.I. and Kung, H.F. 1995. Anti-HIV and anti-tumor activities of recombinant MAP30 from bitter melon. *Gene*, 161: 151-156.

Leng, B., Liu, X.D. and Chen, Q.X. 2005. Inhibitory effects of anticancer peptide from *Mercenaria* on the BGC-823 cells and several enzymes. *FEBS Letters*, 579 : 1187-1190.

Levitz, S. 1992. Overview of host defenses in fungal infections. *Clinical Infectious Diseases*, 14: 37-42.

Li, M.X., Yeung, H.W., Pan, L.P. and Chan, S.I. 1991. Trichosanthin, a potent HIV-1 inhibitor, can cleave supercoiled DNA *in vitro*. *Nucleic Acids Research*, 19: 6309-6312.

Lin, Y.J., Tserng, K.Y. Chen, C.C., Lin L.T. and Tung, T.C. 1970. Abrin and Ricin: New Anti-tumour Substances. *Nature*, 227: 292-293.

Liu, Z., Liu, B., Zhang, Z.T., Zhou, T.T., Bian, H.J., Min, M.W., Liu, Y.H., Chen, J. and Bao, J.K. 2008. A mannose-binding lectin from *Sophora flavescens* induces apoptosis in HeLa cells. *Phytomedicine*, 15: 867-875.

Lobo, D.S., Pereira, I.B., Fragel-Madeira, L., Medeiros, L.N., Cabral, L.M., Faria, J., Bellio, M., Campos, R.C., Linden, R. and Kurtenbach, E. 2007. Antifungal *Pisum sativum* defensin 1 interacts with *Neurospora crassa* cyclin F related to the cell cycle. *Biochemistry*, 46: 987-996.

Lyman, C.A. and Walsh, T.J. 1992. Systemically administered antifungal agents: a review of their clinical pharmacology and therapeutic applications. *Drugs*, 44: 9-35.

Ma, Y.H., Cheng, W.Z., Gong, F., Ma, A.L., Yu, Q.W., Zhang, J.Y., Hu, C.Y., Chen, X.H. and Zhang, D.Q. 2008. Active chinese mistletoe lectin-55 enhances colon cancer surveillance through regulating innate and adaptive immune responses. *World Journal of Gastroenterology*, 14: 5274-5281.

Mavor, A.L., Thewes, S. and Hube, B. 2005. Systemic fungal infections caused by Candida species: epidemiology, infection process and virulence attributes. *Current Drug Targets*, 6: 863-874.

McGrath, M.S., Hwang, K.M., Caldwell, S.E., Gaston, I., Luk, K.C., Wu, P., Ng, V.L., Crowe, S., Danials, J., Marsh, J., Deinhart, T., Lekas, P.V., Vennary, I.C., Yeung, H.W. and Lifson, J.D. 1989. GLQ223: an inhibitor of human immunodeficiency virus replication in acutely and chronically infected cells of lymphocyte and mononuclear phagocyte lineage. *Proceedings of National Academy of Science U.S.A*, 86: 2488-2484.

Mueller, E.A. and Anderer, F.A. 1990. A *Viscum album* oligosaccharide activating human natural cytotoxicity is an interferon gamma inducer. *Cancer Immunology Immunother*, 32: 221-227

Muller, E., and Loeffler, W. 1976. Mycology: an outline for science and medical students. G. Thieme Publishers, Stuttgart, Germany.

Narasimhan, M.L., Coca, M.A., Jin, J., Yamauchi, T., Ito, Y., Kadowaki, T., Kim, K.K., Pardo, J.M., Damsz, B., Hasegawa, P.M., Yun, D.J. and Bressan, R.A. 2005. Osmotin is a homolog of mammalian adiponectin and controls apoptosis in yeast through a homolog of mammalian adiponectin receptor. *Molecular Cell*, 17: 171-180.

Olsnes, S. and Pihl, A. 1978. Abrin and ricin — two toxic lectins. *Trends in Biochemical Sciences*, 3: 7-10.

Osborn, R.W., De Samblanx, G.W., Thevissen, K., Goderis, I., Torrekens, S., Van Leuven, F., Attenborough, S., Rees, S.B. and Broekaert, W.F. 1995. Isolation and characterisation of plant defensins from seeds of Asteraceae, Fabaceae, Hippocastanaceae and Saxifragaceae. *FEBS Letters*, 368: 257-262.

O' Rourke, R.S., Richardson, M. and Selitrennikoff, C.P. 2001. Zeamatin Inhibits Trypsin and α-Amylase Activities. *Applied and Environmental Microbiology*, 67: 2365-2366.

Panday, V.B. and Devi, S. 1990. Biologically active cyclopeptide alkaloids from Rhamnaceae plants. *Planta Medica*, 56: 649-650.

Peumans, W.J., Hao, Q. and Van Damme, E.J.M. 2001. Ribosome-inactivating proteins from plants: more than RNA N-glycosidases? *FASEB Journal*, 15: 1493-1506.

Pfaller, M. and Wenzel, R. 1992. Impact of the changing epidemiology of fungal infections in the 1990's. *European Journal of Clinical Microbiology Diseases*, 11: 287-291.

Pryme, I.F., Pusztai, A.J., Grant, G. and Bardocz, S. 1996. Phytohaemagglutinin-induced gut hyperplasia and the growth of a mouse lymphosarcoma tumour. *Journal of Experimental Therapy and Oncology*, 1: 171-176.

Richardson, M., Valdes-Rodriquez, S. and Blanco-Labra, A. 1987. A possible function for thaumatin and a TMV-induced protein suggested by homology to a maize inhibitor. *Nature*, 327: 432-434.

Roberts, W.K. and Selitrennikoff, C.P. 1990. Zeamatin, an antifungal protein from maize with membrane-permeabilizing activity. *Journal of General Microbiology*, 136: 1771-1778.

Selitrennikoff, C.P., Wilson, S.J., Clemons, K.V. and Stevens, D.A. 2000. Zeamatin, an antifungal protein. *Current Opinion on Anti-infective Investigation Drugs*, 2: 368-374.

Shai, Y. 1995. Molecular recognition between membrane-spanning polypeptides. *TIBS*, 20: 460-464.

Shih, S.F., Wu, Y.H., Hung, C.H., Yang, H.Y and Lin, J.Y. 2001. Abrin triggers cell death by inactivating a thiol-specific antioxidant protein. *Journal of Biological Chemistry*, 276: 21870-21877.

Singh, N.K., Bracker, C.A., Hasegawa, P.M., Handa, A.K., Buckel, S., Hermodson, M.A., Pfankoch, E., Regnier, F.E. and Bressan, R.A. 1987. Characterization of Osmotin A Thaumatin-Like Protein Associated with Osmotic Adaptation in Plant Cells. *Plant Physiology*, 85: 529-536.

Spelbrink, R.G., Dilmac, N., Allen, A., Smith, T.J., Shah, D.M. and Hockerman, G.H. 2004. Differential antifungal and calcium channel-blocking activity among structurally related plant defensins. *Plant Physiology*, 135: 2055-2067.

Stirpe, F., Gasperi-Campani, A., Barbieri, L., Falasca, A., Abbondanza, A. and Stevens, W.A. 1983. Ribosome-inactivating proteins from the seeds of *Saponaria officinalis* L. (soapwort), of *Agrostemma githago* L. (corn cockle) and of *Asparagus officinalis* L. (asparagus), and from the latex of *Hura crepitans* L. (sandbox tree). *Biochemistry Journal*, 216: 617-625.

Stirpe, F. 2004. Ribosome-inactivating proteins. *Toxicology*, 44: 371-383.

Tailor, R., Acland, D., Attenborough, S., Cammue, B.P.A., Evans, I.J., Osborn, R.W., Ray, J., Rees, S.B. and Broekaert, W.F. 1997. A novel family of small cysteine-rich antimicrobial peptides from seed in *Impatiens balsamina*. *Journal of Biological Chemistry*, 272: 24480-24487.

Terras, F.R., Schoofs, H.M., De Bolle, M.F., Van Leuven, F., Rees, S.B., Vanderleyden, J., Cammue, B,P. and Broekaert ,W.F. 1992. Analysis of two novel classes of plant antifungal proteins from radish (*Raphanus sativus* L.) seeds. *Journal of Biological Chemistry*, 267: 15301-15309.

Thevissen, K., Warnecke, D.C., Francois, I.E., Leipelt, M., Heinz, E.,Ott, C., Zahringer,U., Thomma, B.P., Ferket, K.K. and Cammue, B.P. 2004. Defensins from insects and plants interact with fungal glucosylceramides. *Journal of Biological Chemistry*, 279: 3900-3905.

Thevissen, K., Kristensen, H.H., Bart P.H.J. Thomma, B.P.H.J., Cammue, B.P.A. and Francois, I.E.J.A. 2007. Therapeutic potential of antifungal plant and insect defensins. *Drug Discovery Today*, doi:10.1016/j.drudis.2007.07.016.

Trudel, J., Grenier, J., Potvin, C. and Asselin, A. 1998. Several thaumatin-like proteins bind to b-1,3-glucans. *Plant Physiology*, 118: 1431-1438.

Vigers, A.J., Roberts, W.K. and Selitrennikoff, C.P. 1991. A new family of antifungal proteins. *Molecular Plant Microbe Interactions*, 4: 315-323.

Van Loon, L.C., Pierpoint, W.S., Boller, T. and Conejero, V. 1994. Recommendations for naming plant pathogenesis-related proteins. *Plant Molecular Biology Reporter*, 12: 245-264.

Walker, E.M. Jr., Gale, G.R., Glassman, A.B., Atkins, L.M., Bennett, C.E. and Burgis, J.T. 1984. Antitumor effects of abrin and ricin used singly and in combination with cisplatin. *Annals in Clinical Laboratory Science*, 14: 10-20.

Wang, Y, Qian, R.Q., Gu, Z.W., Jin, S.W., Zhang, L.Q., Xia, Z.X., Tian, G.Y., and Ni, C.Z. 1986. Scientific evaluation of Tian Hua Fen (THF) -history, chemistry and application. *Pure and Applied Chemistry*, 58: 789-798.

Wang, Y.X., Neamati, N., Jacob, J., Palmer, I., Stahl, S.J., Kaufman, J.D., Huang, P.L., Winslow, H.E., Pommier, Y., Wingfield, P.T., Lee-Huang, S., Bax, A. and Torchia, D.A. 1999. Solution Structure of Anti-HIV-1 and Anti-Tumor Protein MAP30. *Cell*, 99: 433-442.

Wang Y.Y., Ouyang, D.Y., Huang, H., Chan, H., Tam, S.C. and Zheng, Y.T. 2005. Enhanced apoptotic action of trichosanthin in HIV-1 infected cells. *Biochemical and Biophysical Research Communications*, 331:1075-1080.

Yeung, H.W., Wong, D.M., Ng, T.B. and Li, W.W. 1985. Purification of three isolectins from root tubers of *Trichosanthes kirilowii* (Tianhuafen). *International Journal of Peptide Protein Research*, 27: 325-333.

Yeung, H.W., Poon, S.P., Ng, T.B. and Li, W.W. 1987. Isolation and characterization of an immunosuppressive protein from *Trichosanthes kirilowii* root tubers. *Journal of Immunopharmacology Immunotoxicology*, 9: 25-46.

Yoon, T.J., Yoo, Y.C., Kang, T.B., Song, S.K., Lee, K.B., Her, E., Song, K.S. and Kim, J.B. 2003. Antitumor activity of the Korean mistletoe lectin is attributed to activation of macrophages and NK Cells. *Archieves in Pharmaceutical Research*, 26: 861-867.

Youle, R. and Huang, A. 1976. Protein bodies from the endosperm of castor beans: subfractionation, protein components, lectins, and changes during germination. *Plant Physiology*, 58: 703-709.

Zhang, J.S. and Liu W.Y. 1992. The mechanism of action of trichosanthin on eukaryotic ribosomes-RNA N-glycosidase activity of the cytotoxin, *Nucleic Acids Research,* 20: 1271-1275.

Zinke, H., Eck, J., Langer, M., Möckel, B., Baur, A. and Lentzen, H. 1996. Molecular cloning of the *Viscum album* L (mistletoe) gene for ML-1 and characterization of the recombinant protein. *Phytomedicine,* 1: 25.

Zou, L.B. and Zhan, J.B. 2005. Purification and anti-cancer activity of ricin. *Zhejiang Da Xue Xue Bao Yi Xue Ban,* 34: 217-219.

□□□

25

Medicinal Plant Databases - A Review

P. Shanmughavel

CONTENTS

1.1	Introduction	382
1.2	Electronic Databases	382
	1.2.1 Secondary information database	382
	1.2.2 Tertiary information databases	386
1.3	China Academy of Traditional Chinese Medicine Databases	387
1.4	Dictionary of Natural Products	387
1.5	Florin Medicinal Plants	387
1.6	Biologically Active Phytochemicals and Their Activities; and Phytochemical Constituents of Grass Herbs and Other Economic Plants	388
1.7	Brazilian Medicinal Plants Database	388
1.8	Flavour and Fragrance Materials	388
1.9	Medicinal Plants of Malta	389
1.10	Conclusion	389
	References	390

1.1 Introduction

Concomitant with the growing interest world-wide in the conservation, cultivation and use of medicinal, aromatic and other related groups of plants, there has been a four-fold increase in the volume of literature published on these plants during the past two decades (UNEP, 1992). Until early 1970's, printed publications were the almost exclusive means available for recording and disseminating scientific information. Developments in information technology during the 1980's and 1990's have led to increasing proportion of this pool of information now being held in electronic form in databases which can either be searched online from remote sites or consulted offline at the reader's own desk.

Whereas the bulk of the information held in databases is still copied from or entered simultaneously with its appearance in printed publications, we are now beginning to see documents, which are published exclusively in electronic form (Bhat, 1995; 1997). Recent developments in telecommunications and information technology and the rapidly growing popularity of Internet as the medium of communication of the 1990's suggest that a significant proportion of information dissemination will occur through this medium within the next few years (Balaji, 2001; Neelakandan *et al.*, 2001). Internet resources containing information on medicinal plants, which cannot be found elsewhere have already started to appear. However, the present paper is mainly concerned with electronic databases providing information on medicinal plants, which can be searched without being connected to the Internet.

1.2 Electronic Databases

These may be classified into

a) Secondary information databases (eg. Bibliographic databases) which gives summaries of individual papers contained in primary publications and point the user to the original publication for further information.

b) Tertiary information databases (eg. Electronic compendia, encyclopedia and other reference works) which give detailed information on specific topics which may or may not indicate the source literature from which they are compiled.

1.2.1 Secondary information databases

Information on medicinal and other related groups of plants can be found in bibliographic databases dedicated exclusively to theses groups of plants as well as in botanical, biological, agricultural, chemical, medical, veterinary or multidisciplinary databases with much wider subject coverage. The most important among these are listed below.

MAPA: Medicinal and Aromatic Plants Abstracts (MAPA) published by the National Institute of Science Communication (NISCOM, formerly known as the Publications and Information Directorate, Council of Scientific and Industrial Research), New Delhi, is a bimonthly printed journal in its 19th volume year. It provides good coverage of global literature on medicinal and aromatic plants. Over 55,000 abstracts have been published in the first 18volumes of MAPA covering the period 1979-1996. The data from 1988 onwards (about 30,000 records) are also held in electronic form in the MAPA database and distributed on CD-ROM and other electronic media. Source literature currently scanned for compiling the database includes some 600 periodicals (of which about 200 are Indian) originating from 60 countries and published in 30 different languages. Conference proceedings, books and patents are also covered. Coverage of Indian and South-East Asian literature is particularly good and some records in this database are unique and cannot be found in other databases.

CAB ABSTRACTS: Among multidisciplinary bibliographic databases, CAB ABSTRACTS provides the best all round coverage of world literature on medicinal and related groups of plants. Source literature scanned includes books, reports, conference proceedings and over 12,000 serials published from all over the world in a wide range of languages and covering virtually every discipline involved in the study of medicinal and related plants (botany, agronomy, biotechnology, phytochemistry, economics, medicine and veterinary science). Current rate of new material is over 150,000 records per year.

The database contains over 3 million bibliographic records prepared since 1972, of which approximately 60,000 are concerned with studies on medicinal and related groups of plants. The printed abstract journal, Review of Aromatic and Medicinal Plants, which is in its third volume year, provides excellent coverage of this literature. Earlier literature, dating back from 1931, which is not included in the electronic database, can be found in long-established CABI printed journals including Horticultural Abstracts and Forest Products Abstracts. These are among some 45 printed abstract journals published by CABI in the different disciplines covered in the database.

The database is made available to users through online and CD-ROM vendors such as DIALOG, DIMDI, ESA IRS, CAN-OLE, SILVER PLATTER, etc.

AGRIS : The AGRIS (International System for Agricultural Sciences and Technology) database is managed by the Library and Documentation Systems Division of FAO, Rome. The database is compiled from information provided by 158 national and 28 regional and international input centers around the world and contains nearly 3 million bibliographic records prepared since 1975. Subject areas covered include all aspects of agriculture, fisheries, human nutrition and management of natural resources and environment.

In addition to scientific literature, the database contains material which cannot easily be found in other databases, compiled from less widely known publications of local or regional relevance. The database can be searched using keywords in several languages with the aid of a multilingual vocabulary. Current rate of input of new records is over 150,000 per year. The database is made available to users through online vendors and CD-ROM suppliers such as DIALOG, DIMDI, ESA-IRS, FAXON, MICROINFO and SILVER PLATTER.

AGRICOLA : Bibliographic database compiled and maintained by the US National Agricultural Library, Beltsville, Maryland. The database contains over 3 million records of post-1970 publications covering all aspects of agriculture and provides particularly extensive coverage of US literature not covered by other agricultural databases (e.g. reports from agricultural experiment stations containing material not published elsewhere). It is made available to users through online and CD-ROM vendors DIALOG, DIMDI, FAXON, MICROINFO, SILVER PLATTER and others.

PASCAL : Compiled by the Institut de 1'Information Scientifique et Technique (INIST) of the French Centre National de la Recherche Scientifique, PASCAL is one of the world's largest multidisciplinary bibliographic information databases covering literature published in all areas science and technology, including biology and medicine. This French/English bilingual database contains over 12 million records relating to papers published in over 25,000 serials, 56,000 reports, 60,000 conference proceedings and 100,000 theses covering the period from 1973 to date. Current rate of growth is over 500,000 records per year. It provides better coverage of French literature than any other database. Search aids include a lexicon containing over 80,000 descriptors and the lexicon is available in 3 languages, viz. French, English and Spanish.

A sister database called FRANCIS, also compiled by INIST, covers literature in social sciences (including ethnology) and economics.

PASCAL can be accessed online, or offline on CD-ROM, using the services of online and CD-ROM vendors such as QUESTEL/ORBIT, ESA-IRS and DIALOG/DATA STAR.

BIOSIS PREVIEWS: Probably the world's largest bibliographic information database on biological subjects including medicine, compiled by BIOSIS of Philadelphia, USA. The database contains over 12 million citations and the current rate of addition of new records is 540,000 per year. As with other databases, the source literature includes journals, books, monographs and conference proceedings. In addition to online access, the database is published for offline consultation on CD-ROMs and in printed for in biological abstracts and other abstract journals .Subscriptions to these are available through online vendors and CD-ROM distributors.

CHEMICAL ABSTRACTS: For phytochemical information on medicinal plants and patent-related literature, chemical abstracts are an indispensable resource. The CHEMICAL ABSTRACTS database compiled by CAS (A division of the American Chemical society) in Columbus, Ohio, contains some 13 million abstracts of literature and patents in all areas of chemistry and chemical engineering covering the period from 1970 to date. Over 700,000 new records are added per year. The accompanying CA REGISTRY database contains information on over 16 million chemical substances. The database can be searched online or on CD-ROM. Literature published prior to 1970 can be found in the printed journal chemical abstracts. Search aids include a very detailed list of indexing terms used.

Online access is provided through the STN International online network. Subscriptions are also available through national offices in many countries. e.g. Royal Society of Chemistry in the UK; VCH Verlagsgesellsehaft mbH in Germany, Centre National de 1× Information Chimique in France, Japan Association for International Chemical Information in Japan, etc.

MEDLINE: A limited amount of information relating to medicinal plants (from mainstream medical literature) can be found in this most widely known database of medicine compiled by the National Library of Medicine, USA.

The database is available to users through a large number of online and CD-ROM vendors including: BLAISE LINK, CD-PLUS, DATA-STAR, DIALOG,DIMDI, EURO-CD DIFFUSION, FAXON, INFOPRO TECHNOLOGIES, LEARNED INFORMATION LTD, MIC KIBIC, OPTECH LTD, QUESTEL/ORBIT, STN, TELESAMPO etc.

Although direct online searches are still fee-based, free search facilities are now provided by a number of organizations through their websites on the Internet. The CD-ROM is also becoming widely available in public libraries around the world.

EMBASE: Literature published in some 3,600 biomedical journals is covered in this bibliographic database of medicine produced by Elsevier, Amsterdam. It contains some 6.5 million records covering the period from 1974 to date. Products derived from the database include CD-ROMs and printed journal Excerpta Medica. Like MEDLINE, coverage of literature on medicinal plants is restricted to that appearing in mainstream biomedical journals.

The database also accessible through various online and CD-ROM vendors including CD PLUS, DATA-STAR, DIALOG, DIMDI, FAXON, INFOPRO TECHNOLOGIES, MICROINFO; SILVER PLATTER and STN.

1.2.2 Tertiary information databases

Unlike the secondary information databases listed above, tertiary electronic publications containing information on medicinal and related groups of plants vary widely in their design, structure and contents. A member of them, produced for different purposes by various organizations, are listed below in alphabetical order. While some of them can be accessed online, others are only available on CD-ROM or other offline electronic media.

AHEAD CD-ROM: The Asian Health, Environmental and Allied Databases(AHEAD) CD-ROM series consists of 3 disks containing various databases contributed by the participating organizations based in India, Thailand, Philippines, Malaysia, Singapore and Bangladesh. The project is sponsored by the International Development Research Centre of Canada and co-ordinated by the National Institute of Science Communication (NISCOM), New Delhi.

Disk 1 entitled "Environment Asia" contains full text and bibliographic databases related to water management, recycling of waste water, hygiene education and community participation. Disk 2, named "Wealth Asia", and contains the entire medical and aromatic plants bibliographic database mentioned above and a full text database of Indian plant, animal and mineral resources, based on the well known "Wealth of India" encyclopedic book series. Disk 3 is called "Health Asia" and contains a bibliographic database on tropical (Mosquito-transmitted) diseases and occupational safety and health, a full text database on water-borne (Diarrhoeal) diseases, and a natural toxins database providing text and pictorial information on poisonous plants and animals.

APINMAP: The Asian Pacific Information Network on Medicinal and Aromatic Plants (APINMAP) is a UNESCO-sponsored voluntary network of organizations in 14 Asian and Pacific region countries (Australia, People's Republic of China, India, Indonesia, Republic of Korea, Malaysia , Nepal, Pakistan, Papua New Guinea, The Philippines, Sri Lanka, Thailand, Turkey and Vietnam) with a Secretariat based in the Philippines. Its objective is to promote exchange of information relating to medicinal and aromatic plants between its member organizations.

BACIS: Booleans Aroma Chemical Information Service (BACIS) offers a set of five databases (which can be installed on a desktop PC) mainly aimed at users in the perfumery and flavoring industries. These contain information on volatile compounds in foods, analytical chemical data compiled from published literature on essential oils and other natural compounds, and trade-related data.

1.3 China Academy of Traditional Chinese Medicine Databases

The Institute of Information of the China Academy of Traditional Chinese Medicine has put together several bibliographic and factual information databases. The largest among these is the Traditional Chinese Medical Literature Analysis and Retrieval System (TCMLARS), which contains about 200,000 bibliographic records of literature on traditional Chinese medicine, Chinese herbal medicine, integrated traditional Chinese medicine and Western medicine.

CHINESE UNIVERSITY OF HONG KONG DATABASES: The Chinese Medicinal Material Research Centre (CMMRC) of the Chinese University of Hong Kong has compiled a database of TCM which contains botanical, chemical, pharmacological and clinical information selected from Chinese medical treatises and translated into English. This database is updated by abstracting papers from over 180 medical and other scientific journals in Chinese.

1.4 Dictionary of Natural Products

DNP on CD-ROM published by Chapman and Hall contains chemical, physical, bibliographic and structural data on over 113,000 natural products, organized into over 36,000 entries. Pharmacologically active compounds, food ingredients and many compounds of biochemical significance are well covered. Categories of natural products covered include amino acids, peptides and proteins, antibiotics, nucleosides, alkaloids, terpenoids, flavonoids, lignin's, lichen acids, polyacetylenes, tannins and coumarins.

ETHMED: Ethmed is a database currently being compiled at the Yakko Kaiseki Centre (Analytical research Centre for Ethno medicines) of the Institute for Wakan-Yaku (Traditional Sino Japanese Medicines) affiliated to the Faculty of Pharmaceutical Sciences of Toyama University, Japan. Data on morphology, anatomy, active principles, biological activity and uses of medicinal plants are being recorded.

1.5 Florin Medicinal Plants

FLORIN MEDICINAL PLANTS now contains information on over 5,000 taxa of vascular plants from more than 200 families and is expected to grow to twice its current size in the near future. Taxonomic data, plant parts used bioactive substances they contain and their therapeutic activity or toxicity are among the interactively searchable fields.

1.6 Biologically Active Phytochemicals and Their Activities; and Phytochemical Constituents of Grass Herbs & Other Economic Plants

A set of two databases compiled by Dr James Duke of ARS/USDA. The first of theses contains information on some 3000 biologically active (medicinal, antimicrobial, pesticide and allelopathic) physiochemical, their reported activities and inhibitory concentrations for doses. The second database lists the chemical constituents of approximately 1000 plant species.

1.7 Brazilian Medicinal Plants Databases

This is a database being developed by the Medicine Plants laboratory of the Escola Superior de Agricultura "Luiz de Queiroz", Sao Paulo University, and CIAGRI, the computing centre of the university.

It currently contains the common name in Portuguese, Latin name and synonyms, family, biological activity, and therapeutic uses of over a thousand species.

1.8 Flavour and Fragrance Materials

Is a chemical entry database published by Allured Publishing Corporation. The database contains 2,500 records with CAS registry numbers, structure, molecular formulae and physical constants for flavor and fragrance materials.

FLOTURK: FLOTURK (FLOra of TURKey) is a database compiled and maintained by the Anadolu University Medicinal Plant and Drug Research Centre (TBAM). It contains botanical, phytochemical, chemotaxonomic and pharmacological activity-related data as well as information on production and commercial potential of Turkish flora.

GREEN MEDICINE: Is a Chinese Herbal Medicine database containing information on 390 biomedical syndromes, 257 basic formulas, 490 individual herbs and 600 variations.

INMEDPLAN: INMEDPLAN (Indian Medicinal Plants Network of Distributed Databases) is an initiative of a network of several Indian organizations with expertise on different aspects of medicinal plants to build a multidisciplinary (botanical, horticultural, pharmacological and other) information pool by sharing their resources. The network secretariat is at FRLHT, Bangalore.

MAPI: MAPI (Major Aromatic Plants of India) is a database compiled by the Central Institute of Medicinal and Aromatic Plants, Luck now. It contains very detailed botanical, agronomic, phyto chemical and bibliographic information on 45 major aromatic plants of India. The database has a very elaborate structure with a total of 86 unique data entry fields for each record.

1.9 Medicinal Plants of Malta

An electronic inventory of 300 medicinal and aromatic plants of Malta, compiled by the University of Malta, containing text and images.

MEDICINAL PLANTS OF PAPUA NEW GUINEA: Developed by the Wau Ecology institute Herbarium, this database contains botanical, photochemical and ethno pharmacological information on plants native to PNG.

NAPRALERT, MEDFLOR and DEREP: These are three inter-related databases developed by the department of Medicinal Chemistry and Pharmacognosy of the College of Pharmacy, University of Illinois at Chicago, USA.

The NAPRALERT (Natural Products ALERT) database contains bibliographic and factual data on natural products of plant, microbial and animal origin. It is compiled from ethno medical source literature scanned from some 125,000 journal articles, books, abstracts and patents covering the period from 1975 to data.

NATTS: NATTS is the Natural Products database developed by the central Drug Research institute, Lucknow. It contains factual information on medicinal plants. Detailed data on botanical characters, collection site details, pharmacological screening results, chemical structures of active constituents, uses in folk medicine and in established traditional medicine, information from traditional health systems' literature and from modern scientific literature are recorded in the six sub-files constituting the database.

TRADIMED: TRADIMED is a TCM database developed by the Natural Products research Institute of the Seoul National University. The database contains pharmacological data (efficacy, dosages, and adverse effects) for traditional Korean and Chinese drugs, chemical data (composition and structural formulae) for natural products from botanical, microbial and marine organisms, and color images of medicinal plants.

1.10 Conclusion

The bulk of the literature on medicinal plants currently available in electronic form appears to be inspired by the objective of identifying new plants containing bioactive compounds or isolating and characterizing the active principles from plants already used in herbal therapy in some parts of the world. Only an insignificant fraction of the digital information available on major databases relate to the therapeutic use of theses plants in the traditional system of medicine from which they are identified in the first place (Shanmughavel, 2003; 2004). Yet, the same database contains virtually no information on the conditions under which these plants are already being used effectively in the traditional system of medicine concerned. Literature on established systems of traditional and folk medicine from around the world is yet to make its appearance on electronic media. Further researches and the creation of database are needed on these aspects.

References

Balaji, U. 2001. *Current Science,* 80: 604-605.

Bhat, K.K.S. 1995. *Acta Horticulture,* 390: 11-17.

Bhat, K.K.S. 1997. *Nonwood Forest Products,* 11: 60-77.

Neelakandan, V.N., Mohanan, C.N. and Sukumar, B. 2001. *Indian Catalogue,* 21.

Shanmughavel, P. 2003. *Bioinformatics India,* 3: 14-20.

Shanmughavel, P. 2004. *Bioinformatics India,* 4: 34- 39.

UNEP, 1992. *Conservation on Biological Diversity,* 70-83.

□□□

26

In silico Structural Analysis of Hypothetical Proteins : A Therapeutic Approach

R. Radhika and P. Shanmughavel

CONTENTS

1.1 Introduction 392

1.2 Structure Classification 393
  1.2.1 Primary structure 393
  1.2.2 Secondary structure 394
  1.2.3 Tertiary structure 394
  1.2.4 Quarternary structure 395

1.3 Classes of Protein Structure 395

1.4 Structure Prediction 397

1.5 Computational Prediction of Protein Structure 397

1.6 Sequence Prediction of a Protein by Translation of DNA Sequence 398

1.7 Secondary Structure Prediction 398
  1.7.1 Methods of secondary structure prediction 398

1.8 Tertiary Structure Prediction 399

1.9 Ab-initio Protein Modeling 400

1.10 Comparative Protein Modeling 400

1.11 Protein Threading 400

1.12 Structure Validation 401

1.13 Model Refinement 402

1.14 Structure Function Relationship 406
   1.14.1 Function from structure 406
   1.14.2 Function prediction from sequence 406

1.15 Phylogeny Based Methods 407

1.16 Conclusion 407

References 408

1.1 Introduction

According to a definition of protein, they are large organic compounds made of natural polymers composed of amino acids arranged in a linear chain and joined together by peptide bonds between the carboxyl and amino groups of adjacent amino acid residues. All the amino acids possess common structural features, including α-carbon to which an amino group, a carboxyl group and a variable side chain are bonded.

Proteins undergo a process called 'Protein folding' by which higher structures are formed from primary structure, thereby their biological functions are determined. They fold into one or more specific spatial conformations, driven by a number of non-covalent interactions like hydrogen bonding, ionic interactions, vanderwaals forces and hydrophobic forces. A single polypeptide chain may attain different stable folded conformation having different biological activity but only one conformation is considered to be the native or 'active' conformation. The amino acid of a protein directs the folding pathway, sometimes assisted by proteins called chaperonins (Pattabhi and Gautham, 2002).

A hypothetical protein is a protein whose existence has been predicted, but for which there is no experimental evidence that it is expressed *in vivo*. Bioinformatics analysis intends to predict the secondary and tertiary structures of hypothetical proteins from their primary sequences itself. Hence the field of bioinformatics is highly useful in predicting structures for newly sequenced proteins as there is a large deposition of protein sequences and there is a requirement for three-dimensional structure determination updation accordingly.

A medicinal plant is any plant which, in one or more of its organ, contains substance that can be used for therapeutic purpose or which is a precursor for synthesis of useful drugs (Sofowora, 1982).Medicinal plants possess various therapeutic properties hence plays an important role in the field of pharmacology. Apart from this, they help to boost the economic value of our country. The structure and function prediction of the hypothetical proteins would help us to understand the working machinery of the plant which in turn help us to explore new therapeutic values of the plant.

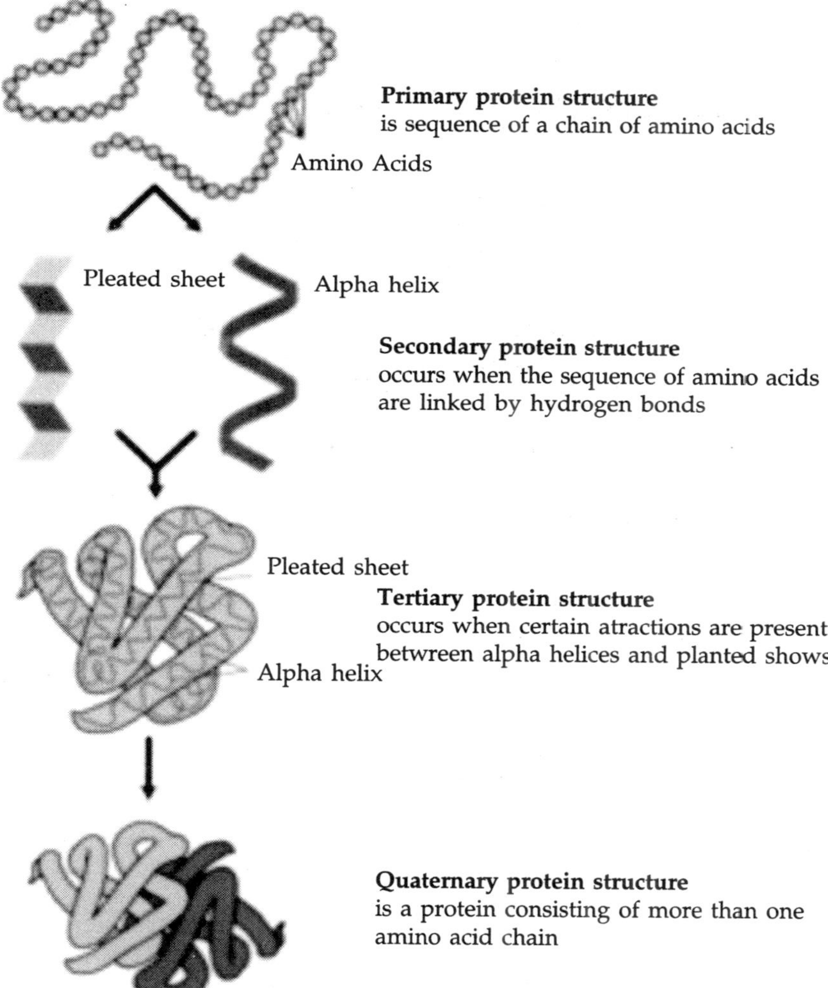

Fig. 1 : The four levels of protein structure

1.2 Structure Classification

1.2.1 Primary structure

The sequence of various amino acids constitutes the primary structure of the peptide or protein. The primary structure is dictated by the order of the amino acids and the genetic information found in a cell. A specific sequence of nucleotides in DNA is transcribed in to mRNA, which is read by the ribosome in a process called translation. The sequence of a protein is unique to that protein, and defines the structure and function of the protein. It is stated that the sequence of a protein is unique to a protein; hence it defines the structure and function of the protein (Branden and Tooze, 1999)

Fig. 2 : Amino acid chain connected by disulphide bond

1.2.2 Secondary structure

The next higher level of protein structure is called the secondary structure. The secondary structure of the protein is the result of interactions of side chains that are located very near to each other. The side chain of the residues have various functional groups that can have different types of forces like hydrophobic forces, hydrophilic forces and the hydrogen bonding resulting in conformations of lower energies. Specifically two secondary structural elements alpha helix and beta sheets have regular geometry. They are constrained to specific values of dihedral angles Ø and Ψ(psi), confined to the specific region of Ramachandran plot. The part of the backbone that is not in the regular secondary structure is said to be random coil.

A motif is a small combination of secondary structural elements (helix-turn-helix), also called as supersecondary structures. Protein structural motifs often include loops of variable length and unspecified structure.

A fold is similar to motif with larger combination of secondary structural units connected with loops like helix bundle or beta-barrel.

1.2.3 Tertiary structure

The secondary structural elements usually fold back on themselves in to a compact shape with the help of turns and loops. It even enables the primary structure residues to interact with each other giving rise to tertiary structure.

The formation of tertiary structure is usually driven by the burial of hydrophobic forces and the stabilization of structure achieved by the involvement of forces such as hydrogen bonding, ionic interactions and disulfide bonds. All non-covalent interactions not involved in secondary structure constitute the overall fold of the protein and they play an important role in the function of the protein. The function of a protein depends upon its tertiary structure.

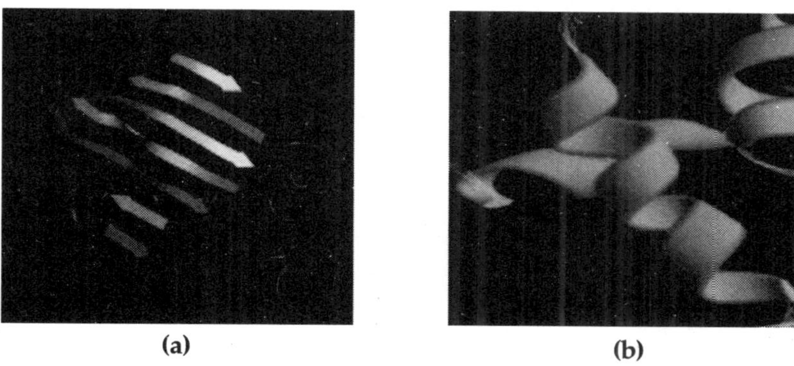

(a) (b)

Fig. 3 : Diagrammatic representation of Alpha helices (a) and Beta sheets (b)

The tertiary structure of many proteins is built from several ***domains***. A structural domain is an element of the protein's overall structure that is self-stabilizing and often folds independently of the result of the protein chain. The same domain tend to appear in different proteins due to the fact of evolution; genes or parts of genes can be doubled or moved around within the genome. To sum up, domains are self-stabilizing units in proteins, which have a biological function.

1.2.4 Quarternary structure

The quarternary structure is formed due to the interaction between several subunits by peptide bonds. The individual subunits or chains may be connected by a disulfide bond or covalent bond. The complexes of two or more polypeptides are called multimers. Multimers made up of identical subunits are called as a dimer, trimer or a tetramer. A multimeric complex may have an important enzyme activity.

1.3 Classes of Protein Structure

Based on the types and arrangements of secondary structural elements four principal class of protein structure or folds are recognized.

(i) Class α comprises a bundle of α helices connected by loops on the surface of the proteins.

(ii) Class β comprises antiparallel â sheets, usually two sheets in close contact forming a sandwich.

(iii) Class α /β comprises mainly parallel β sheets with intervening α helices, but may also have mixed β sheets.

(iv) Class α + β comprises mainly segregated α helices and antiparallel β sheets.

(v) Small disulphide rich proteins, which have few helices or sheets (Mount, 2003).

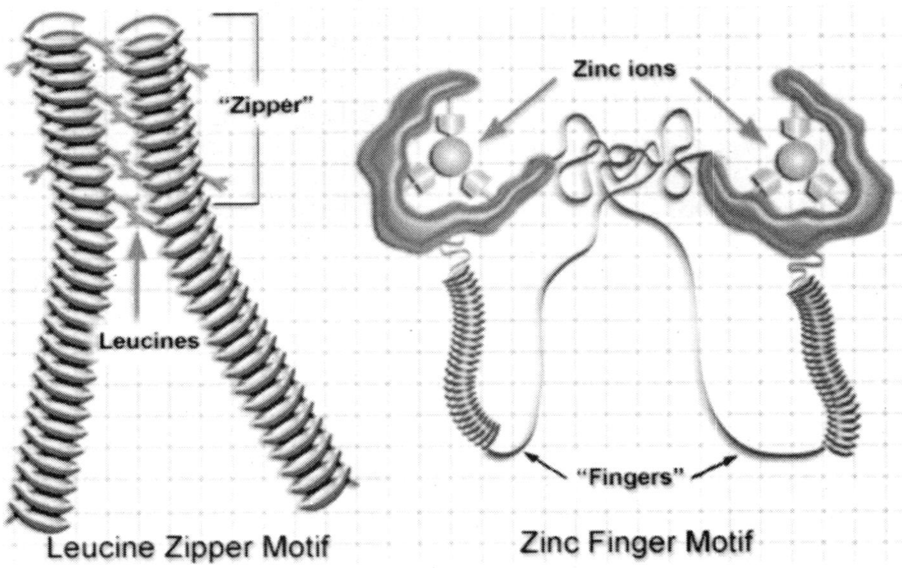

Fig. 4 : Examples of motif: Leucine Zipper motif and Zinc Finger motif

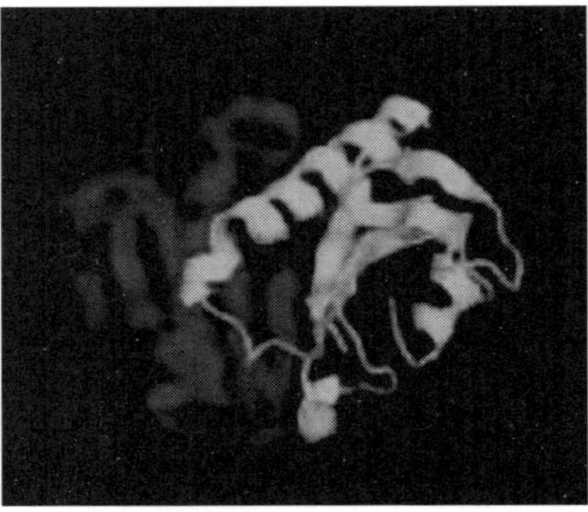

Fig. 5 : Diagrammatic representation of quarternary structure

1.4 Structure Prediction

There are a large number of proteins whose sequences are known, but very few have their structures solved. Experimental techniques like X-ray crystallography, NMR are highly specialized, expensive and time-consuming for solving protein structures. Hence there is an urgent need for the prediction of the three dimensional structure of proteins from the available primary sequence of proteins. The field of bioinformatics fulfils this purpose in an easy and time saving manner. The prediction of three-dimensional structures of proteins from the primary sequences is based on the fold recognition methods. Protein structure prediction is based on an important fact that protein adopt a limited number of three dimensional configurations and the newly determined protein structure of proteins have secondary structural elements having close structural similarity with the previously determined structures. In other words, the folds recognized by computational methods for the experimentally solved structures revealed that the same fold exists between already developed and newly generated protein structure and simultaneously proteins without any sequence similarity can adopt the same fold, which means the process of predicting structure from sequence is a tedious process (Branden and Tooze, 1999).

1.5 Computational Prediction of Protein Structure

The generation of a protein sequence is much simpler than the generation of a protein structure. However, the structure of a protein gives much more insight in the function of the protein than its sequence. Therefore, a number of methods for the computational prediction of protein structure from its sequence have been proposed. Bioinformatics is explained as the application of information technology to the field of molecular biology. The major areas of research in this field include sequence alignment, gene finding, genome assembly, protein structure alignment, protein structure prediction, prediction of gene expression and protein-protein interactions and the study of evolutionary relationships, creating and viewing 3-D models of protein structures. The main applications of bioinformatics are the creation and advancement of databases, algorithms, computational and statistical techniques, and theory to solve formal and practical problems arising from the management and analysis of biological data. Out of these, a major application is to understand the relationship between amino acid sequence and three-dimensional structures in proteins. It employs various structure prediction methods such as *ab initio* prediction methods that uses just the sequence of the protein; threading uses existing protein structures and homology modeling uses a reliable 3D model for a protein of unknown structure from one or more related known protein structure.

1.6 Sequence Prediction of a Protein by Translation of DNA Sequence

Protein sequences are predicted by translating DNA sequences that are cDNA copies of mRNA sequences from a predicted start and end of an open reading frame. Partial sequences like ESTs provide only the carboxy- terminal end of the protein sequence which are usually only 99% accurate. In prokaryotes, bacterial genomes which have few or no introns in their genes, are translated completely. In eukaryotes, which contain both exons and introns in their genes, the protein-encoding exons are predicted and translated.

1.7 Secondary Structure Prediction

The secondary structure prediction methods are based on the assumption that there should be a correlation between amino acids sequence and secondary structure. Structure prediction from amino acid sequence begins with an analysis of a database of known structures. These databases are examined for possible relationships between sequence and structure. The ability to predict secondary structure also depends on identifying types of secondary structural elements in known structures and determining the location and extent of these elements.

1.7.1 Methods of secondary structure prediction

There are several methods employed for the secondary structure prediction of proteins. Of which three widely used methods are as follows (Mount, 2003)

(i) Chou-Fasman/GOR method

The Chou-Fasman method was based on analyzing the frequency of the 20 amino acids in á helices and â sheets and turns of then-known relatively small number of protein structures. The secondary structural elements are identified on the basis of the already determined residual value present in the secondary structure. It is only about 50-60% accurate in predicting secondary structure domains. The known secondary structures are scanned for the occurrence of amino acids in each type of structure.

(ii) Neural network

The most sophisticated methods that have been devised to make secondary structural predictions for proteins use artificial intelligence, or so called neural net algorithms. In the neural network approach, computer programs are trained to be able to recognize amino acid patterns that are located in known secondary structures and to distinguish these patterns from other pattern not located in these structures. Its accuracy level is measured to be about 70%.

(iii) Nearest- Neighbour Method

It is a machine learning method. They predict the secondary structural confirmation of an amino acid in the query sequence by identifying sequences of known structures that are similar to the query sequence. Another exclusively sequence based method used for predicting distant relatives and protein folding patterns is hidden markov model.

(iv) Hidden Markov Model

HMM are used to model alignments or three-dimensional structures in proteins. The models are trained on a pattern of alpha helices, beta strands, tight turns and loops in specific structural classes, which then may be used to provide the most probable secondary structure and structural class of a protein.

1.8 Tertiary Structure Prediction

Protein structure prediction methods can be classified into three categories. They are comparative modeling (Murzin, 2001), threading (Altschul *et al.*, 1990; Jones, 1992; Vingron and Waterman, 1994; Altschul *et al.*, 1997; Karplus *et al.*, 1998) and ab-initio folding (Jones, 1999; David *et al.*, 2000; Lundsro¨m *et al.*, 2001; Zhang, 2004; Kennard, 1977; Tsai, 2001). Comparative modeling and fold recognition methods predict protein structures based on already solved structures. High resolution models are built using template based methods by the identification of close homologues from the PDB. In case of the absence of templates, ab-initio modeling is used.

The three dimensional structure of proteins are formed by the folding of the protein sequences. The different protein sequences can form same folds, thus there are many combinations of amino acids that can fit together into the same three-dimensional conformation, filling the available space and making suitable contacts with neighboring amino acids to adopt a common three dimensional structure. Sequence alignment can be used to identify a family of homologous proteins that have the same sequence, and presumably a similar three-dimensional structure. The sequence alignment searches for families and superfamilies of known structure in various databases.

When the sequence of a protein of unknown structure has no detectable similarity to other proteins, other methods of three dimensional structure predictions may be employed like threading. These methods examine the sequence of a protein for compatibility of the side group with a known protein core. The sequence is threaded into a database of protein cores to look for matches, if a reasonable degree of compatibility is found with a given structural core, the protein is predicted to fold in to a similar three-dimensional configuration.

1.9 *Ab-initio* Protein Modeling

Ab-initio or *de novo* protein modeling methods build three dimensional protein models based on physical principles rather than on the structures of previously solved structures. There are many types of modeling procedures that either attempt to predict three-dimensional structure of tiny proteins using various approaches like protein folding, global optimization techniques etc. For large proteins, powerful algorithms and larger computational resources like supercomputers or distributed computing are used (Altschul *et al.*, 1990). These factors make *ab-initio* structure prediction, an active research field. It is considered as the most difficult category of protein structure prediction (Bonneau *et al.*, 2001; Duan *et al.*, 1998).

1.10 Comparative Protein Modeling

Homology modeling, also known as comparative modeling is a class of methods in protein structure prediction for constructing anatomic resolution model of a protein from its amino acid sequence. It is based upon the fact that homologous proteins share very similar structures and a protein's fold is more evolutionarily conserved than its amino acid sequence. It is based upon previously solved structures as starting points, or templates. It searches for the tertiary structural motifs in the proteins from the protein structural databases based upon the protein folding process. The target sequence is modeled with reasonable accuracy on a distantly related template. Homology modeling is most accurate when the target and template have similar sequences.

1.11 Protein Threading

Protein threading is a method for the computational prediction of protein structure from amino acid sequence. It scans the amino acid sequences of an unknown structure against a database of solved structures. In each case, a scoring function is used to assess the compatibility of the sequence to the structure, thus yielding possible three-dimensional models. This type of method is also known as 3D-1D structures and linear protein sequences. Inverse folding search evaluates the compatibility of a given structure with large database of sequences, thus predicting which sequences have the potential to produce a given fold. Fold recognition methods are divided into two types: methods that derive a 1-D profile for each structure in the fold library and align the target sequence to these profiles and methods that consider the full 3-D structure of the protein template. Algorithms like dynamic programming are used for threading a sequence onto a structure and combinatorial optimization methods such as simulated annealing are used for 3-D threading (Lesk, 2003).

1.12 Structure Validation

Structure validation is carried out for modeled structures for determining their stereochemical quality. A predicted structure confined to the following parameters is considered to be a valid model.

The parameters used for checking the quality of a model are,

(a) *R.M.S.D:* On of the first parameters to evaluate is the root mean square deviation of the ensemble of calculated structures. A small rmsd is considered to be an indicator of precisely determined structure.

(b) *Procheck*: It is a program that checks the stereochemical quality of a protein structure.

(c) *Whatcheck*: It is another collection of protein verification tools that calculates inside-outside and torsion angle errors.

(d) *Structure Z-scores*: Negative Z-scores indicate that the corresponding parameter is worse than average.

(e) *RMS Z-scores*: The RMS Z-scores should be close to 1.0.

Some of the other parameters that are used as good indicators of stereochemical quality include planarity, chirality, phi/psi preferences, chi angles, non-bonded contact distances and unsatisfied donors and acceptors.

The geometric constraints considered for structure validation are,

(a) *Bond Lengths*: In a good protein structure, the internal bond lengths should agree to known stereochemical values. They should lie within a few standard deviations of the mean bond length observed in the PDB. The bond lengths of the residue bonds are observed between the atoms N-CA, CA-CB, CA-C, C-O and C-N.

(b) *Bond Angles*: The internal bond angles within the statistical outliers are normally distributed about the mean angles observed in the PDB. The report contains measurements of the angles C-N-CA, N-CA-C, N-CA-CB, CB-CA-C, CA-C-N, CA-C-O, and O-C-N.

(c) *Dihedral Angles*: The internal dihedral angles, the phi, omega, psi, and chi1 angles are reported along with the more recent OC-CO dihedral.

(d) *Nonbonded Contacts* : It includes all interatomic distances constituting all non-bonded and non-hydrogen bonded contacts where the van der Waals radii overlap by more than a prescribed value.

The Statistical constraints considered for structure validation are,

(a) Omega and Chi1 : Deviations from perfect cis or trans conformations under the assumption of a Gaussian mixture for the omega angle are obtained.

(b) Phi and Psi : The Phi and Psi angles of a proposed model of PDB structures is evaluated by the four arvotated grid squares namely core, Allowed, Generous and Oatride. The 10 degree grid squatres are placedon the 2D Square [-Pio, pi] x [-Pi, Pi]

Ramachandran Plot

The Ramachandran analysis of peptide dihedral angles allows us to classify allowed and nonallowed conformations of the folded protein structural elements (Ramachandran and Sasisekharan, 1968).

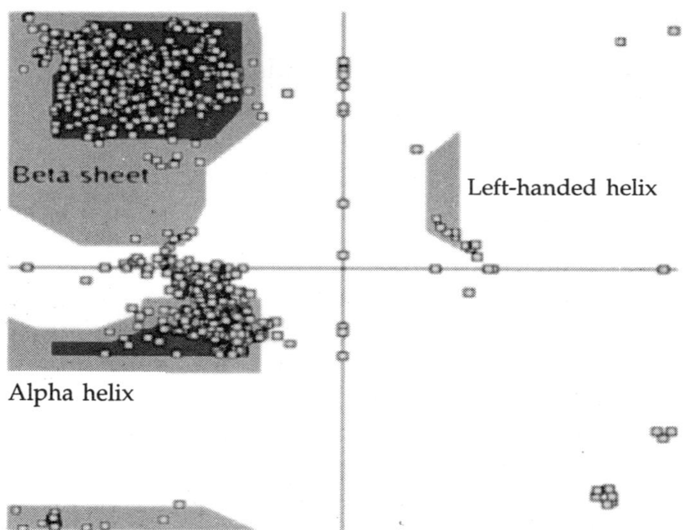

Fig. 6 : A Ramachandran plot also known as a Ramachandran map or a Ramachandran diagram is an easy way to visualize dihedral angles ϕ and ψ of amino acid residues in protein structure and it shows the possible conformations of ϕ and ψ angles for a polypeptide.

1.13 Model Refinement

The refinement of low-resolution predicted models to high-resolution structures that are close to the native state is posing a greatest challenge in protein structure prediction (Jagielska *et al.*, 2008).

A method is described for the refinement of rough protein models into protein models possessing least energy or energy minima. Energy minimization is widely used in molecular modeling and is an integral part of techniques such as conformational search procedures. It is also used to prepare a system for other types of calculation like simulations of complex systems such as macromolecules. It may be used prior to a molecular dynamics or Monte Carlo simulation for removing any unfavorable interactions in the initial configuration of the system (Leach, 2001).

We can classify minimization algorithms into two groups: those which use derivatives of the energy with respect to the coordinates and those which do not use derivatives. Derivatives can be useful because they provide information about the shape of the energy surface, and which can enhance the efficiency with which minimum is located. Most minimization algorithm can go only downhill on the energy surface and so they can enhance the efficiency with which the minimum is located. An ideal minimization algorithm goes only downhill on the energy surface and that locates only the minimum that is nearest and is the one that provides the answer as quickly as possible. The input to a minimization program consists of a set of initial coordinates for the system may be obtained from experimental technique such as X-ray crystallography or NMR or a theoretical method employed such as conformational search algorithm (Leach, 2001).

The derivative methods can be classified according to the highest-order derivative method. First order methods use the first derivatives whereas second-order methods use both first and second derivatives. The non-derivative methods can thus be considered a zeroth- order method as it does not use any derivative.

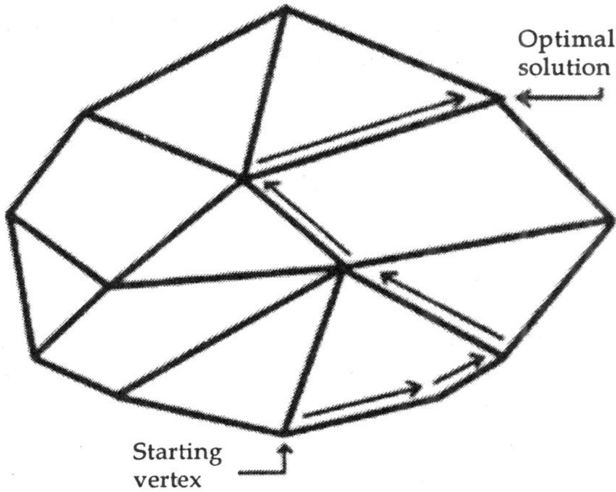

Fig. 7 : Representation of Simplex Method

(i) Non-derivative minimization methods

(a) Simplex method

The simplex method was the first method developed to solve linear programs. The method uses the concept of a simplex, which is a polytope of N + 1 vertices in N dimensions: a line segment in one dimension, a triangle in two dimensions, a tetrahedron in three-dimensional space etc. A system of linear inequalities is used to define a polytope. The simplex algorithm begins at a starting vertex and moves along the edges of the polytope until it reaches the vertex of the optimum solution.

(b) Sequential univariate method

In this method, only one variable is changed at a time and produces a sequence of improved approximation to the minimum point. By starting at a base point X in the ith iteration, the values of n-1 variables are fixed, and the remaining variables are made variable. The problem is called a one-dimensional minimization problem since only one variable is changed. The search is continued in a new direction which is obtained by changing any one of the n-1 variables that were formed in the previous iteration. The search procedure is continued by taking each coordinate direction in turn. The first cycle is completed when all the n directions are searched sequentially; the entire process is repeated for sequential minimization. The procedure is continued for the objective function in the n directions of a cycle, until no function improvement takes place (Rao, 2008).

(ii) First order minimization methods

(a) Steepest descent method

An algorithm for finding the nearest local minimum of a function which presupposes that the gradient of the function can be computed. The method of steepest descent, also called the gradient descent method, starts at a point P_0 and, as many times as needed, moves from P_i to P_{i+1} by minimizing along the line extending P_1 from in the direction of $-\Delta f\ (p_i)$, the local downhill gradient.

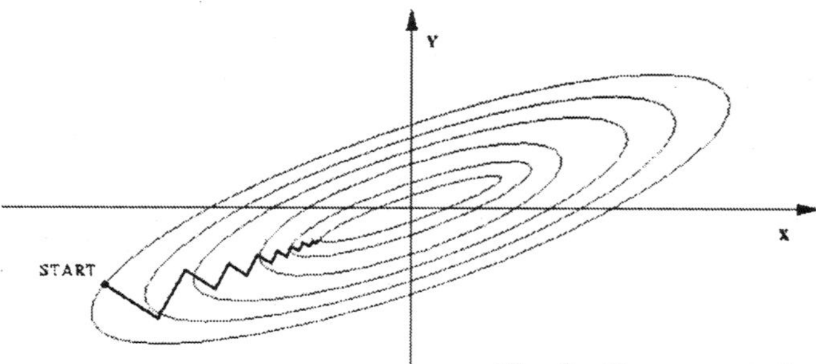

Fig. 8 : Representation of Steepest descent method

The steepest descent method or saddle-point approximation, is a method used to approximate integrals of the form

$$\int_a^b e^{Mf(x)}\,dx$$

where $f(x)$ is some twice-differentiable function, M is a large number, and the integral endpoints a and b could possibly be infinite. The technique is also referred to as **Laplace's method**, which in fact concerns the special case of real-valued functions f admitting a maximum at a real point.

(b) Conjugate gradient method

The conjugate gradient method is an algorithm for finding the nearest local minimum of a function of variables which presupposes that the gradient of the function can be computed. It uses conjugate directions instead of the local gradient for going downhill. A long, narrow valley indicates that the minimum is reached in far fewer steps than by using the method of steepest descent (http://www.mathworld.wolfram.com).

(iii) Second order derivative method

Newton-Raphson method

In numerical analysis, Newton's method or NewtonRaphson method, is considered as the best known method for finding successively better approximations to the zeroes or roots of a real-valued function. Newton's method converges very quickly, especially if the iteration begins near the desired root. Given a function $f(x)$ and its derivative $f'(x)$, we begin with a first guess x_0. A better approximation x_1 is

$$x_1 = x_0 - \frac{f(x_0)}{f'(x_0)}$$

Newton's method can also be used to find a minimum or maximum of a function. The derivative is zero at a minimum or maximum, so minima and maxima can be found by applying Newton's method to the derivative. The iteration becomes:

$$x_{n+1} = x_n - \frac{f'(x_n)}{f''(x_n)}$$

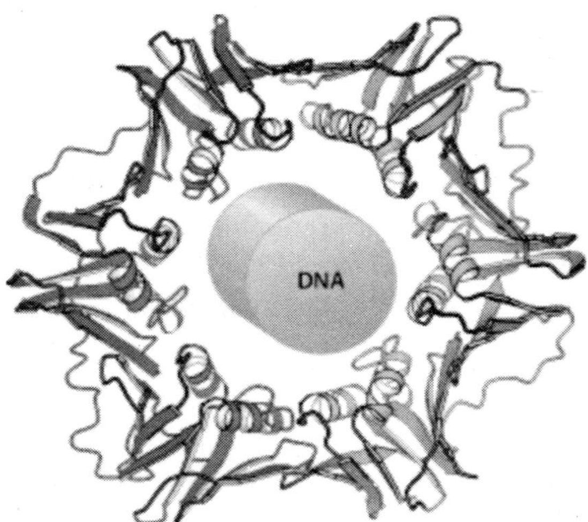

Fig. 9 : A protein component of the DNA replication machinery surrounds a section of DNA double helix allowing DNA to be copied without the replication machinery without dissociation.

1.14 Structure Function Relationship

1.14.1 Function from structure

Functions of a protein are mainly determined by its structure. A three-dimensional protein structure usually contains information about biochemical function e.g. binding sites, catalytic activity, and interactions between the protein and other molecules: such as its interaction with other proteins; with DNA or RNA; and interactions with ligands. Surface cavities, also called pockets or clefts, are ordinarily regarded as potentially active sites where the protein carries out the functions. The size and shape of a protein cavity dictates the three-dimensional geometry of ligands that can bind to together. The accurate identification of ligand binding sites in protein structures can be valuable in determining protein function. Once the binding site is known, it becomes easier to perform *in silico* and experimental procedures that may allow the ligand type and the protein function to be determined. For example, binding pocket shape analysis relies heavily on the correct localization of the ligand binding site. The identification and visualisation of cavities (ligand binding sites) is the starting point for many drug-design projects based on protein structure (Lengauer, 2002).

1.14.2 Function prediction from sequence

This would include "classic" methods such as detection of functional motifs and inferring function from sequence similarity. In this approach to predicting function from sequence, the main thing is to identify the functionally important ***motifs*** or ***patterns***. This can be done by multiple sequence alignment from homologous sequences to the query protein (http://learn.genetics.utah.edu/archive/protein).

1.15 Phylogeny Based Methods

Phylogenomics helps us to represent the molecular function evolution for a cluster of proteins using the phylogenetic tree construction. In addition, it is of interest to understand the evolutionary relationships between the homologous proteins from various organisms, which are based on the evolutionary distance cluster between the homologous proteins from various organisms (Engelhardt, 2006).

1.16 Conclusion

The prediction of protein structure is an active and promising area of research. Protein structure prediction methods attempt to provide a means of generating a plausible structure for proteins whose structures have not yet been experimentally determined. The structure prediction is made possible by sequence analysis between a new protein sequence and a protein of known structure even if there is a marginal similarity between them. This work is made easier with the help of structural databases of already determined protein structures. Meanwhile, the methods for secondary structure and threading analysis can provide useful predictions, although with variable levels of reliability. The concurrence of results of structure prediction from various methods determines the accuracy of the predicted structures. Thus, bioinformatics applications helps to predict protein function either through sequence-to-function or structure-to-function relationship. This facilitates to understand the biochemical pathways and cellular function, and thus aiding to know the field of proteomics and genomics in the case of diseased genes, and leads to improved drug design. In specific, we can conclude that protein structure is the keyhole for protein-ligand docking thereby revolutionizing drug design (Lengauer, 2002).

Some Useful Links

Protein sequence databases

SwissProt: http://www.expasy.ch/sprot/

TrEMBL: http://www.expasy.ch/sprot/

NCBI: http://www.ncbi.nlm.nih.gov/

Secondary structure prediction

GOR: molbiol.soton.ac.uk/compute/GOR.html

Jpred: http://www.compbio.dundee.ac.uk/~www-jpred/

Homology modelling

SWISS-MODEL: http://expasy.hcuge.ch/swissmod/SWISS-MODEL.html

MODELLER: http://salilab.org/modeller/download_installation.html

WHAT IF: http://swift.cmbi.ru.nl/servers/html/index.html

Threading/fold recognition

HHpred: http://toolkit.tuebingen.mpg.de/hhpred

I-TASSER: http://zhang.bioinformatics.ku.edu/I-TASSER/

LOOPP: http://cbsuapps.tc.cornell.edu/loopp.aspx

mGenTHREADER/GenTHREADER: http://bioinf.cs.ucl.ac.uk/psipred/psiform.html

PHDthreader: http://www.embl-heidelberg.de/predictprotein/

T3P2: http://www.mbi.ucla.edu/people/frsvr/frsvr.html

Ab initio structure prediction

HMMSTR/Rosetta: http://www.bioinfo.rpi.edu/~bystrc/hmmstr/server.php

ProFunc: http://www.ebi.ac.uk/thornton-srv/databases/*ProFunc*

PredictProtein: http://www.predictprotein.org/

References

Altschul, S.F., Madden, T.L., Scha¨ffer, A.A., Zhang, J., Zhang, Z., Miller, W. and Lipman, D.J. 1997. Gapped BLAST and PSI-BLAST: a new generation of protein database search programs. *Nucleic Acids Research*, 25: 3389-3402.

Altschul, S.F., Gish, W., Miller, W., Myers, E. and Lipman, D. 1990. Basic local alignment tool. *Journal of Molecular Biology*, 215: 403-410.

Bernstein, F.C., Koetzle, T.F., Williams, G.J.B., Meyer, E.F., Rodgers, M.D.B.J.R., Tsai, R.J., Ruczinski, I. and Baker, D. 2001. Functional inferences from blind ab initio protein structure predictions. *Journal of Structural Biology*, 134: 186-190.

Branden, C. and Tooze, J. 1999. Introduction to Protein structure. Second edition, Garland Publishing, Inc.

David, R., Korenberg, M.J. and Hunter, I.W. 2000. 3D–1D threading methods for protein fold recognition. *Pharmacogenomics*, 1: 445-455.

Duan, Y. and Kollman, P.A. 1998. Pathways to a protein folding intermediate observed in a 1-microsecond simulation in aqueous solution. *Science*, 23: 740-744.

Engelhardt, B.E., Jordan, M.I., Muratore, K.E. and Brenner, S.E. 2005. Protein molecular Functiion Prediction by Bayesian Phylogenomics. *PLoS Computational Biology*, 1: 0432-0445.

Jagielska, A., Wroblewska, L. and Skolnick, J. 2008. Protein model refinement using an optimized physics-based all-atom force field. *Proceedings of the National Academy of Sciences*, 105: 8268-8273.

Jones, D.T., Taylor, W.R. and Thornton, J.M. 1992. A new approach to protein fold recognition. *Nature*, 358: 86-89.

Jones, D.T. 1999. GenTHREADER: an efficient and reliable protein fold recognition method for genomic sequences. *Journal of Molecular Biology*, 287: 797-815.

Karplus, K., Barrett, C. and Hughey, R. 1998. Hidden Markov models for detecting remote protein homologies. *Bioinformatics*, 14: 846-856.

Kennard, O., Shimanouchi, T. and Tasumi, M. 1977. The Protein DataBank: a computer-based archival file for macromolecular structures. *Journal of Molecular Biology*, 112: 535-542.

Lesk, A.M. 2003. Introduction to Bioinformatics. Oxford University Press Publications Indian edition.

Leach, A.R. 2001. Molecular Modelling Principles and Applications, Second Edition, Prentice Hall, Pearson Education Limited UK.

Lengauer, T. 2002. Bioinformatics- From Genomes to drugs.Volume I: Basic Technologies, Wiley-Vch Verlag GmbH, Weinheim.

Lundsro¨m, J., Rychlewski, L., Bunnicki, J. and Elofsson, A. 2001. PCONS: a neural-network-based consensus predictor that improves fold recognition. *Protein Science*, 10: 2354-2362.

Mount, D. 2003. Bioinformatics - Sequence and Genome Analysis. CBS Publishers & Distributors, New Delhi.

Murzin, A.G. 2001. Progress in protein structure prediction. *Natural Structural Biology*, 8: 110-112.

Vingron, M. and Waterman, M.S. 1994. Sequence alignment and penalty choice. Review of concepts, case studies and implications. *Journal of Molecular Biology*, 235: 1-12.

Pattabhi, V. and Gautham, N. 2002. Biophysics, Kluwer Academic Publishers, New York, Boston, Dordrecht, London, Moscow.

Ramachandran, G.N. and Sasisekharan, V. 1968. Conformation of poly-peptides and proteins. *Advances in Protein Chemistry*, 23: 283-437.

Rao, S.S. 2008. Engineering Optimization; New Age International (P) Limited, Publishers.

Sofowora, A. 1982. Medicinal plants and traditional medicine in Africa, John Wiley and Sons Limited, Chichester.

Zhang, Y. and Skolnick, J. 2004. Automated structure prediction of weakly homologous proteins on genomic scale. *Proceedings of National Academy of Sciences USA*, 101: 7594-7599.

http://en.wikipedia.org/wiki/Hypothetical_protein

http://www.strgen.org.

http://www.chemcomp.com/journal/provalid.htm

http://learn.genetics.utah.edu/archive/protein.

http://www.mathworld.wolfram.com.

□□□

27

Herbs as Green Nanofactories

S.Girija

CONTENTS

1.1	Introduction	412
1.2	Nanoparticles Synthesis and Applications	412
1.3	Plants Exploited for Nanoparticle Synthesis	413
1.4	Methods of Synthesizing Metal Nanoparticle using Plant Sources	415
1.5	Role of Plant Extract on Morphology of Nanoparticle	417
1.6	Conclusion	418
	References	420

1.1 Introduction

The utilization of biological systems has emerged as a novel method for the synthesis of nanoparticles. Nanoparticles are defined as particles between 1nm to 100nm in size, some defining them up to 1 micron. Nanoparticles exhibit unique properties in terms of photoemission, conductivity and catalytic activity when compared to their respective bulk materials. The applications of these small particles are defined in a bigger way in different biological studies. Synthesis of nanoparticles using plant extract is a recent advancement in the field of nano biotechnology because it yields an eco-friendly method of nanoparticle synthesis. The currently, marketed nanoparticles are mostly synthesised using chemical methods, where its surface may be coated with toxic chemicals during the synthesis (Girija and Balachandran, 2009). Recently, synthesis of nanoparticles has been reported utilizing several herbal plant extracts and their biomass. It opens up a new venue for using the herbal plants for the synthesis of various metal nanoparticles in large scale. Here, we review the plants which are already exploited for the synthesis of nanoparticles and the possible use of plant resources as nanofactories to synthesize less toxic metal nanoparticles.

1.2 Nanoparticles Synthesis and Applications

Nanoparticles are used in a wide range of applications not only in the optoelectronic devices, chemical and biological sensors, but also in DNA labeling and drug delivery (Kohler *et al.*, 2001; Schatz *et al.*, 2000), biomedicine, pharmaceuticals, cosmetics, energy and environmental detection (Liu, 2006). It also has tremendous application in Food and Agriculture industry. Nanoparticles or nanocapsules could provide a more efficient means to distribute pesticides and fertilizers, reducing the quantities of these chemicals introduced into the environment. There is a wide scope of applying nanoparticles and nanocapsules to plants for agricultural use (Gonzalez-Melemdi *et al.*, 2008). Gene transfer by bombardment of DNA absorbed gold particles has been successfully used to generate transgenic plants in species - independent manner (Christou *et al.*, 1998). Silica nanoparticles were used to deliver DNA in plant cells without the requirement of specialized equipment (Torney *et al.*, 2007). The conventional methods of nanoparticle synthesis are physical and chemical; physical methods include evaporation and laser ablation, whereas, chemical methods involve the reduction of a salt with reducing agents like $NaBH_4$ (Jana *et al.*, 2001; Wang *et al.*, 2005). Synthesis of nanoparticle using either top down involving cutting, carving etc., or bottom up approach, involves manipulation at the molecular level by employing various steps, where the toxic and/or flammable chemicals were used, which can have undesirable side effects in some applications, e.g. in medicine (Bali *et al.*, 2006). The concern for environmental contaminations is also heightened, as the chemical procedures involved in the synthesis of nanomaterials generate a large amount of hazardous byproducts (Sharma *et al.*, 2007). Exploitation of biological organisms in the synthesis of nanoparticle is a biotechnological approach to material science for the low cost production of metal nanoparticles (Joerger *et al.*, 1999). In the last decade, biosynthesis of nanoparticles is an emerging highlight of the intersection of nanotechnology and biotechnology. It has received an increasing attention due to a growing need to develop environmentally benign technologies in material synthesis such as gold and silver (Huang *et al.*, 2007). The biosynthetic method employing plant extracts for synthesizing metal nanoparticles has received attention in recent years, which is a simple and viable method alternative to chemical and physical. The metal nanoparticles like silver and gold were produced when aqueous solutions of Silver nitrate and Chloroauric acid were incubated with boiled plant extract. The studies on silver and gold nanoparticle formation in live alfalfa plant (Gordea *et al.*, 2002) is an embodiment for all subsequent reports on metal nanoparticle synthesis using different plant extracts reported till date. It opens up a new venue for tapping the unexploited plants for synthesis of various metal nanoparticles (Leela and Vivekanandan, 2008), which would have tremendous value in application of nanoparticle in human welfare, which includes Agriculture, food and drug industry. This approach promises synthesis of less toxic nanoparticles using plants, whereas most of the chemicals added during chemical synthesis of nanoparticle would contribute to its toxicity. Thus toxicity can be reduced by using plant extract for nanoparticle synthesis. Analyzing the various factors responsible for nanoparticle synthesis using plants would be helpful in scaling up this technology for mass production.

1.3 Plants Exploited for Nanoparticle Synthesis

Plants being the source in all-human use it doesn't leave the breaking scientific advancement like nanotechnology. Nanoparticles of common heavy metal ions, e.g. Ag, Cu, Co, Zn and Ni, were synthesized by exposing plants to aqueous metal salt solutions (Bali *et al.*, 2006). The utilization of biological systems has emerged as a novel method for the synthesis of nanoparticles. Since, the first report on the ability of plants synthesizing gold nanoparticles (*Medicago sativa*) (Gordea *et al.*, 2002), the green nanotechnology for synthesizing eco-friendly nanoparticle came as a boon to show that plants will be a potential platform to synthesize different metal nanoparticles. Metallic nanoparticles are traditionally synthesized by wet chemical techniques, where the chemicals used are quite often toxic and flammable. The recent research articles published on synthesis of nanoparticles using plant extracts and their dried or fresh biomass gives hope that this technology would be an efficient system to synthesize different nanoparticles for various applications including Agriculture (Gonzalez-Melemdi *et al.*, 2008). Synthesis of gold nanoparticles in live *Medicago sativa* (Gordea *et al.*, 2002), and *Sesbania drummondii* (Sharma *et al.*, 2007) plant species conglomerate the use of plant extracts, as an efficient source for the synthesis of metal nanoparticles like silver and gold. Biosynthetic processes for nanoparticles production would be more useful if nanoparticles were produced extracellularly using plants or their extracts (Kumar and Kumar Yadav, 2008). The extracellular synthesis of silver nanoparticles using Geranium (*Pelargonium graveolens*) was obtained by treating aqueous silver nitrate solution with geranium leaf extract. Rapid reduction of the silver ions is observed leading to the formation of highly stable crystalline silver nanoparticles in solution (Shankar *et al.*, 2003). Similarly the leaf extracts of Neem (*Azadirachta indica*) is found to synthesize silver and gold nanoparticles (Shankar *et al.*, 2004a) when 5 ml of boiled leaf extract were incubated with 45ml of 10^{-3} M of $AgNO_3$ and $HAuCl_4$ separately. The time required for more than 90% reduction of Ag^+ and Au^{3+} was 2 and 4 h respectively. Silver nanoparticles formed were predominantly spherical and polydisperse with diameters in the range 5 to 35 nm. The metal particles were observed to be stable in solution even 4 weeks after their synthesis.

The bioreduction behavior of different leaf extracts such as *Helianthus annus* (*Asteraceae*), *Basella alba* (*Basellaceae*), *Oryza sativa*, *Saccharum officinarum*, *Sorghum bicolour* and *Zea mays* (*Poaceae*) in the synthesis of silver nanoparticles was investigated (Leela and Vivekanandan, 2008). Although all the plants exhibited a strong absorption between 400 to 500 nm, the sunflower leaf extracts showed very strong absorption compared to other leaf extracts used. It was suggested that the differences in the morphology of nanoparticle synthesized may be the reason for the differences in optical properties. Similarly Pine, (*Pinus desiflora*) Persimmon, (*Diopyros kaki*) Ginkgo (*Ginko biloba*), Magnolia (*Magnolia kobus*) and Platanus (*Platanus orientalis*) were used and compared for their extracellular synthesis of metallic silver nanoparticles (Song and Kim, 2008). Gold triangles synthesized using tamarind leaf extracts were found to be useful in sensing of vapor (Ankamwar *et al.*, 2005a). The commercially important medicinal plant *Aloe vera* was successfully used for synthesizing the gold nanoparticles (Chandran *et al.*, 2006) and Indium oxide nanoparticle (Maensiri *et al.*, 2008) it proves that the plants can also be used to synthesize metal oxide nanoparticles. Corriander leaf extract mediated gold nanoparticles were reported (Narayanan and Sakthivel, 2008). In all the above reports the boiled fresh leaf extracts were used for nanoparticle synthesis. In contrast, the dried biomass of *Cinnamomum camphora* leaves were also able to synthesize the silver and gold nanoparticles (Huang et al., 2007). *Avena sativa* stem biomass was successful in synthesizing the gold nanoparticles (Armendariz *et al.*, 2004). With the use of *Emblica officinalis* fruit extract as reducing agent, the extracellular synthesis of highly stable Ag and Au nanoparticles has been achieved (Ankamwar *et al.*, 2005b). Consequently, the methanolic extracts of *Eucalyptus camaldulensis* and *Pelargonium roseum* were successful in the synthesis of Gold nanoparticles (Ramezani et al., 2008). The above reports summarize that the different metal nanoparticles like gold, silver and metal oxide like indium can be successfully synthesized from extracts of leaf, fruit and also by using fresh (stem of oat) and dry (dried leaf of camphor) plant biomass.

1.4 Methods of Synthesizing Metal Nanoparticles using Plant Sources

The first report on synthesis of gold nanoparticle in alfalfa live plants has been reported, where the alfalfa plants has synthesized the gold nanoparticle of 2-20nm when they were grown on agar medium containing gold (III) from potassium tetrachloroaurate. Analyzing the shoot and root using X-ray energy dispersive spectroscopy (EDS), transmission electron microscopy (TEM) and high-resolution transmission electron microscopy (HRTEM) revealed that gold (0) was uptaken by the alfalfa plants and continuous growth and rearrangements of gold nanoparticles has occurred within the plant. Silver nanoparticle formation in alfalfa shoot and root using germinated sprouts was observed by Gordea *et al.* (2003), who explained that alfalfa roots were capable of absorbing silver as Ag (0) from the agar medium and transfer it to the shoot of the plant in the same oxidation state. Silver atoms arrange themselves to form nanoparticles, and these nanoparticles join to form larger arrangements and they also observed that the Ag atoms accumulated inside alfalfa plants underwent nucleation and this can be used in future to synthesize the quantum dots (Semiconductor nanoparticle). This will be useful in synthesizing quantum dots with different optical and fluorescent properties for biological applications like live plant cell staining and imaging, DNA labeling, disease diagnosis and biosensors.

For the synthesis of silver nanoparticles the $AgNO_3$ (Shankar *et al.*, 2004a), Ag_2SO_4 (Ankamwar *et al.*, 2005b) aqueous solutions were used in different molarities (10^{-3} of $AgNO_3$ and $10^{-4}Ag_2SO_4$) and for the synthesis of gold nanoparticle the chloroauric acid (10^{-3} of $HAuCl_4$) is used. The finely cut leaf samples were boiled using sterile distilled water and resulting extracts were used for the synthesis of silver and gold nanoparticles for the reduction of aqueous Ag^+ and Au $^{3+}$ into Ag and Au and subsequent formation of silver and gold nanoparticles in different plant species (Shankar *et al.*, 2004a; Ankamwar *et al.*, 2005a; Leela and Vivekanandan, 2008; Song and Kim, 2008). The Ag nanoparticle formations were confirmed by using UV-vis spectroscopy where it showed strong absorbance at 430 nm (Song and Kim, 2008), whereas, the gold nanoparticle showed strong absorbance at 550 nm (Shankar *et al.*,

2004a). Subsequently, when the *Cinnamomum camphora* sundried leaf biomass were incubated with 50 ml of 1 mM aqueous $AgNO_3$ and $HAuCl_4$ solution in the orbital shaker at 150 rpm at 30°C, the bioreduction of Ag^+ or $AuCl^-_4$ was confirmed by measuring UV-vis spectra of the aliquots of sample suspension (0.2 ml), diluted with 2 ml deionized water periodically (Huang *et al.*, 2007).

Formation of the metal nanoparticles by reduction of the aqueous metal ions during exposure of the broth to boiled neem leaves were measured by UV-vis spectroscopy, where the silver and gold nanoparticles exhibited yellowish-brown and ruby red colors, respectively in water, these colours arised due to excitation of surface plasmon vibrations in the metal nanoparticles (Shankar *et al.*, 2004a). Lemongrass and geranium plant extracts were used to induce the formation of gold nanoparticles when reacted with aqueous chloroauric acid (Shankar *et al.*, 2004b,c). Triangular gold nanoparticles were observed by Ankamwar *et al.* (2005a), when the boiled tamarind leaf broth was incubated with aqueous chloroauric acid solution at room temperature. Consequently the fruit extract of *Emblica officinalis* produced the silver and gold nanoparticle when incubated with boiled pericarp extract with aqueous Ag_2SO_4 and $HAuCl_{4,}$ the solution changed to yellow and pink respectively indicating rapid reduction of Ag^+ and $AuCl^-_4$ ions to yield Ag and Au nanoparticles (Ankamwar *et al.*, 2005b). Similarly when the dried biomass of *Cinnamomum camphora* was used for the silver nanoparticle synthesis, the solution changed from colourless to pale green in about 1 h, after addition of the biomass to the silver nitrate solution, the final colour deepening increased with increase in the amount of dried biomass (Huang *et al.*, 2007). It was observed that the slight size variations has occurred when two different level (0.5g and 1.0g) of biomass were used. These observations deemed that it is not only the silver nanoparticles with concentrated size distribution be fabricated even the different shapes such as triangular or spherical nanoparticles of gold can be produced by simply adjusting the amount of dried biomass. Interestingly, few medicinal plants were screened to use their methanolic extract for gold nanoparticle formation among these the *Eucalyptus camaldulensis* and *Pelargonium roseum* showed enhanced reduction ability compared to neem leaf broth (Ramezani *et al.*, 2008), it shows the reduction ability of metal iron depends on extract of the plant species used.

Studies on understanding the biosynthetic pathway involved in the fabrication of metal nanomaterial mediated by a biological system are important to gain a better control on the process and products. Little is known so far about the interaction of biomolecules and gold nanoparticles. Shankar *et al.* (2004b) believed that the reduction of tetrachloroaurate ions was caused by the reducing sugar present in the lemongrass extract, while the growth of gold nanotriangle crystal was the result of an interaction between *de novo* gold nanoparticles and aldehydes/ketones present in the extract. The reduction of the metal ions and stabilization of the gold/silver nanoparticles were believed to occur by the various terpenoids or alkaloids present in the geranium extract (Shankar *et al.*, 2004c). For the synthesis of silver and gold nanoparticle, aliquots of plant extracts were incubated with Silver nitrate and Chloroauric acid and the bioreduction of silver and gold nanoparticles were confirmed by analyzing UV-visible spectra of plant extract at different time intervals and further characterization of nanoparticles size and shapes were reported using XRD (X-ray diffraction), SEM (Scanning electron microscopy) and TEM (transmission electron microscopy) analysis. FTIR (Fourier transform infrared spectroscopy) absorption spectral analysis was found useful in analyzing the leaf broth before and after bioreduction, it offers information regarding the chemical changes of the functional groups involved in the bioreduction.

As can be seen from the above, the use of plant extract in the synthesis of nanoparticles is a relatively new and exciting area of research with considerable potential for development.

1.5 Role of Plant Extract on Morphology of Nanoparticle

In nanoparticle, the shape and size determines the properties such as biological activities, for example: antimicrobial activity of silver nanoparticle varied with the size of the nanoparticles (Panacek *et al.*, 2006) and the catalytic property was dependent on cubic structure in palladium nanoparticle (Shi and Masel, 1989). The strong NIR absorbance of the gold nanotriangles finds interesting applications in cancer hyperthermia (Kelly *et al.*, 2003).

The consequent reports on nanoparticle synthesis using plant sources shows a different shapes of nanoparticles can be obtained using plant biosynthesis, excellent shape-selective formation of single crystalline triangular gold nanoparticles was observed (Shankar *et al.*, 2004b) using the extract of the lemongrass plant (*Cymbopogon flexuosus*), it was also proved by Chandran *et al.* (2006), where the triangle shaped gold nanoparticles were synthesized using *Aloe vera* plant extract. The biological synthesis of gold nanotriangles using *Aloe vera* extract provides a simple and efficient route for the synthesis of nanomaterials with tunable optical properties directed by particle shape. The size of the gold nanotriangles varied from 50 to 350 nm by merely adjusting the amount of *Aloe vera* extract used in the gold ion reduction. These results throw light that application-based nanoparticle can be synthesized by changing the amount of plant extract. Silver nanoparticles synthesized in *Capsicum annuum* was reported by Li et al. (2007) indicated that the proteins, which have amine groups, played a reducing and controlling role during the formation of silver nanoparticles in the solutions. The particle size could be controlled by changing the reaction temperature, leaf broth concentration and $AgNO_3$ concentration (Song and Kim, 2008). The gold nanoparticle synthesized in roots and shoots of *Sesbania drummondii* plant exhibited fairly well defined dimensions and monodispersity (Sharma *et al.*, 2007). Huang *et al.* (2007) reported that the triangular or spherical shaped gold nanoparticles could be easily modulated by reacting the sundried biomass of *Cinnamomum camphora* leaf with aqueous silver or gold precursors at ambient temperature. In general, the particle size was larger than that reported using Geranium (Shankar *et al.*, 2003) and *Aloe vera* (Chandran *et al.*, 2006) plant extract. However, by switching the amount of the dried biomass from 0.1 to 0.5 g subjected to the same chloroauric acid, the particles shifted from nanotriangles to spherical particles.

The shape and size of the nanoparticles plays very important role in determining the properties of the nanoparticles like photoemission and catalytic activity, which are very important in the application of these particles in different fields especially in biological applications. The above reports summarize that the plant extracts can not only be used as an efficient system for nanoparticle synthesis but also can be used to manipulate the size and shape of the nanoparticle by employing different plant extracts at different concentrations. Armendariz *et al.* (2004) have reported the pH-dependent binding trend of Au (III) ions to *Avena sativa* (Oat) biomass and subsequent formation of Au nanoparticles of variable sizes. They observed the Rod-shaped nanoparticles were common at all pH values used (pH 2-6) during the study but the size of the nanoparticles varied with the pH, this shows the plant based nanoparticle synthesis will compete with chemical and physical synthesis. The plant extracts and biomass offers a great venue for obtaining different shapes and size of nanoparticle by merely adjusting the concentration of plant sample and the environmental conditions like temperature and pH. Plant species plays an important role in determining the morphology of the nanoparticle. The above reports reveal that different plant species have produced different shapes and size of the nanoparticle.

1.6 Conclusion

Nanotechnology is the new and exciting field of research, in which recent advances in nanotechnology are integrated into biological realm, in particular into molecular biology and cell biology. Currently, there are several methods for the production of nanoparticles like chemical and physical method but the report says that they are harmful to the environment. Based on the reports available and the results we obtained (Unpublished) using Neem fresh leaf and dried plant extracts to synthesize, silver nanoparticles for tuning their fluorescent properties shows plants are the best nanofactories for eco friendly synthesis of metal nanoparticles. It will be useful if nanoparticles were produced extracellularly using plants or their extracts in a controlled manner according to their size, dispersity and shape. Although the particle formation and reduction were confirmed by using different analytical studies in all the plant reported, the mechanism behind the particle formation were in scanty and the issues such as monodispersity and shape selectivity for obtaining phase pure monodisperse nanoparticles are yet to be addressed and focused (Shankar *et al.,* 2004a). The reduction of metal ions and the stabilization of the Ag and Au nanoparticles are believed to occur by various hydrolysate tannins present in the *Emblica officinalis* fruit (Ankamwar *et al.,* 2005). This would help in understanding the biochemical and molecular mechanism of the nanoparticle synthesis.

Isolation and testing of individual fractions of metal ion reduction and binding with the nanoparticles is necessary (Leela and Vivekanandan, 2008). The molecular compound responsible for reduction and controlling the size as capping agent would be the near future studies in plant cell biology for efficient manipulation of nanoparticle for desired application with reduced toxicity and to implement their application in various fields including Agriculture. Environmental friendly method of biological silver nanoparticles production provides faster rates of synthesis comparable to those of chemical methods and can be potentially used in various human contacting areas such as cosmetics, foods and medical applications (Song and Kim, 2008).

Development of nano-devices using biological materials and their use in wide array of applications on living organisms has recently attracted the attention of biologists towards nanobiotechnology (Mohanapuria, 2008). Synthesis of nanoparticle using the plant source is a great advantage, which offers eco friendly method of nanoparticle synthesis. The embodiment of reports on nanoparticle formation by different plant species have been reviewed here to excel in this field of research with potential application. The further challenge in this field of research is studying the kinetics of nanoparticle formation using plant extracts and analyzing the biomolecules responsible for the nanoparticle formation. Further this is also, to evaluate the role of plant metabolites in determining the morphology of the nanoparticle formation, whereby nanoparticles with different tunable properties can be produced, which will extend the applications of herbal science in nanotechnology in the future.

References

Ankamwar, B., Chaudhary, M. and Sastry, M. 2005a. Gold Nanotriangles Biologically Synthesized using Tamarind Leaf Extract and Potential Application in Vapor sensing. *Met.- Org. and Nano-Met. Chem.* 35:19–26.

Ankamwar, B., Damle, C., Absar, A. and Murali, S. 2005b. Biosynthesis of gold and silver nanoparticles using *Emblica officinalis* fruit extract, their phase transfer and transmetallation in an organic solution. *J Nanosci.Nanotechnol.* 10:1665–1671.

Armendariz, V., Herrera, I., Peralta-Videa, J. R ., Jose-Yacaman, M., Troiani, H., Santiago, P. and Gardea-Torresdey, J .L .2004. Size controlled gold nanoparticle formation by *Avena sativa* biomass: use of plants in nanobiotechnology. *J. Nano. Res.* 6: 377–382.

Bali, R., Razak, N., Lumb, A. and Harris, A.T. 2006. The synthesis of metallic nanoparticles inside live plants. *Proc.Int.Conf. Nanosci. and Nanotech.*.3-7:224-227.

Chandran, S.P., Chaudhary, M., Pasricha, R., Ahmad, A. and Sastry, M. 2006. Synthesis of Gold Nanotriangles and Silver Nanoparticles using *Aloe vera* Plant Extract. Biotechnol. *Prog.* 22: 577-583.

Christou, P., McCabe, D.E. and Swain, W.F. 1988. Stable transformation of soybean callus by DNA-Coated gold particles. *Plant Physiol.* 87: 671-674.

Gardea-Torresdey, J.L., Parsons, J.G., Gomez, E., Peralta-Videa, J., Troiani, H.E., Santiago, P and Jose Yacaman, M. 2002.Formation and growth of Au nanoparticles inside live alfalfa plants. *Nano Lett.* 2:397–401.

Gardea-Torresdey, L., Gomez, E., Peralta-Videa, J.R., Parsons, J.G., Troiani, H. and Jose Yacaman, M. 2003. Alfalfa spouts: a natural source for the synthesis of silver nanoparticles. *Langmuir.* 19:1357– 1361.

Girija. S. And Balachandran, Y.L.2009. Nanoparticle : For imaging of Biological systems- A review. *Proc. Nat. Wom. Sci. Cong.* (In Press)

Gonanzale – Melendi, P., Fernandez-Pacheco, R., Coronado, M,J., Corredor, E, Testillano, P.S., Risueno, M.C., Marquina, C., Ibarra, M.R., Rubiales, D. and Perez- De- Luque, A. 2008. Nanoparticles as Smart Treatment – delivery Systems in Plants: Assessment of Different Techniques of Microscopy for their Visualization in Plant Tissue. *Ann. of Bot.* 101: 187- 195.

Huang, J., Li, Q., Sun, D., Lu, Y., Su, Y., Yang , X, , Wang, H., Wang, Y., Shao, W., He, N., Hong, J. and Chen, C.2007. Biosynthesis of silver and gold nanoparticles by novel sundried *Cinnamomum camphora* leaf. *Nanotech.*18: 105104 –105114.

Jana, N.R., Gearheart, L. and Murphy, C.J. 2001.Wet Chemical Synthesis of High Aspect Ratio Cylindrical Gold Nanorods. *J. Phys. Chem.* 105:4065-4067.

Joerger, R., Klaus,T. and Granqvist, C.G. 1999.Biologically Produced Silver-Carbon Composite Materials for Optically Functional Thin-Film Coatings. *Adv. Mat.* 12(6): 407 – 409.

Kelly, K. L.,Coronado, E., Zhao, L. L. and Schatz, G. C. 2003.The optical properties of metal nanoparticles: The influence of size, shape, and dielectric environment. *J. Phys. Chem.* 107:668-677.

Kohler, J. M., Csaki, A., Reichert, R., Straube, W. and Fritzche, W. 2001. Selective labeling of oligonuckleotide monolayers by metallic nanobeads for fast optical readout of DNA chips. *Sens.Act.* 76: 166-172.

Kumar, V. and Kumar Yadav. S. 2008. Plant-mediated synthesis of silver and gold nanoparticles and their applications. *J.Chem. Tech. and Biotech.* (In Press).

Leela, A. and Vivekanandan, M. 2008. Tapping the unexploited plant resources for the synthesis of silver nanoparticles. *Afr. J. Biotech.* 17: 3162-3165

Li, S., Shen, Y., Xie, A., Yu, X., Qiu, L., Zhang, L. and Zhang Q.2007. Green synthesis of silver nanoparticles using *Capsicum annuum* L. extract. *Green Chem.* 9: 852 - 858.

Liu, W.T.2006. Nanoparticles and their Biological and Environmental Applications. *J.Biosci. and Bioengg.* l 102: 1-7.

Maensiria, S., Laokula, P., Klinkaewnaronga, J., Phokhaa, S., Promarakc. V. and Seraphin. S.2008. Indium oxide (In_2O_3) nanoparticles using *Aloe vera* plant extract: Synthesis and optical properties. *J.Optoelect and Adv. Mater.* 10:161 165.

Mohanpuria, P., Rana, N. K. and Kumar Yadav , S.2008. Biosynthesis of nanoparticles: Technological concepts and future applications. *J. Nanopart. Res.* 10:507–517.

Narayanan, B. and Shakthivel, N. 2008. Coriander leaf mediated biosynthesis of gold nanoparticles. *Mater. Lett.* 62:4588-4590.

Panacek, A., Kvýtek L., Prucek, R., Kolar, M., Vecerova, R.,Pizurova, N., Sharma, K. V., Nevecna, T. and Zboril, R. 2006. Silver Colloid Nanoparticles: Synthesis, Characterization, and Their Antibacterial Activity. *J. Phys. Chem.* 110: 16248-16253.

Ramezani, N., Ehsanfar, Z., Shamsa, F., Amin, G., Shahverdi, H .R., Monsef Esfahani, H R. Shamsaie, A Bazaz, R. D. and Shahverdi, A.R. 2008. Screening of Medicinal Plant Methanol Extracts for the Synthesis of Gold Nanoparticles by their Reducing Potential. *Z. Naturforsch.* 63: 903 - 908.

Sharma, N.C., Sahi, S. V., Sudip, N. Parsons, J .G. Gardea-Torresdey, J.L. and Tarasankar, P. 2007. Synthesis of plant-mediated gold nanoparticles and catalytic role of biomatrix-embedded nanomaterials. *Env. Sci Technol.* 14: 5137–5142.

Schatz, G.C., Lazarides, A.A., Kelly, K.L. and Jensen, T.R. 2000.Optical properties of metal nanoparticles aggregates important in biosensors. *J.Mol.Struc.* 529: 59-63.

Shankar, S.S., Ahmad, A. and Sastry, M. 2003. Geranium leaf assisted biosynthesis of silver nanoparticles. *Biotechnol.* 6:1627-31.

Shankar, S. S., Rai, A and Ahmad, A. 2004a. Rapid synthesis of Au, Ag, and bimetallic Au core-Ag shell nanoparticles using Neem (*Azadirachta indica*) leaf broth. *J Colloid Inter. Sci.* 2: 496-502.

Shankar, S.S., Rai, A., Ankamwar, B., Singh, A., Ahmad, A. and Sastry, M. 2004b. Biological synthesis of triangular gold nanoprisms. *Nat. Mater.* 3: 482–488.

Shankar, S.S., Rai, A., Ahmad, A. and Sastry, M. 2004c.Biosynthesis of silver and gold nanoparticles from extracts of different parts of the geranium plant. *App. Nano. Sci.* 1: 69–77.

Shi, A.C. and Masel, R. I. 1989. The effects of gas adsorption on particle shapes in supported platinum catalysts. *J. Catal.* 120: 421-431.

Song, J.Y.and Kim, B.S.2008. Rapid biological synthesis of Silver nanoparticle using plant leaf extracts. *Bioproc. and Biosys. Eng.* 1615-7605.

Torney, F., Trewyn, B.G., Lin, V. S.Y and Wang, K. 2007. Mesoporous silica nanoparticles deliver DNA and chemicals into plants. *Nat. Nanotech.* 2: 295-300.

Wang, S.T., Yan,. J and Chen, L. 2005.Formation of gold nanoparticles and self assembly into dimer and trimer aggregates. *Mater. Lett.* 59: 1383–1386.

□□□

28

Status of Biodiversity and Conservation

D.Sathiyan

CONTENTS

1.1. Introduction 424
1.2. Biodiversity of the World 424
1.3. Biodiversity of India 425
1.4. Importance of Biodiversity 425
1.5. Threats to Biodiversity 426
1.5.1. Habitat destruction 427
1.5.2. Alien invasive species 428
1.5.3. Negative effects of climate change 428
1.5.4. Trade in species of plants & animals 429
1.6. Conservation — A Challenge 431
1.7. Approaches to Conservation 431
1.7.1. In-situ conservation 431
1.7.2. Ex-situ conservation 432
1.8 Biodiversity Conservation- The Main Players and Expected Roles 432
1.9 Conclusion 434
References 434

1.1 Introduction

Biodiversity is defined as "the variability among living organisms from all sources including, inter alia, terrestrial, marine and other aquatic ecosystems and the ecological complexes of which they are part; this includes diversity within species, between species and of ecosystems." (CBD)

1.2 Biodiversity of the World

Biodiversity of the world today is a product of about 3.5 million years of evolution. Though 'Earth' has always been endowed with rich variety of life, the word 'Biodiversity' was of recent origin, coined few decades ago. Our knowledge on biodiversity has been only partial. About 1.8 million species only have been described by scientists so far, while the estimated total number of species in the world may range between 5-30 million. Every year, new species are getting discovered by scientists all over the world. On the other hand, several extinctions too are recorded by them. In addition to this recorded extinctions, there may be numerous extinctions too among species unknown to science.

Biodiversity is not uniformly distributed among the countries of the world. About 18 countries together hold about 66 % (2/3rd) of the world's Biodiversity thereby gaining the name 'Mega diversity Countries'. India is one such country. Conservation International (CI) an international NGO, recently identified 34 'Biodiversity Hotspots' of the World, which are areas supporting a higher number of endemic plants (at least 1500) and have already lost about 70 percent extent of their original habitat (Conservation International 2004). In other words these are extraordinary rich areas in terms of biodiversity which are under serious threats, qualifying for the term 'Environmental Emergency Rooms on Earth'. We have the following 3 hotspots in the Indian Region: a) The Himalayas b) The Western Ghats & Sri Lanka (the portion of Western Ghats) and c) Indo – Burma (the portion of Andaman Islands and a part of the North East India).

Similarly, Birdlife international, a global partnership of organisations for conservation of birds, has identified about 218 Endemic Bird areas (EBA) world over, denoting areas of high conservation values (Birdlife international, 2003). Aves are the most well studied group under 'Animalia' with the least percentage of data deficiency against its species. Their status is well known and they are considered to be the indicators to the status of Biodiversity. Endemic Bird Areas are the habitats which support more number of birds with restricted distribution range ('endemic'). India has about 9 such EBAs.

Concepts like Mega Diversity countries, Biodiversity Hotspots and Endemic Bird Areas have been proposed to help prioritize the areas of focus for channelizing our conservation efforts. Though biodiversity found everywhere is important, prioritizing the areas for conservation is certainly a wise strategy considering the paucity of time, funds, focussed scientific research.

1.3 Biodiversity of India

In addition to the above EBAs, Hotspots, etc, India is home to about 3% of the world's mangroves. Indian land mass (2.4% of the world) harbours about 8% of world's biodiversity. (Jagdish Kiswan, Dr Devendra Pandey, AK Goyal and Dr AK Gupta, 2007) India is home to almost all kinds of ecosystems viz., Mountain ecosystem, Desert ecosystem, Marine ecosystem, freshwater ecosystem, Mangrove ecosystem, Riverine ecosystem, Island ecosystem, Forest ecosystem, etc. As per the widely used system of forest classification proposed by Champion and Seth, India has more than 200 forest types, which shows the extend of variety among its forests. India supports about 13.66% of world's birds and about 8.5% of world's mammals, 7.84% of world's reptiles, nearly 11.72% of world's fishes, and 4% of Amphibians. Floral diversity in India is staggering with about 17,000 species of flowering plants of which nearly $1/3^{rd}$ are endemic.

India has long been identified as a centre of diversity and a place of origin of crop plants. It is one of the 8 such centres identified by Russian scientist Nicolay Ivanovich. Many millets (51 species), fruits (104), spices & condiments (27), vegetables and pulses (55), Fibre crops (24), oil seeds etc. that are commonly used today world over have their wild relatives among native Indian biodiversity.

1.4 Importance of Biodiversity

Biodiversity and human life are closely related. Nearly 40 % of world's economy is estimated to be derived from biodiversity. Biodiversity satisfies nearly 80 % needs of the poor lot of the world. As recorded in the Millenium Ecosystem Assessment report (MEA, 2005), biodiversity offers the following ecosystem services of value to humanity:

Provisioning services	Regulating services	Cultural services	Support services
Food	Climate regulation	Aesthetic	Nutrient cycling
Fresh water	Flood regulation	Spiritual	Soil formation
Wood and fiber	Disease regulation	Educational	Primary Production
Fuel	Water purification	Recreational	

The list is not exhaustive. Biodiversity ensures the stability of ecosystems too. In short, biodiversity lays the foundation for the continued existence of a healthy planet and our well being.

1.5. Threats to Biodiversity

The fact that human beings exert tremendous negative pressure on biodiversity has been well established. Our species is the main 'driver' to many factors that hasten the rate of loss of biodiversity. These factors include habitat destruction, habitat fragmentation, Invasive Alien Species (IAS), trade in the species of plants & animals, negative effects of climate change, unsustainable or over exploitation of populations of species, etc.

Species Extinction is the gravest aspect of biodiversity crisis as it is irreversible. The history of Earth shows that there were five (5) periods of 'mass extinctions', in the past. The last such mass extinction (5th) is recorded to have wiped out the population of Dinosaurs. The rate of extinctions being witnessed now is said to be faster by 1000 times or even more, than the natural rate of extinctions. The 'natural extinction rate' is the rate at which species were getting extinct before the onset of influences by the human beings, which, according to some estimates, is a loss of one species in a period of 3-5 years. As such, a section of the scientists warn that 'sixth extinction crisis' is going on at present which needs to be addressed with urgent conservation measures.

The status of biodiversity at the global level is assessed by IUCN (International Union for Conservation of Nature) by assigning threat categories to the assessed individual species of plants and animals. The status is published as 'IUCN RED LIST'. The IUCN threat Categories are given below.

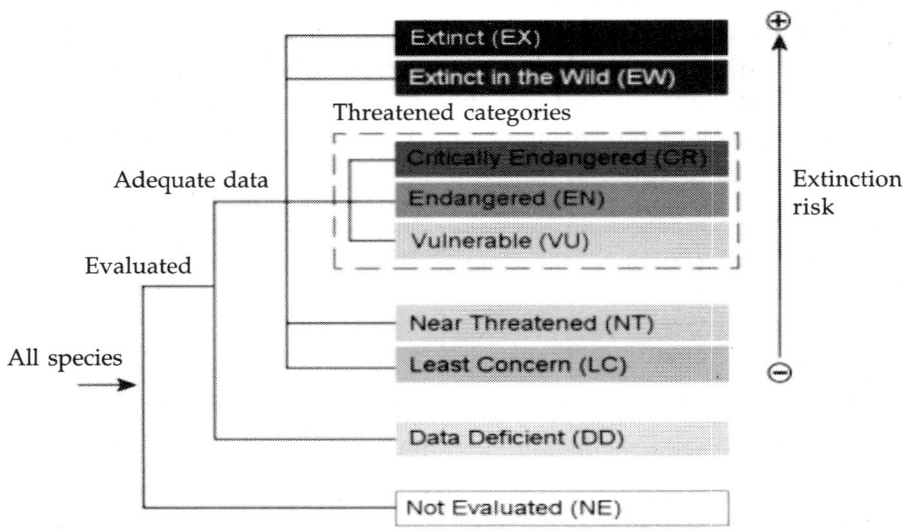

The threat status of species is updated frequently. The latest IUCN update includes 44 838 species in the red list. Of these 869 (2 %) species are classified as extinct or Extinct in the wild. Another 16,928 (16 %) species are classified as 'Threatened with extinction', with a break up of 3246 species as Critically Endangered, 4770 species as Endangered and 8912 species as Vulnerable. In general terms, as per IUCN Red list 2008, about 22 % mammals, 31 % Amphibians, 14 % Birds are included under 'Threatened' Categories. (IUCN Red List 2008)

'The living planet Index (1970-2000)' prepared jointly by the World Wildlife Fund (WWF), United Nations Environment Program (UNEP) and World Conservation Monitoring Centre (WCMC), as an indicator of the state of world's biodiversity, also shows a downward trend in biodiversity. It is recorded that the 'Terrestrial index', which shows the ability of selected terrestrial ecosystems studied for their support to populations of selected species, fell by 30 percent between 1970-2000. Similarly the freshwater index fell by 50 % and marine index by 30 %, during the same period.

1.5.1 Habitat destruction

If we look at the reasons for the decline in biodiversity, habitat destruction ranks first. It is identified as one of the main threats to 85 % of the species included in the IUCN Red list. Habitat fragmentation and habitat deterioration are the other forms of damages to biodiversity due to human causes.

It is worthwhile to note here that while biodiversity is present everywhere, both in macro as well as micro forms, the macro biodiversity that occurs on terrestrial ecosystems has been the most studied. Among terrestrial ecosystems forest ecosystems are estimated to hold nearly 90% biodiversity. So when we talk about biodiversity loss due to loss of habitats it is synonymous with the loss of forests. In a period of about 20 years, we have seen disappearance of about 35% of world's Mangroves. Between 1990-2000, the world had lost forests at a rate of about 8.9 million ha/year. Between 2000-2005 it was estimated to be 7.3 million ha/year, an equivalent of 200 KM^2/day. Among the main losers of forest habitats, Brazil, Indonesia & Sudan ranked at the top during 2000-2005, in terms of extent of forest areas lost (FAO 2006). India fared better, by maintaining its forest cover intact during the same period. The silver lining is the fact that the rate of destruction of forests is coming down the world over and that the plantation area is picking up.

During the period 1951-1976, an area of 4.13 million ha forests have been diverted for other uses, mainly for agriculture, in India. (Jagdish Kiswan, Dr Devendra Pandey, AK Goyal and Dr AK Gupta, 2007). After enactment of the Forest Conservation act in 1980, these diversions were effectively minimised and rationalised. We have about 76.8 million ha area under forest cover at present, accounting for 23.68 % of the total Geographic area of 328.7 million ha. It has been the stated policy of the Government to increase the forest cover to 33%.

1.5.2 Alien invasive species (AIS)

This is rated as the 2nd most serious threat to biodiversity the world over. Favouring introduction of an exotic species based on its performance in its native habitat has been a general phenomenon. Many a times, when such introductions are effected without proper studies, we find ourselves dealing with a new menace in the form of an invasive. High adaptability of the invading species and the high susceptibility of the new site are the two sets of factors which favour an invasive. There are numerous examples of AIS like the case of Indian Mangoose in Hawaii/Fuji islands, cane toad in Fuji, samoa & Australia, Rosy wolf snail in Mauritius & Hawai and Brown tree snake in Guam islands. Alien invasive species have been found to overpower and threaten the native biodiversity often, to cause extinction of species in a few cases, besides causing huge economic losses.

In India we are concerned with the following invasive species, viz., *Lantana camara, Eupatorium glandulosum, Parthenium species, Mimosa species, Eichhornia crassipes, Mikania micrantha, Ulex enropaeus, Prosopis juliflora, Cytisus scoparius, Euphorbia royleana* etc (Jagdish Kiswan, Dr Devendra Pandey, AK Goyal and Dr AK Gupta, 2007). In Keoladeo Ghana National Park at Bharathpur in Rajasthan, the forest department had to spend crores of rupees recently to physically uproot all *Prosopis juliflora* trees that invaded into that wetland affecting the native biodiversity including birds.

1.5.3 Negative effects of climate change

This has become a matter of serious concern for conservationists of biodiversity all over the world. Earth's history reveals that climate change has been a natural phenomenon caused mainly by natural forces, over a period of time at a slow rate. It is believed that during the last million years, the earth has experienced cycles of warming and cooling (approx. 100,000 years for each cycle to complete) by which the global average temperatures fluctuated by about 9°F each time taking Earth into an ice age and then warming it again.

It is feared that due to the deleterious effects of increased human activities during the last 350 years, the climate of the earth is found to change at a faster rate, this time towards the warmer side. In other words the current climate change is indeed a 'Faster global warming'. Studies reveal that since 1990 the global average temperature has gone up by 0.5° C and it is projected to increase by 1° C by 2100 and 1.5° C by 2200. Accumulation of Green house gases (GHG) like Carbon dioxide, Methane, Carbon monoxide, Nitrous oxide, Chloro Fluoro Carbon (CFC), etc in the atmosphere, leads to trapping of heat energy by preventing their escape from the earth's atmosphere, thereby causes 'Global warming'.

Forests play a dual role in the matter of climate change. Due to deforestation and burning of debris huge quantity CO_2 is getting released into the atmosphere which hastens the rate of global warming. On the other hand, existence of green cover that is left intact, acts as a sink which fixes CO_2 from the atmosphere in their biomass, thereby help reducing the rate of global warming.

Changes are predicted in the climate of different regions, which will affect the cropping pattern and yield in that region. As a result, the 'Good' sites of today may turn 'Better' in future and the 'Bad' sites may turn 'Worse'. On the other hand, it is also predicted that some areas currently supporting successful farming may deteriorate and loose their capacity, while few other areas may be blessed with a favourable climate and a higher productive capacity. It is also predicted that global warming will release lots of water from the melting ice bergs into the rivers and seas, causing damages to some of the coastal civilizations and their farming practices. The sea level is projected to rise 0.1m to 0.5m by the year 2200 and 0.8 m by 2300. The pattern and intensity of rainfall may change across regions. It is expected that frequent flooding at one region and an extended dry season at another will be the order of the future.

These extreme events of weather will affect the biodiversity adversely. Ecosystems may shift in their extent and location in tune with the changes in climate. As for as India is concerned, it is predicted that 77% of forest area in India is likely to experience shift in forest type, with the forests in the northeast becoming wetter and those in the northwest drier. Changes in range and distribution of some of the plant and animal species may occur. Forest fires are predicted to increase in frequency and enormity, and the recent fire incidences in Australia and some European countries are in line with this prediction. Cost of pest and disease control measures is poised to increase.

The adverse effects of Climate change are projected to cause extinction of 20-30 percent of the species of plants and animals that exist today. Plants and animals with narrow range of adaptations will be suffering the most due to these changes in climate and 'Endemics' are facing the most serious risks.

1.5.4. Trade in species of plants and animals

This is the next biggest threat to the biodiversity of the world. 'Wildlife trade' is the second biggest illegal international trade after Narcotics (Menon

V and Kumar A., 1998). It has got a huge turnover and clandestine markets. Populations of many plant and animal species have been decimated because of this illegal trade.

Species like tiger, elephant and rhinoceros have seen serious decline in their populations in Asia, because of trade in their body parts which are in demand for their perceived medicinal properties. For making each shahtoosh shawl, 3 or 4 Tibetan antelopes are killed. Musk deer is an endangered animal in India due to the demand for its musk pods. Pangolins and civets are in trade for their use in tribal medicine. Rhesus macaques are traded for their use in biomedical research. Blackbuck, giant squirrel, slow loris, red panda, hoolock gibbon, lion tailed macaque are some of the other animals which are captured and traded (Menon V and Kumar A., 1998).

Besides this, more than 300 species of birds are also captured and traded in India. Horned owls, hornbills, egrets, and hoopoes are in demand for their supposed uses as medicines and in black magic. Turtles and tortoises are slaughtered for meat and for the pet trade business. Python, cobra and sand boa are some of the snakes which are turned into shoes, purses and fashion apparel apart from being used as pets by snake charmers. Some species of butterflies and moths too fetch high value in trade.

Despite international efforts in the form of CITES (Convention on International Trade in Endangered Species of Flora and Fauna) to regulate international trade in endangered species, the trade goes on illegally. Creating awareness and ensuring strict law enforcement are the keys to improve the situation. In India, the provisions of Wildlife Protection Act 1972 have helped a lot, the cause of conservation of many species of plants and animals.

Trade in medicinal Plants is an area of great prospects for economic growth and simultaneously a serious threat to the survival of some species. The World health Organization (WHO) estimates that about 80 % people of the world are dependent on herbal medicines. In the large global markets, medicinal plants are traded to the tune of 60 billion US Dollars/year. (Jagdish Kiswan, Dr Devendra Pandey, AK Goyal and Dr AK Gupta., 2007).

In 1998, the Government of India notified a 'negative list' of 28 plant species, regulating their collection, cultivation and exports. With the creation of National Bio diversity Authority (NBA) and State Bio diversity Boards (SBDB), as provided for by the Bio diversity Act 2002, all activities including research, exchange (import & export) of plants/animals or their derivatives, are to be cleared by these institutions. This act has put in place a mechanism for benefit sharing aimed at protecting the Traditional Indigenous Knowledge (TIK) sources in India. This act also ensures documentation of such knowledge through 'People Biodiversity Registers' at a local level.

1.6 Conservation- A Challenge

The term 'conservation' means 'Offering protection to biological resources while ensuring sustainable utilization at the same time.' Rich Indian bio diversity is facing huge amount of pressures from various sources, both biotic and a biotic. The fact that India accounts for more than 16% of world's human population, 20% of cattle, and a mighty 55% of buffaloes, coupled with high level of dependency on components of biodiversity, explains the enormous pressure on India's biodiversity.

Almost half the forests in India have been degraded. 78% of Indian forests are subjected to grazing. Average productivity of Indian forests is on the lower side and is at 1/3rd of its potential. It is estimated that nearly 1.36 mill. ha of forests have been encroached till 2002 of which only 10% areas have been retrieved back till 2004. All these facts and figures give us an indication that conservation of biodiversity in India is a huge challenge.

1.7 Approaches to Conservation

Biodiversity conservation in a populous and diverse country like India is a complex matter.

The following 2 approaches are generally adopted and practised in India, viz. *In-situ* Conservation and *Ex-situ* conservation.

1.7.1 *In-situ* conservation

It means conservation of biodiversity in its natural habitat. Creation of a wide Protected area network (PAN) is a measure under this approach. This is in conformity with article 7 of the convention of Biological Diversity (CBD) to which India is a signitory.

As per latest information (as on March 2009), India has 99 National parks, 514 WL sanctuaries, 29 Tiger Reserves, 14 Biosphere Reserves, 43 community Reserves and 4 Conservation Reserves, covering nearly 5% of the total geographical area of the country. (National wildlife Database cell, 2009)

Of late the Government of India has laid emphasis on extending measures of conservation to areas outside the traditional forests areas under the forests departments. Creation of Conservation Reserves (conservation in lands under the jurisdiction of other departments of the Government) and Community Reserves (conservation in lands under community ownership, with their participation) are measures taken in this direction.

Under the broader idea of *in-situ* conservation, we may adapt suitable approaches for conservation depending upon the field realities at hand, viz, landscape approach, species approach, ecosystem approach, etc.

1.7.2 *Ex-situ* conservation

It means 'conservation of biodiversity outside its natural habitat. It is not an idea competitive to in-situ conservation. It is only supplementary to *in-situ* conservation and it helps extending the sphere of in-situ conservation. *Ex-situ* conservation is in conformity with article 8 of the convention of Biological Diversity (CBD).

Ex-situ conservation measures are useful in maintaining stock gene pools of threatened species/varieties. This is the last repertoire at our hands to revive a species that is classified as 'Extinct in the Wild', when such stock populations are maintained under safe environments. Ex-situ conservation measures have huge potential in conservation of rare and threatened species of plants and animals. They are also useful for research and education purposes.

Creation of Zoos/ Zoological parks, Arboretums, botanical gardens, seed banks, in vitro cell culture facilities, cryopreservation of living cells like embryo, cells, sperms, are all examples for ex situ conservation measures.

1.8 Biodiversity Conservation - The Main Players and Expected Roles

The following are the main players in biodiversity conservation and their expected roles are mentioned against them.

Players	Expected roles
The Governments - International Initiatives - National - States - Local	• Enabling legal environment • Transparency • Infrastructure: Men & materials • Finance, Accounting & Audit • Delivery mechanism & monitoring
The NGOs - International (WWF, IUCN, CI, TRAFFIC, etc) - National (BNHS, ATREE, WTI, etc) - Regional/ Local	• Technical support/ Specialists • Focussed approach • Awareness creation • Expanding the reach of conservation
The Private entities - Corporates & Companies	• Corporate Social Responsibility • Ecosystem Services Review • Financial support
People - Individuals (Concerned & otherwise; Activist & Passive) - Non formal Associations, Groups, etc	• Awareness & sensitivity • Initiative & local action • Teamwork

In India forests hold maximum biodiversity and the forest departments are the custodians. There are about 30 laws which enable or impact biodiversity conservation in India directly or indirectly. Wildlife Protection Act (WLPA) 1972, Forest Conservation Act (FC Act 1980), Indian Forest Act (IFA 1927), Biodiversity Act 2002 along with their amendments, are the main laws which effectively provide a legal framework for the conservation of biodiversity in India.

Biodiversity conservation has been taken up in a big way by all these players, and the level of awareness about biodiversity conservation & the sensitivity being shown towards biodiversity related issues are on the high in India. The judiciary too has been playing an active part in biodiversity conservation in India, through its judgements and legal interpretations. Formation of a separate 'green bench' by the Hon'ble Supreme Court of India to deal with forestry matters, is indicative of the importance being accorded to conservation. Several corporate houses and companies have set examples by adopting green technologies & non polluting mechanisms in production processes, by resorting to tree planting works and by providing financial assistance to other biodiversity conservation measures, as part of their corporate social Responsibility (CSR). In general, biodiversity conservation is strongly supported by an aware general public and some dedicated NGOs while the leading role is effectively discharged by the Departments of Forests, Wildlife & Environment.

At the global level, it has been realised that conservation will fail without the support of local people who are one of the stack holders of that common resource. India has travelled much in the path of Joint forest Management / participatory forest management, albeit with mixed results. We are not averse to adapt any such measures which give benefits to local people while ensuring biodiversity conservation at the same time. In this regard, promotion of 'Ecotourism' projects in various parts of India is worth mentioning. It means promotion of nature based tourism as a source of income to local people while being sensitive to safety of biodiversity in those areas.

The face of forest management has changed so drastically in the last century from a 'people exclusion approach / Rule of law' to a 'People inclusive approach'. We have also deviated much from the 'Forests for production of goods' approach of yore, in favour of 'Biodiversity Conservation' and 'Forests for their ecosystem services' approaches. The field of conservation has become so dynamic now, with the addition of newer facets very frequently. Aspects like ecosystem services, their valuation and integration of this knowledge in our decision making system are finding more takers, of late. Individuals and companies are aware of and concerned about their ecological footprints on nature. Environmental Impact Assessment, Environmental Auditing, Ecosystem Services Review are increasingly resorted to, to find out where we stand and what our actions have caused and may cause. Environmental equity is expected to take a leap in the form of compensatory mechanisms like PES (Payment for ecosystem Services) in the near future. Higher level of sensitivity is being shown towards issues of pollution too and emission levels are being controlled. There are marked based incentives for pollution control either directly or indirectly. International cooperation in matters of conservation, forms a backdrop for all the above.

1.9 Conclusion

It is clearly established that the global biodiversity is facing serious threats. Increased rate of species extinction remains a matter of serious concern. The human kind continues to be dependent on the components of biodiversity in various ways and human actions have been the prime drivers behind factors deteriorating biodiversity. The silver lining is the fact that there is an increased awareness and initiation of ambitious conservation measures at the global level as well as in India. Conservation of biodiversity is at a decisive phase, characterized by enormous challenges vis-a-vis ambitious conservation initiatives. Success of conservation at the global level in general and India in particular, and how effectively biodiversity could be preserved for the future generations to see, enjoy and benefit, will largely depend on how effectively the main players of conservation discharge their roles in the coming days.

References

Birdlife International (BI), Endemic bird areas of the world, online @ http://www.birdlife.org/datazone/ebas/eba_programme.html

CBD- Convention on Biological Diversity, Text of CBD online @ http://www.cbd.int/convention/convention.shtml

Conservation International, 2004- Hotspots Revisited -2004 online @ http://www.biodiversityhotspots.org/Pages/maps.aspx

FAO, 2006- Global Forest Resources Assessment 2005, FAO Forestry Paper -147, published by FAO, Rome, 2006.

IUCN Red List 2008 - online @ http://www.iucnredlist.org/

Jagdish Kiswan, Dr Devendra Pandey, AK Goyal and Dr AK Gupta (2007) - India's Forests, Ministry of Environment & Forests, Govt. of India publication, New Delhi.

MEA 2005- Millenium Ecosystem Assessment, 2005. Ecosystems and Human Well-being: Synthesis. Island Press, Washington, DC. online @ http://www.millenniumassessment.org/documents/document.356.aspx.pdf

Menon V and Kumar A. (1998)- Wildlife crime: An enforcement guide, Wildlife Protection Society of India, New Delhi.

National wildlife Database cell, 2009, Wildlife institute of India, online @ http://www.wii.gov.in/nwdc/index.html

□□□

29

Intellectual Property Rights : Protection of Herbal Products and Plant Varieties

D.Thiyagaraja Guptha

CONTENTS

1.1 Introduction 438

1.2 Various Kinds of IPRs 438

1.2.1 Copyrights 438

1.2.2 Trademarks 438

1.2.3 Patents 438

1.3 Limitations of Patents Law towards Herbal Plants Protection 439

1.4 Loss of Traditional Knowledge due to Heedlessness in Protection 440

1.4.1 Turmeric 441

1.4.2 Basmati 441

1.4.3 Neem 442

1.5 Plant Breeders' Rights for the Protection of Plant Varieties in India 443

1.6 Procedure Involved in Protection 445

1.7 Cost of Registering a Plant Variety 446

1.8 Infringement – Offences and Legal Actions 446

1.9 Conclusion 447

References 447

1.1 Introduction

Ideas and knowledge are an increasingly important part of trade. Most of the value of new medicines and other high technology products lies in the amount of invention, innovation, research, design and testing involved.

Intellectual Property Rights (IPR) plays an important role in the international trade and commerce and also in the industrial, economical and social development of a nation. It protects the rights of individuals and businesses who have transformed their ideas into property, by granting rights to the owners of those properties. Such legal rights allow the right holders to exclude the unauthorized commercial use of their creations / inventions by third persons. Broadly, intellectual property is divided into two categories. The first category covers industrial property, which includes patents, industrial designs, Geographical Indications and trademarks, all of which have industrial applications. The other refers to copyright laws, which are applied to such things as literary, dramatic and artistic works; rights relating to performing artists, the production of phonograms and rights of broadcasters in their radio and television programs. IPR's that do not fit into this classical division are termed as sui generis, meaning one-of-its-kind. Such sui generis rights include those covering lay out designs of semi conductor chips and plant breeder's rights.

1.2 Various Kinds of IPRs

1.2.1 Copyrights

Copyright is the exclusive right granted by statute to the author of the works to reproduce dramatic, artistic, literary or musical work or to authorize its reproduction by others. The copyright persists for a finite period after the author's death after which it can be sold or inherited. As such it protects the expression of the idea rather than the idea itself. It also extends to films or television.

1.2.2 Trademarks

Trademark means any symbol, logo, or name used to enable the public to identify the supplier of goods. Trademarks can be registered, which gives the holder the exclusive right to use them. Manufacturers, distributors, or importers may register them. They can be sold and are an important form of commercial property.

1.2.3 Patents

A patent is a legal monopoly granted for a limited time to the owner of an invention. In many countries, an inventor of a new product or process can apply for a patent giving the holder the exclusive right for 20 years to produce the good or use the process. This right can be used either through their own business or by charging a license fee.

The history of the development of international patent reforms makes fascinating reading. Although the concept of intellectual property was very old, there appeared the need to harmonize laws to facilitate international trade and the free flow of technology. This became necessary because it was difficult to obtain sufficient protection in other countries of the world, in view of the disparity in the laws in each country. A prime example of this occurred, when the Empires of Austria and Hungary invited several countries to participate in an international exhibition of inventions held in Vienna in 1873. Many countries refused to display their inventions in view of inadequate legal protection. This incident had a far-reaching impact; the Congress of Vienna for patent reforms was convened in 1873, following which an international congress on industrial property was convened in Paris in 1878. A final draft proposing an "International" union of laws was prepared by France and sent to other countries with an invitation to attend the international conference in Paris in 1880. `The Paris Convention' as it was called concluded in 1883. Since then, the Convention has been subject to several revisions. In 1967 in Stockholm, the last revision was made to the Paris Convention by which an international organization was formed to administer and promote intellectual property on an international level – the World Intellectual Property Organization (WIPO). The main principles of the Paris Convention states that nationals, of any country, which is a signatory to the convention, will enjoy the same treatment (with respect to "industrial property" laws) in other countries, as if they were nationals of the respective countries.

The development of the IPR legislation in our country has been in parallel with the international scenario. The first Intellectual Property Rights legislation was enacted in British India in 1856. This Act was amended on the lines of the British Patent Law (1852) and re-enacted in 1859. The monopoly granted to the inventor was known as "exclusive privileges". Subsequently, several changes to the law in this field were introduced through various enactments, which paved the way to the Indian Patents and Designs Act 1911. The realization that the laws were not designed to motivate or protect Indian inventors led to the enactment of the more progressive Indian Patents Act of 1970. Eventually following the TRIPS agreement, Indian Parliament passed the Patents (Amendment) Act of 2005.

1.3 Limitations of Patents Law towards Herbal Plants Protection

Patents are probably the most important IPR today for agricultural goods and services as they provide, wherever these are available, the strongest protection for patentable plants and animals and biotechnological processes for their production. Patents universally give the patentee the right to prevent third parties from making, using or selling the patented product or process. Patents, however, have to be disclosed to the public through the patent documents. This enables researchers to develop further useful products or services. Patentable products and processes have to meet the criteria of patentability, viz., novelty, i.e. that which is not known in the prior art, non-obviousness i.e. that which involves an inventive step and usefulness i.e. that which is industrially applicable. With some nuanced differences the patent laws of all countries follow these criteria. However, not all countries allow the patenting of plants and animals or even microorganisms or biotechnological processes.

At present the Patents (amendment) Act, 2005 too does not allow the patenting of plants or animals as intellectual property protections for life forms have been riddled with many perplexing technical issues. The prerequisites for granting a patent, i.e., novelty and non-obviousness, are difficult to prove in the case of plants. Moreover, plant varieties compared to mechanical inventions, are difficult to replicate as they are reproduced asexually. So India has excluded from patentability the plants and animals in whole or any part of them including seeds, varieties and species and essentially biological processes for production or propagation of plants and animals by section 3(J) of The Patents (amendment) Act, 2005. Thus, having excluded plants from patent protection, as a trade-off, India had to provide an effective sui generis system in place for the protection of new plant varieties. This was achieved with the passing of the Protection of Plant Varieties and Farmers' Rights Act, 2001.

1.4 Loss of Traditional Knowledge due to heedlessness in Protection

Traditional knowledge (TK) associated with biological resource and herbal plants and products are intangible components of the resource itself. TK has the potential of being translated into commercial benefits by providing valuable leads for development of useful products and processes. The valuable leads provided by traditional knowledge save time, money, and investment of modern biotech industry into any research and product development. Hence, a share of benefits must accrue to creators and holders of TK. Due to globalization of production systems, increase in population, destruction of forests; bio-diversity is declining at a rapid pace. Bio-diversity and associated traditional knowledge is also declining due to decreased motivation amongst the local communities to conserve and protect them. Mis appropriation of traditional knowledge not only violates the rights of communities who conserved TK but also adversely affects the conservation and sustainable use of the traditional knowledge and that of bio-diversity. In the recent past, there have been several cases of bio-piracy of Traditional Knowledge about herbal plants from India.

1.4.1 Turmeric

Turmeric is a tropical herb grown in East India, and the powdered product made from the rhizomes of its flowers has several popular uses worldwide. Turmeric powder, which has a distinctive deep yellow color and bitter taste, is used as a dye, a cooking ingredient, and a litmus in chemical tests, and has medicinal uses as well. In the mid-1990s, this product became the subject of a patent dispute with important ramifications for international trade law. A U.S. patent on turmeric was awarded to two Indian-born scientists of the University of Mississippi Medical Center in 1995, specifically for the "use of turmeric in wound healing." This patent also granted them the exclusive right to sell and distribute turmeric. Two years later, a complaint was filed by India's Council of Scientific and Industrial Research, which challenged the novelty of the University's "discovery," and the U.S. patent office investigated the validity of this patent. In India, where turmeric has been used medicinally for thousands of years, concerns grew about the economically and socially damaging impact of this legal "bio-piracy." An internationally enforced patent on the "use of turmeric in wound healing" would mean that suddenly hundreds of millions of Indians (and anybody else) would have to pay in order to apply turmeric to their wounds. Given drug prices today, it might become prohibitively expensive for most of the people.

But in 1997, the patent was revoked on the ground that the healing properties of Turmeric had been 'common knowledge' in India for centuries. But for two years the patent on turmeric had stood, although the process was non-novel and had in fact been traditionally practiced in India for thousands of years, as was eventually proven by ancient Sanskrit writings that documented turmeric's extensive and varied use throughout India's history. The patent on turmeric, had it not been overturned, would have caused an outflow of money from India to the U.S. in the form of licensing fees.

1.4.2 Basmati

Basmati rice means the *"queen of fragrance or the perfumed one."* This type of rice has been grown in the foothills of the Himalayas for thousands of years. Its perfumy, nut-like flavor and aroma can be attributed to the fact that the grain is aged to decrease its moisture content. Basmati, a long-grained rice with a fine texture is the costliest rice in the world and has been favored by emperors and praised by poets for hundreds of years.

On September 2, 1997, an American company RiceTec Inc, was granted a patent by the US patent office to call the aromatic rice grown outside India 'Basmati'. RiceTec Inc, had been trying to enter the international Basmati market with brands like 'Kasmati' and 'Texmati' described as Basmati-type rice but with minimal success. However, with the Basmati patent rights, RiceTec was able to not only call its aromatic rice Basmati within the US, but also label it Basmati for its exports. This had grave repercussions for India and Pakistan because not only India lost out on the 45,000 tonne US import market, which formed 10 percent of the total Basmati exports, but also on its position in crucial markets like the European Union, the United Kingdom, Middle East and West Asia. In addition, the patent on Basmati was believed to be a violation of the fundamental fact that the long grain aromatic rice grown only in Punjab, Haryana, and Uttar Pradesh to be called Basmati.

The invention related to novel rice lines and to plants and grains of these lines. The invention also related to a novel means for determining the cooking and starch properties of rice grains and its use in identifying desirable rice lines. Specifically, one aspect of the invention related to novel rice lines whose plants were semi-dwarf in stature, substantially photoperiod insensitive and high yielding, and produce rice grains having characteristics similar or superior to those of good quality Basmati rice. The invention also provided a method for breeding these novel lines. A third aspect related to the finding that the starch index (SI) of a rice grain could predict the grain's cooking and starch properties, to a method based thereon for identifying grains that could be cooked to the firmness of traditional Basmati rice preparations, and to the use of this method in selecting desirable segregants in rice breeding programs.

The government of India reacted immediately after learning of the Basmati patent issued to RiceTec Inc., by forming a high level inter-ministerial group comprising of representatives of the ministries and departments of commerce, industry, external affairs, Council for Scientific and Industrial Research (CSIR), Agriculture, Bio-technology, All India Rice Exporters Association (AIREA), APEDA, and Indian Council of Agricultural Research (ICAR) to begin an in-depth examination of the case. The rights granted to Rice Tec Inc., on basmati were finally reduced to minimal traits by USPTO as a result of the steps taken by the Indian government.

1.4.3 Neem

In 1971, a timber company in the United States figured out the neem tree's usefulness in acting as a pesticide and began planting neem tree seeds. He received a patent on it and, in 1988, sold the patent to the US based company W.R. Grace. In 1992, W.R. Grace secured its rights to the formula that used the emulsion from the Neem tree's seeds to make a powerful pesticide. It also began suing Indian companies for making the emulsion.

There are approximately 14 million neem trees (*Azadirachta indica*) in India and the access to neem products was very cheap and easy to get. It is a tropical evergreen, related to the mahogany, which mainly grows in arid regions of India and Burma and Southwest Asia and West Africa. When temperatures do not drop below freezing, it may grow up to 50 feet tall. They are estimated to live up to 200 years. The neem tree has many versatile traits that can be traced back to the Upavanavinod, an ancient Sanskrit treatise dealing with agriculture. This treatise cites the neem tree as a cure for ailing soil, plants and livestock. The tree has been referred to as the 'curer of all ailments' and the 'blessed tree' by both the Hindu and Muslim population in India. It has been used to make spermicides and pesticides and as a cure for several diseases. The neem tree is known as the tree for all seasons because of its versatility. Making pesticides emulsion does not take highly sophisticated equipment, as native people have been making it for over 2000 years. Indians have developed their own process of cracking off the top that would then be used on plants as a pesticide. Neem based pesticides, medicines and cosmetics have been produced by some laboratories in India, but there has not been an attempt to make ownership of the formula legal because the then Indian law did not allow agricultural and medicinal products to be patented.

Such bio-piracies are numerous in nature and they steal the traditional knowledge of a country. The presence of effective national laws and varied provisions for the protection of the herbal plants and their products would help the Indian farmers and plant breeders to effectively safeguard the traditional knowledge of this country. On account of these considerations, India has formulated a sui generis legislation for the protection of plant breeders' rights.

1.5 Plant Breeders' Rights for the Protection of Plant Varieties in India

In early history, as part of the process of establishing fixed settlements and becoming a farmer, man selected and kept seed or plants of those species that offered a secure food source. By the end of the eighteenth century, when systematic plant breeding by selection began, the plants grown by farmers were the result of several thousands of years of partly conscious and partly unconscious selection. The art of plant breeding resulted from the realization by innovative farmers in the eighteenth century was possible by systematic selection. In the twentieth century, the rediscovery of Mendel's laws of heredity contributed to the establishment of plant breeding on a scientific basis.

The essence of plant breeding is the discovery or creation of genetic variation in a plant species and the selection from within that variation of plants with desirable traits that can be inherited in a stable fashion. Different types of plant variety have been developed, depending upon the physiology of the plants of each species and the ways in which the plants of the species can be reproduced.

The simple objective of a breeder is to produce a variety which is an improvement on the plants used as the starting point. However, this is a difficult challenge. Many useful characteristics, such as yield and quality, are controlled by the interaction of very large numbers of genes, about most of which, little is known. Very large numbers of plants must be examined by the plant breeder over many different seasons and under different growing conditions. Once a desirable plant has been identified, it is still necessary to fix its genetic structure in order that it can be multiplied into a variety, the individual plants of which perform in the desired way. Thus, the breeding of a plant variety takes place over many years.

Large-scale breeding work calls for significant annual investment in land, specialized equipment (including, for example, greenhouses, growth chambers and laboratories), and skilled scientific manpower, which must continue over several years to find and develop an improved plant variety. Not all plant breeders are successful and, even where successful, changes in market requirements may eliminate the possibility of a return on investment, so there is also risk involved. However, the benefits arising from the combinations of increased output and improved quality made possible by plant breeding are such, that society has good reasons to encourage investment and risk-taking in this field.

The process of plant breeding is long and expensive; however, it can be very quick and easy to reproduce a variety. Clearly, few breeders would spend many years of their life, making substantial economic investment, in developing a new variety if there was no means of being recompensed for this commitment. Hence, sustained breeding efforts are only possible if there is a chance to reward investment. It is, therefore, important to provide an effective system of plant variety protection, with the aim of encouraging the development of new varieties of plants, for the benefit of society.

So in the year 2001 India got a plant varieties protection legislation (The Protection of Plant Varieties and Farmers' Rights Act (PPV & FR), 2001; Act 53 of 2001) heralding agriculture of the 21st century: A 'brand' new age, where 'new' plant varieties can be labelled and sold; the 'creator' of breeding seeds for commercial purposes is granted protection by law through Plant Breeder Rights (PBR). This will enable seed companies and other developers of new varieties, including hybrids and transgenics, to have ownership rights over them, as also exclusive rights to market the seeds of these varieties for the ultimate benefit of the farmers. As such, this will prevent the piracy and misuse of protected material by others. Besides, this will pave the way for healthy growth of the research-based seed industry, providing farmers access to better seed for higher and superior quality production. Such ensuring rights give exclusive commercial control to the breeder over the propagating material of the 'protected' variety. Agriculture, reincarnating as agribusiness in today's global trade system, dares to commodify all plants, seeds, and knowledge related to any of the above.

A farmer who has breed or developed a new variety shall be entitled for registration and other protection under PPV&FR Act, 2001 in the same manner as a breeder of a variety. Further he shall be entitled to save, use, sow, re-sow, exchange and share or sell his farm produce including seed of a variety protected under this act in the same manner as he was entitled before the coming into force of this act provided that the farmer shall not be entitled to sell branded seed of a variety protected under this act. But as such those plant varieties present in wilderness cannot be registered, under PPV&FR Authority. However, any traditionally cultivated plant variety, which has undergone the process of domestication / improvement through human interventions can be registered and protected subjected to fulfillment of the eligible criteria. The protection, provided under the PPV&FR Act of 2001, will be valid for 18 years for trees and vines and 15 years for other plants.

To be eligible for protection, the variety must satisfy the Novelty, Distinctness, Uniformity and Stability (NDUS) criteria. The test for novelty is decided on whether the variety was commercially exploited before seeking protection. Any commercial exploitation in India must be within one year before the date of filing for protection. To qualify as distinct, the new variety must be clearly distinguishable by at least one essential characteristic from any other existing variety of common knowledge in any country. So the filing of application for protection in any country will be treated as a matter of common knowledge of the variety for the purpose of the provision. To qualify as uniform, the variety must be sufficiently uniform in its essential characteristics and a variety exhibits stability if its essential characteristics remain unchanged after repeated propagation and in cases where the propagation is in repeated cycle, the essential characteristics must remain unchanged at the end of each cycle.

1.6 Procedure Involved in Protection

The protection of new variety is based on registration. Any farmer or group or community of farmers claiming to be the breeder of the new variety is entitled to file an application to register the new variety. The application for registration has to contain the details of the variety satisfying the criteria for registration, information regarding the farmer's contribution in the evolution or development of the variety and the declaration that the genetic material or the parental material used to develop the variety has been lawfully obtained. The registrar will examine the applications for its veracity and would conduct NDUS tests and other special tests, if required. The application is advertised in the official journal upon acceptance whereby any person can oppose the application within three months and if no opposition is filed within the notice period then a certificate of registration is issued. Then the breeder must deposit a quantity of seeds or propagating material along with parental line seeds of the registered variety in the National Gene Bank for reproduction within the stipulated time. The Application for registration of varieties has to be submitted to

The Registrar, PPV&FR Authority NASC Complex, DPS Marg, New Delhi - 110012.

1.7 Cost of Registering a Plant Variety

A. Form Charges

1.Application Form Charges (in rupees) 200/-

B. DUS test fee

1.Rice, Wheat, Maize, Sorghum, Pearl millet, Pigeon pea, Mungbean, Urdbean, Chickpea, Fieldpea, Lentil, Rajmash 20,000/- to 50,000/-

C. Annual Fee

The Authority shall notify the amount separately in the official Gazette (in the PVJ of India).

II. Other charges

A. Registration Charges

1.For Extant Variety of Common Knowledge where no DUS test is needed. 2,000/-

2.Registration Certificate charges 5,000/-

B. DUS test site visit charges/person/visit 500/-

(To be retained at DUS test site after issue of receipt)

1.8 Infringement – Offences and Legal Actions

1.If a person who is not a breeder of a variety or a registered agent or a registered license of that variety, sells, exports, imports or produces such variety without the permission of its breeder.

2.If a person uses, sells, exports, imports or produces any other variety giving such variety, the denomination identical with or deceptively similar to the denomination of a variety already registered in such a way that it causes confusion in the mind of general people in identifying the registered variety.

Then it is considered to be infringement and the infringer shall be punishable with imprisonment for a term not less than six months which can be extended up to three years or with a fine not less than one lakh which may be enhanced to five lakhs, or both.

1.9 Conclusion

The farmers and plant breeders have two tools for their benefit. If the research and invention is on a novel herbal product or various processes of extraction of the product then such products and processes are protected by the help of patents and if the innovation lies in the herbal seeds and plants themselves and if they satisfy the NDUS properties then those varieties are protected by plant breeders' rights. Thus with the help of these provisions the entire new seed and plant varieties, products and processes can be amply protected by the breeder for his commercial exploitation , for stimulating research and development and for the improvement of the entire society.

References

http://www.etcgroup.org

http://www.neemfoundation.org

http://www.patentoffice.nic.in/

http://www1.american.edu/projects/mandala/TED/basmati.htm

The article "Intellectual Property? Right!" by T.V. Malavika of DePenning and DePenning.

The article "IPR for plant varieties" by Surinder Sud.

The article "Protection of Plant Varieties and Farmers' Rights Act, 2001 — Just leave the seed alone" by Shalini Bhutani and Kanchi Kohli.

The article "The Protection of Medicinal Plants in India" by Sangeeta Udgaonkar.

The PPV & FR Act of India.

www.ipindia.nic.in

□□□

30

Role of Plant Secondary Metabolites in Ruminant Nutrition

Jadhav Sunil Ekanath, Guru Charan, Tilak Raj
Ashutosh Murkute & Shashi Bala Singh

CONTENTS

1.1	Introduction	450
1.2	Biosynthesis and Storage of PSMs	451
1.3	Saponins	452
	1.3.1 Chemistry	452
	1.3.2 Effect on rumen protozoa	452
	1.3.3 Effect on rumen fermentation	453
	1.3.4 Metabolism	453
1.4	Tannins	454
	1.4.1 Chemistry	454
	1.4.2 Mode of action	454
	1.4.3 Metabolism	455
	1.4.4 Effect of feeding tannins in ruminants	455
1.5	Essential Oils	457
	1.5.1 Chemistry	457
	1.5.2 Effect of feeding essential oils in ruminants	457
1.6	Effect of PSMs on Methane Inhibition	459
1.7	High Altitude Fodder Plants as a Futuristic Source of PSMs	460
1.8	Conclusion	461
	References	461

1.1 Introduction

Plant secondary metabolites (PSMs) are extremely large group of compounds and contain wide variety of phytochemicals which are classified into several major categories like phenolic compounds, glycosides, alkaloids, steroids, triterpenoids etc. PSMs differ from the ubiquitous primary metabolites (carbohydrates, proteins, fats etc.) as their distribution is limited and often being produced only by specific plants or group of plants. Though PSMs do not have vital biochemical role in the process of building and maintaining plant cells, recent research have shown that PSMs play very important role in the eco-physiology of plants. They serve to protect plants from pathogens, herbivores, inter-plant competitions and abiotic stresses and to attract beneficial organisms such as pollinators or symbionts (Kaufman *et al.*, 1999; Wink, 1999). They may also have a possible nutritional role (particularly nitrogen containing PSMs), during germination of seeds.

Use of plants and plant extracts for health improvement, treatments and as performance enhancer in human and animal has received wide acceptance from all over the world. The estimated numbers of species of higher plants on earth are 2,50,000 of which more than 80,000 are medicinal (Kumar *et al.*, 1997). This is the fact that 80% of the world population relies on traditional treatment (Fransworth *et al.*, 1985) and 25% of prescriptions dispensed by pharmacies contain extracts or active principles from herbs (Newman *et al.*, 2000). The database of the world literature of natural products lists over 1,35,000 isolated and characterized natural products (Cordell, 2000) which is from just an estimated 5-15% of the plant species which have been investigated for the presence of PSMs.

Plant secondary metabolites initially were considered as anti-nutritional factors until their scientific exploration for use as performance enhancer in animals was carried out. Use of plants in ethno-veterinary practices can be traced back up to 2600 BC in Mesopotamia. Several medicinal and aromatic plants or extracts are traditionally used to treat various digestive disorders and to enhance productive performance of animals (Jain, 1999). The beneficial effects of medicinal plants or extracts typically results from the individual or combination of secondary metabolites present in these plants.

Main focus of research work being carried out on use of PSMs in animal production is manipulation of rumen ecosystem either by influencing rumen microbial activities or by changing environmental conditions prevailing in rumen. Use of specific PSM acting against or favoring specific ruminal microbe or groups of microbes and its further application in ruminant production and health is being worked out extensively. Ruminants possess unique ability to transform low quality fibre and nonprotein nitrogen into good quality protein sources such as milk and meat for human consumption. The organic matter in the feed is converted to volatile fatty acids, microbial mass and gases through anaerobic microbial fermentation in the rumen. Volatile fatty acids and microbial mass satisfy the major part of the ruminant's energy and protein requirements, respectively. The reason that ruminants are so important to mankind is that much of the world's biomass is rich in fibre and could be converted into good quality protein sources for human consumption only by ruminants. However microbial fermentation in rumen leads to loss of energy in the form of methane. Methane production in the rumen represents loss of energy amounting to 5 to 15 % of ingested gross energy by animal. Moreover methane is one of the important green house gases that contribute to global warming. It is now well recognized that generation of methane in ruminants can be decreased by promoting shift in fermentation towards propionate production. PSMs have been found to possess anti-methanogenic properties. Saponins, tannins and essential oils are the group of secondary metabolites that have been widely investigated in ruminants for their effect primarily on rumen metabolism.

Anti-microbial, immuno-enhancing and anti-parasitic properties of PSMs has attracted researchers and nutritionist. Antibiotics and other compounds like hormones are being used to enhance animal productivity. However, increasing risk of microbial resistance to antibiotics, consumers awareness about the use of non plant origin performance enhancer in animal feeds and hazardous effects of chemical residues due to these compounds on human health are primary reasons behind use of phytochemicals as such or in the form of feed additives in animal diets.

1.2 Biosynthesis and Storage of PSMs

Biosynthesis of PSMs mostly takes place in cytoplasm of the cell. Different organelles of the plant cell such as vesicles, mitochondria, endoplasmic reticulum and chloroplast are site for PSM synthesis. The gene expression or pathways responsible for synthesis of PSMs are regulated by the environment, season and other external factors. Though PSM synthesis occurs in different organs like leaves, fruits, roots etc. they are transported to other parts of plants through different transportation systems and finally stored in different tissues. The site of storage often depends on the polarity of the compounds and varies with the plant types. The hydrophilic compounds viz. alkaloids, glucosinolates and tannins are stored in vacuoles or idioblasts, while lipophilic compounds viz. terpene-based essential oils are stored in trichomes, glandular hairs, resin ducts, thylakoid membranes or on the cuticle (Wiermann, 1981). In annual plants secondary metabolites are stored in fruits, flowers or seeds however in perennial plants they are stored in bulbs, rhizomes or bark of root or stem.

Plants also regulate the synthesis and storage of secondary metabolites so that the more vulnerable tissues such as fruits and young leaves contain higher concentrations of PSMs than senescing tissues (Wink, 2004).

1.3 Saponins

1.3.1 Chemistry

Saponins are high molecular weight surface-active triterpene glycosides containing sugar moiety attached to either sterol or other triterpene. They are composed of two parts viz. glycone (sugar) and aglycone or genin (triterpene). Sugar moiety usually contain glucose, galactose, glucoronic acid, xylose, rhamnose or methyl pentose, glycosidically linked to a fat soluble aglycone (sapogenin) which may be steroid (C_{27}) or triterpenoid (C_{30}) in nature. The aglycone may contain one or more unsaturated C-C bonds. Depending on the sugar moiety attached, saponins can be mono or polydesmodic. The oligosaccharide side chain is normally attached at C_3 position (monodesmosidic), but many saponins have an additional sugar moiety at C_{26} (steroid) or C_{28} (triterpenoid) position (bidesmosidic). Bidesmodic saponins are by far the most common saponins. Some authors distinguish a third group called steroidal amines which are classified by others as steroidal alkaloids (Bruneton, 1995). Generally, immature plants have been found to have higher saponin contents than more mature plants of the same species (Yoshiki *et al.*, 1998). Saponins are characterised by bitter taste and exhibit strong foaming properties in aqueous solutions, from which the name saponin was derived.

1.3.2 Effect on rumen protozoa

Saponins exhibit defaunating abilities. Defaunation is exclusion of protozoa from rumen. In the rumen protozoa play a significant role in methane production and volatile fatty acid production. Rumen protozoa feed upon rumen bacteria and fungi and digest them. Saponins form irreversible complexes with cholesterol. Cholesterol and other sterols are components of cell membranes of protozoa, but not bacteria. Reaction of saponins with protozoal cell membrane causes breakdown of the membrane, cell lysis and death.

Various *in vitro, in sacco* or *in vivo* experiments in ruminants showed that saponins reduces protozoal numbers and inhibits bacteroilytic activity of rumen protozoa. Various saponin containing plants viz. acacia, yucca, quillaja and soap nut have been found to have dose dependent potent anti-protozoal effect. These effects are however dependent on the type of extract, saponin source, dose, type of the diet, animal species involved in study, inactivation of saponins in saliva and resistance of saponins to bacterial degradation. Oral supplementation of *S. sesban* saponins failed to produce effects. However when

S. sesban was placed directly into the rumen through rumen canula, protozoal numbers decreased significantly (Odenyo *et al.*, 1997). This finding suggested that inactivation of saponins might have occurred before reaching the rumen probably in the mouth. Methods of extraction have also been found to affect anti-protozoal activity of saponins in different ways. Methanolic extraction of saponins followed by acid hydrolysis have been reported to yield sapogenins and saponin artifacts (Lu and Jorgenson, 1987) which were found to have more pronounced anti-protozoal effect (Klita *et al.*, 1996). Similarly, incorporation of methanol extract of *Sapindus rarak* at 0.07% of live body weight for every three days in sheep diets reduced protozoal numbers by 57% (Thalib *et al.*, 1996). Inhibitory effect of saponins against protozoa also depends on type of the basal diet as it was observed that hay and concentrate containing diets exhibited more effect of dietary saponins on protozoal population as compared to hay alone diet. However toxic effect of saponins on ruminal protozoa may be transitory as it has been reported by Ivan *et al.* (2004), that after an adaptation period of approximately 21 days in sheep there was no effect of feeding saponin containing *E. cyclocarpum* foliage on protozoal population in sheep.

1.3.3 Effect on rumen fermentation

Reduced ammonia nitrogen concentration and increased molar proportion of propionate are typical in the rumen when rumen protozoa are inhibited (Williams and Coleman, 1992). Increased propionate is desirable since it acts as alternate hydrogen sink in the rumen thereby reducing methane formation and increasing efficiency of microbial protein synthesis. It is also proposed that saponins binds to NH_3 when NH_3 concentration in the rumen is high, and release NH_3 when NH_3 concentration is low thereby increasing the efficiency of nitrogen utilization in ruminants, especially when highly soluble non-protein nitrogen source like urea is available for degradation (Hussain and Cheeke, 1995).

1.3.4 Metabolism

Metabolisms of triterpnoid and steroid saponins have been studied in different ways. Most of the investigations on metabolism of triterpenoid saponins are confined to digestive tract only. Metabolism of steroid saponins in ruminants leads to the secondary photosensitization. Some of the plants responsible for causing secondary photosensitization include *Tribulus terrestris, Agave lecheguilla* and *Narthecium ossifragum*. Miles *et al.* (1994) studied the ruminal and hepatic metabolism of *Tribulus* saponins in sheep and proposed the mechanism by which saponin causes photosensitization. Ingested saponins are hydrolysed in the rumen releasing free sapogenins. Most of the sapogenins are rapidly converted by ruminal microflora to epi-sapogenins which on absorption gets conjugated with glucoronic acid in liver and excreted in the bile as calcium salts of sapogenin glucuronides. These salts are insoluble and precipitate in the bile duct as crystals. The disruption of bile excretion results in retention of phylloerythrin and induction of photosensitization.

1.4 Tannins

1.4.1 Chemistry

Tannins are complex group of water-soluble polyphenolic compounds. There are two biosynthetically-distinct classes of tannins viz. hydrolysable tannins (HT) and condensed tannins (CT; proanthocyanidins). Hydrolysable tannins are composed of esters of gallic acid (gallotannins) or ellagic acid (ellagitannins) with a sugar core which is usually glucose and are readily hydrolysed by acids or enzymes into monomeric units. Condensed tannins are polymers of flavan-3-ol-units. They give coloured products, the anthocyanidins on treatment with acidic alcohol and hence they are also called as proanthocyanidins. Condensed tannins content of different plants refer to both oligomeric (dimmers, trimers, tetramers) and polymeric proanthocyanidins. The word "tannin" was coined because of its ability to tan hide and transform it to leather. Tannins are capable of forming complexes with proteins and other macro molecules including soluble and structural carbohydrates, minerals, vitamins and bacterial cell membranes (Makkar *et al.*, 1987).

1.4.2 Mode of action

By forming hydrogen bonds between the phenolic sub units of the polymer and the carbonyl groups of peptides of the protein, tannins bind with protein at near neutral pH (4.0 to 7.0) to form tannin protein complexes, but dissociate and release protein at acidic pH (<3.5) (Jones and Mankan, 1977). Dietary tannins in ruminants undergo same mechanisms and binds with dietary as well as microbial proteins in the rumen. Tannin–protein complexes formed in the rumen are hydrolysed in the abomasum (Hagerman & Carlson, 1998). Once the complex passes through the rumen, it gets subjected to gastric (pH 2.5 to 3.5) and pancreatic secretions and dissociates making the protein available for absorption from the small intestine. In non-ruminants tannins interfere with nutrient digestion and promote excretion of endogenous nitrogen through formation of tannin-enzyme complexes (Longstaff and McNab, 1991).

Various modes of actions by which tannins produce their effects in the gastro intestinal tract (GIT) of the animals have been studied. Tannins produce their effects by inhibition of microbial action in the rumen, sequestration of minerals in insoluble complexes and producing damaging effects on intestinal function (Makkar *et al.*, 1995). Tannins decreases voluntary feed intake of the animals which may probably be associated with astringency caused by the formation of tannin–salivary protein complexes in the mouth or signals of gut distension resulting from tannin interactions with proteins of the gut wall (D'Mello & Devendra, 1995). Tannins affects gut permeability, causing decreased passage of nutrients through the gut wall (Walton *et al.*, 2001). Free tannins in the lower GIT may bind to endogenous proteins and produce harmful effects causing changes in intestinal structure and inhibiting nutrient digestion and absorption in the small intestine. Histochemical and biochemical investigations revealed that tannins also inhibit enzyme activities in the intestine. When intestinal structure and brush-border enzyme production was investigated in sheep fed a diet of mulga (*Acacia aneura*), striking structural and functional changes were found in the abomasum and small intestine of these animals as compared with animals fed control diets (Robins & Brooker, 2005). Where larger polymers binds with the protein, smaller flavonols and flavonol glycosides are known to produces histopathological, histochemical and biocehimical alterations in the lower GIT. At low to moderate concentrations, tannins especially condensed tannins have been shown to exert positive effect by protecting feed protein from ruminal degradation and subsequently increasing its availability at intestinal level for digestion and absorption. Dietary condensed tannins may also contribute to animal health by reducing the detrimental effects of internal parasites in sheep and the risk of bloat in cattle.

1.4.3 Metabolism

Hydrolysable and condensed tannins behave in different way in the rumen. Several studies have shown that rumen microbes are capable of degrading hydrolysable tannins. The ester bonds and depside linkages are probably cleaved in the rumen by rumen microbes producing esterase and tannin acyl hydrolase to form metabolites viz. gallic acid and ellagic acid. Gallic acid is then decarboxylated to pyrogallol and converted to resorcinol and phlorglucinol ring to acetate and butyrate. However, little evidence exists for degradation of condensed tannins in the rumen and most of the studies reveals that there is no degradation of condensed tannins in the rumen.

1.4.4 Effect of feeding tannins in ruminants

Tannins influence methanogenesis and efficiency of microbial protein synthesis in the rumen. Tannin reduces plant protein degradation in the rumen. The reduction in proteolysis may be attributed to direct effects of tannins on microbial enzyme activity or to indirect effects on rumen metabolite concentrations which may regulate microbial proteolytic activity. Tannins reduce ammonia nitrogen, total volatile fatty acids (TVFA) concentration and methanogenesis in rumen. Bhatta *et al.* (2009) reported that tannins suppress methanogenesis by reducing methanogenic populations in the rumen either directly or by reducing protozoal population, thereby reducing methanogens symbiotically associated with the protozoal population. Moreover there are synergistic effects of HT and CT on ruminal metabolism if fed together as compared to either HT or CT alone. When plants containing either hydrolysable tannins only or both hydrolysable as well as condensed tannins were evaluated *in vitro* to determine the effect on methane production and on ruminal archaeal and protozoa populations, it was observed that tannin sources containing HT and CT were more potent in suppressing methanogenesis than those containing only HT (Bhatta *et al.*, 2009).

It has been seen that consumption of foliages or herbages rich in condensed tannins could interfere with the development of gastro-intestinal nematodes resulting in subsequent reduction in their fecal egg count (Niezen *et al.*, 2000; Min and Hart, 2003). Akkari *et al.*, (2008) found significant reduction in fecal nematode egg count in lambs fed tannin containing *Acacia cyanophylla*. Feeding of different tanniferous plants like *Hedysarum coronarium, Lotus corniculatus,* Cassava, Sainfoin and *Acacia karoo* has shown anti-nematode properties in sheep and goats.

Recent advances in research on tannins include studies on effects of tannins on ruminal biohydrogenation and manipulation of fatty acid profile of products of animal origin. Products of animal origin containing saturated fatty acids are responsible for increasing risk of cardiovascular diseases. However conjugated linoleic acid (CLA) (cis-9, tans-11 C-18) content in ruminant meat and milk has been shown to reduce incidence of cancer, diabetes and atherosclerosis. Biohydrogenation of fat in rumen converts unsaturated linoleic and linolenic fatty acids into saturated stearic acid. Dietary tannins in ruminants reduces *in vitro* biohydrogenation (Khiaosa-Ard *et al.*, 2009; Vasta *et al.*, 2009a) and enhances muscle Ä^{9}- desaturase enzyme expression in sheep (Vasta *et al.*, 2009b) which is responsible for endogenous formation of CLA in the muscle and mammary glands of animals. Very recently Vasta *et al.* (2009) studied metabolic fate of fatty acids involved in ruminal biohydrogenation in sheep fed concentrate or herbage with or without tannins. Supplementation of diet with commercial tannic acid power i. e. quebracho (*Schinopsis lorentzii*) @ of 4.0 % on dry matter basis has resulted in reduced concentration of stearic acid with an increase in vaccenic acid (trans-11, C-18:1) in ruminal fluid of sheep fed concentrate based diet. Tannic acid supplementation also reduced accumulation of saturated fatty acids in blood and resulted in two fold increase in rumenic acid concentration of *longissimus* muscle. These investigations indicated that tannin supplementation could be useful strategy to increase rumenic acid and polyunsaturated fatty acid (PUFA) content and to reduce saturated fatty acids in ruminant meat.

1.5 Essential Oils

1.5.1 Chemistry

Essential oils are complex mixtures of secondary metabolites containing low-boiling point terpenes (derived from the mevalonate and deoxyxylulose pathways), phenylpropenes (derived from the shikimate pathway) found in limited number of plant families. Volatile terpenes can be divided into two groups namely, monoterpenes and sesquiterpenes. Monoterpenes are most common, and the hydrocarbons may be of acyclic, monocyclic and bicyclic in nature with alcohols, aldehydes, ketones, esters, ethers, peroxides or phenols as functional group. The structural variations in sesquiterpenes are of the same nature as in monoterpenes, with hydrocarbons, alcohols and ketones being the most common. Phenylpropanoids are far less common than terpenes. They are mostly as allyl and propenylphenols and sometimes as aldehydes. Aromatic essential oils possess various biological activities including antibacterial, antifungal, antiviral and anti-inflammatory effects. Even these oils can also be active against higher organisms such as nematodes, helminthes, insects etc.

1.5.2 Effects of feeding essential oils in ruminants

Generally, essential oils have terpenoid structures and their bioactivity is the result of the combination of all their constituents. In ruminants, essential oils affect rumen fermentation. They interact with microbial cell membranes and inhibit their growth and reduce colonization of substrate by rumen bacteria (Castillejos *et al.*, 2005). As a result there may be decrease in the rate of deamination of amino acids, reduced methanogenesis and decreased degradation of protein resulting in lower ammonia nitrogen, methane and acetate and higher propionate and butyrate concentrations. Results have indicated that garlic oil, cinnamaldehyde (the main active component of cinnamon oil), eugenol (the main active component of the clove bud), capsaicin (the active component of hot peppers), and anise oil, may increase propionate production, reduce acetate or methane production, and modify proteolysis, peptidolysis, or deamination in the rumen. However, the effects of some of these essential oils are pH and diet dependent, and their use may be beneficial only under specific conditions and production systems. For example, capsaicin appears to have small effects in high-forage diets, whereas the changes observed in high-concentrate diets (increases in dry matter intake and total volatile fatty acids, and reduction in the acetateto-propionate ratio and ammonia nitrogen concentration) may be beneficial.

Essential oils have profound effect on nitrogen metabolism in the rumen. Thymol, a major component of some essential oils has been reported to be a strong deaminase inhibitor (Van Nevel and Demeyer, 1988). Essential oils were observed to selectively inhibit the population of *C.sticklandii* and *P. anaerebius* which are involved in rumen nitrogen metabolism and responsible for producing more ammonia and thus essential oils can also be looked into to enhance nitrogen utilization efficiency of ruminants. Essential oils also have inhibitory effect on rumen microbes. This inhibitory effect of essential oils is more on fungi than bacteria. This selective activity of essential oils can be used to manipulate rumen fermentation selectively. It is seen that Thymol is a suitable compound as it remained undegraded in an anaerobic conditions for 60 days (Varel, 1999) and it inhibited the growth of *Streptococcus bovis* and *Selenomonas ruminantium* (Evans and Martin, 2000). Ando *et al.* (2003) have reported decrease in rumen protozoal numbers in steers fed peppermint as additive. This anti-protozoal effect was attributed to L-Menthol, the main component of peppermint essential oil. Different plants rich in essential oils like *Achillea millefolium, Cymbopogan citratus* (herbage), *Eucalyptus globulus* (leaves), cymbopogan (herbage) and eucalyptus (leaves) have been evaluated as feed additives in animals.

Dietary terpenes have also been found to appear in milk and meat of animals grazed on natural pastures and rangelands rich in terpenes and directly affect the development of their milk products like cheese flavor (Carpino *et al.*, 2004) or causes changes in microbial proteolysis taking place during cheese maturing (Buchin *et al.*, 2003). Thus there is growing interest among nutritionist in the post ingestion metabolism of these compounds. Malecky and Broudiscou (2009) studied disappearance of nine different monoterpenes exposed *in vitro* to the rumen microflora of dairy goats grown in either starch or fiber rich cultures. Different terpenes tried were (*E*)-β-ocimene, (Z)-β-ocimene (linear alkenes), alpha terpinene and α-phellandrene (monocyclic alkenes), (-)-α-pinene, (+)-α-pinene, (-)-β-pinene and sabinene (bicyclic alkenes), and linalool (linear terpene alcohol). Under presence of fiber adapted microflora degradation of α-phellandrene and all the isomers of β-ocimene and pinene reached a plateau within 3 hours, however α-terpinene showed same trend under fiber as well as starch adapted microflora. With the only starch adapted microflora this steady state was reached for most other terpenes within 6 hours of incubation and sabinene was the only compound still disappearing after an incubation period of 6 hours. Influence of degradation patterns of dietary terpenes in ruminants can be exploited to influence animal product characteristics in desired way.

Essential oils have been documented as having varying antimicrobial properties (Hammer *et al.,* 1999). In few studies it was also observed that feeding of essential oils resulted into reduction of liver abscess in animals exhibiting anti-microbial property. Meyer *et al.*, (2009) reported that total liver abscess tended to be lower by 39% due to supplementation of essential oil mixture in steers.

1.6 Effect of PSMs on Methane Inhibition

Methane is one of the potent green house gases. The domestic livestock is responsible for production of enteric methane as high as 80 Tg per year and 25 Tg per year methane is produced from manure of livestock (Cynoweth, 1996). Recent studies have shown that plant secondary metabolites could be used to manipulate rumen fermentation in the desired way. Considerable amount of work has been carried out by different researchers on effect of PSMs on inhibition of methane production by archaea in rumen. Most of the studies involved use of plant extract in suitable solvent (methanol, ethanol, water etc.) in assessing their anti-methanogenic properties in ruminants. Kreuzer *et al.* (2009) has reviewed anti-methanogenic properties of tropical plant extracts. Table 1. depicts the details of tropical plant extracts containing saponins, tannins and essential oils and their effects on methane inhibition under different experimental conditions.

Table 1 : Tropical plants and their extracts with potent anti-methanogenic effects.

Plant (Latin) name	Plant part	Methane (CH_4) inhibition (%)	Experimental model	Reference
Plants rich in saponins				
Acacia angustissima	Leaf	50[c]	*in vitro*	Soliva *et al.*, 2008
		13	*Rusitec*[d]	Bekele *et al.*, 2009
Camelia sines	Seed	26	*in vitro*	Hu *et al.*, 2005
		17	*in vitro*	Hu *et al.*, 2006
Sapindus mukorossi	Seed	96[a]	*in vitro*	Agarwal *et al.*, 2006
Sapindus saponaria	Seed	41[c]	*in vitro*	Soliva *et al.*, 2008
		15	*Rusitec*	Hess *et al.*, 2003
		4	*Lambs*	Hess *et al.*, 2004
Sesbania sesban	Leaf	55[c]	*in vitro*	Soliva *et al.*, 2008
		33	*Rusitec*	Bekele *et al.*, 2009
Yucca shidigera	Leaf	54	*in vitro*	Pen *et al.*, 2006
Plants rich in tannins				
Acacia meansii	Bark	7	Lambs	Carulla *et al.*, 2005
Agelaea obliqua	Leaf	84	*in vitro*	Hayler *et al.*, 1998
Leucaena leucocephala	Leaf	33	*in vitro*	Hayler *et al.*, 1998
Mangifera indica	Leaf	50	*in vitro*	Hayler *et al.*, 1998
Phyllantus discoideus	Leaf	42	*in vitro*	Hayler *et al.*, 1998
Terminalia chebula	Fruit pulp	100[b]	*in vitro*	Patra *et al.*, 2006

Contd...

Table 1. Contd...

Plants rich in essential oils				
Allium sativum	Bulbs	70[b]	*in vitro*	Kamra *et al.*, 2008
	Bulbs	12	*in vitro*	Garcia-Gonzales *et al.*, 2008
	Oil	75	*in vitro*	Busquet *et al.*, 2005
Foeniculum vulgare	Seeds	71[b]	*in vitro*	Kamra *et al.*, 2008
Populus deltoids	Leaf	86	*in vitro*	Kamra *et al.*, 2008
Psidium guayaba	Leaf	82	*in vitro*	Kamra *et al.*, 2008
Syzygium aromaticum	Flower buds	86[b]	*in vitro*	Kamra *et al.*, 2008

[a]Ethanolic extract, [b]Methanolic extract, [c]Compared with most inefficient plant tested.
[d]*Rusitec*: Rumen simulation technique
Source: Kreuzer. M., Kamra, D. N., Soliva, C. R. 2009. Utilizing the natural resources of the tropics: plants and plant extracts mitigating methane in ruminants. Proceedings of Animal Nutrition World Conference: 14-17 Feb 2009, New Delhi, India: 96-98.

1.7 High Altitude Fodder Plants as a Futuristic Source of PSMs

High altitude cold arid regions of Trans-Himalaya are characterized by adverse climatic conditions viz. prolonged extreme winter, subzero temperatures, high wind velocity, low humidity, intense ultra violet radiations and hypoxic air. However this region possess diverse flora and fauna. It is well established fact that PSMs synthesis is regulated by environmental conditions and other external factors. Extremely harsh climatic conditions and stressful environment prevailing in Ladakh (one of the cold, arid, high altitude region) results in development of unique, potent bioactive molecules in plants and fodders growing in this region. Livestock graze on scarce but nutritive wild fodder plants possessing medicinal importance & growing on high mountains even in winter (temperature up to -30°C) conditions. It is believed that feeding of animals on these wild fodder plants results in intake of PSMs having nutritional and medicinal implication and are able to produce and reproduce normally even in extreme environmental conditions. Different wild plants identified as fodder for animals include *Aconogonum tortuosum, Cicer microphyllum* (Wild gram), *Ephedra gernrdiana, Hippophae rhamnoides* (Seabuckthorn), *Prangos pabularia* (Prangos), *Lepidium latifolium* (Dittander) and *Iris lacteal* (Chaurasia *et al.*, 2003). Beside these plants, local unconventional feed ingredients in Ladakh (Willow and poplar tree leaves, apricot cake, seabuckthorn pomace etc.) are futuristic potential feed resources that could be exploited for balanced nutrition of animals in this region (Jadhav *et al.*, 2009). However nutritional attributes and medicinal importance of these fodder plants needs to be investigated in animal production and veterinary practices.

Seabuckthorn is a wonder plant of Ladakh region and leaves of this plants are very good source of fodder for animals in this region. Seabuckthorn is rich source of PSMs that could be used in animal feeding system. Studies on evaluation of these feed resources in the animal ration in high altitude cold conditions of Ladakh are scanty. However few studies reports nutritive evaluation of seabuckthorn leaves in sheep. Jadhav *et al.* (2008a) and Jadhav *et al.* (2009b) studied effect of feeding Seabuckthorn leaves in complete feed of sheep in Ladakh. Singh *et al.* (2005) studied DM intake potential and degradation parameters of Seabuckthorn leaves by *in sacco* as well as *in vitro* gas production technique and indicated potential of Seabuckthorn leaves as cattle feed. Further studies are essentially required for identification of active components in these fodder plants and their efficient utilization in veterinary and animal production practices.

1.8 Conclusion

The effectiveness of PSMs is such that even at concentrations as low as 0.1 g/kg feed may be sufficient to bring about conspicuous changes in animal performance. Livestock sector contributes immensely to economy of any nation. An endeavour for increasing animal productivity has often resulted into deterioration of environment either due to emission of green house gases of animal origins or polluting environments due to animal excretions. PSMs however are preferred to overcome these issues. Future threats of side effects on human health due to indiscriminate use of antibiotics, hormones and other synthetic compounds for enhancing animal productivity can be minimised very efficiently by the judicious use of PSMs. Alteration of nutritional profile of animal products in desired way by the use of specific PSMs is very much possible and should get more focus in research and development activities in near future. However, although PSMs look promising in many ways for their use as performance enhancer in animals, issues such as toxicity, photosensitivity, residual effects, taint, allergic reactions and cost effectiveness needs to be addressed more effectively so that these compounds will gain widespread acceptance.

References

Akkari, H., Draghouth, M. A. and H. Ben Salem. 2008. Preliminary investigations of the anti-nematode activity of *Acacia cyanophylla* Lindl.: Excretion of gastrointestinal nematode egg in lambs browsing *A. cyanophylla* with and without PEG of grazing native grass. *Small Ruminant Research*. 74, 78-83.

Ando, S., Nishida, T, Ishida, M, Hosoda, K. and Bayaru, E. 2003. Effect of peppermint feeding on digestibility, ruminal fermentation and protozoa. *Livestock Production Science*, 82:245-248.

Agarwal, N., Kamra, D. N., Chudhary, L. C. and Patra, A. K. 2006. Effect of *Sapindus mukorossi* extracts on *in vitro* methanogenesis and fermentation characteristics in buffalo rumen liquor. *Journal of Applied Animal Research*, 30: 1-4.

Bhatta, R., Y. Uyeno, K. Tajima, A. Takenaka, Y. Yabumoto, I. Nonaka, O. Enishi and M. Kurihara. Difference in the nature of tannins on in vitro ruminal methane and volatile fatty acid production and on methanogenic archaea and protozoal populations. *J. Dairy Sci.* 92: 5512-5522.

Bekele, A. Z., Clement, C., Kreuzer, M. and Soliva, C. R. 2009. The efficiency of *Sesbnia sesban* and *Acacia angusissima* in limiting methanogenesis and increasing ruminally available nitrogen in a tropical grass-based diet depends on accession. *Animal Production Science.*

Busquet, M., Calsamiglia, S., Ferret, A., Carro, M. D. and Kamel, C. 2005. Effect of garlic oil and four of its compounds on rumen microbial fermentation. *Journal of Dairy Science*, 88: 4393-4404.

Bruneton, J. 1995. Pharmacognosy, phytochemistry, medicinal plants. Lavosier publishing, Paris. pp. 538-544.

Buchin, S., B. Martin and A. Hauwuy. 2003. Do terpenes influence flavour of cheese? In Flavour Research at the Dawn of 21st Century - Proceedings of the 10th Weurman Flavour Research Symposuim, Beaune, France. Tec' & Doc, Paris, France. Pp-753-756.

Castillejos L, Calsamiglia S, Ferret, A. and Losa, R. 2005. Effects of a specific blend of essential oil compounds and the type of diet on rumen microbial fermentation and nutrient flow from a continuous culture system. *Animal Feed Science and Technology* 119, 29–41.

Carpino, S., S. Mallia, S. La Terra, C. Melilli, G. Licitra, T. E. Acree, D. M. Barbano and P. J. Van Soest. 2004. Composition and aroma compounds of Ragusano cheese. Native pastures and total mixed rations. *J. Dairy Sci.* 87:816-830.

Carulla, J. E., Kreuzer, M., Machmûller, A., and Hess, H. –D. 2005. Supplementation of *Acacia meansii* tannins decreases methanogenesis and urinary nitrogen in forage-fed sheep. *Australian Journal of Agricultural Research*, 56: 961-970.

Chaurasia, O. P., Basant Ballabh, Verma Anju, Ahmed Z. and Raut, B. 2003. Potential fodder plants of Ladakh. Extension folder number 9. Publisher: Field Research Laboratory, Defence Research Development Organization (DRDO), Leh (Ladakh), 194101, J&K.

Cynoweth, D. P. 1996. Environ. Monit. *Assess.* 42: 3-18.

Cordell, G. A. 2000. Biodiversity and drug discovery - a symbiotic relationship. *Phytochemistry*. 55, 463-480.

D'Mello, J. P. F. and Devendra, C. 1995. Tropical Legumes in Animal Nutrition. Wallingford, Oxon.: CAB International.

Evans, J.D. and Martin, S.A. 2000. Effects of thymol on ruminal microorganisms. *Current Microbiology*, 41:336-340.

Farnsworth, N. R., Akerele, O., Bingel, A. S., Soejarto, D. D. and Guo, Z. 1985. Medicinal plants in therapy. *Bulletin of the World Health Organization* 63, 965–981.

Garæia-Gonzáles, R., López, S., Fernández, M., Bodas, R. and Andzález, J. S. 2008. Screening the activity of plant and spices for decreasing methane production *in vitro*. *Animal Feed Science and Technology*. 147: 36-52.

Hammer, K. A., C. F. Carson and T. V. Riley. 1999. Antimicrobial activity of essential oils and other plant extracts. *J. Appl. Microbiol.* 86:985-990.

Heyler, R., Steingass, H. and Drochner, W. 1998. Effect of various feed stuffs rich in tannin content on rumen methanogenesis in vitro using the Hohenheim Gas Test (in German). *Proceedings of the society of nutritional physiology,* 7: 35 (Abstr).

Hagerman, A. E. & Carlson, D. M. 1998. Biological responses to dietary tannins and other polyphenols. *Recent Research Developments in Agricultural and Food Chemistry* 2, 689–704.

Hess, H. -D., Kreuzer, M., Díaz, T. E., Lascano, C. E., Carulla, J. E., Soliva, C. R. and Machmûller, A. 2003. Saponin rich tropical fruits affect fermentation and methanogenesis in faunated and defaunated rumen fluid. *Animal Feed Science and Technology*, 109: 79-94.

Hess, H. -D., Beuret, R., Lötscher, M., Hindrochsen, I. K., Machmûller, A., Carulla, J. E., Lascano, C. E. and Kreuzer, M. 2004. Kueuzer, M., 2004. Ruminal fermentation, methanogenesis and nitrogen utilization of sheep receiving grass hay-concentrate diets offered with *Sapindus saponaria* fruits and *Cratylia argentea* foliage. *Animal Science*, 79: 177-189.

Hu, W., Liu, J., Ye, J., Wu, Y. and Guo, Y. 2005. Effect of tea saponin on rumen fermentation in vitro. *Animal Feed Science and Technology*, 120: 333-339.

Hu, W., Liu, J., Wu, Y., Guo, Y. and Ye, J. 2006. Effect of tea saponin on in vitro ruminal fermentation and growth performance in growing Boer goats. *Archives of Animal Nutrition*, 60: 89-97.

Hussain, I and Cheeke, P.R. 1995. Effect of dietary *Yucca schidigera* extract on rumen and blood profiles of steers fed concentrate or roughage based diets. *Animal Feed Science and Technology*, 51: 231-242.

Ivan, M. Koenig, K.M., Teferdegne, B, Newbold, C.J., Entz, T., Rode, L.M and Ibrahim, M. 2004. Effects of the dietary *Enterolobium cyclocarpum* foliage on the population dyanamics of rumen ciliate protozoa in sheep. *Small Ruminant Research*, 52: 81-91.

Jadhav, S. E., Tilak Raj, Guru Charan Singh and Shashi Bala Singh. 2009b. Effect of feeding Seabuckthorn leaves and Lucerne hay based complete diets on growth and blood biochemical profile of sheep at high altitude. In: Advances in Agriculture, Environment and Health, Satish Serial Publisher, New Delhi, pp.415-421.

Jadhav, S. E., Singh, S. B. Tilak Raj and Guru Charan. 2009a. Comparative evaluation of Seabuckthorn leaves and lucerne hay feeding in complete feed of lambs at high altitude cold arid conditions of Ladakh. Proceedings of Animal Nutrition Association World Conference, February 14-17, 2009, New Delhi, India, pp.2.

Jadhav, S. E., Guru Charan, Tilak Raj, K. and Shashi Bala Singh. 2009. Seabuckthorn leaves and various non conventional feed ingredients: Futuristic animal feed resources of Ladakh. Souvenir of National Conference on Seabuckthorn and Environment: High Altitude Perspectives organized by Defence Institute of High Altitude Research, DRDO, September 25-27, 2009, Leh (Ladakh), India, pp-33-34.

Jain, S.K. 1999. Dictionary of ethnoveterinary plants of India. Deep Publications New Delhi.

Jones, W.T. and Mankan, J. L. 1977. Complexes of the condensed tannins of sainfoin with fraction 1 leaf protein and with submaxillary mucoprotein, and their reversal by polyethyleneglycol and pH. *Journal of the Science of Food and Agriculture*, 28: 126-136.

Kamra, D. N., Patra, A. K., Chaterjee, P. N., Kumar, R., Agarwal, N. and Chaudahary, L. C. 2008. Effect of plant extracts on methanogenesis and microbial profile of the rumen of buffalo: a brief overview. *Australasian Journal of Experimental Agriculture*, 48: 175-178.

Kumar, S., Shukla, Y.N., Lavania, U.C., Sharma, A. and Singh, A.K. 1997. Medicinal and aromatic plants: Prospects for India. *Journal of Medicinal and Aromatic Plant Sciences*. 19: 361-365.

Khiaosa-Ard, R., S. F. Bryner, M. R. L. Scheeder, H-R. Wettstein, F. Leiber, M. Kreuzer and C. R. Solvia. 2009. Evidence of inhibition of terminal step of ruminal á-linolenic biohydrogenation by condensed tannins. *J. Dairy Sci.* 92:177-188.

Kaufman, P.B., Cseke, L. J.; Warber, S., Duke, J. A., Brielmann, H. L. 1999. Natural products from plants. CRC Press, Boca Raton, FL.

Klita, P.T., Mathison, G.W., Fenton, T.W. and Hardin, R.T. 1996. Effect of alfalfa root saponins on digestive function in sheep. *Journal of Animal Science*, 74: 1144-1156.

Kreuzer. M., Kamra, D. N., Soliva, C. R. 2009. Utilizing the natural resources of the tropics: plants and plant extracts mitigating methane in ruminants. *Proceedings of Animal Nutrition World Conference*: 14-17 Feb 2009, New Delhi, India: 96-98.

Longstaff, M., and McNab, J. M., 1991. The inhibitory effects of hull polysaccharides and tannins of field beans (*Vicia faba* L.) on the digestion of amino acids, starch and lipids and on digestive enzyme activities in young chicks. *Br. J. Nutr.* 65, 199-216.

Lu, C. D. and Jorgensen, N. A. 1987. Alfalfa saponins affect site and extent of nutrient digestion in ruminants. *Journal of Nutrition* 117, 919–927.

Min, B. R. and Hart, S. P. 2003. Tannins for suppression of internal parasites. *J. Anim. Sci.* 81, 102-109.

Meyer, N. F., Erickson, G. E., Klopfenstein, T. J., Greenquist, M. A., Luebbe, M. K., Williams, P and Engsrom, M. A. 2009. Effect of essential oils, tylosin and monensin on finishing steers performance, carcass characteristics, liver abscesses, ruminal fermentation and digestibility. *J Anim. Sci.* 87:2346-2354.

Miles, C.O., Wilkins, A.L., Erasmus, G. L. and Kellwerman, T. S. 1994. Photosensitivity in south Africa.VIII. Ovine metabolism of Tribulus terrestris saponins during experimentally induced geeldikkop, *Onderstepoort. Journal of Veterinary Research,* 61:351.

Makkar, H. P. S., Singh, B. and Dawra, R. K. 1987. Tannin-nutrient interaction- a review. *International Journal of Animal Sciences,* 2:127-140.

Makkar H. P. S., Blummel, M. and Becker, K. 1995. Formation of complexes between polyvinyl pyrrolidones or polyethylene glycols and tannins, and their implication in gas production and true digestibility in *in vitro* techniques. *British Journal of Nutrition* 73, 897–913.

Malecky, M. and Broudiscou, L. P. 2009. Disappearance of nine monoterpenes exposed in vitro to the rumen microflora of dairy goats: Effects of inoculum source, redox potential, and vancomycin. *J. Anim. Sci.* 87: 1366-1373.

Newman, D. J., Cragg, G. M. and Snader, K. M. 2000. The influence of natural products upon drug discovery. *Natural Product Reports* 17, 215–234.

Niezen, J. H., Robertson, H. A., Shelton, D., Waghorn, G. C., Green, R., 2002. The effect of feeding sulla (*Hedysarum coronarium*) or lucerne (*Medicao sativa*) on lamb parasite burdens and development of immunity to gastrointestinal nematodes. *Vet. Parasitology.* 105, 229-245.

Odenyo, A. A., Osuji, P. O. and Karanfil, O. 1997. Effect of multipurpose tree supplements on ruminal ciliate protozoa. *Animal Feed Science Technology,* 67: 169-180.

Patra, S. K., Kamra, D. N., and Agarwal, N. 2006. Effect of plant extracts on *in vitro* methanogenesis, enzyme activities and fermentation of feed in rumen liquor of buffalo. *Animal Feed Science and Technology,* 128: 276-291.

Pen, B., Sar, C., Mwenya, B., Kuwaki, K., Morikawa, R. and Takahashi, J. 2006. Effects of *Yucca schidigera* and *Quillaja saponaria* extracts on in vitro ruminal fermentation and methane emission. *Animal feed science and technology,* 129: 175-186.

Robins, C. and Brooker, J. D. 2005. The effects of Acacia aneura feeding on abomasal and intestinal structure and function in sheep. *Animal Feed Science and Technology.*

Singh, B., Bhat, T. K. and Singh, Virendra. 2005. Seabuckthorn leaves as cattle feed-A comparative evaluation. *Indian J. of Animal Nutrition.* 22(1):16-20.

Thalib, A., Widiawati, Y., Hamid, H., Suherman, D. and Sabrani, M. 1996. The effects of saponin from *Sapindus rarak* fruit on rumen microbes and performance of sheep. *Jurnal-Iimu-Ternak-dan-Veteriner* 2: 17-21.

Vasta, V., Mele, M., Serra, A., Scerra, M., Liciano, G., Lanza, M and Priolo, A. 2009. Metabolic fate of fatty acids involved in ruminal biohydrogenation in sheep fed concentrate or herbage with or without tannins. *J. Anim. Sci.* 87:2674-2684.

Vasta, V., H. P. S. Makkar, M. Mele and A. Priolo. 2009a. Ruminal biohydrogenation as affected by tannins *in vitro*. *Br. J. Nutr.* Dio:10.1017/S0007114508137898.

Vasta, V., A. Priolo, M. Scerra, K. G. Hallett, J. D. Wood and O. Doran. 2009b. Ä^9 desaturase protein expression and fatty acid composition of *longissimus dorsi* in lambs fed herbage or concentrate with or without tannins. *Meat Sci.* 82:357-364.

Van Nevel, C. J. and Demeyer, D. I. 1998. Manipulation of rumen fermentation. In The Rumen Microbial Ecosystem, pp. 387-443 (P N Hobson, editor). London and New York: Elsevier Applied Science.

Varel, V. H. 1999. Livestock manure odor abatement with plant derived oils and nitrogen conservation with urease inhibitors: A review. *Journal of Animal Science*, 80 (E.Suppl. 2):E1-E7.

Walton, J. P., Waghorn, G. C., Plaizier, J. C., Birtles, M. and McBride, B. W. 2001. Influence of condensed tannins on gut morphology in sheep fed Lotus pedunculatus. *Canadian Journal of Animal Science* 81, 605–607.

Wink, M. 1999. Introduction: biochemistry, role and biotechnology of secondary products. In:Biochemistry of secondary product metabolism (M. Wink, ed.) CRC Press, Boca Raton, FL. Pp 1-16.

Wink, M. 2004. Evolution of toxins and antinutritional factors in plants with special emphasis on Leguminosae. In Poisonous Plants and Related Toxins, pp. 1–25 [T Acamovic, CS Stewart and TW Pennycott, editors]. Wallingford, Oxon.: CABI publishing.

Williams, A.G. and Coleman, G.S. 1992. The rumen protozoa. Springer-verlag, London.

Wiermann, R. 1981. Secondary plant products and tissue differentiation. In The Biochemistry of Plants, vol. 7, pp. 85–116 [EE Conn, editor]. New York: Academic Press.

Yoshiki, Y., Kudou, S and Okubo, K. 1998. Relationship between chemical structures and biological activities of triterpenoid saponins from soybean (Review). *Bioscience Biotechnology and Biochemistry*, 62: 2291-2299.

□□□

ABOUT THE EDITORS

Dr.T.Parimelazhagan, is a Reader in Department of Botany, School of Life Sciences, Bharathiar University, Coimbatore. He did his graduation from St. Joseph's College, Trichirappalli and M.Sc and Ph.D degree from Bharathiar University, Coimbatore. He has worked as Scientist in Defence Institute of High Altitude Research, Defence Research & Development Organization (DRDO), Ladakh. Dr.Parimelazhagan has received National Science Day Medal and Laboratory Scientist of the Year award 2005 (DRDO). He has also been conferred with Young Scientist award 2008 in the International Seminar on Medicinal and Aromatic Plants held at Sri Venkateswara University, Tirupathi for his contributions in herbal product development. He is handling two major research projects funded by DRDO and UGC. He has filed three patent and more than 20 research publications to his credit. His current research focuses on the Phytomedicine and pharmacology of medicinal plants for value addition.

Prof. Dr. S. Manian, a specialist in herbal science and plant pathology, had his higher education in St. Joseph's College, Trichirappalli and Centre for Advanced Studies in Botany, University of Madras, Chennai. In the past three decades, he has built an illustrious research and teaching career with 150 original research articles and reviews published in reputed journals, 15 Ph.Ds and 32 M.Phils guided and innovative academic programmes started. He was instrumental in introducing a Master degree programme in Bioinformatics in the country by initiating it in Bharathiar University in 2001 and serving as its founder course director. He has handled many funded projects which include application oriented industrial projects and has patented the novel inventions. Prof. Manian has widely travelled in Europe and East Asian countries. He was a Commonwealth Academic Staff Fellow to Horticultural Research International, UK (1995-96), INSA -JSPS Visiting Professor to Kyushu University, Japan (1998) and INSA-CAS International Visiting Scientist to Xishuangbanna Tropical Botanical Garden, Chinese Academy of Sciences, China (2005). At the moment, he serves as Professor and Head, Department of Botany, Director, School of Life Sciences and Dean, Faculty of Research in Bharathiar University, Coimbatore. He is also serving as expert member in various government funding agencies.

Dr. M. Pugalenthi is a Assistant Professor, Department of Botany, Government Arts College, Coimbatore. He has 13 years of research and 10 years of teaching experience of undergraduate and postgraduate classes. He did M. Sc., Botany from Saraswathi Narayanan College, Madurai and Ph.D. in Botany from Bharathiar University, Coimbatore. He has been awarded Gold medal and many endowment prizes for securing the highest marks in M. Sc., Botany in Madurai Kamaraj University, Madurai. He has published more than 35 research papers in various Journals of National and International repute. Dr. Pugalenthi has completed a UGC funded Major Research Project and currently operating another Major Research Project funded by UGC. His thrust area of research is "Exploitation of under-utilized legume seeds as an alternative protein source". He has guided a Ph.D. and more than 15 M. Phil students.